FUNDAMENTALS OF MEDICAL VIROLOGY

FUNDAMENTALS OF MEDICAL VIROLOGY

Second Edition

LOUIS S. KUCERA, PH.D.
Professor
Department of Microbiology and Immunology
Bowman Gray School of Medicine
Wake Forest University
Winston-Salem, North Carolina

QUENTIN N. MYRVIK, PH.D.
Professor
Department of Microbiology and Immunology
Bowman Gray School of Medicine
Wake Forest University
Winston-Salem, North Carolina

LEA & FEBIGER
Philadelphia 1985

Lea & Febiger
600 Washington Square
Philadelphia, PA 19106-4198
U.S.A.
(215) 922-1330

First Edition, 1974
Reprinted, 1976
Second Edition, 1985

Library of Congress Cataloging in Publication Data

Kucera, Louis S.
Fundamentals of medical virology.

Rev. ed. of: Fundamentals of medical virology/
Jean D. Acton . . . [et al.] 1974.
Bibliography: p.
Includes index.
1. Virology. 2. Virus diseases. I. Myrvik,
Quentin N., 1921– . II. Fundamentals of medical
virology.
III. Title. [DNLM: 1. Virus diseases. 2. Viruses.
WC 500 K95f]
QR360.K83 1984 616'.0194 83-26831
ISBN 0-8121-0921-X

PRINTED IN THE UNITED STATES OF AMERICA

Print No. 4 3 2 1

Preface

Since the first edition of this book, the field of virology has exploded into the new areas of biotechnology and monoclonal antibody technology. Viruses provide unique tools for the advancement of biotechnology, particularly in the areas of recombinant DNA research and the application of modern genetics in the development of newer vaccines for the control of some human infectious diseases. Monoclonal antibodies provide a source of powerful reagents for more specific and rapid procedures in the diagnosis of viral diseases. Virology offers a major challenge to the epidemiologist, diagnostician, clinician, molecular biologist and geneticist. Recent major discoveries have linked viruses with acute immunodeficiency disease syndrome (AIDS) and human cancer. For example, it appears that a variant of the human T-cell leukemia-lymphoma virus (HTLV) may be the etiologic agent of AIDS. Another exciting and significant discovery of this decade has revealed that the retrovirus oncogene originates from normal cellular DNA sequences. These rapid advancements have generated a wealth of new information that is highly relevant to the study of health related sciences. Therefore, the intent of this second edition is to provide the student with concise, up-to-date coverage of the essential fundamentals of both basic and applied virology that have application to these advancements. We hope that this new edition will be particularly useful to students taking a first course in virology as well as to beginning graduate students, medical students, public health officers, clinicians, epidemiologists and researchers.

This edition comprises two major sections—basic and clinical virology. The basic virology section addresses each major virus family and includes electron micrographs of representative viruses, data on biochemical structure and antigenic properties, host range, culture conditions and their respective replicative cycles. Chapters describing arenaviruses, coronaviruses and hepatitis viruses have been added.

The clinical virology section deals with medically relevant viruses that infect major organ systems. Each chapter describes the etiologic agent(s), clinical symptoms, pathogenesis, immunity, laboratory diagnosis, epidemiology, treatment and prevention of viral diseases. Also included are individual chapters on classification, growth and assay, viral replication cycles and genetics of animal viruses, as well as on viral pathogenesis, control of viral infections, principles of laboratory diagnosis, viruses infecting the fetus and newborn and viral carcinogenesis.

This edition has been expanded from 24 to 34 chapters by including individual chapters for each major virus family. The bibliography at the end of each chapter includes reviews or important original papers. The literature citations have been chosen to provide an introduction to a topic or additional insight into virology that the reader may want to pursue. Some references are included for historical interest.

Special acknowledgments are given to the contributors of this text and to authors who allowed us to use original illustrations and photomicrographs. We also gratefully acknowledge the various publishers who gave us permission to use copyrighted material.

Winston-Salem, North Carolina

LOUIS S. KUCERA
QUENTIN N. MYRVIK

Contributors

Chapter 4	Genetics of Viruses
Chapter 12	Orthomyxoviruses
Chapter 13	Paramyxoviruses

Dr. Douglas Lyles
Associate Professor
Department of Microbiology and Immunology
Bowman Gray School of Medicine
Wake Forest University
Winston-Salem, North Carolina

Chapter 33 Viral Diseases of the Fetus and Newborn

Dr. Andre J. Nahmias
Professor of Pediatrics and Chief of Infectious Diseases and Immunology
School of Medicine
Emory University
Atlanta, Georgia

Contents

Chapter 1

Characteristics and Classification of Viruses

A. HISTORY

The word virus (L.) was used in ancient times to note any noxious agent or poison. Before the discovery of filterable viruses, the word *virus* was commonly used nonspecifically to designate any infectious microbe. In this regard, several infectious diseases were recognized long before the nature of the etiologic agents was understood. For example, in 1798 Jenner introduced the practice of vaccination against smallpox with exudate from cowpox lesions without recognizing that he was dealing with a viral agent. In 1884, 14 years before the isolation of the first animal virus, Pasteur developed a vaccine for rabies. By the middle of the last century, scientists were making major efforts to determine the cause and effect of many biologic phenomena, including infectious disease. It was evident to those pioneer investigators that disease could be caused by several different classes of harmful agents including poisons, toxins and pathogenic bacteria. By the 1880s, Koch, Pasteur, Ehrlich and others made remarkable progress in describing, isolating and culturing pathogenic bacteria and relating them to specific diseases. However, it soon became obvious that there were many infectious diseases which were not caused by bacteria. Organisms could not be cultured from some specimens obtained from the lesions of some diseases that could be serially passed from one animal to another and were, therefore, infectious. Furthermore, such material remained infectious after passage through bacteriologic filters. Eventually agents of this type were given the name *filterable viruses* and later, simply *viruses.*

Illnesses now known to be caused by viruses were recognized for thousands of years. Hippocrates described the fever blister as a herpes (meaning to creep), hence the name herpes simplex virus. A Chinese description of a disease dating back to the tenth century B.C. is reminiscent of smallpox. Yellow fever was known for centuries in tropical Africa.

Iwanowski is generally recognized as the "father" of the science of virology for his discovery of the tobacco mosaic virus (TMV) in 1892. He reported that the agent which produced the mosaic disease of tobacco plants passed freely through "bacteriological filters." However, his conclusions concerning the nature of the etiologic agent of the disease were erroneous. He suggested that the disease was caused either by a toxin elaborated by some bacterial agent or by a bacterium which passed through the filters he used. The significance of Iwanowski's findings was not recognized until 1899 when Beijerinck proved the serial transmission of TMV by bacteria-free filtrates in which no bacteria could be detected.

The introduction of the chick embryo for virus propagation (1928–1930), followed by improvements in existing cell-culture techniques for growing viruses (1945–1950), marked the beginning of the highly productive era of modern virology. A list of discoveries which have contributed most importantly to the development of virology is presented in Table 1–1.

Table 1–1. Some Important Milestones in the Development of Medical Virology

Approximate Date	Principal Contributors	Contribution
1892	Iwanowski	Recognized that a "filterable agent" was responsible for mosaic disease in tobacco plants
1898	Loffler and Frosch	Discovered the foot-and-mouth disease virus, the first animal virus.
1898	Beijerinck	"Rediscovered" tobacco mosaic virus and called it *contagium vivum fluidum.*
1902	Reed	Discovered the cause of yellow fever, the first human virus to be described.
1907	von Prowazek	Described the first insect virus.
1908	Ellerman and Bang	Demonstrated that chicken leukemia could be transferred with cell-free extracts.
1911	Rous	Described the transmission of Rous sarcoma virus in chickens.
1915	Twort	Discovered bacteriophage.
1917	D'Herelle	"Rediscovered" bacteriophage and developed the plaque assay for quantifying infectious virus; coined the term *bacteriophage.*
1931	Woodruff and Goodpasture	Introduced embryonated eggs for propagating viruses.
1933	Shope	Described papilloma virus of rabbits.
1934	Lucke	Described a renal carcinoma with a suspected herpes-type virus etiology in frogs.
1935	Stanley	Obtained tobacco mosaic virus in crystalline form.
1936	Bittner	Described mouse mammary tumor virus and its transmission by milk.
1939	Delbruck	Began systematic studies of bacteriophage.
1941	Hirst	Demonstrated that influenza virus would agglutinate red blood cells (hemagglutination test).
1949	Enders	Showed that poliovirus multiplies in and destroys non-neural tissue in culture.
1951	Gross	Induced lymphoid leukemia in mice with a cell-free extract of tumors from leukemic mice.
1952	Dulbecco	Developed the plaque technique for assaying animal viruses.
1952	Hershey and Chase	Showed that only the DNA of bacteriophage is required for replication.
1953	Salk	Developed inactivated poliovirus vaccine.

Table 1–1. *Continued*

Approximate Date	Principal Contributors	Contribution
1954	Sigurdsson	Distinguished slow viruses from more familiar chronic and latent viruses.
1955	Sabin	Developed an active, attenuated poliovirus vaccine.
1957	Isaacs and Lindenmann	Discovered interferon production by influenza virus infected chicken cells.
1957	Colter	Extracted infectious nucleic acid from animal viruses.
1957	Stewart, Eddy, Gochenour, Borgese, and Grubbs	Isolated polyoma virus in tissue culture.
1959	Friend, C.	Reported cell-free transmission of reticulum cell leukemia in mice.
1958	Burkitt	Reported on a lymphoma involving the jaw in African children.
1960	Enders	Developed an active, attenuated measles virus vaccine.
1962	Rauscher	Described a virus-induced lymphoid leukemia in mice.
1962	Trentin, Yabe, and Taylor	Reported that adenoviruses of human origin induced tumors in newborn hamsters.
1964	Temin	Postulated that viral RNA can direct the synthesis of DNA provirus
1964	Epstein and Barr	Described the presence of a herpes-type virus associated with Burkitt's lymphoma.
1969	Huebner and Todaro	Presented the viral oncogene hypothesis that most or all vertebrate species contain genomes of RNA tumor viruses that are vertically transmitted from parent to offspring.
1970	Baltimore; Temin	Independently found evidence of RNA-directed DNA polymerase (reverse transcriptase) in virions of RNA tumor viruses.
1970	Spiegelman	Established that the RNA-directed DNA polymerase in oncogenic RNA viruses catalyzes the synthesis of an RNA:DNA hybrid.
1971	Danna and Nathans	Used restriction endonucleases to analyze the genome of SV40.
1972	Hehlmann, Kufe, and Spiegelman	Demonstrated that human leukemic cells, but not normal cells, contain RNA sequences homologous to those found in Rauscher leukemia virus.
1973	Gajdusek	Described kuru and Creutzfeld-Jakob diseases as models for slow virus diseases of the CNS; concept of viroids in human diseases.
1975	Blumberg, Larouze, London, Werner, Hesser, Milman, Saimot, and Payet	Relationship of hepatitis B virus to hepatocellular carcinoma in humans.

Table 1–1. *Continued*

Approximate Date	Principal Contributors	Contribution
1977	Berk and Sharp	Reported that some adenovirus transcripts consist of three colinear transcripts spliced together.
1977	Sanger	Used restriction enzymes to sequence the DNA of ΦX 174 bacteriophage.
1978	Collett and Erickson; Levinson, Oppermann, Levintow, Varmus, and Bishop	Produced evidence that the transforming src gene of RNA tumor viruses coded for a polypeptide with protein kinase activity.
1978	Halgerman, Rogers, Van de Voorde, Heuverswyn, Herreweghe, Volchaert, and Ysebaert	Reported the complete nucleotide sequence of SV40 DNA.
1979	Enquist, Madden, Schiop-Stansly, and Woude	Cloned herpes simplex virus-DNA fragments in a bacteriophage lambda vector.
1979	Roussel	The myc gene was defined as the unique transforming sequence of retroviruses which cause myeloid leukemia. The myc gene represents a transduced cellular gene.
1980	Taniguchi, Fujii, Kuriyama, Muramatsu	Molecular cloning of human interferon cDNA.
1981	Poiesz, Ruscetti, Reitz, Kalyanaroman, Gallo	Isolation of a new type C retrovirus (HTLV) from human T-cell leukemias.
1982	Prusiner	Description of a small proteinaceous infectious particle (Prions) associated with slowly progressive neurologic diseases.
1983	Land, Parada, Weinberg	Showed that cooperation between polyoma virus oncogene and a cellular proto-oncogene virus was required to produce a tumor.

B. DEFINITION AND ORIGIN

Because of the many specialized virus-host cell relationships which have been discovered, there is no completely satisfactory definition of a virus. The following definition proposed by Lwoff in 1957 is presented in an attempt to differentiate viruses from rickettsia, chlamydia and cell organelles which contain nucleic acid: *Viruses are obligate intracellular parasites that contain either DNA or RNA; they depend on the synthetic machinery of the cell for replication of specialized elements that can transfer the viral genome to other cells.* The complete infectious virus particle is called the *virion.*

How viruses reached the state of obligate intracellular parasitism poses an interesting, if presently unanswerable, question. Probably, the various groups of viruses arose by parallel evolution rather than by degeneration of a complex common ancestor. The fact that a number of cytoplasmic organelles contain functional DNA which serves as template for synthesis of messenger RNA (mRNA) and codes for protein suggests that the DNA in these organelles was originally derived from nuclear DNA and acquired the capability of autonomous replication. It is conceivable that a further step in evolution gave rise to progenitors of the various groups of DNA viruses.

Concomitantly, the RNA viruses may have evolved from cellular RNA-replicating systems. Cellular mRNA could have acquired the ability to initiate its own replication. In this regard, double-stranded RNA has been detected in a number of species of uninfected animal cells. However, the presence of a latent RNA virus in these cells has not been ruled out. Another alternative is that RNA viruses were derived from DNA viruses whose mRNA acquired the attribute of self-replication thus eliminating the requirement for transcription from DNA.

The development of experimental procedures for determining the "relatedness" of species of DNA and RNA by complementation or hybrid-

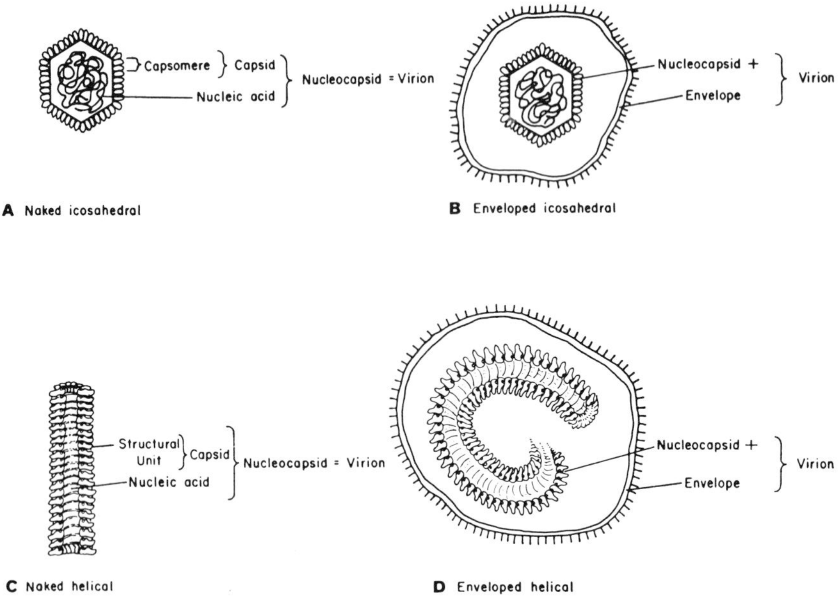

Fig. 1–1. Schematic representation of the structure and symmetry of four types of virions. The capsids of icosahedral viruses are composed of aggregates of structural units called *capsomeres*. The capsids of helical viruses are composed of identical repeating structural units. The capsid plus the nucleic acid comprises the nucleocapsid.

ization, "nearest neighbor" base sequence frequencies, and base ratios has provided approaches for studying the origin of viruses. Nearest neighbor base sequence analysis between the nucleic acids of certain viruses and the DNA of mammalian cells or of bacteria is compatible with the hypothesis that some viruses arose from the DNA of animal cells and others from bacteria. Although the viruses within a given major group, such as the adenoviruses, appear to have only small regions of viral genome which are homologous, they probably evolved from a common ancestor. Refinement and extension of existing analytical procedures may yield more definitive information about the phylogenetic relationship of the various groups of viruses.

C. COMPOSITION AND STRUCTURE OF THE VIRION

All virions are composed of a single species of nucleic acid, either single-stranded, or double-stranded RNA or DNA, enclosed in a protein coat, the *capsid*. The capsid is composed either of similar repeating protein molecules called *morphologic units* or *capsomeres* (aggregates of morphologic units). The structure composed of the nucleic acid surrounded by the capsid is the *nucleocapsid*. Some viruses are naked nucleocapsids, whereas others are enclosed in a phospholipid bilayer *envelope* or *peplos* of cellular origin and, thus, are said to be "enveloped" (Fig. 1–1). Some envelopes are covered with "spikes" or surface projections of varying lengths spaced at regular intervals. The spikes are presumably viral coded and are incorporated into the host cell membrane prior to virion maturation and release by a budding process.

Initial observations with the electron microscope have indicated that viruses can be classified into one of four broad morphologic classes: spherical, rod-shaped, complex, or tadpole-shaped. The viruses infecting animals belong to one of the first three morphologic classes, whereas those infecting bacteria (bacteriophage) belong in the fourth class.* Refinement of techniques for negative staining of purified virus preparations and for fixing and staining virus-infected cells for electron microscopy has permitted a more precise visualization of virus symmetry and structure.

All of the virions which originally appeared to be spherical have been shown to have nucleocapsids with either icosahedral or helical symmetry. Viruses with icosahedral symmetry contain capsomeres which may be round or prismatic in shape and are arranged in equilateral triangles. The icosahedron has 12 vertices, 20 triangular faces and 30 edges (Fig. 1–2). Diagrams of viruses with icosahedral symmetry are shown in Figures 1–1A and 1–1B. It is apparent from Figure 1–1B and

*For more information readers should see Q.N. Myrvik, A. Kreger and R.S. Weiser, Fundamentals of Bacteriology and Mycology, 2nd ed. (Philadelphia: Lea & Febiger, in press), Ch. 3.

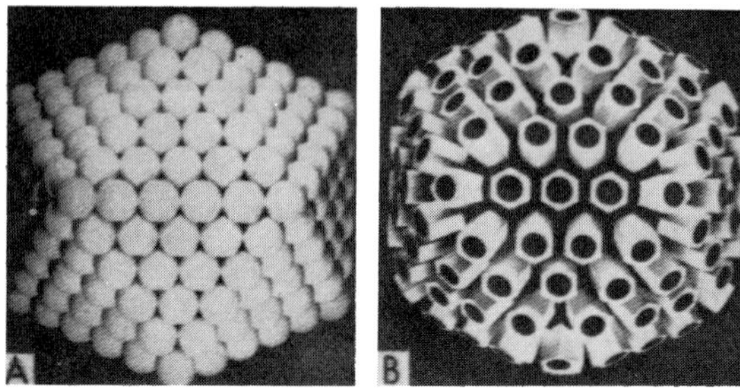

Fig. 1–2. Models of icosahedral viruses showing (A) round and (B) prismatic capsomeres arranged in equilateral triangular faces. (Reproduced with permission from Wildy, P., and Watson, D.H.: Electron microscopic studies on the architecture of animal viruses. (Symposia on Quantitative Biology. Vol. 27. Cold Spring Harbor, New York, The Biological Laboratory, Long Island Biological Association, 1962.)

1–1D that an enveloped icosahedral or helical virus can appear as a spherical particle.

The elucidation of the structure of rod-shaped viruses indicated that they all possess helical symmetry and may be rigid or flexible. The helical viruses which infect vertebrates all appear to be flexible and to be surrounded by an envelope (Fig. 1–1D). When the helical nucleocapsid is coiled in an envelope, the virion may assume either the shape of a rod or a sphere. Some viruses, such as influenza virus, possess virions of both shapes and, therefore, are said to be pleomorphic.

The best-studied helical virus is a tobacco mosaic virus, a non-enveloped rigid RNA plant virus (Fig. 1–1C). Fragments of TMV rods observed in the direction of the virion axis have a central channel readily observed in electron micrographs of negatively stained preparations. The nucleocapsid is composed of identical structural units arranged in a helix with the nucleic acid coiled between the turns of the helix.

The nucleocapsids of the enveloped animal viruses with helical symmetry have been observed following treatment with ether or with other substances which disrupt the lipid-containing envelope. The internal components of most animal viruses with helical symmetry resemble TMV. It has been suggested that the pleomorphism of some viruses (such as parainfluenza virus type 1) is, in part, a consequence of envelope and nucleocapsid flexibility which permits considerable variability in their configuration. It is also possible that pleomorphism may result from the incorporation of multiple nucleocapsids into a single envelope.

Not all viruses exhibit classical icosahedral or helical symmetry. Those which do not fit into these two categories, because of their unusual complexity, have been arbitrarily classified as *complex* or *binal* viruses. The symmetry of the poxviruses, the largest and most complex viruses which infect vertebrates, has not been determined with certainty; it has been suggested that the internal component be called a "core" until agreement can be reached on the question of what constitutes the nucleocapsid.

Although members of the poxvirus group vary with respect to size and morphology, vaccinia virus can be considered to be representative of this group. Electron microscopic examinations of shadowed preparations and thin sections of the mature vaccinia virion have shown it to be a brick-shaped particle with an outer multilayered membrane surrounding a biconcave core (Fig. 1–3). A pair of lens-shaped lateral bodies lie in the concavities of the core and are responsible for the central thickening of the particle.

In contrast to viruses (e.g., herpes simplex) which derive their envelope from the cell membrane, the envelope of the vaccinia virus is acquired while the virus is within the cytoplasm of the host cell. Occasionally, a surface membrane is acquired by vaccinia virus upon release from the cell.

D. CLASSIFICATION SCHEMA

Ideally, a classification scheme should reflect the evolutionary and phylogenetic relationships between groups of organisms. Since such relationships have not been established between the major virus groups, the existing classification schema have evolved primarily to serve as useful keys for identifying and classifying viral isolates.

Unfortunately, the viral nomenclature in current use is the result of free-lance labeling of viruses by investigators often with disregard for developing a systematic approach to classification. For example, many viruses were named for the geographic location from which they were first isolated or where the first case of a disease occurred; e.g., O'Nyong-yong virus initially was isolated in Africa, and the first diagnosed coxsackievirus infection occurred in Coxsackie, New York. The practice of incorporating into the virus name the disease associated with it has some value because it relates significant information about the virus; for example, poliovirus causes poliomyelitis. However, the discovery of increasing numbers of viruses posed the need for grouping them in some logical taxonomic order.

Among the criteria which have been used to classify viruses into groups are the nature of the disease and the organ system most frequently involved. Thus, the dermatropic viruses included the measles, smallpox, and herpes agents, whereas poliovirus and the equine encephalitis agents were classified as neurotropic viruses. Other schema were based on the mode or route of transmission. Whereas such systems are useful for associating certain viruses with specific syndromes, they fail to take into account the wide variability in the characteristics of

DNA VIRUSES

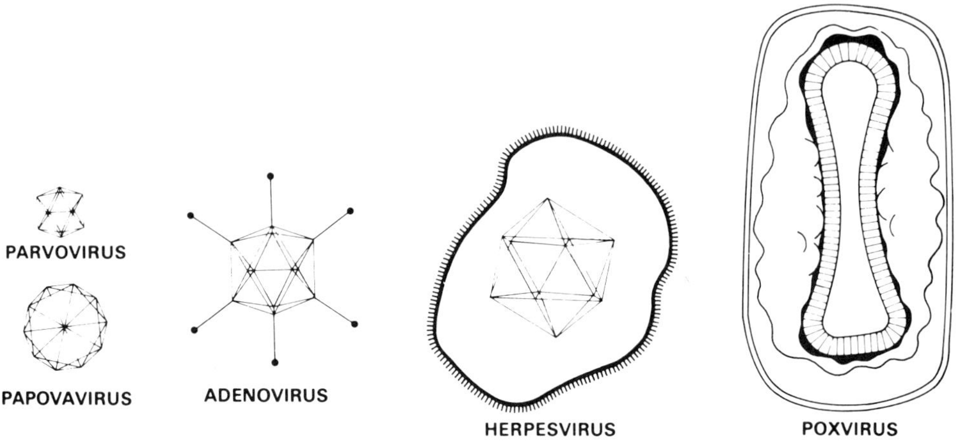

RNA VIRUSES

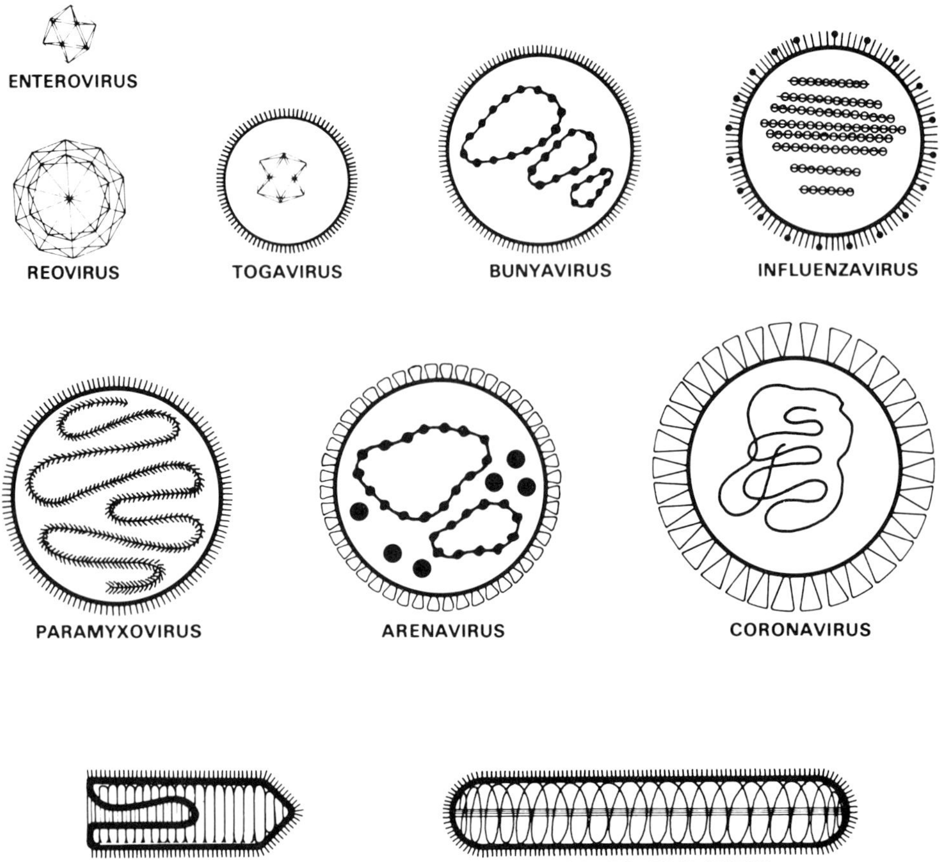

individual viruses within a given group and the fact that viruses can replicate in many different tissues without producing clinical disease.

During recent years, an International Committee on Taxonomy of Viruses has recommended that viruses be given genus and latinized family names. Thus, viruses are currently divided into families and genera on the basis of their physical, biochemical, morphologic, and antigenic properties and mode of genome replication. In the most generally accepted scheme of classification, the viruses are divided into two major divisions: the *riboviruses* and the *deoxyriboviruses*, based on the type of nucleic acid present in the virion. Criteria used for subdividing these two divisions into the major virus families include (1) symmetry of the nucleocapsid, (2) presence or absence of an envelope, (3) number of strands of nucleic acid, (4) size and shape of the virion, and (5) number of capsomeres on icosahedral viruses or the diameter of the helix in helical viruses. A simplified classification, based on the 1982 meetings of the International Committee on Taxonomy of Viruses (ICTV) is presented in Table 1–2. Diagrams of viruses representing the major virus groups are illustrated in Figure 1–3.

E. CHARACTERIZATION OF MAJOR GROUPS

Usually more than one type of disease entity can be caused by viruses within each major group. Conversely, the same clinical picture often can be caused by different viruses or by other microorganisms. For example, meningitis can be a consequence of infection with picornaviruses, togaviruses, herpesviruses, as well as a number of species of bacteria. The clinical symptoms produced by a given agent also can vary. For example, poliovirus infection can be inapparent or it can result in mild illness, aseptic meningitis, or paralytic poliomyelitis. The virus groups and important viral agents, together with the common diseases they cause, are summarized in Table 1–3. This list serves to introduce the major viral diseases of man and animals which will be discussed in subsequent chapters.

Fig. 1–3. Comparative size and shape of virions representing the major virus groups. (Reprinted with permission from Palmer, E.L., and Martin, M.L.: An Atlas of Mammalian Viruses. Boca Raton, Florida, CRC Press, Inc., 1982.)

Table 1–2. General Properties of Major Virus Groups

Nucleic Acid Type	No. Strands Nucleic Acid	Nucleocapsid Symmetry	Virion Naked or Enveloped	Shape of Virion	Size of Virion (nm)	Family	Typical Agents
DNA	1	Icosahedral	Naked	Spherical	18–24	Parvoviridae	Parvovirus, Adeno-associated virus
	2	Icosahedral	Naked	Spherical	40–55	Papovaviridae	Papilloma (Wart) virus; Polyoma; Simian virus 40
				Spherical	70–80	Adenoviridae	Adenovirus
			Enveloped	Roughly Spherical	110	Herpesviridae	Herpes simplex virus; Cytomegalovirus; Varicella-Zoster virus; Epstein-Barr virus
		Complex	Naked	Complex	130	Iridoviridae	Frog virus
				Brick Shaped	230 × 300	Poxviridae	Smallpox; Variola; Vaccinia
RNA	2	Icosahedral	Naked	Spherical	54–75	Reoviridae	Orbivirus; Reovirus; Rotavirus
	1	Icosahedral	Naked	Spherical	18–30	Picornaviridae	Enterovirus; Rhinovirus; Cardiovirus
			Enveloped	Spherical	35–40	Togaviridae (Arbovirus)	Alphavirus (Eastern equine encephalitis virus); Flavivirus (Yellow fever virus)
				Helical	90–100	Bunyaviridae	California encephalitis virus
				Roughly Spherical	100	Retroviridae	Leukosis viruses (RSV); Mammary tumor viruses; Visna virus
		Helical	Enveloped	Roughly	80–120	Orthomyxoviridae (Myxovirus)	Influenza virus
				Pleomorphic	100 × 300	Paramyxoviridae	Parainfluenza virus
				Bullet-shaped	60 × 225	Rhabdoviridae	Rabies virus; Vesicular stomatitis virus
				Oval or Pleomorphic	110–130	Arenaviridae	Lassa fever virus; Lymphocytic choriomeningitis virus
				Roughly Spherical	80–160	Coronaviridae	Human strains

Table 1–3. Properties and Diseases Associated with Selected Viruses within Major Virus Families

1. PAPOVAVIRIDAE—Cyclic, closed circular, double-stranded DNA; replicate in nucleus; cause cell transformation and tumors in animals.
   A. Papillomavirus: *Warts*
   B. Polyomavirus: *No disease in man;* oncogenic mouse virus
   C. Simian vacuolating virus (SV40): primate virus, oncogenic in rodent and human cells
2. ADENOVIRIDAE—Linear, double-stranded DNA; replicate in nucleus; 31 serotypes known
   Adenoviruses: *Respiratory tract, eye infections*
3. PARVOVIRIDAE—Linear, single-stranded DNA; replicate in nucleus; both defective and standard virus types exist.
   A. Adeno-associated (Satellite) viruses: *No disease in man*
   B. H1 and H3 viruses: *Isolated from human malignancy; significance unknown*
4. HERPESVIRIDAE—Linear, double-stranded DNA; replicate in nucleus; acquire envelope by budding from nucleus into cytoplasm.
   A. Alphaherpesvirinae
      a. Herpes simplex; *"Fever blisters," cervical cancer*
      b. Varicella-Zoster: *Chickenpox, shingles*
   B. Betaherpesvirinae
      a. Cytomegalovirus: *Cytomegalic inclusion disease*
   C. Gammaherpesvirinae
      a. Epstein-Barr (EB): *Burkitt's lymphoma, mononucleosis*
   D. B virus: *Ascending myelitis*
   E. Lucke: *Frog adenocarcinoma*
   F. Marek's disease: *Lymphoproliferative disease in chickens*
5. POXVIRIDAE—Linear, double-stranded DNA; replicate in cytoplasm; membrane is viral coded and is acquired within the cytoplasm. DNA-dependent RNA polymerase and DNA-dependent DNA polymerase associated with virion.
   A. Variola: *Smallpox*
   B. Vaccinia: *Complications following vaccination*
   C. Molluscum contagiosum: *Benign skin nodules*
   D. Orf: *Contagious pustular* dermatitis
6. PICORNAVIRIDAE—Linear, single-stranded, infectious RNA; replicate in cytoplasm.
   A. Enterovirus
      1. Poliovirus: *Poliomyelitis*
      2. Coxsackieviruses: *Aseptic meningitis, herpangina, pleurodynia, myo-* and *pericarditis, common cold*
      3. Echoviruses: *Aseptic meningitis, febrile illness with or without rash, common cold*
   B. Rhinoviruses: *Common cold*
   C. Cardioviruses: *Myocarditis*
7. REOVIRIDAE—Linear, double-stranded RNA; RNA in segments; replicate in cytoplasm; virion contains RNA polymerase which transcribes mRNA from the double-stranded genome.
   A. Reovirus: May cause *minor febrile illness, diarrhea,* and *upper respiratory disease,* but relationship to clinical disease in man is not clear.
   B. Orbivirus: *Colorado tick fever*
   C. Rotavirus: *Infantile diarrhea*
8. TOGAVIRIDAE (ARBOVIRUS)—Linear, single-stranded RNA; replicate in cytoplasm; acquire envelope by budding from plasma membrane. Many transmitted by arthropod vectors.
   A. Alphaviruses: Group A. Eastern and Western encephalitis viruses: (equine encephalitis viruses): *Encephalitis*
   B. Flaviviruses: Group B. St. Louis encephalitis virus: *Encephalitis* yellow fever virus: *Yellow fever*
   C. Rubivirus: Rubella virus: *German measles*

Table 1–3. *Continued*

9. BUNYAVIRIDAE—Segmented, negative-stranded RNA genomes in helical nucleocapsids; arthropod borne.
   A. (California Group) California encephalitis viruses: *Encephalitis*
10. ORTHOMYXOVIRIDAE (MYXOVIRUS)—Linear, single-stranded, segmented "negative-strand" RNA; phases of replicative cycle in nucleus and cytoplasm; envelope acquired by budding from plasma membrane. RNA-dependent RNA polymerase associated with virion; contain neuraminidase and hemagglutinin on separate proteins.
    A. Influenza types A, B, and C viruses: *Influenza*
11. PARAMYXOVIRIDAE—Linear, single-stranded, "negative-strand" RNA; some may have phases of replicative cycle in nucleus and cytoplasm; envelope acquired by budding from plasma membrane. RNA-dependent RNA polymerase associated with virion.
    A. Paramyxoviruses: Contains envelope-associated neuraminidase and hemagglutinin on a single protein.
       a. Parainfluenza virus: *Severe respiratory infections, croup, bronchiolitis, bronchitis, pneumonia in children, common cold in adults.*
       b. Mumps virus: *Mumps (epidemic parotitis)*
    B. Pneumovirus: Lacks envelope-associated neuraminidase; possesses a hemagglutinin
       a. Respiratory syncytial virus: *Severe respiratory infections (bronchiolitis and pneumonia in children; common cold in adults)*
    C. Morbillivirus: Lacks envelope-associated neuraminidase; possesses a hemagglutinin
       a. Measles virus: *Measles (rubeola)*
12. RHABDOVIRIDAE—Linear, single-stranded "negative-strand" RNA; replicate in cytoplasm. "Bullet-shaped" virion; acquire envelope by budding from plasma membrane.
    A. Lyssavirus:
       a. Rabiesvirus: *Rabies*
    B. Vesiculovirus
       a. Vesicular stomatitis virus: *Infects insects and mammals (cattle)*
13. RETROVIRIDAE—Contains 2 identical linear, single-stranded RNA molecules hydrogen bonded together to form 70S RNA plus small RNA from the host; RNA-directed DNA polymerase (reverse transcriptase) associated with the virion; multiplies by integration into host DNA; acquires an envelope at the plasma membrane.
    A, Leukosis viruses: *Cause leukemias and sarcomas in animals*
    B. Mouse mammary tumor (Bittner) virus: *Cause mammary carcinoma in mice*
    C. Visna virus: *Causes slowly progressive neurologic disease in sheep resulting in demyelination of CNS*

14. ARENAVIRIDAE—Single-stranded RNA; helical nucleocapsid; replicates in cytoplasm; pleomorphic; interior of particle appears unstructured and contains a variable number of electron-dense granules; acquire envelope by budding, chiefly from plasma membrane.
    A. Lymphocytic choriomeningitis (LCM) virus: *Virus infects mice but may spread to humans causing aseptic meningitis*
    B. Lassa fever virus: *Rare serious generalized disease in humans spread from rodents.*
15. CORONAVIRIDAE—Single-stranded RNA; helical nucleocapsid; replicate in cytoplasm; surface carries characteristic pedunculated (petal) projections; maturation by budding into cytoplasmic vesicles.
    A. Human respiratory viruses: *Respiratory tract infections*

Table 1–3. *Continued*

16. Tentatively Unclassified Viruses
    Hepatitis A (infectious) and hepatitis B (serum) viruses; associated with infectious and serum hepatitis, respectively.
    A. Hepatitis A virus—Single-stranded RNA; icosahedral nucleocapsid; nonenveloped; four major polypeptides in capsid.
    B. Hepatitis B virus—Partially double-stranded circular DNA; double-shelled particle with central core; contains DNA polymerase; also incomplete forms (spheres and filaments); polypeptide covalently attached to 5′ end of DNA, protein kinase in virion core.

REFERENCES

Fenner, F.: The Biology of Animal Viruses. Vol. I., New York, Academic Press, 1968.
Kurstak, E.: Classification of Human and Related Viruses. *In* Comparative Diagnosis of Viral Diseases. Vol. I., New York, Academic Press, 1977.
Lechevalier, H.A. and Solotorovsky, M.: Three Centuries of Microbiology. New York, McGraw-Hill Book Co., 1965.
Luria, S.E., Darnell, J.E., Jr., Baltimore, D. and Campbell, A.: General Virology. 3rd Ed., New York, John Wiley & Sons, Inc., 1978.
Matthews, R.E.F.: Classification and nomenclature of viruses. Intervirology, *12*:129, 1979.
Matthews, R.E.F. (ed.): A Critical Appraisal of Viral Taxonomy. Boca Raton, Florida, CRC Press, Inc., 1983.
Melnick, J.L.: Taxonomy of viruses. Prog. Med. Virol., *25*:160, 1979.
Melnick, J.L.: Taxonomy and nomenclature of viruses. Prog. Med. Virol., *28*:208, 1982.
Myrvik, Q.N., Kreger, A., and Weiser, R.S.: Fundamentals of Bacteriology and Mycology. 2nd Ed., Philadelphia, Lea & Febiger, (in press).

Chapter 2

Growth and Assay of Viruses

A. METHODOLOGY FOR VIRUS CULTIVATION

Since viruses are obligate intracellular parasites, the techniques for their propagation are more complex than those employed to culture bacteria. The three most common procedures used to cultivate animal viruses are (1) animal inoculation, (2) embryonated egg inoculation, and (3) inoculation of cell cultures. The presence of infectious virus is usually recognized by some abnormal manifestation, e.g., disease in animals, lesion (pock) formation on the chorioalloantoic membrane of the chick embryo or formation of a plaque (altered appearance of cells) in cell monolayers. The types of responses produced in host cell systems are usually characteristic for certain viruses.

1. Animal Inoculation

In the earliest attempts to propagate viruses, the definitive animal host of the agent was utilized. Such efforts were hampered by the complexity of the intact animal and the wide animal-to-animal variation in susceptibility, even within a single species. With the development of refined techniques for in vitro cultivation of cells, capable of supporting virus replication, animals are no longer selected for the routine propagation of most viral agents. However, animal inoculation is still invaluable for studying viral oncogenesis, pathogenesis of virus diseases, the immune response to viruses, the effect of environmental factors on virus infections, and the primary isolation of some viruses (e.g., coxsackie A viruses).

The method used to inoculate animals depends on the anatomic site of the target cells and the nature of the virus. For example, enveloped viruses are rapidly inactivated by acid pH and would not survive inoculation via the alimentary tract. The following routes of inoculation are commonly employed: (1) intravenous, (2) intracerebral, (3) intraperitoneal, (4) intranasal, (5) intratracheal, (6) intradermal, and (7) subcutaneous. After virus replication has occurred, tissue from infected areas of the body can be removed, minced or homogenized, and stored (usually in the frozen state) for subsequent use as a source of infectious virus.

2. Egg Inoculation

Several viruses have been cultured in the cells lining the cavities of the embryonated egg or in the developing embryo itself (Fig. 2–1). Although duck and turkey eggs have been used to cultivate some viruses (e.g., rabiesvirus), the chicken egg is usually employed.

The site of inoculation and the age of the embryo employed depend upon the virus being cultured. For example, primary isolation of influenza virus from throat washings is best accomplished by inoculation into the amniotic cavity of 7-hour to 8-day-old embryos; however, this

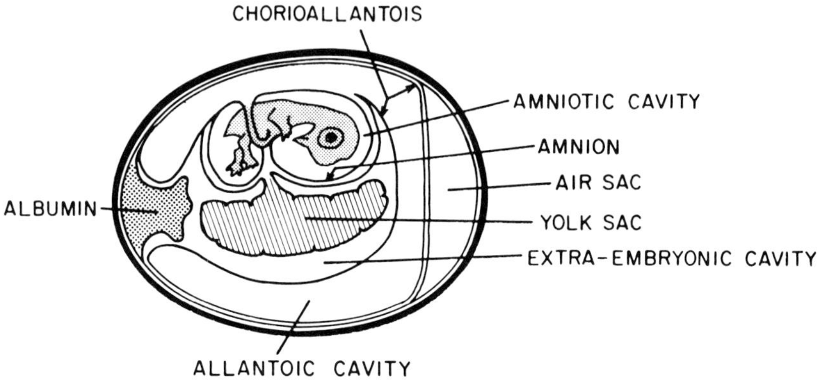

Fig. 2–1. Embryonated chicken egg. A longitudinal section through an embryonated egg showing the arrangement of the embryo and embryonic membranes.

virus replicates readily in the allantoic membrane of 12- to 13-day-old embryos after several passages ("adaptation") in the amnion. The most frequently employed sites of virus inoculation, other than the amniotic and allantoic cavities, include the chorioallantoic membrane and the yolk sac. In special circumstances, virus may be injected directly into the developing embryo by either intravenous, intraperitoneal or intracerebral inoculation. After virus replication has occurred in the cells of the membranes or embryo, virus may be released into the surrounding fluids which can be used as a source of virus. For example, influenza virus, which replicates in the cells of the amniotic membrane, is released into the amniotic fluid which can be readily harvested with a syringe and needle. Other viruses, such as poxvirus and herpesvirus, which replicate in the chorioallantoic membrane remain cell-associated in the lesions or "pocks" produced on the membrane. The infectious viruses can be released by grinding the infected membranes in an isotonic salt solution.

3. Cell-culture Inoculation

Cell-culture techniques have largely replaced the cumbersome and expensive procedures which used animals and embryonated eggs as means of propagating many viruses. Although there are inherent problems in the preparation and maintenance of tissue cultures suitable for virus propagation, their value is evidenced by the fact that more than 100 new viruses of human origin have been described since tissue-culture techniques have found routine application. The use of cell cultures not only has uncovered many new agents, but also has greatly facilitated the study of long-recognized viruses such as poliovirus and has led to the development of vaccines for poliomyelitis and other virus diseases. Many details of host-cell virus relationships, mechanisms of pathogenesis and an understanding of the molecular events involved in virus replication have been elucidated by the use of cell cultures.

There are three basic types of animal cell cultures: (1) primary and secondary cultures, (2) diploid cell strains, and (3) continuous cell lines. A *primary culture* is derived directly from the tissues of an animal or embryonated egg. The dissociated cells (usually obtained by treatment of tissue with trypsin) are suspended in a fluid growth medium and allowed to settle on solid surfaces such as those presented by a test tube, Petri dish or flask. The cells attach to the solid surface and divide mitotically to produce a *monolayer* of cells which constitutes the primary culture. A confluent primary culture of normal tissue may be subcultured to initiate a limited number of *secondary cultures.* Both primary and secondary cultures retain the normal diploid number of chromosomes characteristic of the parent tissue but usually undergo degeneration and cease to divide after 30 to 50 passages in culture or following repeated subculture. However, some primary or secondary cultures undergo a change which permits prolonged serial culture for up to 50 subcultures or more. They usually become altered morphologically even though they usually maintain the original number of chromosomes. Such altered cells are called *cell strains.*

During the course of culturing cell strains, *continuous cell lines* may emerge. Continuous cell lines represent cells which have assumed characteristics associated with transformed cells. These cell lines grow faster, form multilayers of cells and invariably are aneuploid (altered numbers of chromosomes). Continuous cell lines also can be obtained directly from primary cultures of malignant tissue or can be produced by infecting primary cultures or cell strains with oncogenic viruses. Many continuous cell lines derived from viral transformal primary cultures produce tumors when transplanted into susceptible animals.

During serial culture of cell lines, transformed cells often emerge spontaneously which do not attach firmly to solid surfaces but can be propagated in suspension cultures. Suspension cell cultures are useful for obtaining high yields of some viruses and for experiments which require repeated sampling of a homogeneous cell population.

The type of cell culture selected for propagating viruses is determined by the sensitivity of the cells to infection by the virus and by the specific requirements of the experiments or diagnostic information needed.

B. DETECTION OF VIRAL REPLICATION IN CELL CULTURES

The effects of virus replication within a cell can range from complete destruction of the cell to no visible effect. The type of cytopathogenic effect (CPE) depends upon the virus and the host cell system employed; however, morphologic changes that occur in a given system are characteristic for a virus and can be used as a basis for diagnosis of some virus diseases.

The following are among the most frequently encountered morphologic changes that occur following virus infection: (1) Cell lysis or necrosis

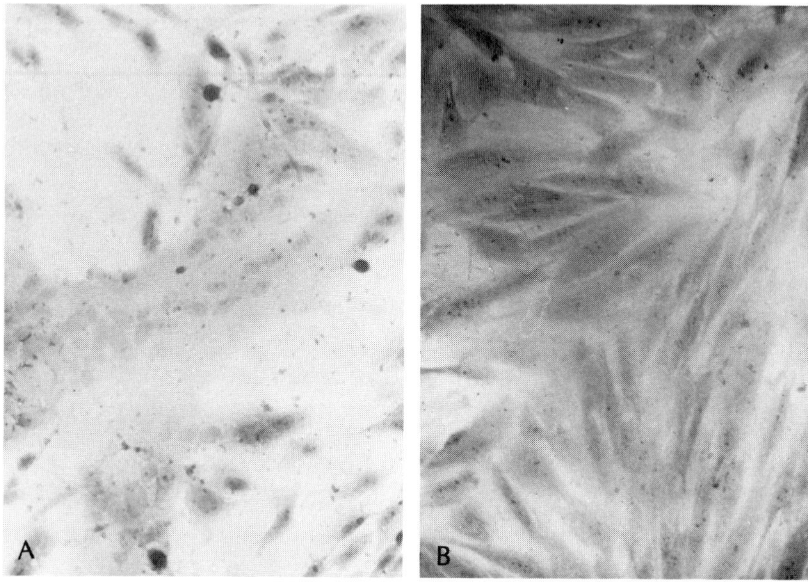

Fig. 2–2. Viral cytopathic effect showing herpesvirus type CPE with cell lysis. (A) Human cell culture infected with herpes simplex virus for 24 hours. Note large multinucleated giant cell (syncytium) in the center. Periphery shows dark, rounded, and lysed cells. (B) Uninfected human cell culture. (Crystal-violet stained preparations)

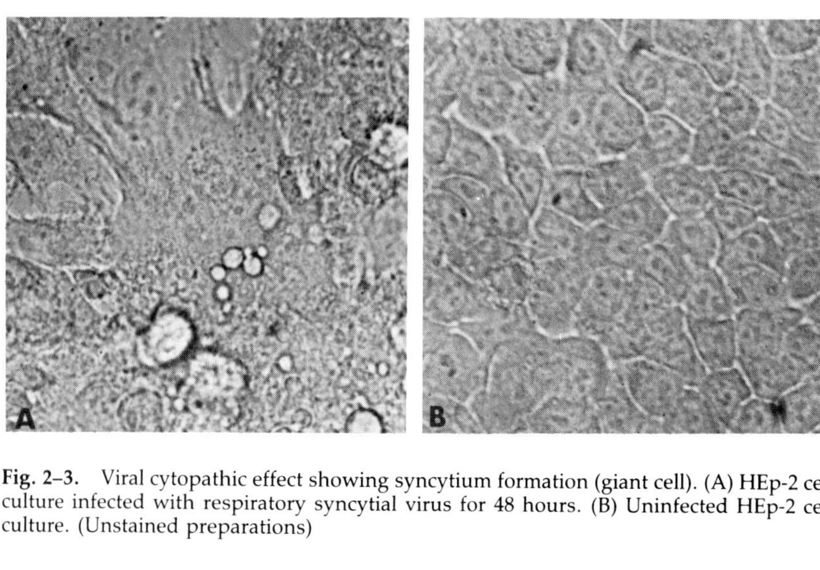

Fig. 2–3. Viral cytopathic effect showing syncytium formation (giant cell). (A) HEp-2 cell culture infected with respiratory syncytial virus for 48 hours. (B) Uninfected HEp-2 cell culture. (Unstained preparations)

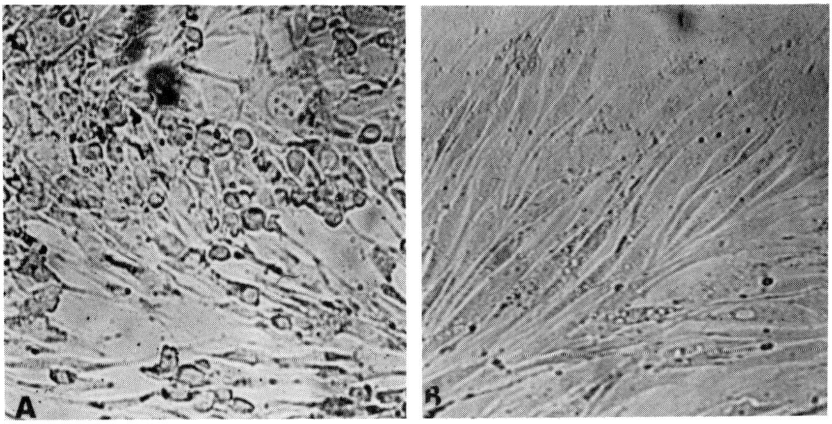

Fig. 2–4. Viral cytopathic effect showing cell clumping. (A) Rabbit kidney cells infected with adenovirus type 3 for 3 days. (B) Uninfected rabbit kidney cell culture. (Unstained preparations)

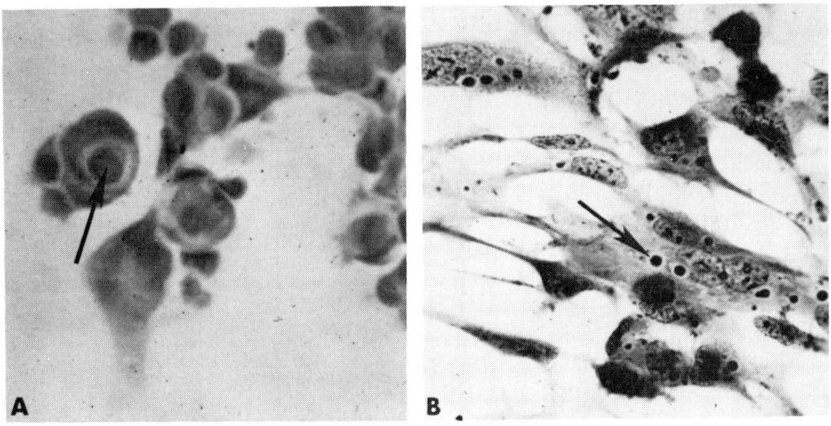

Fig. 2–5. Inclusion-body formation. (A) Intranuclear inclusion bodies in cells infected with herpes simplex virus. (B) Intracytoplasmic inclusion bodies in cells infected with vaccinia virus. The arrows indicate the inclusion bodies. (Stained preparations)

(Fig. 2–2). (2) Formation of multinucleate "giant cells" called *syncytia* (Figs. 2–2, 2–3). These are formed when viral-induced alterations in the cell membrane result in fusion of contiguous cells which initially are not destroyed but appear as a continuous cytoplasm containing accumulated nuclei (Fig. 2–2). (3) Clusters or clumps of cells which occur when the membranes are altered and the cells adhere but do not fuse (Fig. 2–4). (4) *"Inclusion-body"* formation (Fig. 2–5). Intranuclear or intracytoplasmic (Fig. 2–5) structures may appear in virus-infected cells and can be of considerable diagnostic aid; for example, intracytoplasmic inclusions in neural cells are diagnostic for rabies. In some cases, inclusion bodies

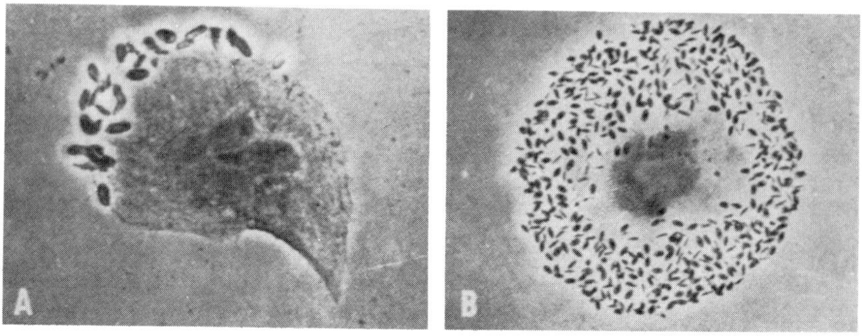

Fig. 2–6. Hemadsorption of HeLa cells infected by Newcastle disease virus. HeLa cells exposed to x-irradiation continue to increase in size but fail to multiply. These "giant" HeLa cells were infected with Newcastle disease virus and incubated. The hemadsorption test was performed after 5 hours (A) and 6½ hours (B) using chicken erythrocytes. (A) The initial point of erythrocyte attachment represents the first area of the plasma membrane to be modified by the incorporation of viral hemagglutinin. (B) As incubation is continued, viral hemagglutinin is synthesized and inserted into the cell membrane. Eventually, the entire circumference of the cell membrane is modified. The area over the nucleus is the last region to become hemadsorption positive. (Reprinted with permission from Marcus, P.I.: Dynamics of surface modification in myxovirus-infected cells. Symp. Quant. Biol., *27*:351, 1962.)

may be the site of virus replication. In others, they may not be sites of replication but nevertheless contain masses of virus particles or viral components. In still other infections, the inclusion bodies appear to be remnants of virus multiplication since they develop after virus replication has terminated.

In the absence of or prior to morphologic changes, the membranes of virus-infected cells may contain either viral components or partially extruded virus particles which have affinity for red blood cells. Such cells are said to be *hemadsorption positive* (Fig. 2–6). In some cases, hemadsorption tests are the sole means of detecting virus replication in cell cultures.

Cellular metabolic alterations are a frequent consequence of virus infections and can result either in an increase or decrease in the pH of the culture fluid. In the case of some viruses that cause cell degeneration, there is a decrease in the accumulation of acid end-products as cell degeneration progresses; the pH is not lowered as it is in uninfected cell cultures. In contrast, other viruses (e.g., adenoviruses) stimulate the production of acid end-products, and the pH attained is lower than the pH of uninfected culture fluids.

Cells infected with some viruses, especially some of the oncogenic agents, stimulate cell division which results in the formation of foci composed of heaps of "transformed" cells (Fig. 2–7). Normal cells possess the property of contact inhibition (blocked in cell division) which

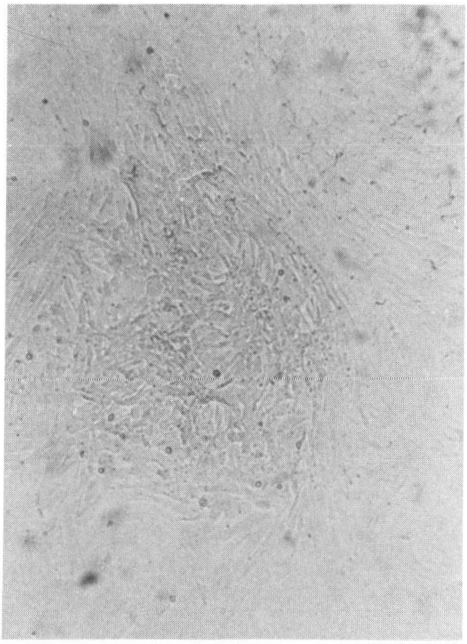

Fig. 2–7. Focus of transformed cells. Culture of rat embryo fibroblasts transformed with herpes simplex virus type 2. The transformed cells (morphologically altered) are in the center of the figure.

causes the cells to grow until a monolayer is formed; transformed cells have lost this property and pile up to form random multilayers of cells.

C. PRINCIPLES EMPLOYED IN THE TITRATION OF VIRUSES

Titrations of viruses can involve either a determination of the number of particles without regard to infectivity or an assay of the amount of infectious virus present.

A. Virus Particle Enumeration

a. Electron Microscopy. The electron microscope can be used to count the number of virus particles in a highly concentrated purified suspension. One method involves adding a known number of latex particles, approximately the size of the virus, to the preparation containing the virus to be counted. By determining the virus to latex particle ratio, the total number of virus particles in the original suspension can be calculated with a reasonably high degree of accuracy.

b. Hemagglutination. Viruses which possess a surface hemagglutinin and agglutinate red blood cells can be titrated by special hemagglutination tests. Both infectious and inactivated virus can cause hemagglutination, thus this assay can be used to enumerate the total number of viral particles present, but not those that are viable. Serial dilutions (usually twofold) of the virus are prepared, and red blood cells

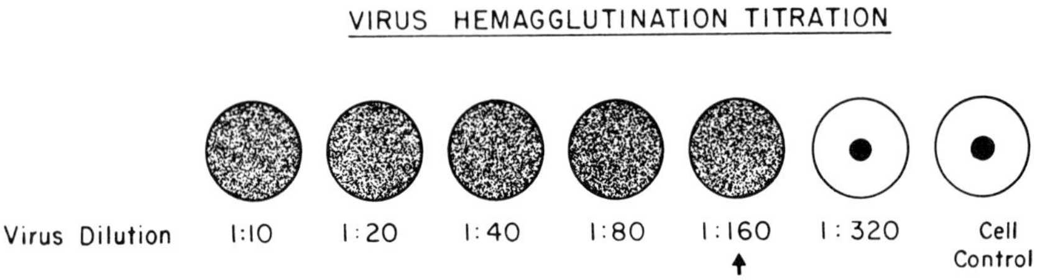

Fig. 2–8. Diagram of virus hemagglutination titration. Red blood cells are added to serial dilutions of virus. Hemagglutination results when sufficient virus binds red blood cells to form a lattice. In the control tube, the red cells settle in a characteristic "button."

are added. If enough virus is present to bind red cells in the form of a lattice, the virus-cell complexes will settle in the tube in a characteristic pattern. After appropriate incubation, the tubes are examined from the bottom to determine the highest dilution of virus which causes hemagglutination (Fig. 2–8). This dilution is referred to as one *hemagglutination unit.*

2. Virus Infectivity Assays

a. Plaque or Focal Assays. The most precise measure of virus infectivity is achieved by procedures which determine the capacity of a suspension of virus to establish a detectable focus (lesion) in a monolayer of host cells. Serial dilutions of virus are added to monolayers of susceptible cells. Following adsorption of the virus to the cells, an agar medium is added to the cultures to localize the virus infection and incubation is carried out for several days to several weeks depending on the rate of virus replication. If virus is transferred from cell to cell without being released into the medium, a fluid medium without agar can be used. As the virus in the initially infected cells replicates, its progeny infect neighboring cells and eventually produce a detectable focus of infection called a *plaque.* Each plaque represents infection of a cell by one infectious virus particle, which is commonly called a *plaque-forming unit* (PFU). Viruses that cause lysis or necrosis of cells produce clear plaques which can be counted after the addition of a stain to differentiate living from dead cells (Fig. 2–9). Viruses, such as tumor viruses, which cause cell proliferation will produce areas of living "transformed" cells. Noncytocidal and nononcogenic viruses may produce plaques which can be detected indirectly. For example, cells infected with viruses that induce production of hemagglutinin which inserts into the cell membrane can be detected by their ability to bind red blood cells. The assay is called the hemadsorption test. After unattached red cells are washed from the cell monolayer, the foci with absorbed red cells can be counted, the number of foci are equivalent to the number of infectious particles in the original inoculum.

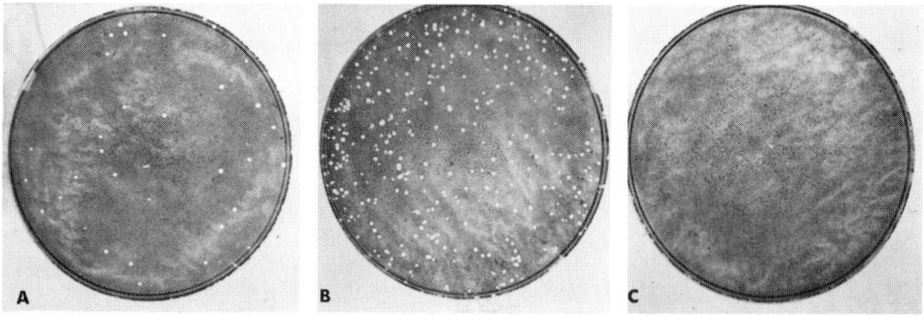

Fig. 2–9. Plaque assay for virus infectivity. Sample A represents a tenfold dilution of sample B. Note small clear areas (plaques) in the monolayer where cells have been destroyed. (C) Uninfected cell control.

b. Pock Assays. Some viruses form localized lesions on the chorioallantoic membrane of the embryonated egg. The titer of infectious virus is determined by counting the lesions, called "pocks," which develop after the addition of known dilutions of virus. The number of pocks is equivalent to the number of infectious particles in the original inoculum.

c. Quantal Assays. Quantal assays (dilution to extinction) depend on an "all-or-none" response in which viruses are allowed to multiply for a period of time sufficient to permit detection of the effect of a single infectious virus particle by the amplification effect of its replication. Serial dilutions of virus are prepared, and susceptible cells are incubated with a measured volume of each of several dilutions of virus. After appropriate incubation, each culture, egg, or animal is examined to determine whether it exhibits an effect attributable to virus replication. The criteria most frequently employed for determining the end point in quantal titrations are (1) development of CPE in cell cultures, (2) death or characteristic disease in animals, (3) appearance of lesions on embryonated egg membranes or the embryo, and (4) development of effects detected by in vitro procedures, such as hemagglutination or hemadsorption tests. The highest dilution of virus which causes a visible effect must contain at least one infectious virus particle. The titer is often expressed as the 50 percent infectious dose (ID_{50}), which is the reciprocal of the highest dilution of virus that produces an effect in 50 percent of the inoculated cell cultures, eggs, or animals. The precision of quantal assays can be improved by using large numbers of test subjects. However, a reasonably accurate estimate of the concentration of infectious virus can be obtained by statistical analysis of the data obtained from a relatively small number of subjects.

REFERENCES

Fenner, F.J. and White, D.O.: Cultivation and assay of viruses. *In* Medical Virology. New York, Academic Press, 1970.

Gospodarowicz, D. and Moran, J.S.: Growth factors in mammalian cell cultures. Amer. Rev. Biochem., *45*:531, 1976.
Habel, K. and Salzman, N.P.: Fundamental Techniques in Virology. New York, Academic Press, 1969.
Holley, R.W.: Control of growth of mammalian cells in cell culture. Nature, *258*:487, 1975.
Jakoby, W.B. and Pastars, I.H. (eds): Cell culture. *In* Methods in Enzymology. Vol. VLIII, New York, Academic Press, 1979, pp. 3–590.
Luria, S.E., Darnell, J.E., Jr., Baltimore, D. and Campbell, A.: General Virology. 3rd Ed., New York, John Wiley & Sons, Inc., 1978.
Rafferty, K.A., Jr.: Epithelial cells: growth in culture of normal and neoplastic forms. Adv. Cancer Res., *21*:249, 1975.

Chapter 3

Pathways and Classes of Virus Replication

The viral replicative cycle is studied most conveniently in cell cultures; however, the events observed in vitro may differ from those which occur in intact animals because in vivo homeostatic control mechanisms and the genetic make-up and immune status of the host can influence the outcome of a viral infection.

A typical viral replicative cycle can be divided into 5 sequential phases: (1) attachment, (2) penetration, (3) uncoating, (4) biosynthesis, and (5) maturation and release. The events which occur during each phase and the duration of the cycle vary depending on the type of nucleic acid in the viral genome, the group of virus being studied, the temperature of incubation and the host cell type. A defect in any step can result in an abortive or defective replicative cycle.

Students of medicine should know the molecular aspects of viral replication in order to understand the pathogenesis of virus diseases and the principles that relate to the action of chemotherapeutic agents. Understanding of the virus-specific events which occur in infected cells has led to the development of some potentially useful chemotherapeutic agents for virus infections (Chapter 23).

A. ATTACHMENT

The first step in infection has traditionally been termed attachment. About one in every 10^3 to 10^4 collisions between virions and the cell surface leads to a physically complementary union between a site on the cell and a site on the virion, probably with the aid of electrostatic forces; in many instances this interaction can be highly specific. The *surface site* on the virion may be a special virus protein (e.g., spike or fiber protein) projecting from the virus particle or mosaic of several capsid proteins. The *receptor site* on the cell is probably a surface glycoprotein in most cases. Different receptor sites exist on the cell for different viruses. Viruses which share common attachment sites usually belong to the same virus group.

Collision of virus particles with a cell does not always result in attachment or infection. The initial binding can be reversible, i.e., the virion can leave the cell surface. Some attached viruses proceed to an irreversible binding.

An important aspect of cellular receptors is that virus disease is often determined by the expression of different genes in different cell types and within the same cell types at different stages of development. During cell differentiation a different set of receptors can be expressed. The receptor sites appear to have a role other than just serving as sites of attachment. For example, in humans poliovirus can infect cells of the nasopharynx, gut and anterior horn cells of the spinal cord; the latter infection can cause poliomyelitis. However, other tissues do not possess receptors for poliovirus. Some viruses (e.g., measles virus) may have receptors in virtually every tissue. Measles virus may infect epithelial cells of the nasopharynx and lungs, spread via lymphatics and the blood to skin and subcutaneous tissues and eventually may infect the central nervous system.

Tissue cells that have receptors during early life may not possess such receptors during adult life. For example, coxsackie A virus receptors are absent in most mouse embryo tissues except differentiating myoblasts correlating with the observation that muscle infection (i.e., myositis) is the primary disease caused by coxsackie A virus in newborn mice.

The envelopes of orthomyxoviruses and some of the paramyxoviruses contain two spike proteins, a hemagglutinin and a neuraminidase capable of destroying cell receptor sites. The multiple hemagglutinating spikes in a single virion are able to bind two red blood cells together; many virions can create multiple bridges between red blood cells leading to hemagglutination. The cell receptor sites contain neuraminic acid. The virion neuraminidase catalyzes the removal of N-acetylneuraminic acid and the bridges between red cells disappear. During infection, orthomyxoviruses encounter mucopolysaccharides in nasal and respiratory tract secretions and it probably uses this enzymatic capacity to enter plasma membranes of cells. Immunization against neuraminidase or hemagglutinin spikes on virions protects against infection.

B. PENETRATION

There are at least 4 mechanisms by which viruses can penetrate cells: (1) On the basis of electron microscopic studies, it has been observed that some viruses are engulfed by a process of phagocytosis called "viropexis" (Fig. 3–1). (2) Penetration of viruses can involve the "fusion" or "melting" of the viral lipoprotein envelope with the cell membrane (Fig. 3–2). This fusion results in alteration of both viral envelope and the cell membrane at the point of contact and permits passage of the naked nucleocapsid directly into the cytoplasm. (3) Penetration also can involve the interaction of the virion with receptor sites on the cell membrane. It has been postulated that uncoating is initiated at the cell surface by host cell enzymes, and as a consequence the virion nucleic acid, and in some cases, the virion-associated RNA or DNA polymerase, are released directly into the cell. The residual capsid material which remains attached to the surface of cells infected with high multiplicities of virus may inhibit the attachment of additional virions. (4) Some viruses become intracellular by a process involving direct penetration of the intact virion through the cell membrane into the cytoplasm.

C. UNCOATING

Before viral nucleic acid can be replicated and viral coded proteins can be synthesized, the envelope and all or parts of the capsid material surrounding the genome must be removed. The initial steps leading to

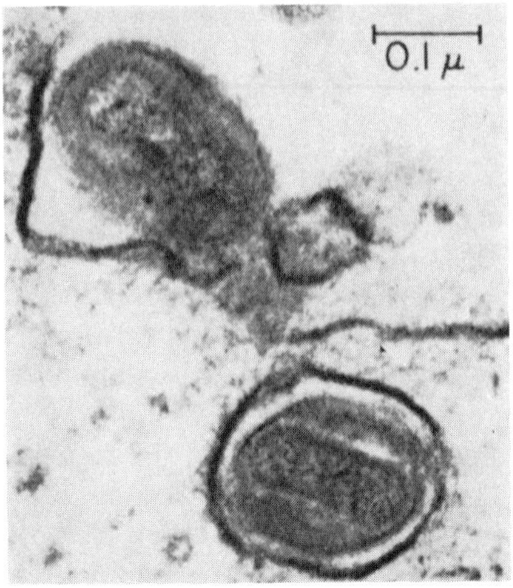

Fig. 3–1. Phagocytosis of vaccinia virus. A particle contacting the surface and another in a vesicle are evident (× 120,000). (Reproduced with permission from Dales, S.: The uptake and development of vaccinia virus in strain L cells followed with labeled viral deoxyribonucleic acid. J. Cell Biol., *18*:51, 1963.)

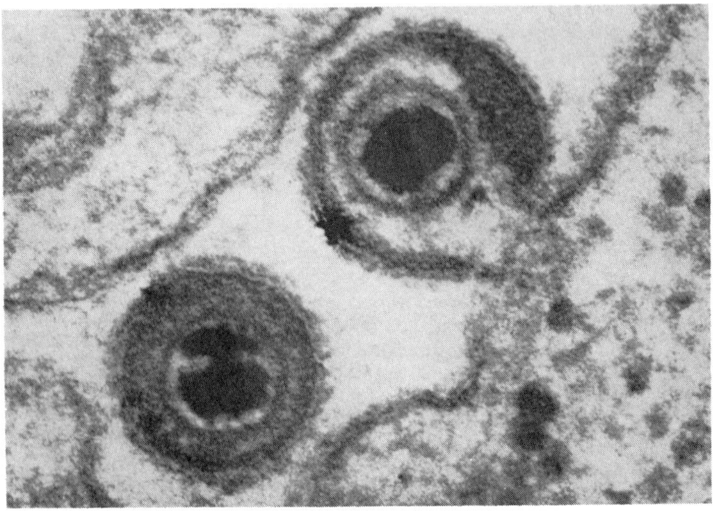

Fig. 3–2. Fusion of virus envelope with cell membrane (× 180,000). (Reproduced with permission from Morgan, C., Rose, H.M., and Mednis, B.: Electron microscopy of herpes simplex virus. I. Entry., J. Virol., *2*:507, 1968.)

capsid disruption can begin during adsorption and penetration and may result in release of RNA into the cytoplasm (e.g., poliovirus). In the case of certain enveloped viruses, the envelope is removed at the cell surface, and the intact nucleocapsid penetrates the cell.

Those viruses which gain entrance to the cell by viropexis are exposed to the lysosomal enzymes that are transferred to the virus-containing vacuole. The uncoating of some viruses (e.g., adenovirus) does not require protein synthesis; therefore, the capsids must be disrupted either by pre-existing cellular enzymes or by physical forces. In contrast, the uncoating of vaccinia virus, which gains entrance by viropexis, is a two-stage process. The first stage, during which the outer membrane and some protein are removed, occurs within the phagocytic vacuole and does not require de novo protein synthesis. As this first stage of uncoating occurs, the phagocytic vacuole breaks down and virus cores, which are still resistant to DNase, are released into the cytoplasm. In order that uncoating be completed, mRNA and protein must be made. It has been established that poxvirus cores can initiate mRNA synthesis as a result of the activity of a DNA-dependent RNA polymerase associated with the partially uncoated virion. Evidently mRNA released from the cores is translated into the enzyme(s) required for the final stage of poxvirus uncoating.

In the case of reovirus, the genome is never uncoated completely. The outer capsid is removed, and transcription takes place within the core particles which appear to remain intact throughout the replicative cycle. At no time is free double-stranded parental RNA found in the cytoplasm.

DNA extracted from DNA viruses and RNA extracted from (+) stranded RNA viruses (virion RNA has the same polarity as messenger RNA) are infectious. Other RNA viruses (− stranded) require a virion-associated RNA polymerase to synthesize mRNA. Removal of the RNA polymerase by extraction of the virion RNA prevents infectivity. The host cell range for infectious nucleic acid is usually much wider than for complete virus because nucleic acid penetration does not require specific receptor sites. However, infectious nucleic acid has a lower efficiency of infection than the corresponding virions, because it lacks the protection provided by the capsid. The role and importance of infectious nucleic acid in the spread and transmission of viral natural diseases is unknown.

D. BIOSYNTHESIS

The site of viral synthesis varies depending upon the nucleic acid composition of the virus and the group to which the virus belongs; however, the following generalizations can be made: (1) most of the deoxyriboviruses (e.g., herpes simplex) synthesize their DNA in the nucleus of the host cell and their protein components in the cytoplasm; exceptions are the poxviruses which synthesize all of their components in the cytoplasm; (2) most of the riboviruses synthesize all viral components in the cytoplasm; exceptions are the orthomyxoviruses and some of the paramyxoviruses and retroviruses in which part of the replicative cycle takes place in the nucleus.

1. Processing and Modification of Viral mRNAs.

Production and transcription of viral mRNA in the nucleus produces large heterogeneous nuclear RNA (HnRNA) that is processed by nucleases to messenger-size molecules before they are transferred to the cell cytoplasm. Adenovirus HnRNA appears to be processed by endonucleolytic segmentation and the splicing of leader segments onto the 5′ end of mRNA. Except for paramyxoviruses, no such processing occurs for viral mRNA synthesized in the cytoplasm. All known viral mRNAs with the exception of reovirus are polyadenylated (50–200 nucleotides) at their 3′-termini. The 3′ polyadenylation can occur by transcription of 5′-terminal polyuridylate sequences in complementary negative strands (e.g., picornaviruses) or by posttranscriptional modification. Also, with the exception of picornaviruses and "late" reovirus mRNA, viral mRNAs are capped at 5′-termini principally with the methylated $m^7G(5')ppp(5')$ N^mpnp(cap I structures) or $m^7G(5')ppp(5')N^mpN^mpNp$(cap II structures). Polyadenylation is necessary for polyribosome formation and capping is needed by some viruses for efficient translation. The processing of most DNA virus mRNA is carried out by preexisting host cell enzymes and appears to be essential for regulating the synthesis of functional viral messages.

2. Protein Synthesis.

Viral replication involves not only the synthesis of virion nucleic acid and capsid proteins but also the enzymes needed to catalyze the synthesis of viral components and maturation protein(s) required for virus assembly. In addition, some virus-induced regulator proteins are needed, either to inhibit certain cellular processes or to accomplish the cascade synthesis of viral-specific products. In some systems, (e.g., (−)stranded RNA viruses, influenza virus) mRNA must be synthesized before viral-specific protein can be made; in others, parental RNA can act directly as mRNA (e.g., (+)stranded RNA viruses, poliovirus).

The new proteins observed following virus infection may be virus coded or they may be host coded but virus induced. Since the genomes of many viruses do not contain enough genetic information to code for all of the newly synthesized proteins present in infected cells, it must be assumed that the mRNA for some virus-induced proteins is transcribed from the host cell genome.

Posttranslational cleavages and modification mechanisms play an important role in the formation of viral proteins. The genome of picornaviruses is translated as a single polyprotein that undergoes cleavage into three smaller proteins during translation. Further cleavages into distinct capsid and polymerase units occur during virus maturation. For some retroviruses, two large polypeptides corresponding to the entire coding capacity of the genome are produced. These precursor polypeptides are subsequently cleaved to functional proteins. In adenovirus, conversion of newly synthesized empty capsids to mature virus is accomplished by cleavage of 5 different polypeptides to smaller proteins. Other posttranslational modifications of proteins in several other virus groups include phosphorylation, glycosylation and sulfation. Also, hydrophobic domains of membrane proteins and glycoproteins are inserted into cellular lipid bilayers during assembly of enveloped viruses (e.g., herpes simplex virus).

3. Mechanisms of Viral Nucleic Acid Transcription and Replication.

DNA viruses can be divided into three classes while RNA viruses are divided into five classes on the basis of (1) genome structure, and (2) mechanism of gene transcription. Virus groups that replicate by similar pathways are grouped together into classes. Corona-arena- and bunyaviruses are not included because too little information is available concerning their mechanism of replication.

a. Class I Viruses have a Double-Stranded DNA Genome: The DNA of These Viruses is Transcribed Asymmetrically to Give Rise to mRNA. This class of viruses includes adenoviruses, herpesviruses, poxviruses, and papovaviruses.

Adenoviruses replicate in the cell nucleus. The ends of the double-stranded DNA contain an inverted repeat sequence suggesting that initiation of DNA synthesis could occur with equal rates at either end of the viral DNA molecule (Fig. 3–3). An adenovirus specific protein (MW = 72,000) that binds to both strands of DNA at the 5′ terminal is thought to be involved in DNA replication. The inverted repetitive base sequences permit the formation of single-stranded circles closed by a duplex stem "panhandle" structure (Fig. 3–3). Viral DNA is transcribed clockwise on the (+) L (late) strand and counterclockwise on the (−) E (early) strand resulting in (−) mRNA and (+) mRNA, respectively. The formation of mRNAs by splicing of noncontiguous RNA segments is now recognized as a general future of gene structure and function for adenoviruses. The E strand and L strand of adenovirus DNA are transcribed into "early" and "late" viral mRNAs, respectively. Early viral mRNAs direct the synthesis of T-antigens required for DNA replication and late mRNA transcription. Late mRNAs direct the synthesis of adenovirus capsid proteins.

Herpesviruses contain linear, double-stranded DNA in virions. Extracts of herpes simplex virus DNA contain four equimolar populations differing in the orientation of a long (l) and a short (s) segment relative to each other. Each segment contains terminal inverted repeat sequences

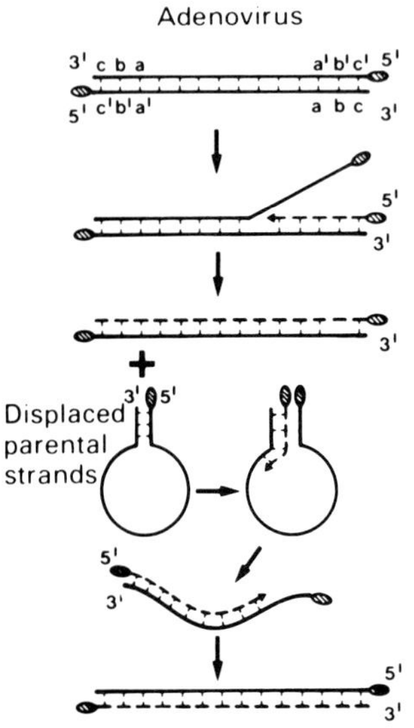

Fig. 3–3. Model of replication of adenovirus DNA. Initiation of DNA synthesis occurs at the 3′ end of one strand. Both strands of the DNA molecule have identical sequences at the two ends. The repetitive base sequences at the ends of the DNA strand permit the formation of single-stranded circles closed by a duplex panhandle. DNA replication is considered asymmetric. Parental nucleic acid is represented by the heavy lines; newly synthesized nucleic acid is represented by the light or broken lines. An adenovirus protein complex at the 5′ end of each DNA chain is probably required for initiation of DNA replication. (From Dulbecco, R. and Ginsberg, H.S.: Virology. Hagerstown, Harper and Row, 1980.)

(Chapter 8). The enzymes involved in herpes simplex virus DNA replication are partly viral in origin (e.g., DNA polymerase, thymidine kinase). The terminal inverted repeat sequences are complementary and join to form circles and concatemers during viral DNA replication (Fig. 3–4). The early-late control characteristic of transcription for adenoviruses also exists in herpesviruses. A temporal regulation of viral transcription involving specific regions of viral DNA leads to the sequential appearance of different classes of mRNAs: immediate early (α), delayed early (β), and late (γ) (Chapter 8). The formation of viral mRNA by splicing noncontiguous segments of mRNA also occurs with herpesviruses.

Poxviruses are composed of linear double-stranded DNA having covalently joined ends permitting the molecule to denature to a single-stranded circle (Fig. 3–5). The virus core contains RNA polymerase, poly (A) polymerase, protein kinase and guanylyl- and methyltransferase activities essential to virus replication. Infection induces a DNA polymerase and polynucleotide ligase. Primary transcription of parental DNA by the virion RNA polymerase reportedly produces early enzymes and structural proteins. Viral DNA synthesis occurs in discrete regions

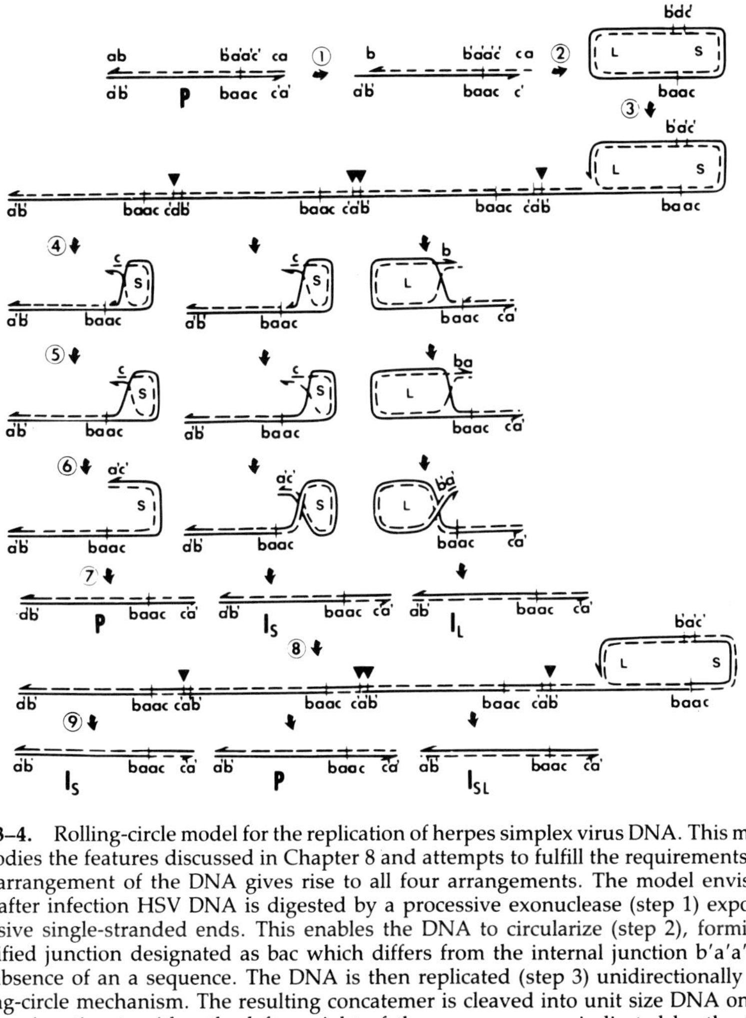

Fig. 3–4. Rolling-circle model for the replication of herpes simplex virus DNA. This model embodies the features discussed in Chapter 8 and attempts to fulfill the requirements that one arrangement of the DNA gives rise to all four arrangements. The model envisions that after infection HSV DNA is digested by a processive exonuclease (step 1) exposing cohesive single-stranded ends. This enables the DNA to circularize (step 2), forming a modified junction designated as bac which differs from the internal junction b'a'a'c' in the absence of an a sequence. The DNA is then replicated (step 3) unidirectionally by a rolling-circle mechanism. The resulting concatemer is cleaved into unit size DNA only at the bac junction to either the left or right of the a sequence as indicated by the filled inverted triangles. Steps 4 through 7 are diagrammatic representations of the regeneration of the missing a sequence. These steps are shown in three different columns to demonstrate how the replacement of these sequences could lead to the production of progeny in P (left column), I_s (center column), and I_L (right column) arrangements. Replication of I_s progeny (steps 8 and 9) would lead to the production of progeny I_s, P, and I_{SL} arrangements by the mechanism outlined in steps 4 through 6. (Reprinted with permission from Jacob, R.J., Morse, L.S., and Roizman, B.: Anatomy of herpes simplex virus DNA. XII. Accumulation of head-to-tail concatemers in nuclei of infected cells and their role in the generation of the four isomeric arrangements of viral DNA. J. Virol., *29*:448, 1979.)

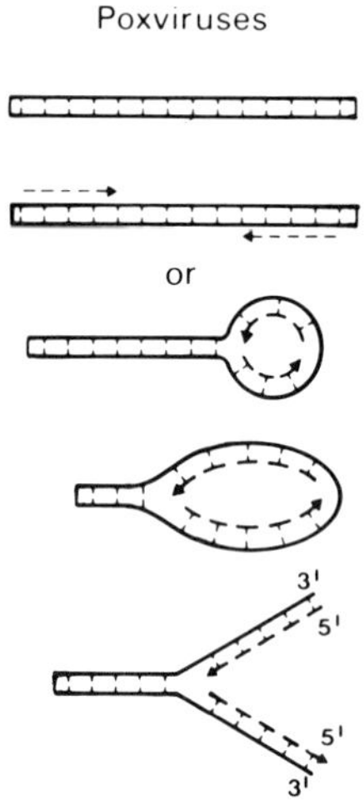

Fig. 3–5. Model for poxvirus DNA replication. The linear double-stranded DNA has covalently joined ends that permit the molecule to denature to a single-stranded circle. Semi-conservative replication initiates at one end and proceeds bidirectionally. A lariat-like structure increases with time and an endonuclease cleaves the molecule at a specific site, allowing separation of the two strands of DNA. (From Dulbecco, R. and Ginsberg, H.S.: Virology. Hagerstown, Harper and Row, 1980.)

in the cell cytoplasm by a process involving semi-conservative replication. Late transcription produces virion enzymes and late structural proteins. Both monocistronic and polycistronic messages are translated in proteins. Polypeptide precursors from the polycistronic messages undergo proteolytic cleavages.

In summary, it can be concluded that double-stranded viral DNA serves as a template, not only for the synthesis of new viral DNA, but also for transcribing into mRNA the genetic information controlling the synthesis of capsid proteins and probably some of the enzymes involved in the viral replicative cycle.

Papovaviruses contain a double-stranded covalently closed, circular DNA. The covalently closed circular DNA of papovaviruses is capable of integration into the genomes of permissive and nonpermissive cells. Viral DNA is replicated as a covalently closed circular DNA molecule (Fig. 3–6). There is a single origin for initiation of viral DNA synthesis located at about 0.67 on the physical map. DNA synthesis is bidirectional from this point until a point of termination is reached about 180 degrees away from the point of initiation. The growing DNA molecules separate from each other while there are gaps in the growing strand; the gaps are later filled and the molecules are converted to closed circular duplexes. Messenger RNA synthesis is divided into early and late mRNAs. The two classes of mRNA are copied from opposite strands and consequently are transcribed in different directions. The strand of DNA

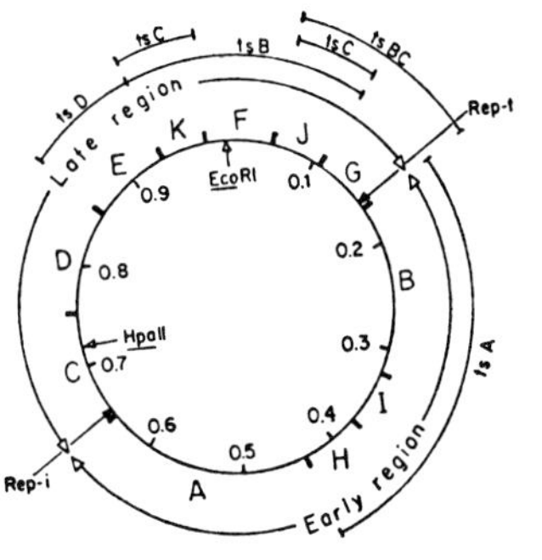

Fig. 3–6. Map of SV40 DNA. The restriction enzyme Eco R1 site represents 0.0 on the map. Initiation of DNA replication (Rep i) begins at map unit 0.67 and proceeds in opposite directions on the two DNA strands. Replication terminates (Rep t) at map unit 0.17. Early transcription starts at 0.65 and terminates at 0.18 (early region). The tsA is a temperature-sensitive mutant in early transition. The late region codes for viral capsid proteins. Several ts mutants have been mapped in the late region. (From Brockman, W.W.: Evolutionary Variants of Simian Virus 40. *In* Progress in Medical Virology. Vol. 23. Edited by J.L. Melnick. Basel, S. Karger, 1977.)

encoding early mRNA is called the E (early) strand; the other is called the L (late) strand. The transcripts are equal to or larger than the size of a single DNA strand. Messenger RNA is derived from splicing or cleavage of the large transcripts.

b. CLASS II VIRUSES HAVE A SINGLE-STRANDED DNA GENOME. This class of viruses includes the family parvoviridae. Parvoviruses are comprised of nondefective and defective viruses. The *nondefective* viruses are capable of autonomous replication and can use host cell enzymes for replication, transcription and other needed functions. *Defective* parvoviruses require an adenovirus helper which provides critical functions needed for replication. The genome of autonomous viruses is a unique single-strand of DNA but the DNA of defective viruses consists of equimolar quantities of complementary single-stranded DNA molecules packaged in separate virions. Cells infected with either type of parvovirus accumulate double-stranded forms of DNA that function as intermediates in replication. The DNA of parvoviruses have inverted terminal repetitive base sequences which are identical at each end of the DNA molecule (Fig. 3–7). The identical inverted repetitive sequences at both ends of the single-stranded (+) and (−) DNAs allow formation of intramolecular single-stranded circles closed by a duplex stem. During DNA synthesis, the 3′-terminal inverted sequences in the single-stranded DNAs of both defective and autonomous viruses unite to form intramolecular terminal hairpin duplexes that self-prime the 5′ to 3′ synthesis of linear concatemers (Fig. 3–7). A single mRNA is transcribed from DNA. In general, three structural proteins have been found in parvoviruses.

In summary, parvoviruses are either nondefective or defective in rep-

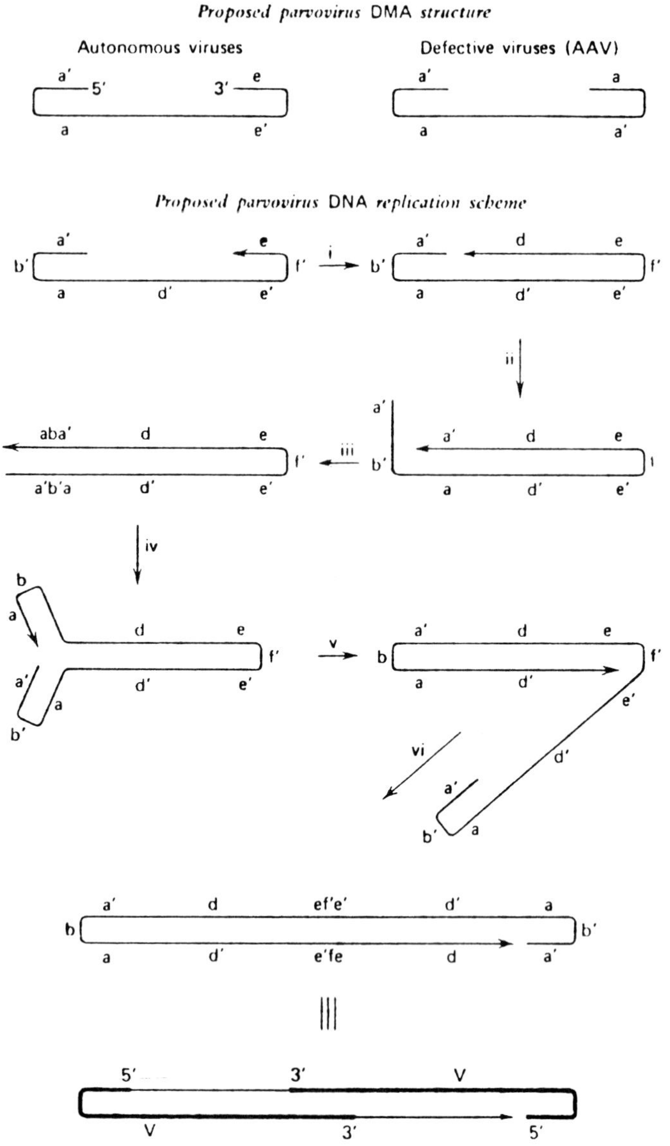

Fig. 3–7. Structure and replication of parvovirus DNA. The upper portion diagrammatically represents the structures of the DNA from autonomous and defective (adenovirus-associated) parvovirus. The designations a and a′ are meant to represent DNA sequences that have exact Watson-Crick base complementarity. The sequences a and a′ represent the stable 5′ hairpin, sequences e and e′ represent a transient 3′ hairpin and d′ represents 95% of the length of the molecule. On the model of replication the 3′ hairpin is shown initiating DNA synthesis (i). When the growing strand reaches the 5′ hairpin, it is dissociated and replicated (ii and iii). The 5′ hairpin then reforms generating a complementary 3′ hairpin (iv) that can reinitiate replication (v), ultimately producing a covalent dimer (vi) consisting of two virion strands and two complementary strands. In the bottom diagram, V signifies the two virion DNA's in a dimer duplex. Repetition of steps (iv), (v), and (vi) would generate a tetramer and continued initiation could produce higher multimers. (From Luria, S.E., Darnell, J.E., and Baltimore, D., et al.: General Virology. 3rd Ed. New York, John Wiley & Sons, Inc., 1978. Adapted from Tattersall, P., and Ward, D.C.: Nature, *263*:106, 1976.)

lication. In either case, single-stranded DNA is replicated via a double-stranded DNA intermediate. The ends of the DNA molecule can form intramolecular single-stranded circles that can self-prime the synthesis of 5′ to 3′ linear concatemers. A single mRNA molecule is transcribed from DNA.

c. Class III Viruses have a Double-Stranded RNA Genome. Reoviruses are the most thoroughly studied of class III viruses. The nucleoprotein core of the virus contains 10 double-stranded RNA segments and 5 shell proteins possessing transcriptase, nucleotide phosphorylase and guanylyl- and methyltransferase activities. Each of 10 (−) RNA strands is asymmetrically transcribed simultaneously to (+) mRNA via virion transcriptase molecules (Fig. 3–8). A single (+) mRNA encoding a single protein is transcribed from each segment of double-stranded RNA. The proteins vary in molecular weight from 34,000–155,000.

In summary, only one strand (− strand) of the parental genome is transcribed. Some of the newly synthesized strands of RNA function as mRNA; the remaining strands function as a template for the synthesis of a complementary strand resulting in the production of new double-stranded viral genomes.

d. Class IV Viruses are (+) Strand Viruses. Virions Contain a (+) RNA Genome Which also Functions as mRNAs on Large Membrane-Bound Polyribosomes. The class includes poliovirus (Class IVa, Fig. 3–9) and togavirus (alphavirus) (Class IVb, Fig. 3–10). The (+) mRNAs are polyadenylated at 3′-termini (Figs. 3–9 and 3–10) but capping at the 5′-terminus occurs only with Class IVb alphaviruses (Fig. 3–10).

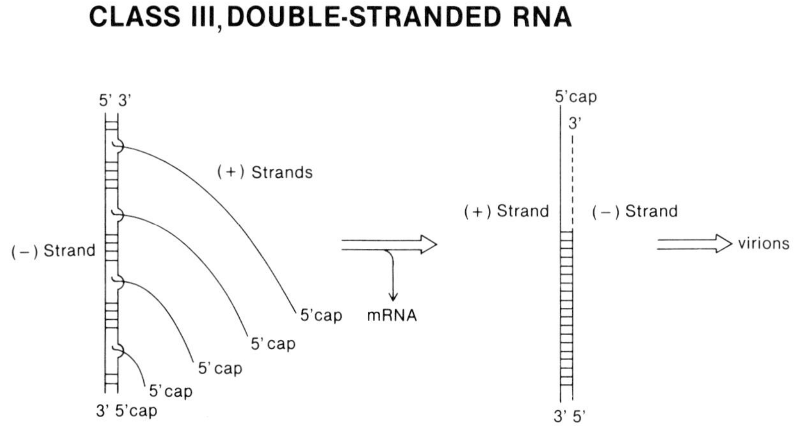

Fig. 3–8. Synthesis of RNA by reoviruses. The (−) strand of virion-type RNA is used as a template to make new (+) strands which contain a 5′ cap but no poly A sequences. These newly made (+) strands function as mRNA or as templates for making new (−) strands and double-stranded virion RNA.

CLASS IV_a, SINGLE-STRANDED (+) RNA

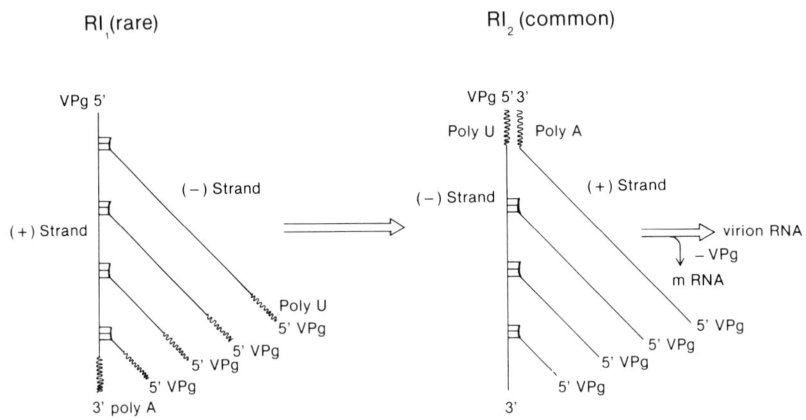

Fig. 3–9. Replication of poliovirus RNA. Virion type (+) RNA is used as template for making (−) strand RNA in replicative intermediates (RI_1). The (−) strand RNA is used as template for making new (+) strands in RI_2. All RNA species in RI_1 and RI_2 have the genome linked VPg at the 5′ end. The 3′ end contains poly A sequences. New (+) strand RNA can be encapsidated into virions or used as mRNA following removal of VPg.

Picornaviruses are an exception to the rule that animal virus mRNAs are capped and methylated. Early translation of poliovirus (+) RNA produces virus-specific polymerases required for genome replication via negative and positive replicative intermediates (RI) (Fig. 3–9). Evidence suggests that a virus protein (VPg) in the replicative intermediate at the 5′ end is a noncapsid protein. The translation product of poliovirus is a large molecular weight polyprotein corresponding to the entire coding capacity of the genome. Specific protease cleavage gives rise to four viral capsid proteins (designated VP_1, VP_2, VP_3, and VP_4) and two noncapsid proteins ($NCVP_2$, $NCVP_4$).

Alphaviruses (+) strand RNA codes only for early synthesis of virus-specific polymerase and serves as template for making (−) strand RNA. The (−) strand RNA is used for making a full length (+) strand RNA and a shorter (−) strand RNA (Fig. 3–10).

In summary, class IV (+) single-stranded viral genomes can function as mRNA or as template for the synthesis of (−) strands. The (−) strands in replicate intermediates serve as templates for the production of new (+) strand viral genomes or as mRNA for making viral structural and nonstructural proteins. The (+) strands have a 5′ cap and 3′ poly A sequences.

e. Class V Viruses, Represented by Orthomyxoviruses, Rhabdoviruses and Paramyxoviruses, are all (−) Strand Viruses. Most of the (−) Strand RNA Genome is Transcribed into (+) mRNA by

CLASS IV_b, SINGLE-STRANDED (+) RNA (ALPHAVIRUSES)

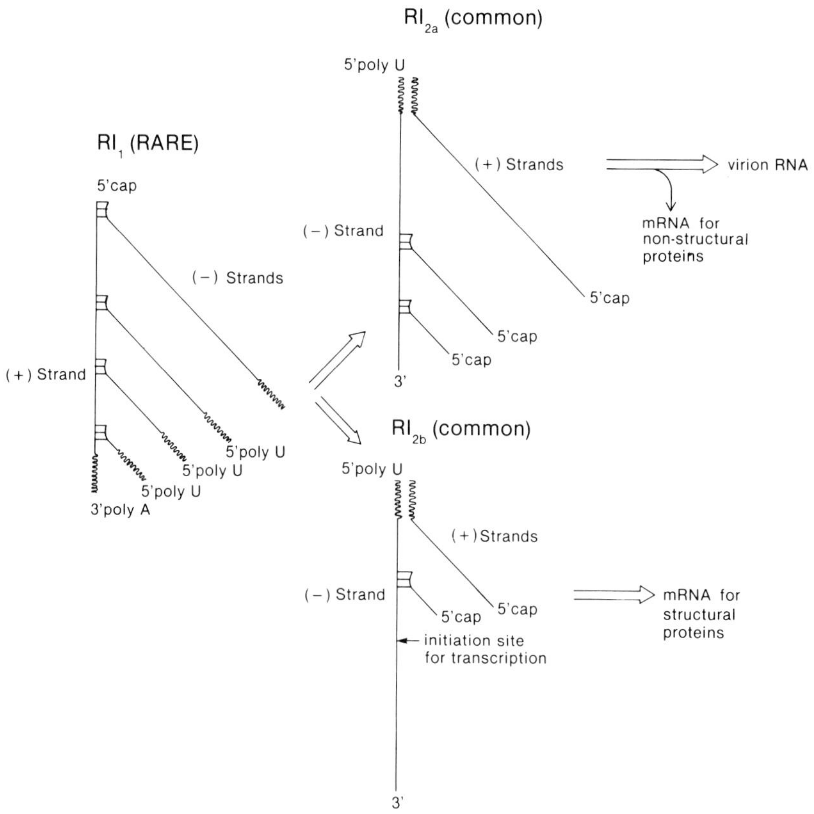

Fig. 3–10. Replication of togavirus (alphavirus) RNA. Virion type (+) RNA is used as template for making (−) strand RNA in replicative intermediates (RI). The (−) strand RNA is used for making two forms of (+) RNA. One form of (+) RNA made on RI_{2a} can be encapsidated into virions or used as mRNA for making nonstructural proteins. The other form of (+) RNA made on RI_{2b} is shorter than virion RNA and is used as mRNA to make virus structural proteins. Virion (+) RNA has 5′ cap and 3′ poly A sequences.

A VIRION-ASSOCIATED TRANSCRIPTASE. It is noteworthy that free viral (−) RNA is not infectious.

Orthomyxoviruses consist of a virus core comprising eight different hairpin double-helical (−) RNA segments. Each (−) strand RNA segment is copied into (+) RNA by the core transcriptase (Fig. 3–11). After capping and polyadenylation each (+) RNA serves as monocistronic message for virus-specific protein synthesis and as templates for the synthesis of (−) RNA genomes via negative replicative intermediates. Early transcription of cellular DNA by RNA polymerase II is necessary for (+) mRNA synthesis. Mixed infection with orthomyxoviruses can

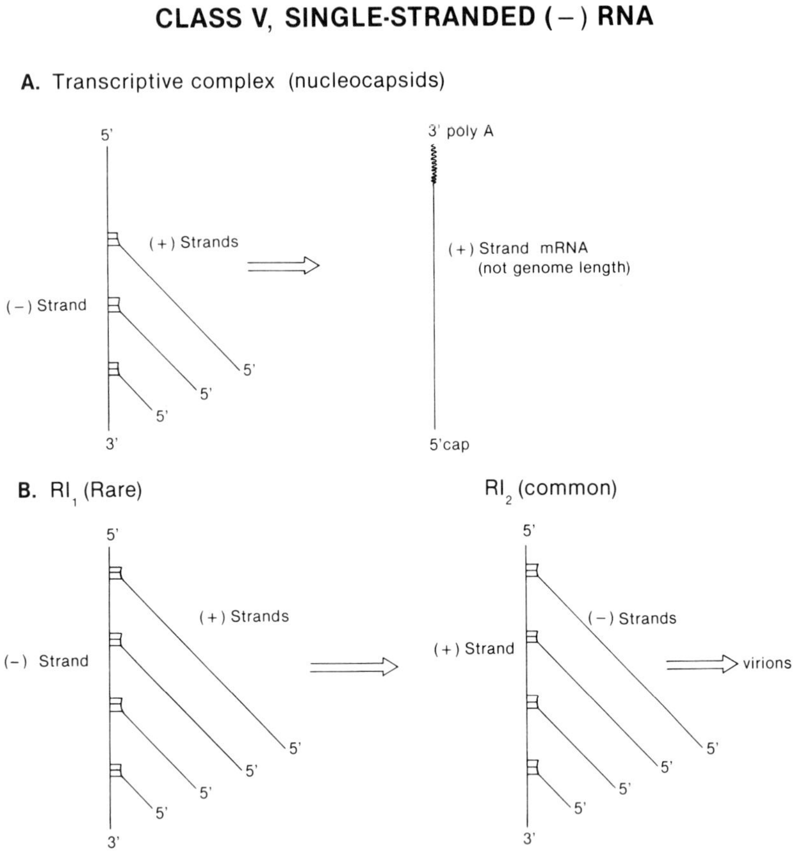

Fig. 3–11. Synthesis of RNA by (−) strand viruses. A. Transcription of (−) strand virion RNA uses a virion-associated transcriptase to make subgenome length (+) strand RNA that contains a 5′ cap and 3′ poly A sequences. B. Virion (−) strand RNA also is used as template to make full length (+) strands in RI_1. The full length (+) strands are used as templates to make new virion (−) strand RNA in RI_2. All (−) strand RNA lack 5′ cap and 3′ poly A sequences.

lead to a high reassortment of (−) RNA segments in progeny viruses (Chapter 12).

Rhabdoviruses contain an intact (−) RNA which is not segmented. The (−) RNA genome is transcribed into 5 monocistronic (+) RNAs via a virion-associated transcriptase. Recent data indicated that transcription is initiated at a single site and messenger size (+) RNA is generated by posttranscriptional cleavages. Each mRNA has a 5′ terminal cap and 3′ terminal poly A sequence and is subgenomic in length. Virion transcriptase is involved in both transcription and posttranscriptional 3′ polyadenylation. Following transcription and translation, synthesis of full length (−) RNA genomes via negative replicative intermediates occurs (Fig. 3–11).

Paramyxoviruses contain either full length (+) RNA or (−) RNA genomes complexed to helically arranged protein subunits in rod-shaped nucleocapsid particles. Therefore, when purified virion RNA is extracted, it self-anneals to form genome length (+) and (−) RNA. The (+) RNA genomes probably serve as intermediates in genome replication. The mRNAs are subgenomic in size similar to the mRNAs of rhabdoviruses (Fig. 3–11). Each mRNA is transcribed via a virion-associated transcriptase and it has a 5′ terminal cap and 3′ terminal poly A sequence. The virion codes for six major proteins.

In summary, the replicative cycle of Class V viruses involves the synthesis of (+) strands of RNA complementary to the virus genome via a virion-associated transcriptase. The (+) strands can function as mRNA or can serve as template for the production of (−) progeny strands.

f. CLASS VI VIRUSES COMPRISE RETROVIRUSES (FORMALLY NAMED ONCORNAVIRUSES, RNA TUMOR VIRUSES, LEUKEMIA VIRUSES). THE GENOME IS A SINGLE STRANDED (+) RNA (70S) THAT CONTAINS TWO IDENTICAL 35S RNAS PLUS STRUCTURAL PROTEINS, REVERSE TRANSCRIPTASE, RNAASE-H AND HOST CELL t-RNA. The two 35S RNA segments are noncovalently linked in the region of their 5′ terminus. The 3′ terminus contains poly A and the 5′ terminus is capped. The RNA primes the initiation of RNA-directed (−) strand DNA synthesis via the virion-associated reverse transcriptase to form a 35S (−) DNA intermediate (Fig. 3–12). The RNAase serves to remove the RNA strand from the RNA:DNA hybrid molecule. The (−) DNA serves as template for making shorter 10S (+) DNA segments. These (+) DNA segments are ligated to the 35S (−) DNA strand to form a double-stranded DNA (provirus). The proviral DNA circularizes, migrates to the cell nucleus and becomes integrated into the host cell genome. Isolated proviral DNA strands are infectious. The integrated proviral DNA is transcribed asymmetrically to the genome size 35S (+) RNA, and subgenomic 21S mRNAs. All mRNAs are translated into virion proteins. 35S (+) RNA also serves as virion RNA to make new progeny virus (Fig. 3–12).

The discovery that the virion RNA of retroviruses can direct the synthesis of DNA has implications far beyond the immediate province of molecular biology. For example, DNA produced by mammalian oncogenic virus RNA may be transmitted indefinitely without evidence of cell alteration or malignancy; alternatively, the regions of the cell genome carrying virus-coded information could become activated many years later and cause "spontaneous" cancer in man.

In summary, these RNA viruses replicate by way of an RNA:DNA hybrid and a double-stranded DNA intermediate which is integrated into the host cell genome. Progeny (+) RNA is transcribed from the integrated viral DNA. The specialized event which permits integration

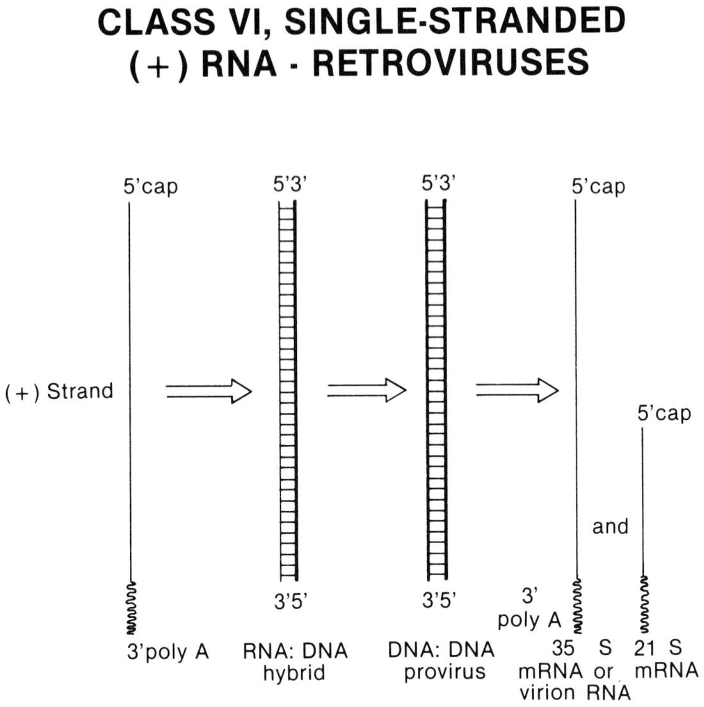

Fig. 3–12. Retroviruses contain two identical (+) strands (segments) of single-stranded RNA with a 5′ terminal cap and 3′ terminal poly A sequence. The (+) RNA is transcribed by a virion-associated reverse transcriptase to a RNA:DNA hybrid and a double-stranded DNA provirus. The DNA:DNA provirus serves as template for making 35S and shorter 21S mRNA. The 35S RNA also is used as virion RNA.

of viral DNA into the host cell genome provides a mechanism by which viral genetic information can be transmitted genetically without phenotypic expression.

E. MATURATION AND RELEASE

The process of virion maturation is determined in part by the site of virus replication and whether or not the nucleocapsid is surrounded by an envelope. The maturation (assembly) or representative naked and enveloped DNA and RNA viruses will be described. A prototype virus is indicated for each category.

1. Naked DNA Virus (Adenovirus).

The polypeptide precursors of adenovirus proteins are synthesized on cytoplasmic polyribosomes and are transported rapidly to the nucleus where they are incorporated into the capsid and internal components of the virus. The production of arginine-rich maturation factor(s) is required for the assembly of the structural proteins and DNA into infectious virus. The virions accumulate in the nucleus and remain cell-associated until they are released gradually as a consequence of cell death and autolysis of infected cells.

2. Enveloped DNA Virus (Herpesvirus).

Proteins synthesized in the cytoplasm are transferred to the nucleus where the nucleocapsids are assembled. The nucleocapsids migrate to the nuclear membrane which has incorporated virus-specific Ags during the course of the infection. By a budding process, the particles become enclosed in an envelope composed of the inner nuclear membrane which contains viral Ags. The enveloped virions leave the nucleus enclosed within cytoplasmic vesicles or vacuoles which appear to open at the cytoplasmic membrane and permit virus release from the cell (Fig. 3–13). If the nucleocapsid leaves the nucleus through a break in the nuclear membrane, it may become enveloped at cytoplasmic or plasma membrane sites. An alternate mechanism for virus release is by way of the perinuclear cisternae, which are continuous with the endoplasmic reticulum and allow the gradual release of virions from the intact cell.

3. Complex DNA Virus (Vaccinia Virus).

Both viral DNA and protein components are synthesized in the cytoplasm and appear as dense granules and fibrils in foci called *factories* or *viroplasm*. At maturation, randomly arranged filaments become surrounded by a newly synthesized multilayered membrane to produce spherical "immature particles." The particles undergo internal differentiation to form the inner membrane, the lateral bodies, and the nucleoid or core which contains the DNA. The external membrane of the

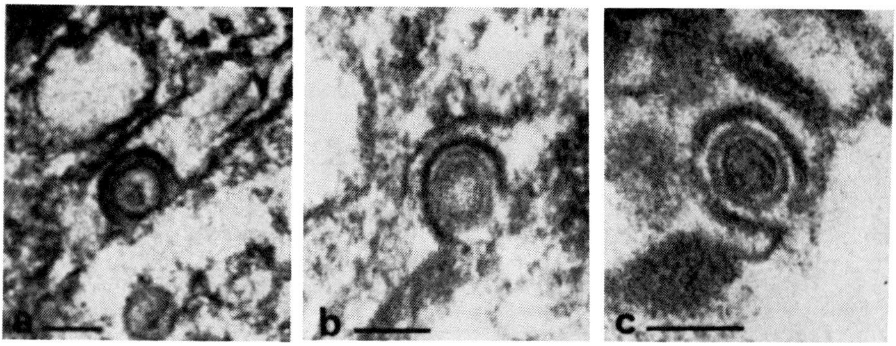

Fig. 3–13. Stages in the budding process of herpesviruses from the inner nuclear membrane of BHK 21/13 cells. Note the thickening of the nuclear membrane as it surrounds the viral particle. Nucleus is at lower right in all micrographs. Bar indicates 0.1 μ (a) H4 virus 21 hr after infection. × 90,000. (b) EAV 12 hr after infection. × 110,000. (c) Pr virus 12 hr after infection. × 144,000. (From Darlington, R.W., and Moss, L.H., III: Herpesvirus Envelopment. J. Virol., 2:48, 1968.)

virion develops its characteristic mature appearance, and the particles accumulate in the cytoplasm. Some virions may be released slowly from microvilli on the cell surface, but most of the virus particles remain within the cell. The virus can spread directly from cell to cell through intercytoplasmic bridges even in the presence of specific neutralizing Ab.

4. Naked RNA Virus (Poliovirus).

Poliovirus RNA and protein precursors are synthesized in membrane-bound cytoplasmic structures, the virus-synthesizing bodies (VSB), in which virion maturation occurs. The capsomeres are formed by self-assembly from monomers of precursor proteins (structural units) which are present in pools within the VSB. As viral RNA is synthesized, it is rapidly enclosed in capsids which are formed by the assembly of capsomeres. Mature virus particles accumulate in the cytoplasmic matrix or within cytoplasmic vacuoles and may resemble intracellular crystals. Lysis of the cell results in the simultaneous and rapid release of large amounts of infectious virus.

5. Enveloped RNA Virus (Parainfluenza Virus).

These viruses are assembled from RNA and proteins produced in the cytoplasm. Their organization into helical nucleocapsids occurs rapidly in the cytoplasmic matrix, presumably by self-assembly. Some virus Ags are incorporated into the cell plasma membrane to form "spikes." The nucleocapsids become aligned in close proximity to the altered cell membrane. Virus maturation is completed when one or more of the nucleocapsids becomes encompassed by a portion of the "spike-containing cell membrane" (Fig. 3–14). Infectious virus is released gradually by a process of exocytosis (budding).

F. ABNORMAL REPLICATIVE CYCLES

1. Abortive Infections.

The infection of cells with virus does not always result in the production of infectious progeny. In nonpermissive cells, some or all of the viral components may be produced, but virus assembly or maturation either does not take place or yields only noninfectious virus particles. This type of virus-host cell interaction is referred to as an *abortive infection.* A virus which undergoes an abortive replicative cycle in one cell type may replicate normally in another. Thus, the defect which prevents infectious virus production in these systems is usually a property of the cell and not the virus.

2. Defective Viruses.

In other systems, the virus is genetically defective and is unable to cause the synthesis of some product essential for the production of

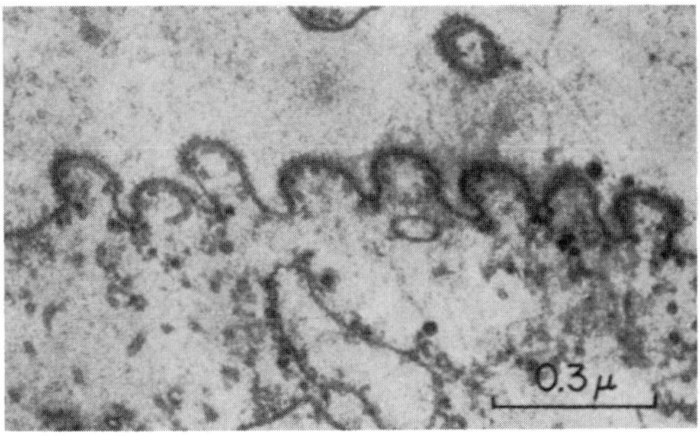

Fig. 3–14. A row of eight budding particles showing many cross-sections of nucleocapsid aligned under the cell membrane (× 63,000). (Reproduced with permission from Compans, R.W., Holmes, K.V., Dales, S., and Choppin, P.W.: An electron microscopic study of the moderate and virulent virus-cell interactions of the parainfluenza virus SV5. Virology, *30*:411, 1964.)

infectious progeny. These viruses can replicate normally if the cells are infected with a "helper virus" which can code for the missing product. The Bryan high titer strain of Rous sarcoma virus (RSV) is an example of a virus defective in its capacity to synthesize essential envelope components. Infectious RSV particles are produced only in cells coinfected with Rous-associated virus (an avian leukosis helper virus) or in cells which contain *"chick helper factor."* The helper factor is endogenous leukosis virus genetic material that is integrated into the cell genome and is transcribed and translated into the necessary envelope components (Chapter 19). Rous sarcoma virus can be surrounded by envelopes containing Ags determined by any avian leukosis virus (Phenotypic mixing, Chapter 4). Thus, immunologically distinct pseudotypes of RSV containing identical genomes can be produced depending upon which helper virus is involved.

The oncogenic DNA viruses (e.g., SV40) exemplify the extreme state of defectiveness. In transformed cells, the virus genome may be integrated into the host cell genome and replicated with it during mitosis. Although viral-specific mRNA and Ags are present in transformed cells, no infectious virus is produced. In some cases the integrated virus can be "rescued" by culturing the nonproducing cells with cells that support the normal replication of the virus (Chapter 34).

3. Viral Interference.

Cells infected by more than one virion may support the normal replication of both viruses or one virus may inhibit or enhance the multiplication of the other. Virus-induced inhibition of viral replication is called the *interference phenomenon* (Chapter 4).

There are three major categories of interference depending upon the relatedness of the viruses involved. (1) *Autointerference* may be observed when a cell is infected with one type of virus at high multiplicities. (2) *Homologous interference* occurs between viruses in the same taxonomic group and can be heterotypic or homotypic. Homologous *heterotypic interference* occurs between two different types of virus belonging to the same group, for example, influenza type A and type B. Homologous, *homotypic interference* occurs between viruses in the same taxonomic group. *Homotypic interference* also occurs between different strains of a single type of virus; for example, between influenza types A_1 and A_2. (3) *Heterologous interference* occurs between a wide range of unrelated viruses.

Viral interference can be affected at the level of adsorption or, more commonly, at some later intracellular step in the replicative cycle. The most important mediator of interference in natural virus infections probably is interferon (Chapter 23). Interferon is a virus-induced, cell-coded protein which mediates inhibition of virus replication at the intracellular level and may be involved in any or all of the types of interference named above. It undoubtedly is responsible for most heterologous interference. However, since some viral interference occurs when interferon production is absent or suppressed, factors other than interferons must be involved in the phenomenon of interference.

Interference which blocks virus adsorption or penetration results from competition for or destruction of receptor sites and thus can occur only between viruses which have common receptor sites. This type of interference, called *viral attachment interference,* is involved in some kinds of autointerference as well as some types of homologous interference.

Homologous interference sometimes occurs only if the interfering virus is allowed to replicate for a short time before the challenge virus is added. The most likely explanation for this observation is that there is competition for substrates or viral replicating sites which are required by both viruses for multiplication.

A special type of autointerference occurs when concentrated preparations of influenza virus are passaged serially. Increasing amounts of noninfectious (defective) virus are produced with each passage. The defective particles are deficient in one particular segment of RNA. The "incomplete" influenza virus interferes with the production of infectious virus; therefore, each passage contains decreasing amounts of infectious virus.

The interference phenomenon undoubtedly has practical significance in natural virus infections. On the one hand, it may be important in containing virus infections and mediating recovery from clinical illness before protective levels of Ab are produced. In contrast, its effect may be detrimental when polyvalent vaccines are administered. For example, homologous heterotypic interference may develop if equal amounts of all three types of attenuated poliovirus are administered simultaneously. This problem has been overcome by applying knowledge of the dynamics of viral interference. The composition of the vaccine is adjusted so that the interfering type is present in lower concentration than the other types. Another problem which may be encountered in administering active vaccines is that viruses present in the host may interfere with the replication and effectiveness of the vaccine.

Much effort has been expended in attempts to utilize interferon prophylactically or therapeutically for human virus infections. This special example of viral interference will be discussed at greater length in Chapter 22.

REFERENCES

Bachrach, H.L.: Comparative strategies of animal virus replication. Adv. Virus Res., *22*:163, 1978.

Ball, L.A. and White, C.N.: Order of transcription of genes of vesicular stomatitis virus. Proc. Natl. Acad. Sci. (U.S.A.), *73*:442, 1976.

Ben-Porat, T. and Rixon, F.J.: Replication of herpesvirus DNA IV: Analysis of concatemers. Virology, *94*:61, 1979.

Hayward, W.S. and Neel, B.G.: Retroviral gene expression. Curr. Top. Microbiol. Immunol., *91*:217, 1981.

Jacob, R.J., Morse, L.S., and Roizman, B.: Anatomy of herpes simplex virus DNA. XII. Accumulation of head-to-tail concatemers in nuclei of infected cells and their role in the generation of the four isomeric arrangements of viral DNA. J. Virol., *29*:448, 1979.

Kaplan, A.S. (ed.): Organization and Replication of Viral DNA. Boca Raton, Florida, CRC Press, Inc., 1982.

Levine, A.J., van der Vliet, P.C., Rosenwirth, B., Rabek, G., Frenkel, G. and Ensinger, M.: Adenovirus-infected, cell-specific, DNA-binding proteins. Cold Spring Harbor Symposia Quant. Biol., *39*:559, 1974.

Levintow, L.: The reproduction of picornaviruses. *In* Comprehensive Virology. Vol. 2. H. Fraenkel-Conrat and R.R. Wagner (eds.). New York, Plenum Press, 1974, pp. 109–169.

Luria, S.E., Darnell, J.E., Jr., Baltimore, D., and Campbell, A.: General Virology. 3rd Ed. Animal Virus Multiplication: DNA Viruses and Retroviruses. New York, John Wiley & Sons, Inc., 1978, pp. 343–389.

Nomoto, A., Kitamura, N., Golini, F., and Wimmer, E.: The 5′-terminal structures of poliovirion RNA and poliovirus mRNA differ only in the genome-linked protein VPg. Proc. Natl. Acad. Sci. (U.S.A.), *74*:5345, 1977.

Palese, P.: The genes of influenza virus. Cell, *10*:1, 1977.

Roizman, B. (ed.): Herpes Viruses. Vol. 1. New York, Plenum Publishing Corp., 1982.

Roizman, B. (ed.): Herpes Viruses. Vol. 2. New York, Plenum Publishing Corp., 1983.

Roizman, B.: The structure and isomerization of herpes simplex virus genomes. Cell, *16*:481, 1979.

Rueckert, R.R.: On the structure and morphogenesis of picornaviruses. *In* Comprehensive Virology. Vol. 6. H. Fraenkel-Conrat and R.R. Wagner (eds.). New York, Plenum Press, 1976, pp. 131–213.

Salzman, N.P. and Khoury, G.: Reproduction of papovaviruses. *In* Comprehensive Virology. Vol 3. H. Fraenkel-Conrat and R.R. Wagner (eds.) New York, Plenum Press, 1974, pp. 63–141.

Sawicki, D.L., Kaariainen, L., Lambek, C., and Gomatos, P.J.: Mechanism for control of synthesis of Semliki Forest virus 26S and 42S RNA. J. Virol., *25*:19, 1978.

Sussenback, J.S., and van der Vliet, P.C.: The mechanism of adenovirus DNA replication and the characterization of replication proteins. Curr. Top. Microbiol. Immunol., *109*:53, 1984.

Taylor, J.M.: DNA intermediates of avian RNA tumor viruses. Curr. Top. Microbiol. Immunol., *87*:23, 1979.

Ward, D.C., and Tattersall, P. (eds.): Replication of Mammalian Parvoviruses. Cold Spring Harbor Laboratory, Cold Spring Harbor, New York, 1978.

Chapter 4

Genetics of Viruses

A. MUTATIONS

The two major mechanisms for genetic modification of viruses are *mutation* and *recombination*. It is important to recognize that the interaction of viruses or their products may also result in nonheritable phenotypic changes in the progeny. Some of the phenomena now known to be the consequence of gene product interactions were once thought to have a genetic basis. An understanding of the possible interactions between viruses is essential for interpreting results obtained when cells are infected with more than one virus.

1. Types of Mutations.

Mutations can arise from changes affecting a single nucleotide (a point mutation) or from larger alterations such as deletions or inversions which may involve hundreds or thousands of nucleotides. Spontaneous mutations occur in animal viruses at approximately the same rate as in other organisms (e.g., mutation rates per gene range from 10^{-5} to 10^{-8} each time a nucleic acid molecule is replicated).

A common cause of point mutations is incorrect base-pairing during either replication or repair of nucleic acid. Spontaneous mispairing can occur when the bases shift from their common tautomeric form to a more rare form. Mutation rates can be increased by treating viruses or their nucleic acid with various physical or chemical agents called *mutagens*. Mutagens mediate mutagenesis either by altering a base so that mispairing occurs or by provoking the repair of nucleic acid which has been damaged by the mutagen. Some analogues of the bases which normally are found in nucleic acid can pair with the normal nucleotides well enough to be substituted for the natural base but poor enough to cause incorrect base-pairing at a later time.

Mutations can occur in practically any measurable detectable phenotypic property of viruses. Variations in plaque type, cytopathic effect, host range, antigenic composition, and drug resistance have been observed. When viruses are serially passed in a host different from the host of origin, certain mutants are selected which express a loss of virulence for the original host. These mutations are referred to collectively as "attenuation." They are usually poorly defined, but are of major importance in the production of live virus vaccines, which retain the immunogenicity but lack the virulence of wild type virus. Attenuated virus vaccines, such as the Semple rabies vaccine, which was empirically developed by Pasteur in 1885, were used long before the molecular mechanism involved in mutation and attenuation were understood.

Analysis of viral genomes by techniques that can detect changes in the primary sequence of nucleotides reveals that many mutations occur that do not result in a measurable change in the phenotype of the virus and are therefore "silent." These mutations occur frequently during the natural (and laboratory) evolution of viruses. The number of such mutations can be used as a measure of the relatedness of viral isolates and thus may be used in analysis of the epidemiology of virus infections.

2. Conditional Lethal Mutations.

Viruses that have mutations which are lethal under certain conditions but which can replicate and yield mutant progeny under other conditions are referred to as *conditional lethal mutants*. Perhaps the most useful conditional lethal mutants are temperature sensitive (ts) mutants, which have been isolated from many animal viruses. These mutants will grow normally at a low (permissive) temperature but not at a high (nonpermissive) temperature, while the wild type virus grows at both temperatures. The basis of temperature sensitivity resides in the instability (or denaturation), at the nonpermissive temperature, of some viral protein(s) that is reasonably stable at the permissive temperature. The difference between the permissive and nonpermissive temperatures may be quite small (5–7° C) because of the cooperative nature of protein denaturation, which can occur over a narrow temperature range. However, temperature sensitivity is rarely an all-or-none phenomenon. The extent of production of progeny by a ts mutant at the nonpermissive temperature is referred to as the "leakiness" of the mutant. The leakiness of a variety of mutants from the same virus usually varies over a wide range.

Temperature sensitive mutants are powerful tools in analyzing the mechanisms of virus replication. Replication of ts mutants at the nonpermissive temperature usually occurs normally until a stage is reached at which the mutant protein is required. Analysis of such an abortive infection can often provide information about the function of the protein that has undergone mutation. Temperature sensitive mutants are also of potential use as vaccine strains of virus, especially for respiratory pathogens such as influenza virus. These vaccine strains would presumably be able to replicate at the lower temperatures of the upper respiratory tract while not being able to replicate and cause disease at the higher temperature of the lower respiratory tract.

3. Deletion Mutations.

Viable *deletion mutants* arise by deletion of genes or parts of genes that are not essential for virus replication. For example, tumor viruses may undergo deletion of genes involved in oncogenic transformation, such as the src gene of avian sarcoma viruses (Chapters 19, 34) or the small t antigen of SV40 (Chapters 6, 34), without losing the ability of the virus to replicate.

Special types of deletion mutations can occur experimentally when viruses are serially passed at high multiplicities of infection. These mutants are referred to as *defective interfering* (DI) particles because they interfere with the replication of the parental virus but are incapable of independent replication. These defective viruses acquire the ability to effectively compete with parental infectious virus for virus-coded replicatory enzymes. It is of special significance that DI particles may play a role in slow or persistent virus infections (Chapter 21).

B. GENETIC INTERACTIONS INVOLVING RECOMBINATION

1. Mechanisms of Recombination.

Recombination is the exchange of genetic material between two viruses which have infected the same cell. It can occur when (1) both viruses are active (i.e., infectious), (2) one virus is active and the other is inactive or (3) both viruses are inactive. Recombination may lead to the production of genetically stable progeny *(recombinants)*, which possess characters not found together in either parent. The mechanism for recombination between viruses with nonsegmented genomes involves the breakage and reunion (crossovers) of the nucleic acid molecules from the two viruses (Fig. 4–1). In the case of viruses which have segmented genomes, recombination can result from the encapsidation of progeny nucleic acid segments produced by different parents (reassortment).

Recombination by crossing over occurs within almost all groups of DNA viruses and within the retrovirus group at the DNA provirus stage (Chapter 19). The genomes of picornaviruses are the only RNAs known to undergo crossovers among animal viruses. The frequency of recombination between two genetic markers depends upon the distance they are separated in the genome; the probability of a crossover event in a given region of nucleic acid strand is proportional to its length. Thus, three or more genetic markers can be ordered (mapped) on the viral genome by analysis of their relative recombination frequencies.

Recombination by reassortment only occurs between viruses with segmented genomes which includes only RNA viruses (e.g., influenza viruses, reoviruses, bunyaviruses). Recombination by this mechanism

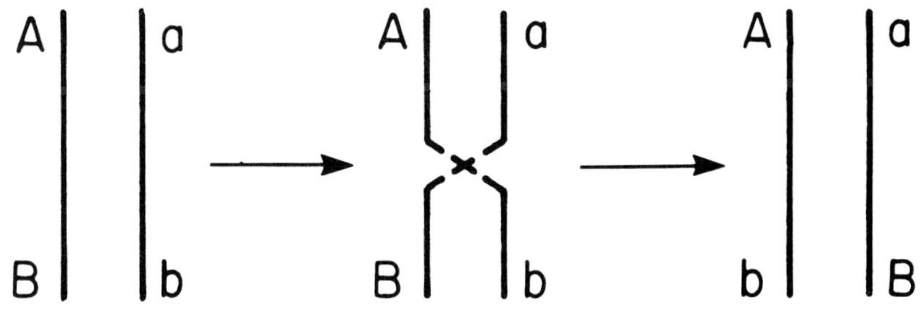

Fig. 4–1. Recombination between viruses results from the breakage and reunion (crossing over) of homologous regions in the nucleic acid molecules from two viruses.

occurs with high frequency. The genetic "map" consists of assigning a genetic marker to a specific genome segment based on the segregation of the marker and its genome segment in many recombinants independent of the remaining segments. Successful mapping requires that the genome segments of the two parents be distinguishable which can usually be accomplished based on size or primary sequence.

2. Recombination in Nature.

Recombination between viruses appears to be an important mechanism for the generation of pathogenic virus strains in nature. For example, pandemic strains of influenza A virus have appeared 3 to 4 times since 1900. These strains have immunologically distinct surface Ags and are thus unaffected by existing host immunity (antigenic shift, Chapter 12) to the parental strain. Antigenic shifts appear to result from recombination between animal influenza viruses, which results in new antigenic surface molecules, and human influenza viruses, which contribute the genes for virulence in humans. This process can be "reversed" in the laboratory. For example, influenza virus vaccine strains can be generated by recombination between the currently circulating pathogenic strain and previously generated strains that lack virulence for humans. Recombinants are selected which possess the desired Ags of the current strains and the avirulence of the vaccine strain.

Pathogenic strains of retroviruses may also arise by recombinational events. For example, viruses which rapidly produce leukemias or sarcomas, such as the Rous sarcoma virus, may have arisen by recombination between retroviruses of low oncogenicity and normal host cellular genes, presumably of a regulatory nature. The inappropriate expression of these genes following virus infection then leads to malignant transformation of these cells (Chapter 19).

3. Recombination Involving Inactive Viruses.

a. MULTIPLICITY REACTIVATION is the production of infectious virus by a cell co-infected with two or more inactive viruses of the same strain. Active virus is produced as a result of recombination between the damaged nucleic acids of inactive viruses which have suffered lethal mutations in different genes; e.g., following exposure to UV-irradiation. Since the administration of vaccines produced by UV-irradiation of viruses could result in the production of infectious virus by way of multiplicity reactivation, this method of inactivation is not currently used for the production of vaccines.

b. CROSS-REACTIVATION OR MARKER RESCUE is a consequence of genetic recombination between an active and a different, but related, inactive virus. This phenomenon results in the production of active progeny possessing one or more genetic traits from the inactive parent.

C. NON-GENETIC INTERACTIONS IN MIXED INFECTIONS

In addition to interactions involving genetic recombination, there can be interactions between viral-induced gene products (proteins) produced in cells infected with more than one virus.

1. Phenotypic Mixing.

This is the incorporation of the genome of one virus into a capsid or envelope which contains some components produced by another virus. The altered phenotype is not a stable change because, upon subsequent passage, the phenotypically mixed virus will produce progeny enclosed in capsids or envelopes coded by the parental genome (Fig. 4–2).

a. Transcapsidation is the mismatching of entire capsids between two viruses replicating in the same cell. This phenomenon is commonly observed when cells are infected with two different enteroviruses. For example, progeny virus released from cells infected with poliovirus and coxsackievirus or echo and coxsackieviruses often have the genome of one virus enclosed within the capsid of the other. In some

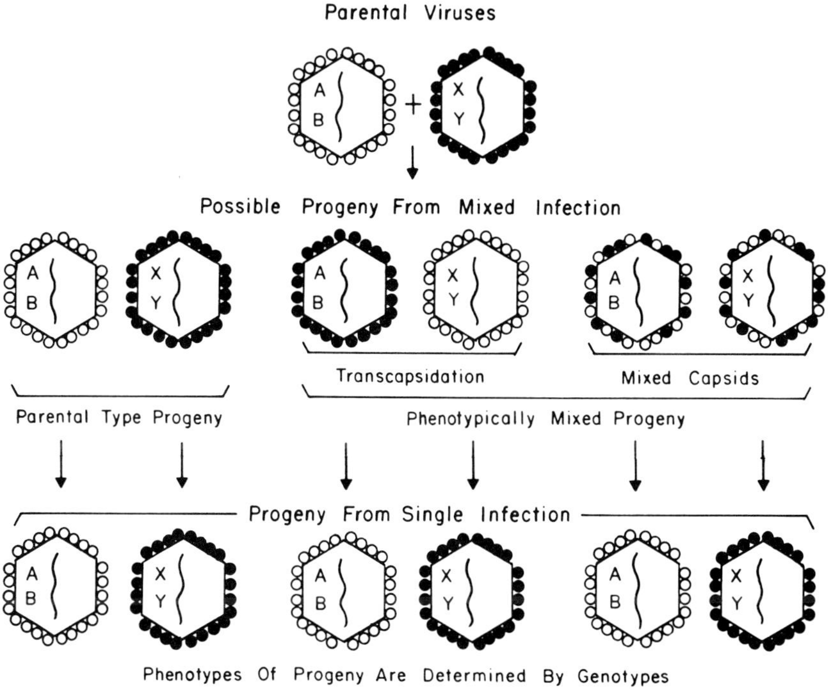

Fig. 4–2. Phenotypic mixing. When cells are simultaneously infected with more than one virus, the progeny from the mixed infection may include (1) viruses identical to the parental type, (2) viruses which contain the genome of one virus and the capsid of another (transcapsidation), and (3) viruses with capsid components of both viruses (mixed capsids). When cells are infected with high dilutions of the progeny from mixed infections, the phenotypes of the viruses produced in the singly infected cells are determined by the genotypes of the infecting virus.

instances, genomes also may be enclosed in "mixed" capsids composed of capsomeres produced by both viruses. This latter type of interaction has been observed following infection of cells with different serotypes of poliovirus or different adenoviruses (Fig. 4–2).

b. Mosaic Envelopes contain a mixture of viral surface glycoproteins produced by at least two viruses. This type of phenotypic mixing occurs between viruses that acquire their envelopes by budding from host cell membranes. The glycoproteins of both parental virus types are inserted into the cell membrane and can be incorporated into the envelope during the budding process. This type of phenotypic mixing can occur between widely divergent virus types such as rhabdoviruses and herpesviruses as well as between closely related virus types.

2. Polyploidy and Genotypic Mixing (Heteropolyploidy).

Polyploidy and *genotypic mixing* result from the incorporation of more than one complete genome into the same virus particle. There is no recombination between the two genomes and progeny of both types are produced upon subsequent passage (Fig. 4–3).

In cells infected with genetically different forms of a virus, "heteropolyploid" particles may be produced by the accidental incorporation of two different nucleocapsids into the same envelope at maturation. Cells singly infected with the genotypically mixed viruses will yield progeny identical to both original parents as well as more genotypically mixed viruses. The production of polyploid particles is a common oc-

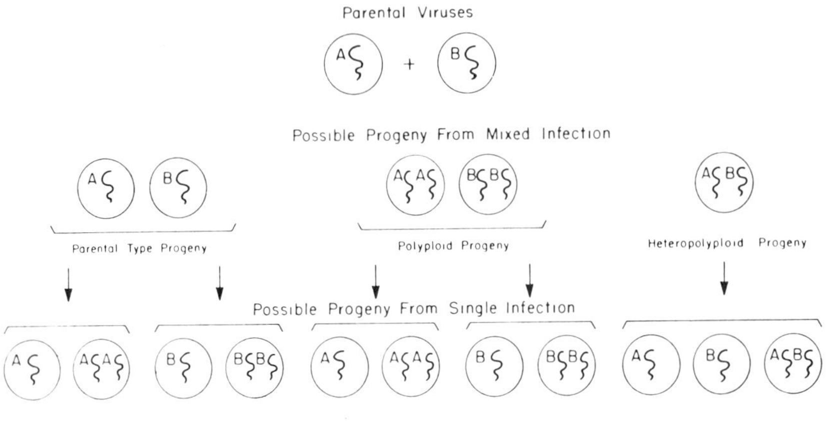

Fig. 4–3. Polyploidy and genotypic mixing. In cells simultaneously infected with variants of a virus, the progeny from the mixed infection may include (1) viruses identical to the parental types, (2) polyploid particles produced by the incorporation of two nucleocapsids of the same virus into one envelope at maturation, and (3) heteroploid particles produced by the incorporation of two different nucleocapsids into one envelope. Cells singly infected with the progeny from a mixed infection may produce more parental type viruses, polyploid particles, or heteropolyploid particles, depending upon the genotype of the infecting virus particle.

currence in cells singly infected with some viruses, for example, the paramyxoviruses. Enveloped virions which contain multiple nucleocapsids are readily detected with the electron microscope.

3. Complementation.

Complementation refers to the interaction of two viruses which results in the production of infectious progeny of one or both types under conditions in which normal replication ordinarily would not occur. Complementation can involve (1) an active and a defective virus, (2) an active and an inactive virus, (3) two defective viruses or two conditional lethal mutants (Fig. 4–4). The viruses may be related or unrelated. Since there is no exchange of nucleic acid between the viruses the individual genotypes are not altered.

In all cases of complementation, one virus induces the production of some essential gene product which the other requires and is unable to synthesize. Since the genotype of the viruses is not altered, the progeny of defective viruses will also be defective and will have the same gene

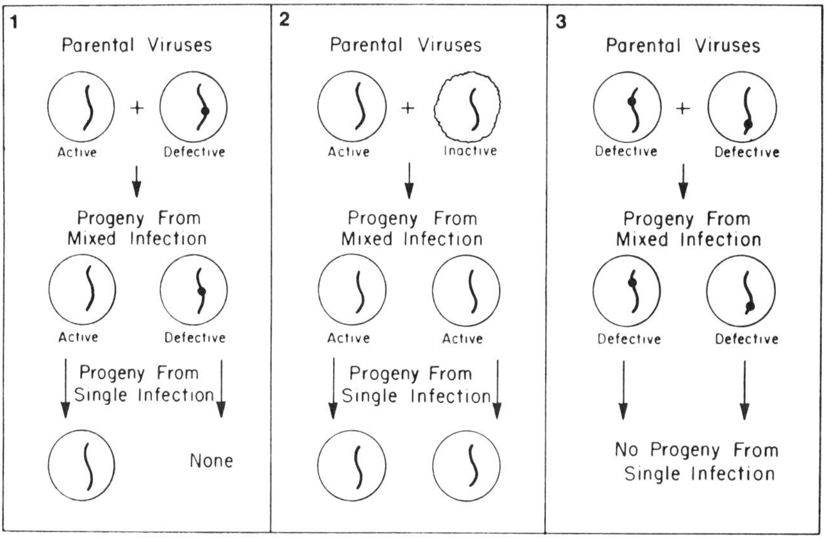

Fig. 4–4. Complementation patterns. Mutations which result in defectiveness are indicated by (ʃ). A defective capsid is indicated by (ʃ). (1) In cells infected with an active and a defective virus, the gene product produced by the active virus can be utilized by the defective virus so that progeny defective viruses are produced as well as active viruses. However, in cells singly "infected" with the defective progeny no virus is produced because no wild-type allele of the active virus was present to complement the mutant cistron in the defective virus. (2) Cells infected with active virus and virus inactivated by methods which damage the capsid but not the nucleic acid yield active virus progeny of both types which will replicate normally in singly infected cells. (3) Two viruses defective in different cistrons may complement each other in mixed infections so that defective progeny of both types are produced. In single infections, neither defective virus will replicate because each still requires the gene product produced by the other virus.

product requirement as the parental virus. For example, most strains of Rous sarcoma virus are defective because they are unable to code for the synthesis of envelope protein(s) necessary for infectivity. The avian leukosis viruses, which act as helper viruses, provide the missing envelope protein, and complete Rous sarcoma virus is produced by unilateral phenotypic mixing. However, the progeny are incapable of independent replication upon subsequent passage unless a helper virus is present (case 1, Fig. 4–4).

An example of complementation of an inactive virus containing a nondefective genome by another active virus (case 2, Fig. 4–4) occurs with poxviruses that have been heat-inactivated. The heat-inactivation denatures a virion RNA polymerase but leaves the viral genome intact. Upon co-infection with both a heat-inactivated virus and a related active poxvirus, the RNA polymerase from the active virus can serve in the replication of the inactive virus, and thus progeny of both parental types are produced.

Complementation between two defective viruses (case 3, Fig. 4–4), can occur, for example, between two ts mutants at the nonpermissive temperature. Co-infection with the two parental ts mutants is performed at the nonpermissive temperature; the progeny that arise as a result of complementation can only be assayed at the permissive temperature, because no exchange of genetic information occurred. Consequently, progeny virus are still defective. Complementation between two defective viruses can only occur if they are defective in different genes. Thus, ts mutants which cannot complement each other are said to belong to the same "complementation group," and are presumed to be defective in the same gene unless proven otherwise. The number of complementation groups into which the ts mutants of a virus fall provides a rough estimate of the minimum number of genes the virus possesses.

REFERENCES

Huang, A.S., and Baltimore, D.: Defective interfering animal viruses. *In* Comprehensive Virology. Vol. 10. H. Fraenkel-Conrat and R.R. Wagner, (eds.). New York, Plenum Press, pp. 73–116, 1977.

Palese, P.: The genes of influenza virus. Cell, *10*:1, 1977.

Palese, P. and Young, J.F.: Variation of influenza A, B, and C viruses. Science, *215*:1468, 1982.

Young, J.F., Elliott, R.M., Berkowitz, E.M., and Palese, P.: Mechanisms of genetic variation in human influenza viruses. Ann. N.Y. Acad. Sci., *354*:135, 1980.

Chapter 5

Parvoviruses

A. PARVOVIRIDAE FAMILY

1. Nondefective and Defective Parvoviruses.

The family Parvoviridae (parvo = small) are the smallest of all known DNA containing viruses; most parvoviruses have been discovered in the last 20 years. Whereas many animals harbor their own latent parvoviruses (Table 5–1), only the adeno-associated virus (AAV) serotypes 2 and 3 have man as a natural host. There is suggestive evidence that at least two other human viruses, acute infectious gastroenteritis virus (Chapter 31) might also belong to the family Parvoviridae. There are two major groups of parvoviruses, (1) nondefective viruses (autonomous replication) such as rat virus, and, (2) defective viruses such as AAV (Table 5–1). AAV requires the presence of a helper virus, either a nondefective adenovirus or a herpesvirus, to initiate their replication cycle. Parvoviruses are considered potential contaminants of cells, virus stocks, and vaccines because of their ambiguous nature, their tendency to be latent in animal tissues, and their remarkable stability in the outside environment.

Structure and Antigenic Properties. Parvoviruses have a particle size of about 18–26 nm, approximately the size of a ribosome (Table 5–2). Virions have icosahedral symmetry (Fig. 5–1) with 32 capsomeres

Table 5–1. Source of Some Parvoviruses

Virus	Source
Mink enteritis virus (MEV)	Mink (liver, spleen)
Rat virus (RV)	Rat (tumor)
H-1	HEp-1 cells
Hemadsorbing enteric virus (HADEN)	Cattle
Densonucleosis virus (DNV)	Insect (Galleria mellonella)
Adenovirus-associated virus (AAV)	
Serotype 1	SV15 contaminant
Serotypes 2 and 3	Adenovirus contaminants
Serotype 4	SV15 contaminant
Feline panleukopenia virus (FPV)	Cat
Minute virus of mice (MVM)	Mouse
KBSH	Pig (KB cells)

Table 5–2. General Properties of Parvoviruses

Properties of nucleic acid: single-stranded DNA; 1.5–2.2 × 10^6 daltons
No. of genes: 7
Virion: nonenveloped; 32 capsomeres
Capsid symmetry: icosahedral
Size (nm): 18–26
Site of replication: nucleus

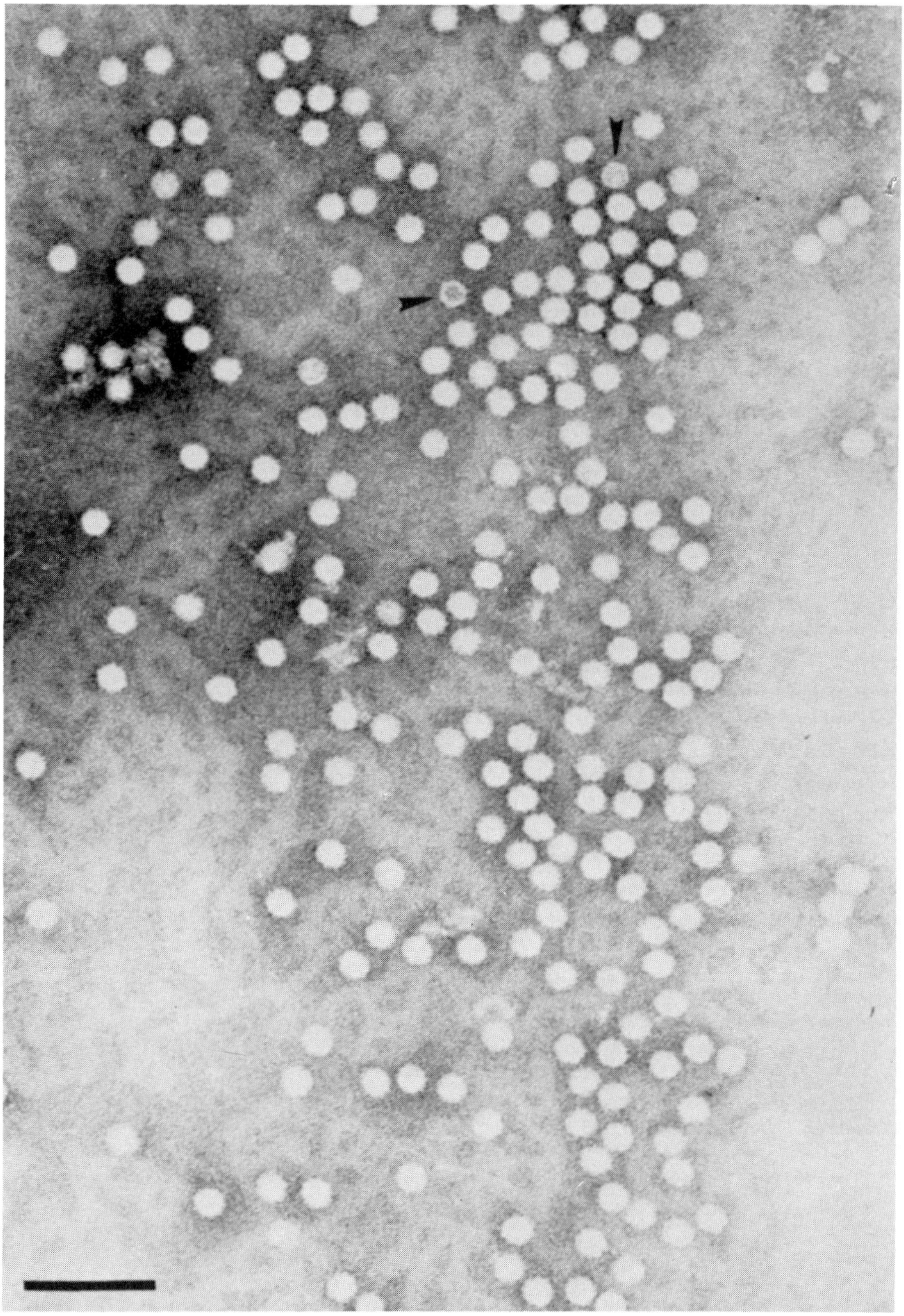

Fig. 5–1. Electron micrograph of nondefective parvoviruses. (Reprinted with permission. Palmer, E.L., and Martin, M.L.: An Atlas of Mammalian Viruses. Boca Raton, Florida, CRC Press, Inc. 1982.)

and no envelope. Their single-stranded DNA has a molecular weight of 1.5 to 2.2 $\times$ 10^6 daltons. All parvovirus DNAs are linear; however, they have inverted terminal repetitions and are capable of forming circular molecules by small hydrogen-bonded duplex segments (Chapter 3). AAV encapsidates complementary strands of DNA separately. Upon DNA extraction of virions in vitro, the complementary strands hybridize to form double strands of DNA. Extracted double-stranded DNA of AAV is infectious.

All parvovirus capsids contain three different sizes of polypeptides with the largest having a molecular weight of 90,000 daltons. The tryptic peptides of the two smaller proteins are subsets of the tryptic peptides of the 90,000 dalton polypeptide. Conceivably, parvovirus mRNA codes for a 90,000 dalton capsid protein which is subsequently cleaved to two smaller proteins which make up the rest of the capsid. It was reported that purified preparations of rat virus has a virion-associated DNA polymerase.

Reportedly, Abs to rat virus are acquired by suckling animals from their mothers. Infection occurs between 2–7 months of age and at least 60 percent of the rats can become infected in a closed laboratory colony. Antibodies to AAV serotypes 2 and 3 were detected in a high percentage of the human population. Humans receiving virus vaccines prepared in simian cells often have Abs directed against a simian AAV serotype 4 (Table 5–1).

Host Range and Culture. The host range for parvoviruses in vivo is restricted but in vitro a wider host range can occur. They have often been isolated from many different permanent cell lines (e.g., HeLa). Commercial preparations of trypsin used for cell cultures has been found contaminated with porcine parvovirus.

Replication Cycle. Growth of parvoviruses is highly dependent on rapidly proliferating host cells and cellular functions supplied during late S phase. This may explain why tumor cells have been the source of many parvovirus isolates and why certain parvoviruses cause developmental abnormalities in embryos. Approximately 75 percent of virions attach to a host cell monolayer after 2 hours of incubation. The eclipse period of about 10–16 hours is followed by an exponential growth phase which occurs between 16 and 24 hours after virus attachment. Viral DNA is converted to a double-stranded replicative form (RF) and after duplication of the RF, single strands are produced (Chapter 3) which subsequently become encapsidated. Virus replication occurs in the cell nucleus and is characterized by the formation of dense DNA-positive inclusions. Nondefective parvoviruses are capable of autonomous replication. Defective AAV depend completely on helper viruses. Adenoviruses are complete helper viruses, whereas herpesviruses are incomplete helpers, suggesting at least two defective steps in AAV replication.

Virus replication is measured by CF, direct particle counting, HA or fluorescent AB tests.

REFERENCES

Berns, K.I.: Molecular biology of the adeno-associated viruses. Curr. Top. Microbiol. Immunol., *65*:1, 1974.

Berns, K.L. (ed.): The Parvoviruses. New York, Plenum Publishing Corp., 1984.

Hoggan, M.D.: Adenovirus associated viruses. Prog. Med. Virol., *12*:211, 1970.

Luria, S.E., Darnell, J.E., Jr., Baltimore, D. and Campbell, A.: General Virology, 3rd Ed., New York, John Wiley & Sons, Inc., 1978, pp. 343–349.

Rhode, S.L., III: Defective interfering particles of parvovirus H-1. J. Virol., *27*:347, 1978.

Rhode, S.L., III and Klaassen, B.: DNA sequence of the 5′-terminus containing the replication origin of parvovirus replicative form DNA. J. Virol., *41*:990, 1982.

Rose, J.A.: Parvovirus reproduction. *In* Comprehensive Virology, Vol. 3. H. Fraenkel-Conrat and R.R. Wagner (eds.), New York. Plenum Press, 1974, p. 1–61.

Young, J.F. and Mayor, H.D.: Studies on the defectiveness of adeno-associated virus (AAV). 1. Effects of phosphonoacetic acid and 2-deoxy-D-glucose on the replication of AAV. Virology, *94*:323, 1979.

Chapter 6

Papovaviruses

A. PAPOVAVIRIDAE FAMILY

The word "papovavirus" was coined in 1962 when Melnick considered that the three viruses, human *pa*pilloma (wart virus), mouse *po*lyomavirus and *va*cuolating virus or simian virus 40, shared common biologic and physical properties (Table 6–1). Papovaviruses have been subdivided into two major genera; Papillomaviruses and Polyomaviruses (Table 6–2). Biologically, the members of the Papovaviridae family have a restricted host range; they characteristically cause persistent infections in the natural host. Papillomaviruses are of special interest to viral oncologists because of their tumorigenic potential either in their natural hosts or in other animal species. Two new polyoma-like viruses have been isolated from man and designated BK and JC virus, respectively. Morphologically, both viruses are identical but they are antigenically distinct. These new viruses are associated with progressive multifocal leukoencephalopathy (Chapter 32) which may be a late sequela of Hodgkin's disease in man.

1. Papillomavirus.

Papillomaviruses induce warts (papillomas) in their natural hosts (e.g., man, cows, dogs). The warts are considered benign tumors but sometimes they may become malignant (e.g., condylomata acuminata, Chapter 34).

Structure and Antigenic Properties. The human papillomavirus is

Table 6–1. General Properties of Papovaviruses

Properties of nucleic acid: double-stranded, circular DNA; 3 to 5 × 10^6 daltons
No. of genes: 4
Virion: nonenveloped; 72 capsomeres
Capsid symmetry: icosahedral
Size (nm): 43–55
Site of replication: nucleus

Table 6–2. Papovavirus Genera

Papillomaviruses	Polyomaviruses
Rabbit	Mouse
Human	Simian virus 40
Bovine	Rabbit kidney vacuolating virus
Canine	K virus (mouse)
Rabbit, oral	BK virus (man)
Canine, oral	JC virus (man)
Equine	
Deer fibroma	

a naked icosahedral particle that contains a circular double-stranded DNA genome. The genome is slightly larger than the genome of polyomaviruses. The capsid is about 53 nm in diameter and contains 72 capsomeres in an outer shell that surrounds a dense core (Fig. 6–1).

Little is known about the antigenic properties of the wart virus, since the only source of material is tissue excised from patients. Antiserum prepared against partially purified virus contains Abs that fix complement and agglutinate virus particles. Virus extracted from one form of wart will produce clinically different types of warts when inoculated in different sites on the body. Although this observation suggests that a single serotype can produce different clinical forms of warts, it does not preclude the existence of more than one serotype of human papillomavirus.

Host Range and Culture. Man is the only known host for human papillomavirus. It has been serially passaged from person to person but has not been transmitted to laboratory animals. Many attempts have

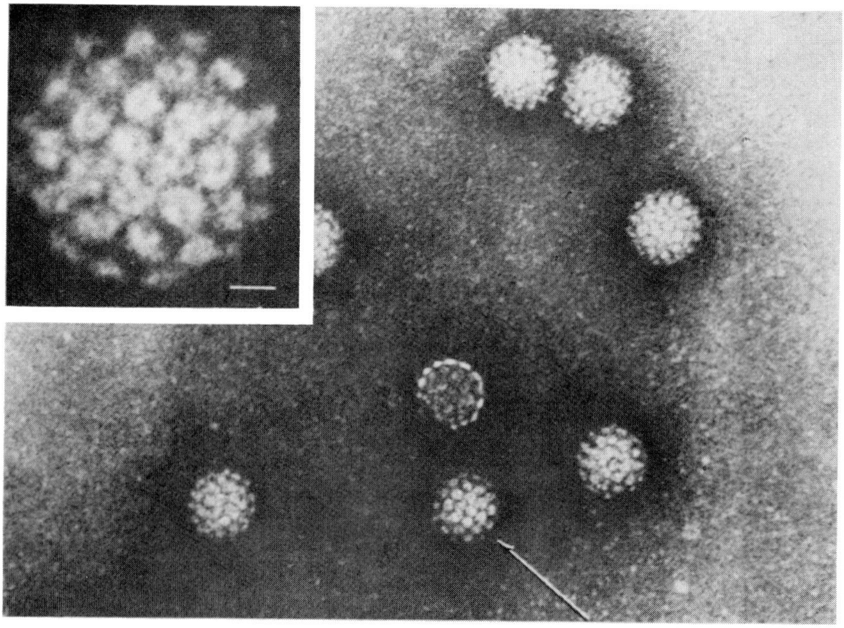

Fig. 6–1. Electron micrograph of human papillomavirus, negatively stained with phosphotungstate. The virus particles are intact and well preserved (arrow) with the exception of one damaged particle. Bar indicates 0.1 nm. Inset is higher magnification of particle at arrow. Bar indicates 100 Å. (Reprinted with permission from Noyes, W.F.: Structure of the human wart virus. Virology, 23:65, 1964.)

been made to propagate the virus in tissue cultures; however, reports of apparent replication of the virus in vitro await confirmation.

2. Polyomavirus.

Polyomavirus causes latent infections in laboratory and wild mice. Infection may occur within the first 2 weeks of life due to contaminated urine or saliva shed by infected adult mice. These viruses are highly tumorigenic when inoculated into newborn mice or hamsters; the tumors produced are of various histologic types, hence the name polyoma.

Structure and Antigenic Properties. Polyomaviruses are morphologically indistinguishable from papillomaviruses; however, they are antigenically distinct and have the capacity to agglutinate red blood cells.

Polyoma virions consist of three major capsid proteins designated VP-1, VP-2 and VP-3 ranging in molecular weight from 20,000–43,000. Capsid proteins are distinct from T (for tumor) and TSTA (tumor-specific transplantation) Ags associated with polyoma virus transformed cells.

Host Range and Culture. Polyoma virus causes latent infections in laboratory and wild mice. Whereas polyomaviruses grow and induce CPE in mouse embryo fibroblast cells, inoculation of hamsters results in tumor formation. The tumor cells can be cultivated in vitro and retain malignancy. The 3T3 cell line of mouse fibroblasts and baby hamster kidney cells can be readily transformed in vitro with these viruses (Chapter 34). Infectious DNA isolated from virions is also tumorigenic. Tumors include spindle cell sarcomas and epithelial tumors which can appear in a number of sites. Infectious virus cannot be recovered from tumors.

Replication. Polyoma virus replication occurs in the cell nucleus. Viral DNA integrates into cellular chromosomes or remains extrachromosomal as supercoiled double-stranded DNA. About 30 percent of the viral genome is transcribed "early" into large (900,000 daltons) transcripts covalently linked to cellular mRNA. The remaining 70 percent of the genome is transcribed "late." All capsid proteins are late viral gene products. During replication some linear host cell DNA is found encapsidated in viral particles called *pseudovirions*.

3. Simian Virus 40.

SV40 is one of the best known and widely studied oncogenic DNA viruses. The virus was originally identified as a latent infection in monkey kidney cells used for production of poliovirus vaccine.

Structure and Antigenic Properties. The virion consists of a major protein, VP-1 and two minor proteins, VP-2 and VP-3, which consist of identical amino acid sequences. Cellular histones are associated with the single, closed-circular duplex DNA molecule. The DNA codes for virion proteins and early proteins (T-antigen, t-antigen or gene A products). Viral DNA can exist in three physical forms. Form I is a supercoiled, covalently closed circular duplex DNA. It is converted to form II by a single-strand break and to form III by a double-strand break. Using restriction endonucleases it is possible to cut SV40 DNA into a large number of defined fragments. The entire nucleotide sequence of the SV40 genome is known. The virus is antigenically distinct from papillomaviruses and polyomaviruses.

Host Range and Culture. SV40 replicates in monkey kidney cells to produce new progeny virus. Infection of rat, hamster or human cells results in *abortive infection* with transformation and little or no new progeny virus production.

Replication. SV40 replication is similar to polyomavirus. Early transcription starts about 6 hours after infection from the minus strand. Late transcription from the plus strand starts about 12 hours. DNA replication is initiated at map coordinate 0.67 units from the Eco R1 restriction endonuclease cleavage site (Chapter 3, Fig. 3–6) and proceeds bidirectionally with production of 4S DNA fragments. These DNA fragments are later joined together by the enzyme ligase to form the complete virus genome. Virus maturation occurs in the cell nucleus.

REFERENCES

Brockman, W.W.: Evolutionary variant of simian virus 40. Prog. Med. Virol., *23*:69, 1977.

Clayton, C.E., Lovett, M., and Rigby, P.W.J.: Functional analysis of a simian virus 40 super T-antigen. J. Virol., *44*:974, 1982.

Crawford, L.V., Cole, C.N., Smith, A.E., Paucha, E., Tegtmeyer, P., Rundell, K., and Berg, P.: Organization and expression of the early genes of SV40. Proc. Natl. Acad. Sci. (U.S.A.), *75*:117, 1978.

Fried, M. and Griffin, B.E.: Organization of the genomes of polyoma virus and SV40. Adv. Cancer Res., *24*:67, 1977.

Graessmann, A., Graessmann, M., and Mueller, C.: Regulation of SV40 gene expression. Adv. Cancer Res., *35*:111, 1981.

Graessmann, A., Graessmann, M., and Mueller, C.: Simian virus 40 and polyoma virus gene expression explored by the microinjection technique. Curr. Top. Microbiol. Immunol., *87*:1, 1980.

Lebowitz, P. and Weissman, S.M.: Organization and transcription of the simian virus 40 genome. Curr. Top. Microbiol. Immunol., *87*:43, 1980.

Levine, A.J.: SV40 and adenovirus early functions involved in DNA replication and transformation. Biochemica et Biophysica Acta, *458*:213, 1976.

Levine, A.J., van der Vliet, P.C., and Sussenback, J.S.: The replication of papovavirus and adenovirus DNA. Curr. Top. Microbiol. Immunol., *73*:68, 1976.

Martin, M.A. and Khoury, G.: Integration of DNA tumor virus genomes. Curr. Top. Microbiol. Immunol., *73*:35, 1976.

Padgett, B.L. and Walker, D.L.: New human papovaviruses. Prog. Med. Virol., *22*:1, 1976.

Tooze, J.: DNA Tumor Viruses. Part 2 of Molecular Biology of Tumor Viruses. 2nd. Ed. Cold Spring Harbor Laboratory, Cold Spring Harbor, New York, 1980.

Chapter 7

Adenoviruses

A. ADENOVIRIDAE FAMILY

1. Adenovirus.

The adenoviruses were discovered in 1953 during a search for a type of cell culture that could be used to isolate the common cold virus. It became apparent that prolonged culture of tonsil and adenoid tissue in vitro would activate viruses that were present as latent infections in healthy individuals. During an epidemic of influenza-like illness in army recruits, several similar agents were isolated in cultures of human cells inoculated with throat washings. These "new" viruses were officially designated adenoviruses to indicate the original tissue location of this group of agents.

The adenoviruses cause only a small proportion of the respiratory illnesses in man. However, some human adenoviruses produce cancer following inoculation into infant hamsters, mice, or rats. These viruses also can transform rodent cells to the malignant state in vitro. Although there is no evidence that adenoviruses are oncogenic in man, they provide an excellent system for studying the mechanism of viral tumorigenesis (Chapter 34).

Structure and Antigenic Properties. The adenoviruses are medium-sized (70–80 nm in diameter) naked viruses that contain double-stranded DNA (Table 7–1). The virions have a dense central core (the nucleoid) surrounded by an outer icosahedral capsid composed of 252 capsomeres (Fig. 7–1). Of these, 240 are called *hexons* because each is surrounded by 6 neighboring capsomeres. The hexons are the major morphologic subunits of the capsid. The capsomeres situated at the 12 vertices of the icosahedron are designated *pentons* because each is bounded by 5 neighbors. A *fiber* containing a terminal knob projects from the base of each penton (Fig. 7–2).

At least 31 serotypes of adenovirus have been isolated from man. The human isolates have been divided into subgroups on the basis of (1) their ability to agglutinate rhesus monkey or rat erythrocytes, (2) their oncogenic potential for hamsters, and (3) the molecular weight and guanine-cytosine content of their DNA. All adenoviruses possess a common group-specific Ag associated with the hexon, as well as subgroup-specific

Table 7–1. General Properties of Adenoviruses

Properties of nucleic acid: double-stranded linear DNA; 20–30 $\times$ 10^6 daltons
Virion: nonenveloped; 252 capsomeres (240 hexons; 12 pentons and fiber proteins)
Capsid symmetry: icosahedral
Size (nm): 70–80
Site of replication: nucleus

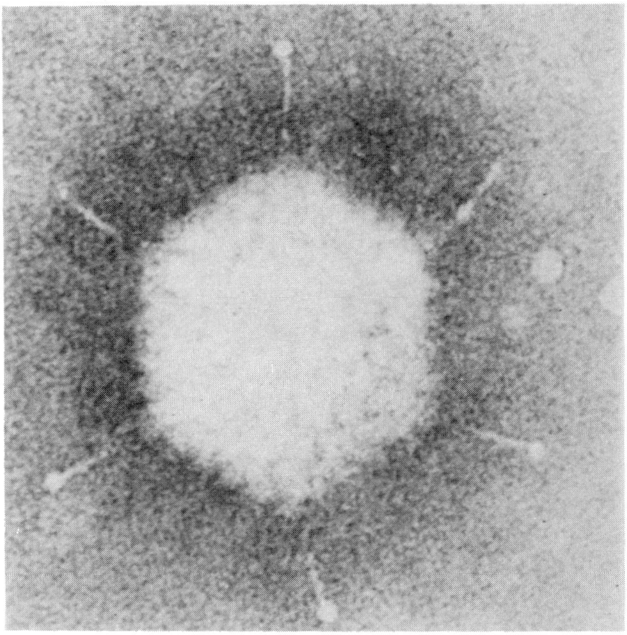

Fig. 7–1. Electron micrograph of adenovirus. Magnification × 500,000. (Reproduced with permission from Valentine, R.C. and Pereira, H.G.: Antigens and structure of the adenoviruses. J. Mol. Biol., *13*:13, 1965.)

determinants on the penton. Type-specific Ags are associated with both the fiber and hexon; type-specific Abs against the fiber inhibit hemagglutination, whereas those against the hexon neutralize infectivity.

Host Range and Culture. Adenoviruses have been isolated from many animal species. With the exception of the human virus isolates which are oncogenic for certain rodents, most are highly species-specific. The host range is reflected in their specificity for cell cultures of their natural host or a closely related species. Primary cultures of human embryonic kidney or amnion and a number of continuous cell lines of human origin support the growth of human adenoviruses. The viruses will multiply in monkey kidney cells only if the cells are also infected with a latent simian virus (SV40) that provides a missing gene product for adenovirus. Adenoviruses grow slowly in cell cultures, and the typical cytopathic effects may not become apparent for 2–4 weeks. The cells, which become rounded and greatly enlarged, usually clump to form grape-like clusters. Nuclei of infected cells may contain characteristic inclusion bodies that are composed of aggregates of virions.

Replication. Complete replication of adenoviruses requires 24 hours (Chapter 3). Infecting virions adsorb to specific receptors and enter cells by pinocytosis or direct penetration. Uncoating occurs in three

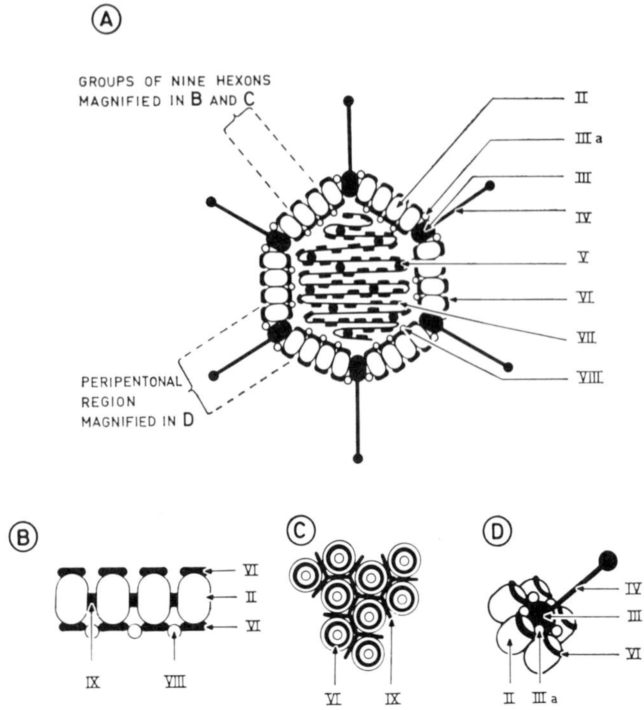

Fig. 7–2. A hypothetical model of the location of the adenovirus proteins. (A) A schematic view of a vertical section of the virion. The different polypeptides are indicated by their Roman numerals as indicated to the left. Magnified views of groups of nine hexons and the peripentonal region are given in B, C, and D. (B) A vertical section through a group of nine hexons showing the tentative location of proteins VI, VIII and IX. (C) A horizontal view from the outside of the group of nine hexons showing the tentative location of the proteins VI and IX. (D) A magnification of the peripentonal region showing the proteins II, III, IIIa, IV and VI. (With permission from Everitt, E., Sundquist, B. Pettersson, U., and Philipson, L.: Structural proteins of adenoviruses. X. Isolation and topography of low molecular weight antigens from the virion of adenovirus type 2. Virology, *52*:130, 1973.)

steps: (1) removal of pentons, (2) removal of hexons from DNA-protein core and (3) release of naked DNA into the cell nucleus. Early and late mRNA's involve several different reactions including polyadenylation at the 3′ terminus, a cap structure at the 5′ end, processing and splicing long transcripts prior to transporting them into the cytoplasm. Early proteins (T-antigen, P-antigen, thymidine kinase, DNA polymerase) and late proteins (hexon, penton, fiber, core proteins) are synthesized on nonmembrane bound polysomes. Fiber proteins are glycosylated and phosphorylated, whereas nonstructural polypeptides are only glycosylated. Viral proteins are transported to the cell nucleus for final assembly into new virus particles.

2. Adenovirus-SV40 Hybrids.

As mentioned above, adenoviruses will grow in monkey kidney cells only if the cells are co-infected with a latent SV40. Progeny adenoviruses contain SV40 genetic information covalently linked to the adenovirus genome. These "hybrid viruses" contain an adenovirus capsid and can be neutralized by specific adenovirus antiserum. There are two types of adenovirus-SV40 hybrids: PARA (particle aiding replication of adenoviruses)-adenovirus and $Ad2^{+}ND$ (nondefective). PARA-adenoviruses consist of two particles, a nonhybrid adenovirion and a defective adenovirus-SV40 genome packaged into an adenovirus capsid. The second hybrid virus $Ad2^{+}ND$ consists of nondefective adenovirus serotype 2 particles containing various amounts (5–44 percent) of the SV40 genome. These hybrid viruses have been useful in analyzing the genetics of SV40.

3. Adenosatellite Virus.

These viruses were originally isolated from adenovirus stocks. Four antigenic types exist in nature. They are antigenically distinct from adenoviruses. Adenosatellite viruses are classified as parvoviruses (Chapter 5).

REFERENCES

Binger, M.H., Flint, S.J., and Rekosh, D.M.: Expression of the gene encoding the adenovirus DNA terminal protein precursor in productively infected and transformed cells. J. Virol., *42*:488, 1982.

Chow, L.T. and Broker, T.R.: The spliced structures of adenovirus 2 fiber message and the other late mRNAs. Cell, *15*:497, 1978.

Nevins, J.R. and Chen-Kiang, S.: Processing of adenovirus nuclear RNA to mRNA. Adv. Virus Res., *26*:1, 1981.

Persson, H. and Philipson, L.: Regulation of adenovirus gene expression. Curr. Top. Microbiol. Immunol., *97*:157, 1982.

Philipson, L.: Structure and assembly of adenoviruses. Curr. Top. Microbiol. Immunol., *109*:1, 1984.

Philipson, L. and Lindberg, U.: Reproduction of adenoviruses. *In* Comprehensive Virology. Vol. 3. H. Fraenkel-Conrat and R.R. Wagner (eds.). New York, Plenum Press, 1974, pp. 143–227.

Philipson, L., Pettersson, U. and Lindberg, U.: Molecular biology of adenoviruses. Virol. Monogr., *14*:1, 1975.

Rapp, F. and Melnick, J.L.: Papovavirus SV40, adenovirus and their hybrids: Transformation, complementation and transcapsidation. Prog. Med. Virol., *8*:349, 1966.

Sussenback, J.S., and van der Vliet, P.C.: The mechanism of adenovirus DNA replication and the characterization of replication proteins. Curr. Top. Microbiol. Immunol., *109*:53, 1984.

Taylor, P.: Adenoviruses: Diagnosis of Infections. *In* Comparative Diagnosis of Viral Diseases. Vol. 1. E. Kurstak and C. Kurstak (eds.). New York, Academic Press, 1977, pp. 85–170.

Zain, S., Sambrook, J., Roberts, R.J., Keller, W., Fried, M. and Dunn, A.R.: Nucleotide sequence analysis of the leader segments in a cloned copy of adenovirus-2 fiber mRNA. Cell, *16*:851, 1979.

Chapter 8

Herpesviruses

A. HERPESVIRIDAE FAMILY

More than 25 viruses comprise the herpesviridae family. Members of the herpesviridae family are morphologically indistinguishable (Table 8–1). The virion is made up of three parts: the core that contains the DNA is surrounded by the capsid, which is enclosed in an envelope composed of one, two, or more concentric membranes (Fig. 8–1). The core is 75 nm in diameter, is usually spherical, and is comprised of the nucleoid (primary body) and an area of low electron density. The icosahedral capsid has a diameter of 100–110 nm and is made up of 162 elongated hollow capsomeres, of which 150 are hexagonal and 12 pentagonal. The diameters of the virion range from 145–200 nm due to variation in the size of the envelope. At the periphery, the envelope is condensed as a membrane 4–5 nm thick from which protrude periodic structures 8–10 nm long and 8 nm thick. Within a population of virions many are "naked" without an envelope. The enveloped virus particle is considered to be infectious (Fig. 8–2).

Table 8–1. General Properties of Herpesviruses

Properties of nucleic acid: linear, double-stranded DNA; 85–150 × 10^6 daltons
Virion: enveloped; 162 capsomeres (24 structural polypeptides)
Capsid symmetry: icosahedral
Size (nm): 145 to 200
Site of replication: nucleus

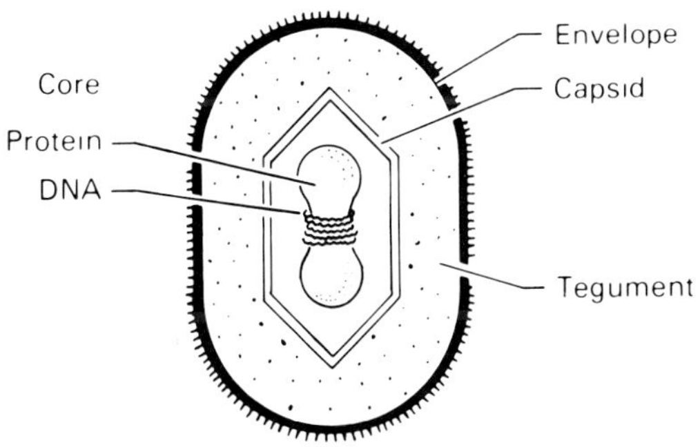

Fig. 8–1. Diagram of a herpes simplex virus particle. The DNA of the virus is wrapped up in a central protein core and capsid proteins. The capsid is composed of 162 elongated hexagonal capsomeres with a central hole. The surrounding tegument between the capsid and envelope consists of globular proteins. The encompassing envelope possesses lateral short projects. (From Dulbecco, R., and Ginsberg, H.S.: Virology. Hagerstown, Harper and Row, 1980.)

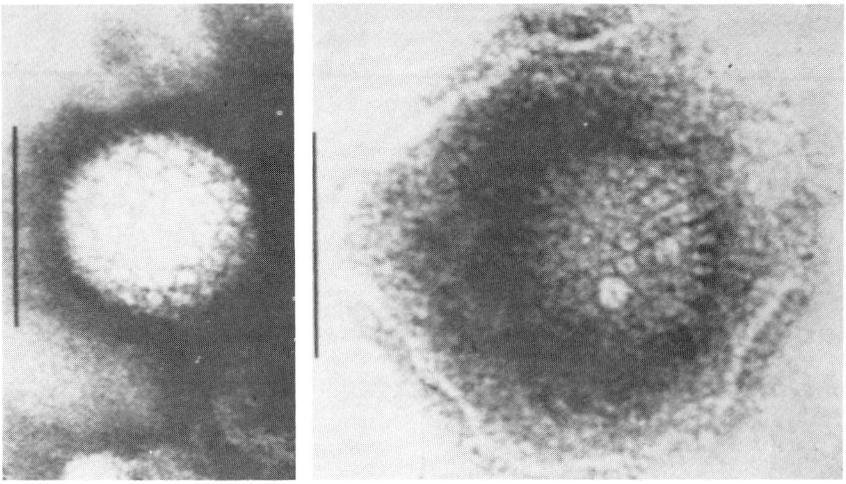

Fig. 8–2. Electron micrographs of herpes simplex virus particles. A. "Naked" and B. enveloped virion. Phosphotungstic acid negative contrasts; 100 μM (bars). (Reprinted with permission from Goodheart, C.R.: Herpesviruses and Cancer. J.A.M.A., *211*:91, 1970.)

1. Herpes Simplex Virus Types 1 and 2 (HSV-1, HSV-2).

Using electron microscopy, nucleotide sequence-specific endonucleases (restriction enzymes; Eco R1, Hind III) which cut DNA at specific sites, and nucleic acid hybridization, the structure of herpes simplex virion DNA was found to have unusual features. The DNA is composed of two unique components designated "long" (L) and "short" (S) each bounded by terminal repetitions and internal inverted repetitions of the terminal sequences (Fig. 8–3). The genome can exist in four possible isomeric arrangements differing in the relative orientation of the two sets of unique components. The sequence arrangement and genomic complexity varies among several investigated herpesviruses (Fig. 8–4).

Serial undiluted passage of HSV results in production of "defective" virus which can interfere with replication of "standard" virus (Chapter 4). The origin of defective virus DNA which contains tandem repetitions of selected sequences has been mapped in the right side of the S component and near the middle of the L component (Fig. 8–3). Defective virus may be important in development of cancer (Chapter 34).

Antigenic Properties. There are two serotypes of HSV. Type 1 herpes simplex can be differentiated from type 2 on the basis of several minor antigenic differences and also by the G + C content of the DNA, by cultural characteristics and restriction enzyme cleavage patterns of viral DNA. HSV-2 has been found in association with carcinoma of the cervix; this aspect of the infection will be discussed in Chapter 34. The types can be differentiated into strains by the analysis of Ab neutral-

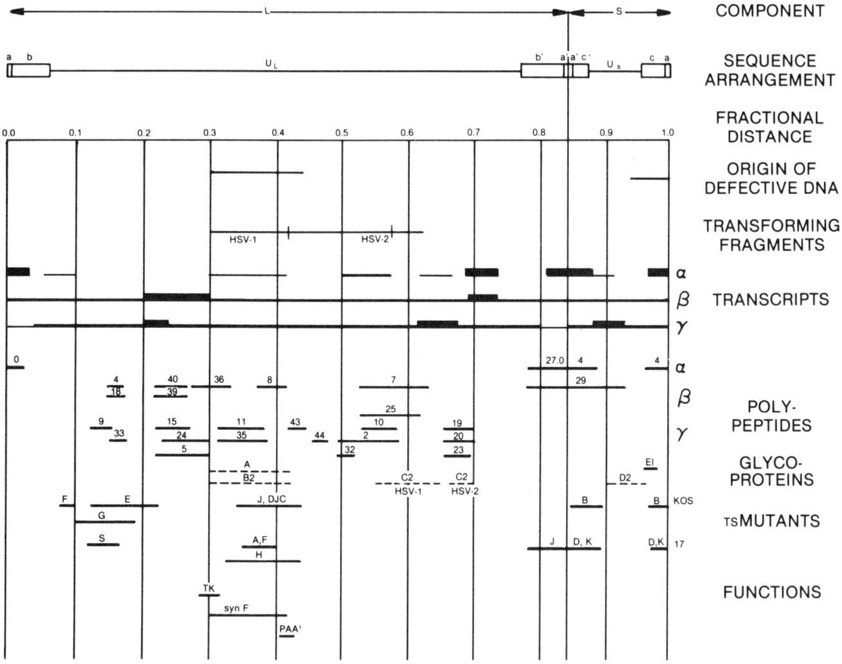

Fig. 8–3. Herpes simplex virus DNA map. The DNA is shown in a conventional orientation with the two components and arrangement of the terminal inverted repeat sequences. Shown are the map positions of defective DNA, transforming fragments, three classes of transcripts and corresponding polypeptides. The map positions of the major glycoproteins are given for HSV-1 and HSV-2. The functions of the polypeptides where known are indicated. (From the back cover of the Cold Spring Harbor Laboratory Abstracts, Herpesvirus Meeting, 1979.)

ization kinetics and changes in restriction enzyme cleavage patterns among the virus strains. These analyses are important in the epidemiology of herpesvirus diseases.

Serologic cross-reactions between HSV-1 and HSV-2 occur. Serologic cross-reactions do not occur between HSV and other human herpesviruses.

Host Range and Culture. Man is the natural host for HSV. However, most strains are pathogenic for conventional laboratory animals, including rabbits, mice, guinea pigs, hamsters, rats, day-old chicks, and owl monkeys, depending upon the route of inoculation.

The virus grows well in a wide variety of tissue culture cells. Rabbit kidney cells are commonly used for the initial isolation of the virus. Human embryo and baby hamster kidney cells also are highly susceptible. Various continuous cell lines and some strains of human diploid cells are less susceptible. Monkey kidney cells support the growth of

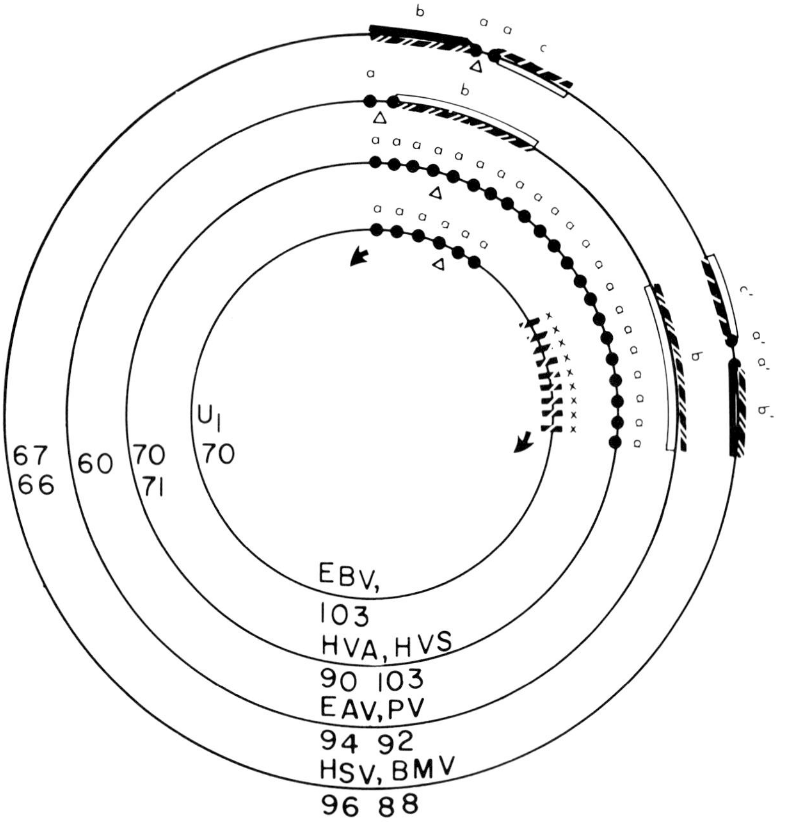

Fig. 8–4. Anatomy of DNA. U_S = unique short component; U_l = unique long component. Lower case letters indicate regions of terminal redundancy. EBV = Epstein-Barr virus, HVA = herpesvirus atelas, HVS = herpesvirus simairi, EAV = equine abortion virus, PV = pseudorabies virus, HSV = herpes simplex virus, BMV = bovine mamillitis virus. The figure includes mol. wt. × 10^6 of DNA. (From the cover of the Abstracts of papers presented at the 4th Cold Spring Harbor Meeting on Herpesviruses, 1979.)

HSV with difficulty, although they are susceptible to other members of the herpesvirus group.

The cytopathic characteristics of HSV infections can take two forms: Either the cells become rounded, swollen, and clumped or they fuse to become multinucleated giant cells (Chapter 22). Both forms usually are observed on primary isolation. The nuclei of infected cells contain inclusion bodies that give unfixed preparations a refractile appearance. Herpes simplex virus can cause tumorigenic transformation of animal cells in tissue culture; there are immunologic, biochemical and molecular hybridization data supporting the view that herpes simplex virus may be involved with the etiology of cervical carcinoma (Chapter 34).

Replication. HSV enters host cells by fusion or viropexis. Uncoating occurs in the cell cytoplasm and virion DNA is transported to the nucleus by an unknown mechanism. Switch-off of cellular macromolecular synthesis occurs via a virion-associated polypeptide or de novo synthesis of an immediate-early virus-coded protein. Viral mRNA and protein syntheses are coordinately regulated and sequentially ordered into α, β and γ. Alpha proteins are made first and reach maximum synthesis in the cytoplasm 2 to 4 hours after infection. They induce the second or β-group proteins at maximum rates between 5 and 7 hours after infection. These shut off synthesis of α-proteins and induce synthesis of γ-proteins. About 51 virus specific polypeptides, including 24 structural and 27 nonstructural polypeptides, can be identified by polyacrylamide gel electrophoresis of infected cell extracts. The physical map of several viral induced polypeptides are shown in Figure 8–3. Viral proteins synthesized on polyribosomes in the cytoplasm are transported to the nucleus where virus assembly occurs. Maturation of nucleocapsids takes place at altered sites on the inner nuclear membrane by a process of budding. Virions are released from infected cells by tubular structures continuous with extracellular spaces or from vacuoles released by infected cells.

2. Varicella-Zoster Virus (V-Z).

Structure and Antigenic Properties. V-Z virus is morphologically identical to other herpesviruses (Fig. 8–2). The virus is the etiologic agent of chickenpox (varicella) and shingles (zoster). Viruses isolated from clinical cases of chickenpox and shingles have identical antigenic and biologic properties; there is only one serotype of V-Z virus. The virus is difficult to grow to high titer in conventional tissue culture systems. Furthermore, animals are not susceptible to infection by V-Z virus. Lack of high titered virus stocks has seriously hampered progress in understanding the molecular virology of the agent. The virus can be propagated in primary or secondary human or monkey cell cultures; human diploid cell strains or cell lines also may be employed. The virus cannot be grown in embryonated eggs.

Cytopathic effects are focal and slow to develop. The virus replicates in the cell nucleus and remains cell-associated; extension of individual lesions occurs by the direct passage of virus to contiguous cells. Infected cells become rounded, swollen, and refractile. In addition, multinucleate giant cells may develop. Radiating cytoplasmic processes extend from the infected cells. Acidophilic inclusions are commonly observed in the nucleus and occasionally in the cytoplasm of stained cells.

3. Cytomegalovirus (CMV).

Structure and Antigenic Properties. Complete virus particles are surrounded by double-membrane envelopes. The virus core is composed of globular subunits. Electron micrographs of infected cells have revealed capsids with or without cores. An additional viral structure, a dense body, is composed of a central sphere enclosed in a double membrane envelope similar to mature virus particles. Dense bodies measure 250–500 nm in diameter compared to about 180 nm for virus particles. Both types of structures are observed in the cytoplasm of infected cells.

Cytomegalovirus (CMV) has a structure common to all herpesviruses. The DNA of the virus has a molecular weight of 142×10^6, a density of 1.716 g/cm^3 in CsCl and a G + C content of 58 percent. The viral DNA does not share sequence homology with HSV. Cytomegalovirus of animals are antigenically distinct. There are several strains of human CMV that share many common antigenic properties; however, it is difficult to separate them into individual serotypes. There are indications that CMV shares some common Ags with HSV, V-Z and Epstein-Barr virus (all herpesviruses). In humans CMV is associated with salivary gland disease, mononucleosis (Chapter 29), congenital abnormalities (Chapter 33) and is considered the most common cause of kidney transplant failures.

Host Range and Culture. Since CMV are species-specific with respect to host range, the human types are grown in cultured human cells. Primary or serially propagated strains of human fibroblasts (e.g., WI-38) are routinely utilized to isolate the virus, to prepare virus stocks for study or to provide antigenic materials for serologic tests. Treatment of human fibroblast cells with 5-iododeoxyuridine (IUdR) enhances replication of CMV in these cells. Infectivity is primarily cell-associated and spread of infectious virus is from cell to cell. Infected cells become swollen, rounded, and light-refractile and contain intranuclear and intracytoplasmic inclusions. Cytopathology may be recognized within 24 to 72 hours when high-virus-titer inocula are used. With low-titered inocula, lesions may not develop in the culture for several weeks. Serial passage of infectious virus to new cultures is best accomplished by transfer of infected cells freed by trypsinization or mechanical scraping. Cytomegalovirus loses infectivity when heated at 56°C for 30 minutes or when exposed to ether or to acid pH.

Data show that ultraviolet-inactivated CMV can transform hamster embryo fibroblasts in vitro, and that the transformed cells induce tumor formation when inoculated into newborn hamsters (Chapter 34).

Replication. Virus-cell interaction depends on whether the virus strain is laboratory adapted or a recent clinical isolate. Clinical isolates of CMV may be serially passaged in tissue culture to adapt the virus for growth in tissue culture cells. Virions and dense bodies enter host cells by fusion of the envelope with the cell membrane or by phagocytosis. Nucleocapsids rapidly accumulate in the area of the nuclear membrane. Virus replication occurs in nuclear inclusions. Nucleocapsids with or without central cores appear 48 to 72 hrs after infection. Envelopment of virions and dense bodies occurs at the inner nuclear membrane or by budding into cytoplasmic vacuoles. Release of virions and dense bodies apparently takes place by a process of reverse phagocytosis from cytoplasmic vacuoles and connecting microtubules.

4. Epstein-Barr Virus (EBV).

Epstein-Barr virus (EBV), an antigenically distinct herpesvirus, was originally observed in cell cultures of Burkitt's lymphoma (BL) (a tumor originally described in Central African children). The virus has also been found in many lymphoid cell cultures derived from patients with BL, nasopharyngeal carcinoma or infectious mononucleosis (IM) as well as from normal individuals. Epstein-Barr virus is considered the etiologic agent of IM and is suspected of being the cause of BL and nasopharyngeal carcinoma (Chapter 34). Infection of man by this virus is common and generally inapparent.

Structure and Antigenic Properties. EBV has a structure typical of other herpesviruses. Progress on determining the molecular composition of purified virions has been slow due to lack of high productive host cells for EBV. Two strains of EBV exist in nature, lymphoid-cell transforming and nontransforming strains. Typical EBV particles in infectious mononucleosis cell lines are shown in Figures 8–5 and 8–6. EBV-DNA is a double-stranded, linear molecule with a molecular weight of 1×10^8. The linear molecule is terminally redundant. Purified virion preparations are composed of 33 polypeptides, 12 of which are glycosylated.

Although EBV is antigenically distinct from other human herpesviruses, it reportedly shares a common precipitating Ag with herpesviruses that cause the Lucke tumor (a renal adenocarcinoma of frogs), Marek's disease (Chapter 34), infectious bovine rhinotracheitis and CMV. EBV infected lymphoblastoid cells in culture contain EBV early Ag (EA), Epstein-Barr nuclear Ag (EBNA), virus capsid Ag (VCA) and a virus

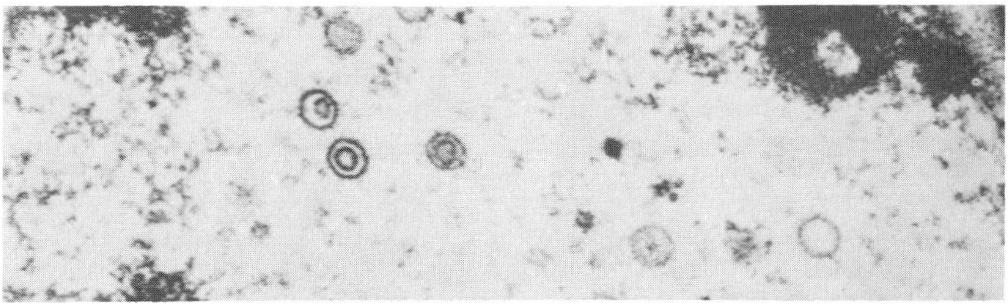

Fig. 8–5. Incomplete herpes-type virus particles located within a nucleus of a peripheral blood cell in a cell culture established from a patient with heterophil-positive infectious mononucleosis. × 50,000. (Reprinted with permission from Moses, H.L., Glade, P.R., Kasel, J.A., Rosenthal, A.S., Hirshaut, Y., and Chessin, L.N.: Infectious mononucleosis: Detection of herpeslike virus and reticular aggregates of small cytoplasmic particles in continuous lymphoid cell lines derived from peripheral blood. Proc. Natl. Acad. Sci. (U.S.A.), *60*:492, 1968.)

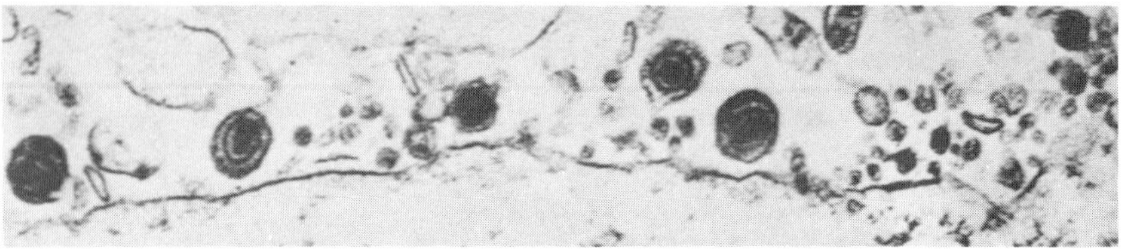

Fig. 8–6. Extracellular herpes-type virus particles that have an internal core with the same dimensions as the nuclear particles (Fig. 8–5) but appear to be surrounded by an envelope giving an overall diameter of 160 nm. × 50,000. (Reprinted with permission from Moses, H.L., Glade, P.R., Kasel, J.A., Rosenthal, A.S., Hirshaut, Y., and Chessin, L.: Infectious mononucleosis: Detection of herpeslike virus and reticular aggregates of small cytoplasmic particles in continuous lymphoid cell lines derived from peripheral blood. Proc. Natl. Acad. Sci. (U.S.A.), *60*:492, 1968.)

specified membrane Ag (MA). These Ags are measured by IF or CF tests using hyperimmune serum.

Host Range and Culture. Unlike other human herpesviruses, EBV does not infect fibroblast cells. The host cell type sensitive to EBV lytic infection is the B-type lymphoblast and human epithelial cells. Infection results in establishment of latent or persistent EBV genome-containing lymphoblast cell lines. Both adult and fetal lymphoblasts are susceptible to infection in vitro. There are two types of EBV genome-containing lymphoblastoid cell lines, i.e., virus producer cell lines like HR-1 and nonproducer cell lines such as Raji. Whereas, cell lines express EBNA, less than 10 percent of producer cell lines spontaneously express other viral Ags or virus particles. Recent evidence suggests that the tumor promoter, phorbol myristic acetate, can induce EBV-specified Ags, viral DNA and virion replication in producer cell lines.

Three enzymatic activities, ribonucleotide reductase, thymidine kinase and DNA polymerase, can be induced in EBV-producer cell lines by treating them with halogenated pyrimidines (e.g., IUdR) or superinfecting the cells with nontransforming EBV. The thymidine kinase activity appears to be a host cell coded enzyme. The other two enzymes may represent host enzymes induced by EBV or may be encoded by the viruses themselves.

Replication. The events leading to productive EBV infection are difficult to study and to interpret because of differences in virus strains and target cells and the nearly complete repression of viral DNA expression by lymphoblastoid cells. Nevertheless, some progress has been made. Using fetal cells infected with nontransforming EBV, the virus-specific EBNA is expressed about 12 to 16 hours after infection. The bulk of EBV-DNA remains extrachromosomal as covalently closed, supercoiled molecules. Only a small portion of virion 2X linear DNA is closely associated with cellular DNA. Infection with EBV stimulates two cycles of cellular DNA synthesis. Induction of EA synthesis precedes viral DNA replication. Expression of VCA and MA can be detected after viral DNA replication. The events leading to virus assembly occur in the cell nucleus. Virus matures by budding through the nuclear membrane.

REFERENCES

Butterson, W., Furlong, D., and Roizman, B.: Molecular genetics of herpes simplex virus. VIII. Further characterization of a temperature-sensitive mutant defective in release of viral DNA and in other stages of the viral reproductive cycle. J. Virol., *45*:397, 1983.

Glaser, R., and Rapp, F.: Biological properties of the Epstein-Barr virus and its possible role in human malignancy. Prog. Med. Virol., *21*:43, 1975.

Hill, T.M., Sinden, R.R., and Sadler, J.R.: Herpes simplex virus types 1 and 2 induce shutoff of host protein synthesis by different mechanisms in friend erythroleukemia cells. J. Virol., *45*:241, 1983.

Hudewentz, J., Bornkamm, G.W. and zur Hausen, H.: Effect of the diterpene ester TPA on Epstein-Barr virus antigen and DNA synthesis in producer and nonproducer cell lines. Virology, *100*:175, 1980.

Kaplan, A.S. (ed.): The Herpesviruses. New York, Academic Press, 1973.

Kucera, L.S.: Herpes simplex virus-host cell interactions. CRC Crit. Rev. Microbiol., *7*:215, 1979.

Roizman, B.: Structural and functional organization of herpes simplex virus genomes. *In* Oncogenic Herpesviruses. Vol. I. F. Rupp (ed.). Boca Raton, Florida, CRC Press, 1980, pp. 19–48.

Smith, J.D. and de Harven, E.: Herpes simplex virus and human cytomegalovirus replication in WI-38 cells. I. Sequence of viral replication. J. Virol., *12*:919, 1973.

Stinski, M.E.: Human cytomegalovirus: Glycoproteins associated with virions and dense bodies. J. Virol., *19*:594, 1976.

Sugden, B., Kintner, C., and Mark, W.: The molecular biology of lymphotropic herpesviruses. Adv. Cancer Res., *30*:239, 1979.

Chapter 9

Poxviruses

A. POXVIRIDAE FAMILY

The poxviridae family includes the largest and structurally most complex viruses that infect man. They are classified as complex viruses because the nucleocapsids do not conform to either helical or icosahedral symmetry. The family is subdivided into 7 genera (*Orthopoxvirus, Leporipoxvirus, Avipoxvirus, Capripoxvirus, Suipoxvirus, Parapoxvirus,* and *Ungrouped Poxviruses)* on the basis of antigenic relatedness and the ability of the viruses in each genus to recombine. There is extensive cross-neutralization between viruses in each genus but none between viruses in different genera. However, all poxviruses share a common nucleoprotein (NP) Ag that can be extracted from the core. Poxviruses multiply in the cell cytoplasm (Table 9–1).

There are 8 poxviruses that can infect man: variola, vaccinia, monkeypox, cowpox, orf, pseudocowpox, molluscum contagiosum and Yaba monkey viruses. The remaining unmentioned viruses are of importance in their respective animal hosts including buffalo, swine, camels, sheep, rabbits, mice, goats, cows, pigeons, turkeys, canaries and other birds.

Smallpox, which is caused by variola virus, has been recognized as a life-threatening disease for many centuries, but now is considered eradicated from the world by active immunization. It was the first disease against which active immunization was widely practiced. The first vaccine used contained variola virus from lesions of smallpox patients. Material from a smallpox lesion was inoculated by incision or puncture into the skin. Although the illness that developed was usually milder than naturally acquired smallpox, variolation was sometimes fatal. This dangerous vaccine was replaced by Jenner's cowpox vaccine taken from cowpox lesions. Jenner's procedure was based on an observation common at that time, namely, that milkmaids who became infected with cowpox developed resistance to smallpox. The cowpox virus was passed from arm to arm for many years with the occasional introduction of new material from infected cows; later the "lymph" used as vaccine was produced by serial passage of the virus in calves, sheep, or rabbits. In the course of time, the virus underwent minor antigenic changes, and the laboratory strains presently used can be differentiated from fresh isolates of cowpox virus. For this reason, strains recently isolated from

Table 9–1. General Properties of Poxviruses

Properties of nucleic acid: double-stranded linear DNA; 150×10^6 daltons
Virion: enveloped; inner core or capsid, two lateral bodies surrounded by external membranes
Capsid symmetry: Complex
Size (nm): approximately 250×400
Site of replication: cytoplasm with inclusion body formation

cowpox are called cowpox virus and the vaccine strains used to immunize against smallpox are called vaccinia virus.

The World Health Organization initiated an immunization program in 1967 to eradicate smallpox from the world. Except for an accident involving variola virus infection of a laboratory technician in England in 1978, the last naturally occurring case of smallpox was recorded in Somalia in October, 1977. For this reason, routine vaccination for smallpox is no longer recommended.

1. Variola, Vaccinia, Monkeypox, and Cowpox Viruses.

STRUCTURE AND ANTIGENIC PROPERTIES. Variola, vaccinia, monkeypox and cowpox viruses are morphologically indistinguishable. The virions measure about 250 × 400 × 100 nm and appear brick-shaped when purified virus is viewed in the electron microscope (Fig. 9–1). However, in ultrathin sections of infected tissue, the particles are ellipsoid. An outer membrane composed of tubular strands of lipoprotein encloses a layer of soluble protein Ags, surrounding the dense dumbbell-shaped nucleoid (core) and the lateral bodies (Fig. 9–2). The nucleoid contains the high-molecular-weight double-stranded DNA genome in association with an inner membrane (Table 9–1).

The viruses differ from each other by only 1 or 2 of the 20 or more Ags that can be detected by immunodiffusion tests. Serologic tests used in the diagnosis of smallpox depend upon a complex soluble Ag (LS) that contains heat-labile (L) and heat-stable (S) components. The LS Ag

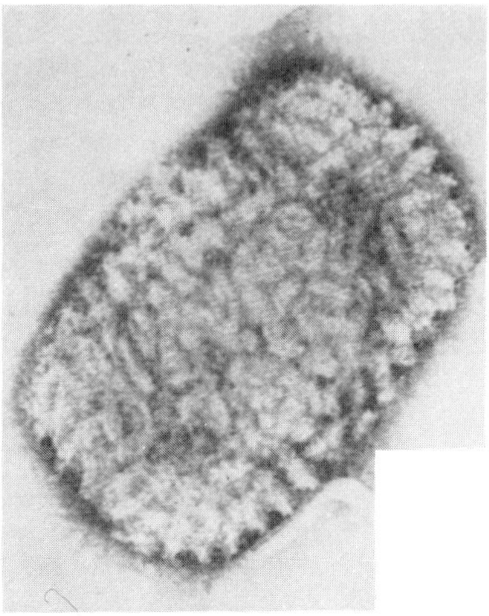

Fig. 9–1. Electron micrograph of purified poxvirus, phosphotungstate stain (superficial penetration). Note the short surface filaments. × 150,000. (With permission from Long, G.W., Noble, J., Jr., Murphy, F.A., Herrmann;, K.L., and Lourie, B.: Experience with electron microscopy in the differential diagnosis of smallpox. Appl. Microbiol., *20*:497, 1970.)

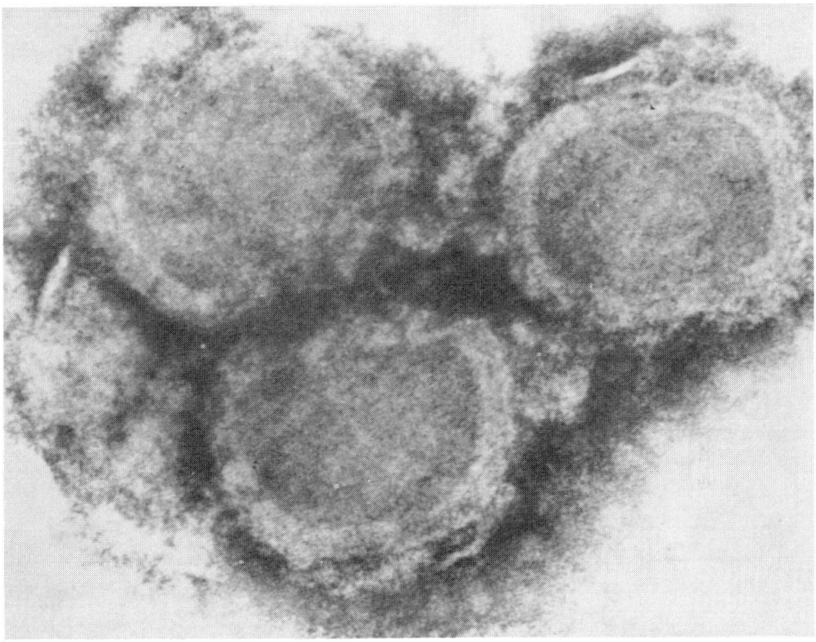

Fig. 9–2. Poxvirus from specimen material, phosphotungstate stain. The deep penetration of stain permits observation of three concentrically laminated zones around the core. × 150,000. (Reprinted with permission from Long, G.W., Noble, J., Jr., Murphy, F.A., Herrmann, K., and Lourie, B.: Experience with electron microscopy in the differential diagnosis of smallpox. Appl. Microbiol., *20*:497, 1970.)

is present in both the virion and extracts of infected tissue or fluid from smallpox vesicles. It is distinct from the high-molecular-weight structural Ag that induces the formation of neutralizing Ab.

Hemagglutinins are produced in infected cells but are not an integral part of the virion; i.e., purified poxvirus does not agglutinate red blood cells.

Host Range and Culture. Man is probably the only natural host and reservoir of variola virus. Monkeys infected experimentally develop a mild generalized disease with a rash. The virus also can be serially passed in suckling mice inoculated intracerebrally. Although lesions can be produced in rabbits, serial passage is achieved with difficulty. Vaccinia and cowpox viruses readily infect cows and sheep, as well as a number of laboratory animals.

Inoculation of embryonated chicken eggs is the most reliable laboratory test for this group of viruses. Variola virus grows readily on the chorioallantoic membrane and produces small dome-shaped pocks that are easily differentiated from the large centrally depressed lesions of vaccinia and cowpox.

The growth of poxviruses in cell cultures reflects their host range in vivo. Vaccinia and cowpox readily produce cytopathic effects in many types of cells, including chick embryo fibroblasts, monkey kidney cells, and a number of continuous cell lines. Variola virus also grows in various mammalian tissue cultures, but cytopathic changes appear more slowly than those produced by vaccinia virus. Although the poxviruses are DNA viruses, they replicate in the cytoplasm and produce characteristic cytoplasmic inclusions called Guarnieri bodies (Table 9–1); the inclusions are composed of a dense aggregation of many virus particles. Virus particles are also referred to as "elementary bodies" or "Paschen bodies." The cytoplasmic inclusion bodies observed in specimen material are useful for establishing a diagnosis of infections caused by members of the poxvirus group.

Replication. Most of the information concerning the replication of poxviruses was obtained with vaccinia virus infected cells. Entry of vaccinia virus into host cells occurs by viropexis (phagocytosis) and formation of phagocytic vesicles. An alternate method of virus entry occurs through virus fusion with the plasma membrane. Uncoating of infecting virus particles and release of virion DNA occurs in two stages. *First stage* uncoating occurs in the phagocytic vesicle where the virion envelope is digested; subsequently, the phagosomal membrane is ruptured resulting in the release of the virus core and lateral bodies into the cytoplasm. *Second stage* uncoating proceeds after immediate-early transcription of virion DNA via a virus core-associated RNA polymerase and translation of protein. Approximately one-half of the viral genome is transcribed early in infection. Although apparently not spliced, the large transcripts are processed to monocistronic capped and polyadenylated mRNA's. Proteins translated from this mRNA population include the "uncoating" protein, several enzymes (thymidine kinase, DNA polymerase) and structural proteins. Uncoated vaccinia DNA initiates the formation of viral inclusions in the cytoplasm which are visible by histochemical staining and light microscopy. Viral DNA synthesis starts about one and one-half hours after infection in cytoplasmic inclusions and ends about 3½ hours later; this is followed by late mRNA and protein synthesis. Some of these late proteins are responsible for "switch-off" of early mRNA translation (e.g., thymidine kinase synthesis). Assembly of progeny virus occurs between 5 and 24 hours after infection in intracytoplasmic inclusions. Mature virus is liberated as the cell disintegrates.

2. Orf (Contagious Pustular Dermatitis) and Pseudocowpox (Milker's Nodule) Viruses.

Structure and Antigenic Properties. The virions of orf and pseudocowpox are indistinguishable morphologically. They are slightly narrower and less brick-shaped than those of the variola-vaccinia subgroup; the tubular strands of the surface membrane are spirally arranged, giving the virus a "ball of yarn" appearance (Fig. 9–3). These viruses are an-

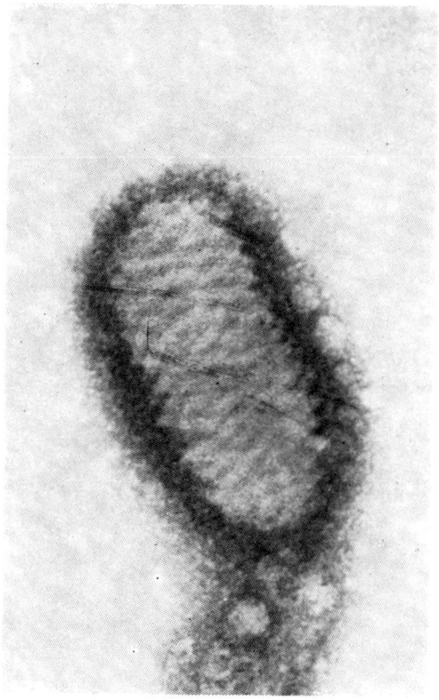

Fig. 9–3. Electron micrograph of Milkers' Nodule virus from specimen material, phosphotungstate stain (superficial penetration). Note the cylindrical shape and spiral arrangement of surface filaments. × 150,000. (Reprinted with permission from Long, G.W., Noble, J., Jr., Murphy, F.A., Herrmann, K., and Lourie, B.: Experience with electron microscopy in the differential diagnosis of smallpox. Appl. Microbiol., *20*:497, 1970.)

tigenically related but share no immunogenic Ags with variola or vaccinia. Neither orf virus nor pseudocowpox virus produces a hemagglutinin.

Host Range and Culture. Sheep and goats are the natural hosts of orf virus; the disease affects mainly the mouth and lips of young lambs and kids, but adults also can be infected. Man may develop localized skin lesions as a result of contact with infected animals.

Orf virus grows poorly, if at all, in most routine laboratory animals or in embryonated eggs but can be cultured in the scarified skin of lambs. Laboratory strains can be propagated in primary human amnion cells as well as in sheep, goat, or bovine cells. The virus causes focal areas of cellular degeneration and destruction after 5–18 days of incubation.

The cow is the principal natural host of pseudocowpox virus, which causes a cowpox-like disease on the teats and udders. The virus may be spread to milkmaids or farmworkers and causes lesions on the hands.

Pseudocowpox virus does not produce lesions in chick embryos or other laboratory animals but can be propagated in cultures of bovine tissue or primary human amnion. Cytopathic effects may not appear for 10–12 days on primary isolation, but, on second passage, cytotoxicity may be evident in 2–3 days.

3. Molluscum Contagiosum and Yaba Monkey Tumor Viruses.

Structure and Antigenic Properties. Morphologically molluscum contagiosum and Yaba viruses resemble the members of the variola-vaccinia genus. A soluble, heat-labile Ag has been prepared from molluscum contagiosum infected skin lesions; this Ag is distinct from Ags produced by other members of the poxvirus group. The properties of Yaba virus have been described elsewhere (Chapter 34).

Host Range and Culture. Man is the only known natural host for molluscum contagiosum virus. The virus has not been cultured in experimental animals or in chick embryos. Although cytopathic effects have been observed in some primary human cell cultures and in HeLa cells (a continuous human cell line), the virus has not been serially passed.

REFERENCES

Baxby, D.: Identification and interrelationships of the variola/vaccinia subgroup of poxviruses. Prog. Med. Virol., *19*:215, 1975.

Cooper, J.A. and Moss, B.: Transcription of vaccinia virus mRNA coupled to translation in vitro. Virology, *88*:148, 1978.

Fenner, F.: The eradication of smallpox. Prog. Med. Virol., *23*:1, 1977.

Hruby, D.E., Guarino, L.A., and Kates, J.R.: Vaccinia virus replication. I. Requirement for the host-cell nucleus. J. Virol., *29*:705, 1979.

Ichihashi, Y. and Oie, M.: Adsorption and penetration of the trypsinized vaccinia virion. Virology, *101*:50, 1980.

Lake, J.R., Silver, M. and Dales, S.: Biogenesis of vaccinia. Complementation and recombination analysis of one group of conditional-lethal mutants defective in envelope self-assembly. Virology, *96*:9, 1979.

Menna, J.H.: Poxviruses. *In* Microbiology: Basic Principles and Clinical Applications. N.R. Rose and A.L. Barron (eds.). Macmillan Publishing Co., New York, 1983, pp. 360–369.

Moss, B.: Reproduction of poxviruses. *In* Comprehensive Virology. Vol. 3. Fraenkel-Conrat, H., and Wagner, R.R. (eds.). New York, Plenum Press, 1974, pp. 405–474.

Moss, B.: Poxviruses. *In* The Molecular Biology of Animal Viruses, Nayak, D.P. (ed.), Vol. 2. New York, Marcel Dekker, 1978, p. 849.

Paoletti, E., Lipinskas, B.R., and Panicali, D.: Capped and polyadenylated low-molecular-weight RNA synthesized by vaccinia virus in vitro. J. Virol., *33*:208, 1980.

Venkatesan, S., Gershowitz, A., and Moss, B.: Complete nucleotide sequences of two adjacent early vaccinia virus genes located within the inverted terminal repetition. J. Virol., *44*:637, 1982.

Chapter 10

Picornaviruses

A. PICORNAVIRIDAE FAMILY

The picornaviruses are the smallest of the RNA viruses, hence the name "pico (small) RNA virus." All members are naked icosahedral viruses, 25–30 nm in diameter, and all contain positive-stranded RNA (Table 10–1); there is no common group Ag. The picornaviruses are subdivided into 4 genera: *Enterovirus* and *Cardiovirus* members are acid-stable and *Rhinovirus* and *Aphthovirus* members are acid-labile. The enteroviruses are found mainly in the gastrointestinal tract; whereas the rhinoviruses are found primarily in nasal passages. Aphthovirus is the genus name for foot and mouth disease virus of cattle, swine, sheep and goats. Cardiovirus is the genus name for mengovirus and encephalomyocarditis (EMC) virus of rodents; these viruses can cause mild febrile illness and mild meningoencephalitis in humans.

1. Enterovirus.

The genus *Enterovirus* is subdivided into 3 groups, (1) poliovirus, (2) coxsackievirus, and (3) echovirus. Poliovirus was the first of these viruses to be recognized. The other 2 groups of viruses were discovered during the course of intensive epidemiologic studies of poliomyelitis. The foundation of modern virology was laid in 1949, when Enders, Weller, and Robbins showed that poliovirus could be isolated and propagated in cultures of nonneural cells. They were awarded the Nobel Prize for their discovery.

a. Poliovirus. Poliovirus is responsible for an acute infectious disease that occasionally involves the CNS. Infection and destruction of motor neurons in the spinal cord may lead to flaccid paralysis.

Structure and Antigenic Properties. Polio virions possess a positive-stranded RNA genome. The nucleocapsid has cubic symmetry, is nonenveloped, and measures about 25–30 nm in diameter (Fig. 10–1). Infectivity of poliovirus is stable to many physical (e.g., freezing) and chemical (e.g., ether or deoxycholate) agents.

Three distinct antigenic types of poliovirus exist that can be identified by an Ab neutralizing test for infectivity; there is essentially no cross-neutralization. Inactivation of virions by formalin, heat, or ultraviolet

Table 10–1. General Properties of Picornaviruses

Properties of nucleic acid: single, positive-stranded, linear RNA; 2.6×10^6 daltons
Virion: naked nucleocapsid
Capsid symmetry: icosahedral
Size (nm): 25–30
Site of replication: cytoplasm

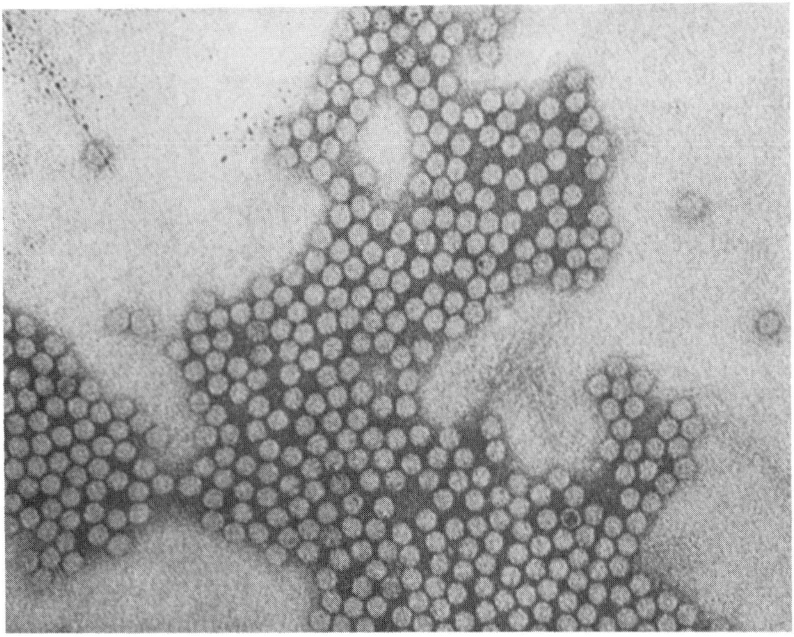

Fig. 10–1. Electron micrograph of typical poliovirus particles. × 141,500. (Reprinted with permission from Mayor, H.D., and Jamison, R.M.: Morphology of small viral particles and subunit components. Prog. Med. Virol., *8*:183, 1966.)

light releases a soluble complement-fixing (CF) Ag that cross-reacts with heterotypic poliovirus Abs. In addition, two type-specific Ags called "N" (native) and "H" (heated) are detectable by precipitin or CF tests. The N Ag is associated with virus infectivity, whereas the H Ag is found in ruptured or incomplete virus particles.

Host Range and Culture. Man is the only natural host for poliovirus. The virus infects only cells lining the intestinal tract or cells of the central nervous system. Nonhuman primates are susceptible to paralysis following intraspinal or intracerebral inoculation. Poliovirus is pathogenic only for man and closely related species because of the requirement for specific attachment sites in the membrane of susceptible cells. Poliovirus can be isolated and grown in rhesus monkey kidney or human cells because in tissue culture the host cells express the receptor required for virus attachment. The virus multiplies in the cytoplasm of cells and is rapidly cytocidal; the cytopathic changes produced include cell rounding, increased refractivity, nuclear pyknosis and finally lysis (Chapter 2).

Replication. Poliovirus attaches to specific receptor sites on the surface of host cells. Structural changes in the virion occur during the first hour of infection at the site of virus attachment. Complete uncoating of the viral RNA occurs in the cytoplasm. The parental virus genome functions as mRNA and is translated into a single large polyprotein (200,000–300,000 daltons). The polyprotein is subsequently cleaved into smaller proteins including an RNA polymerase and capsid proteins. Accumulation of RNA polymerase is followed by poliovirus RNA replication (Chapter 3). The parental (+)-strand is transcribed into a (−)-strand. The (−)-strand serves as a template for making progeny positive strands. RNA strands of both positive and negative polarity are covalently linked at their 5′-termini to a 5000 dalton protein via a tyrosine residue. The protein may serve as primer for transcription (Chapter 3).

Assembly of progeny poliovirus particles occurs via intermediate structures. The first step involves aggregation of the N Ag protein to form a pentamer. The cleaved pentamers form three capsid proteins (VP0, VP1 and VP3) which aggregate to form the procapsid. Viral RNA is then added to the procapsid to form the provirion. Complete virions appear in the cytoplasm after VP0 is cleaved to VP2 plus VP4. Cleavage of VP0 changes the provirion into the virion. Release of progeny poliovirions occurs during cell lysis.

b. Coxsackievirus. The coxsackieviruses were discovered in 1948 when a new virus was isolated from the feces of 2 children in Coxsackie, N.Y., during unsuccessful attempts to grow poliovirus in neonatal mice. Subsequently, the coxsackieviruses were subdivided into groups A and B on the basis of their effects on suckling mice: Group A coxsackieviruses cause flaccid paralysis and death within a week, whereas the group B viruses produce spastic paralysis that develops more slowly.

Antigenic Composition. There is no single Ag common to all group A and B coxsackieviruses. There are at least 24 group A and 6 group B serotypes that are distinguished by type-specific Ags on the virion surface. Although there is no common group A Ag, heterotypic cross-reactions have been observed between some of the group A viruses. All group B viruses possess a common group Ag.

Host Range and Culture. Man appears to be the only natural host for the coxsackieviruses. However, both group A and group B serotypes are readily isolated and propagated in suckling mice. All coxsackie B viruses, but only a few coxsackie A serotypes, grow well in human or monkey kidney cell cultures. Some of the group A viruses that formerly could be propagated only in neonatal mice have been found to multiply in diploid strains of human fibroblasts. The cytopathic effects caused by coxsackieviruses are similar to those caused by other enteroviruses. The cells first become granular and then rounded. Cell degeneration proceeds rapidly, and within 24–48 hours after the effect first becomes apparent, the cells are completely lysed and released from the surface of the culture flask.

c. Echovirus. The echoviruses, like the coxsackieviruses, were discovered accidentally during epidemiologic studies of poliovirus. The viruses were isolated from the feces of apparently healthy individuals. Since they were not known to be associated with any specific disease, they were called "orphan" viruses. Subsequently, they were named *e*nteric *c*ytopathic *h*uman *o*rphan (ECHO) viruses because they were found in the intestinal tract of man and caused cytopathic changes in cell cultures.

Antigenic Composition. There are at least 34 serotypes of echoviruses. They do not share a common group Ag, but some immunologic cross-reactions have been observed between certain serotypes. In addition, some echoviruses cross-react with certain enteroviruses.

The similarity between the coxsackieviruses and echoviruses prompted the adoption of the proposal that all new serotypes be assigned an enterovirus type number rather than attempting to subclassify them as coxsackieviruses or echoviruses. The designations of agents that have already been described was retained. These include poliovirus types 1, 2, and 3; the 24 types of group A coxsackieviruses; the 6 types of group B coxsackieviruses; and the 34 types of echoviruses.

Host Range and Culture. Man is the only known natural host for the echoviruses. However, monkeys develop a viremia and brain lesions following the intraspinal or intracerebral inoculation of some serotypes. The echoviruses can be isolated and propagated only in cell cultures. Primary human or monkey kidney cells are usually used, but some serotypes grow better in diploid strains of human fibroblasts. The echoviruses cause cytopathic effects that are like those of other enteroviruses except that they develop more slowly and are usually less complete.

2. Rhinovirus.

The goal of the long search for "the" common cold virus appeared to have been reached when the first rhinovirus (nose virus) was isolated in 1956. Since then more than 100 serologically distinct viruses possessing the properties of rhinoviruses have been isolated, and the end probably is not in sight.

Structure and Antigenic Properties. The rhinoviruses that have been examined appear to resemble the enteroviruses morphologically (Fig. 10–2). However, the rhinovirus RNA has a higher molecular weight than the RNA of the enteroviruses. There are at least 100 serotypes of rhinoviruses, each possessing a type-specific Ag; there are other Ags that show variable patterns of cross reactivity. However, no common Ag is found in all serotypes.

Host Range and Culture. Rhinoviruses related to human types have been isolated from natural infections in a number of domestic animals; chimpanzees can be infected experimentally with human isolates.

The rhinoviruses are more difficult to grow in tissue culture than enteroviruses. Some serotypes (H strains) grow readily only in cells of

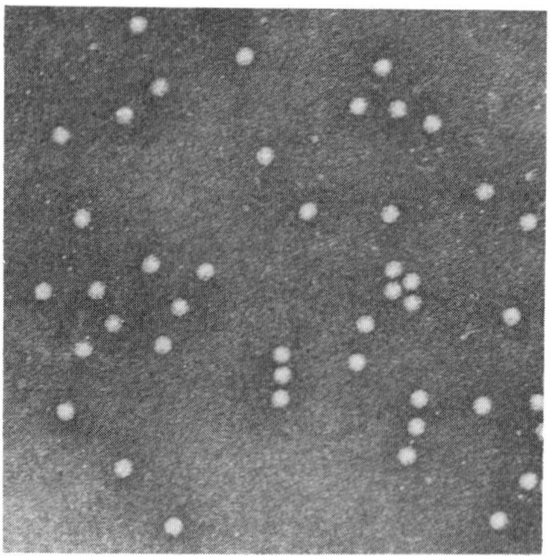

Fig. 10–2. Electron micrograph of rhinovirus. Magnification × 75,000. (Reproduced with permission from Hamre, D.: Rhinoviruses. Monogr. Virol., *1*:1, 1968.)

human origin; other strains will multiply in both monkey and human kidney cell cultures and have been designated "M" strains. Diploid strains of human embryonic lung fibroblasts also support the growth of rhinoviruses; their use has alleviated the special cultural conditions required for isolation in primary cell cultures. The cytopathic effects observed in tissue cultures are similar qualitatively to those produced by enteroviruses but develop more slowly and may be localized and incomplete.

REFERENCES

Butterworth, B.E., Shimshick, E.J., and Yin, F.H.: Association of the polioviral RNA polymerase complex with phospholipid membranes. J. Virol., *19*:457, 1976.

Flanegan, J.B. and Van Dyke, T.A.: Isolation of a soluble and template-dependent poliovirus RNA polymerase that copies virion RNA in vitro. J. Virol., *32*:155, 1979.

Guttman, N. and Baltimore, D.: Morphogenesis of poliovirus. IV. Existence of particles sedimenting at 150S and having the properties of provirion. J. Virol., *23*:363, 1977.

Humphries, S., Knauert, F. and Ehrenfeld, E.: Capsid protein precursor is one of two initiated products of translation of poliovirus RNA in vitro. J. Virol., *30*:481, 1979.

Korant, B., Chow, N., Lively, M., and Powers, J.: Virus specific protease in poliovirus infected HeLa cells. Proc. Natl. Acad. Sci. (U.S.A.), *76*:2992, 1979.

Levintow, L.: Reproduction of picornaviruses. *In* Comprehensive Virology. Vol. 2. Fraenkel-Conrat, H., Wagner, R.R. (eds.). New York, Plenum Press, 1975, pp. 109–169.

McGregor, S. and Rueckert, R.R.: Picornaviral capsid assembly: similarity of rhinoviruses and enteroviruses precursor subunits. J. Virol., *21*:548, 1977.

Nomoto, A., Kitamura, N., Golini, F., and Wimmer, E.: The 5′-terminal structures of poliovirion RNA and poliovirus mRNA differ only in the genome-linked protein VPg. Proc. Natl. Acad. Sci. (U.S.A.), *74*:5345, 1977.

Palmenberg, A.C.: In-vitro synthesis and assembly of picornaviral capsid intermediate structures. J. Virol., *44*:900, 1982.

Rekosh, D.M.K.: The molecular biology of picornaviruses. *In* The Molecular Biology of Animal Viruses. Vol. 1. Nayak, D.P. (ed.). New York, Marcel Dekker Press, 1977, p. 63.

Rueckert, R.R.: On the structure and morphogenesis of picornaviruses. *In* Comprehensive

Virology. Vol. 6. Fraenkel-Conrat, H. and Wagner, R.R. (eds.). New York, Plenum Press, 1976, pp. 131–213.

Yogo, Y. and Wimmer, E.: Sequence studies of poliovirus RNA. III. Polyuridylic acid and polyadenylic acid as components of the purified poliovirus replicative intermediate. J. Mol. Biol., *92*:467, 1975.

Chapter 11

Reoviruses

A. REOVIRIDAE FAMILY

The first double-stranded (ds) RNA virus recognized, a reovirus, was isolated from man. Similar viruses have since been found in plants and insects as well as a number of ruminants, ungulates, and other mammals. The existence of viruses of similar structure in animals, plants and other insects appears to be a phenomenon unique to the reoviridae family.

The primary consideration for inclusion in this family is possession of a linear ds RNA genome composed of 10, 11 or 12 segments. Of six genera included in the reoviridae family, three genera (e.g. *Reovirus, Orbivirus* and *Rotavirus)* contain members infectious for humans. The genera differ widely in morphology (Table 11–1) and host range.

1. Reovirus.

The genus *Reovirus* consists of 3 serotypes. The term reovirus (*r*espiratory *e*nteric *o*rphan virus) was coined to denote their dual respiratory and enteric trophism and their isolation in the absence of known disease. The first recognized reovirus was isolated from feces and was classified as an echovirus (Chapter 10). However, it became apparent that the virus possessed physical and biologic properties that set it apart from other known enteric or respiratory agents.

Structure and Antigenic Composition. Reoviruses are naked icosahedral particles 70 to 75 nm in diameter (Fig. 11–1). The genome consists of 10 loosely linked segments of ds RNA that are surrounded by 2 protein layers. The outer layer appears to contain 92 "holes"; the holes are surrounded by capsomeres composed of either 5 or 6 structural units. The inner protein layer and the RNA genome make up the subviral particle (SVP). The SVP contains the RNA-dependent RNA polymerase that initiates virus replication.

All reoviruses share a group-specific soluble Ag. By hemagglutination inhibition or neutralization tests, 3 distinct serotypes can be distinguished.

Host Range and Culture. The reoviruses can be isolated and prop-

Table 11–1. General Properties of Reoviruses

Properties of nucleic acid: double-stranded, linear RNA consisting of 10, 11 or 12 segments
Virion: nonenveloped; reovirus consists of 2 well-defined capsid shells; orbivirus has an outer shell and inner shell composed of 32 large ring capsomeres; rotavirus is a wheel-like structure.
Capsid symmetry: icosahedral
Size (nm): 70–75
Site of replication: cytoplasm

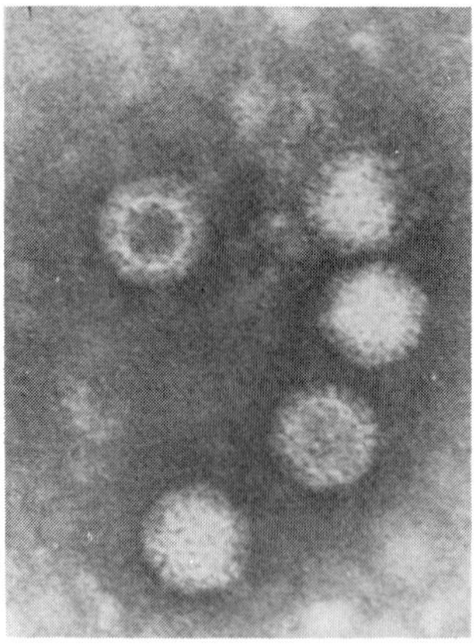

Fig. 11–1. Electron micrograph of reovirus. Magnification × 96,000. (Reprinted with permission from: Spendlove, R.S.: Unique reovirus characteristics. Prog. Med. Virol., *12*:161, 1970.)

agated in primary cultures of epithelial cells from many animal species as well as continuous human cell lines. Infected cells undergo cytopathic changes that are slow to develop and are rather nondescript in appearance. However, distinctive acidophilic inclusion bodies are present in the perinuclear region of the cytoplasm and may eventually encircle the nucleus completely.

Replication. Attachment of reovirus occurs via specific cell receptors. The virus is phagocytized into lysosomal vesicles that contain digestive enzymes. Initial digestion of the virion yields a subviral particle or core containing a RNA polymerase. Each of the 10 segments of virion RNA is transcribed into a monocistronic mRNA which is capped but not polyadenylated (see Chapter 3). The ds RNA segments each code for a separate polypeptide. The core of the virus containing the parental ds RNA is not digested and remains in the infected cell cytoplasm throughout the replication cycle. The mRNA molecules are copied into minus strands of RNA giving rise to progeny ds RNA (conservative replication). Progeny genomes are transcribed into additional mRNAs and viral proteins. The viral proteins are cleaved into smaller polypeptides before they recombine with progeny ds RNA to form immature virus particles. Maturation of virus occurs in several stages in the cell cytoplasm.

2. Orbivirus.

The etiologic agent of Colorado tick fever in humans is classified in the genus *Orbivirus*. The genus is comprised mainly of animal viruses (e.g., blue tongue virus of sheep) transmitted by an insect vector. Colorado tick fever virus shares many properties in common with other

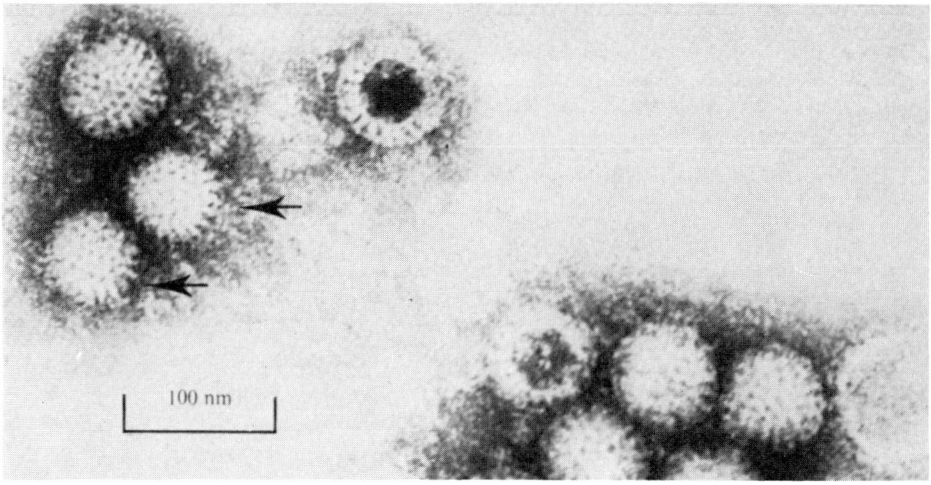

Fig. 11–2. Mature rotavirus particles from an experimentally infected lamb showing single (arrows) and double-shelled rotavirus particles. (With permission from McNulty, M.S.: Rotaviruses. J. Gen. Virol., *40*:1, 1978.)

reoviruses (Table 11–1). The wood tick constitutes both the vector and major reservoir for the virus. The virus is passed transovarially in the tick.

Colorado tick fever virus can be grown in insect tissue culture cells and baby hamster kidney cells. The virus gives distinguishable neutralization reactions but shares common CF Ags with other orbiviruses. The replicative cycle of orbiviruses is similar to reovirus.

3. Rotavirus.

The human rotaviruses (rota = wheel-like) are associated with acute enteritis in infants and young children (Chapter 33), piglets, calves and rhesus monkeys. The virus shares many properties in common with reoviruses (Table 11–1). Virus particles appear as single or double shelled in electron micrographs (Fig. 11–2). Rotaviruses have been successfully grown in cell culture after multiple passages in piglets. The human rotavirus shares common Ags with a rotavirus which causes calf diarrheal disease but not with reovirus. Abs to human rotavirus can be measured by CF and indirect IF tests.

REFERENCES

Fields, B.N.: Genetics of Reovirus. Curr. Top. Microbiol. Immunol., *91*:1, 1981.

Flores, J., Myslinski, J., Kalica, A.R., Greenberg, H.B., Wyatt, R.G., Kapikian, A.Z. and Chanock, R.M.: In vitro transcription of two human rotaviruses. J. Virol., *43*:1032, 1982.

Gorman, B.M.: Variation in Orbiviruses. J. Gen. Virol., *44*:1, 1979.

McNulty, M.S.: Rotaviruses. J. Gen. Virol., *40*:1, 1978.

Schroeder, B.A., Street, J.E., Kalmakoff, J., and Bellamy, A.R.: Sequence relationships

between the genome segments of human and animal rotavirus strains. J. Virol., *43*:379, 1982.
Shatkin, A.J., Sipe, J.D., and Loh, P.: Separation of ten reovirus genome segments by polyacrylamide gel electrophoresis. J. Virol., *2*:986, 1968.
Skup, D. and Millward, S.: mRNA capping enzymes are masked in reovirus progeny subviral particles. J. Virol., *34*:490, 1980.
Verwoerd, D.W., Huismans, H. and Erasmus, B.J.: Orbivirus. *In* Comprehensive Virology. Vol. 14. H. Fraenkel-Conrat and R.R. Wagner (eds.). New York, Plenum Press, 1979, pp. 285–345.
Zarbl, H., Skup, D. and Millward, S.: Reovirus progeny subviral particles synthesize uncapped mRNA. J. Virol., *34*:497, 1980.

Chapter 12

Orthomyxoviruses

A. ORTHOMYXOVIRIDAE

1. Influenza Virus

The Orthomyxoviridae family consist of three genera, *Influenza type A, B,* and *C,* based on antigenic specificity of internal antigens (i.e., M = matrix and NP = nucleoprotein). Influenza A virus is responsible for the periodic pandemics of influenza and for most epidemics during the interpandemic period. Influenza B virus is responsible for occasional influenza epidemics. Influenza C virus is a minor cause of influenza.

Structure and Antigenic Properties. Influenza virions are enveloped, usually spherical, and measure 80–120 nm in diameter (Fig. 12–1); the virion is represented diagrammatically in Figure 12–2. The envelope contains surface projections (spikes) composed of two major glycoproteins, the hemagglutinin (HA) and neuraminidase (NA) which are responsible for eliciting protective immunity against influenza virus (Table 12–1). The glycoproteins are embedded in the viral envelope lipid bilayer by a short hydrophobic segment. The internal surface of the envelope contains a non-glycosylated protein which forms a matrix that lends structural rigidity to the virus envelope. The envelope encloses helical nucleocapsids composed of a major nucleoprotein and three minor components (P1, P2, P3) which are probably subunits of the virion RNA-dependent RNA polymerase. The virion RNA consists of eight unique segments of single-stranded RNA of negative polarity (i.e., they are complementary to viral mRNAs). Each RNA segment encodes a different protein (the seven virion proteins and the nonstructural (NS) protein which is only found in infected cells). The size of the RNAs vary roughly according to the size of the proteins for which they code (Table 12–2).

The emergence of pandemic strains of influenza A virus is due to "antigenic shifts" in the HA and NA surface glycoproteins. Antigenic shift is responsible for pandemics of influenza. The mechanism of antigenic shift occurs when human type A and animal type A influenza viruses coinfect the same cell (human or animal origin). If genetic reassortment (Chapter 4) occurs between the RNA segments of the two viruses a new hybrid of influenza A virus emerges. The genes for HA and NA are contributed by the animal virus parent while the genes for virulence in humans are contributed by the human virus parent. Humans become infected because they have no immunity to the HA or NA glycoproteins of the hybrid virus. Antigenic shifts have occurred 3 or 4 times in this century, with new strains designated by the antigenic specificity of the HA (H1, H2, or H3) and NA (N1 or N2). For example, the H1N1 type arose in 1918, the H2N2 (Asian) in 1957, and the H3N2 (Hong

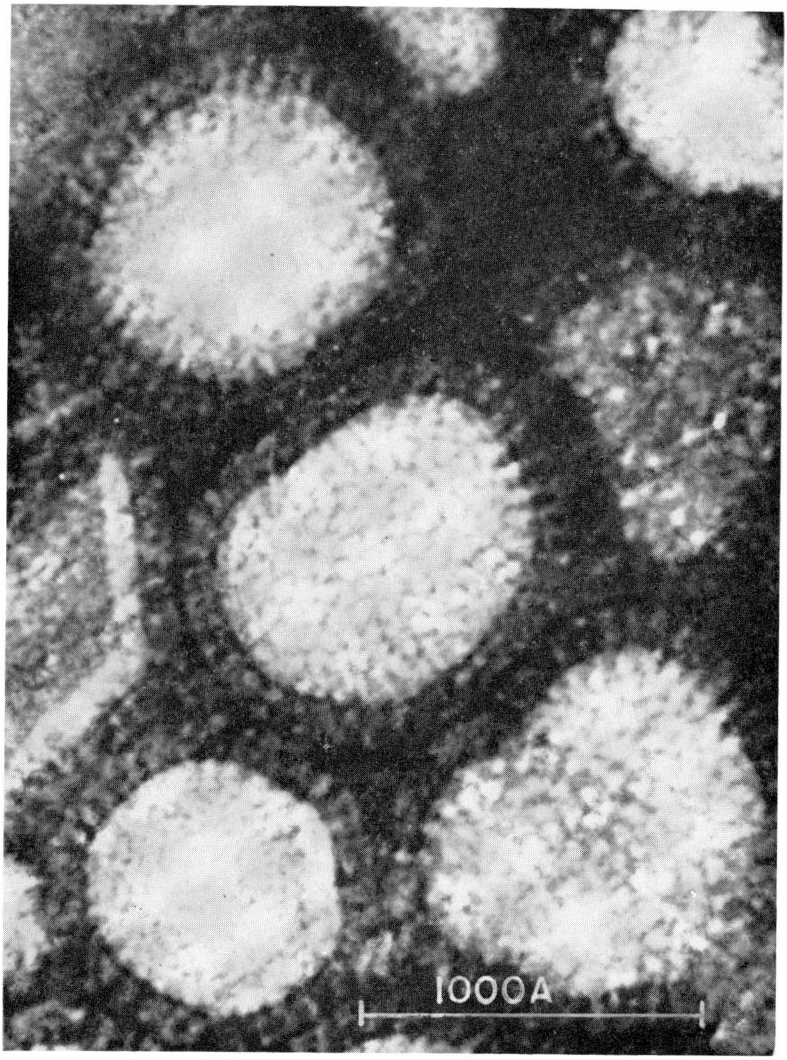

Fig. 12–1. Electron micrograph of phosphotungstate-stained influenza A virus. The particles are of variable size and shape and show evenly spaced projections perpendicular to the virus envelope. (From Virology, *11*:79, 1960.) (× 300,000.)

Kong) in 1968. In 1978 the H1N1 type re-emerged and circulated with the H3N2 virus among the non-immune population.

Between pandemic periods, the prevailing influenza A virus undergoes minor antigenic changes every 1–2 years referred to as *"antigenic drift."* The minor changes appear to be due to mutations in the genes for HA and NA. These glycoproteins are highly immunogenic and are able to undergo mutation without affecting the function of the protein.

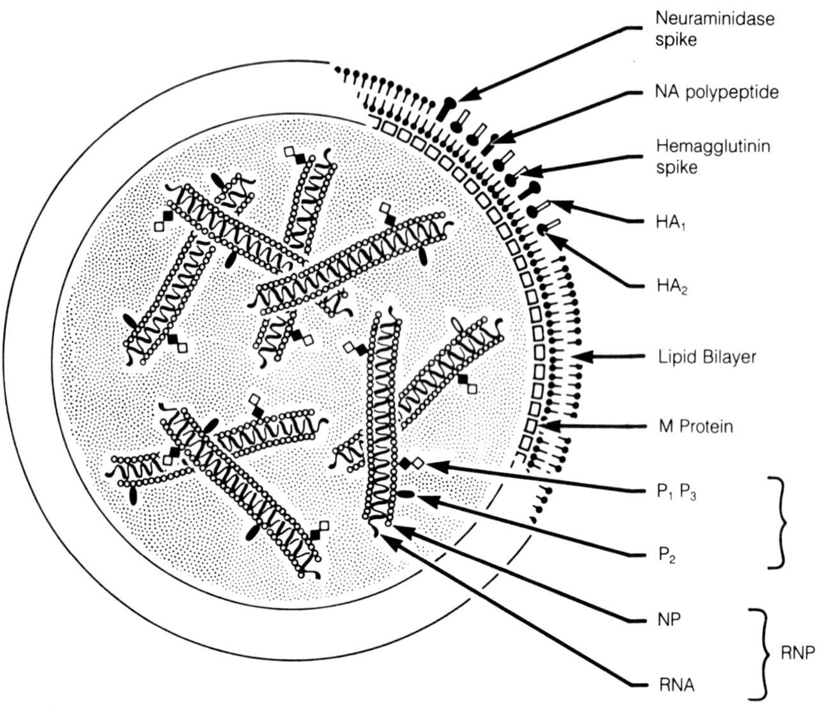

Fig. 12–2. Schematic diagram of an influenza virus particle. The eight nucleocapsid fragments are segmented and have a helical configuration due to association of internal proteins (NP, P_1, P_2 and P_3) with the single-stranded (−) RNA. The hemagglutinin (HA) spike is composed of three sets of HA_1 + HA_2 polypeptides. The neuraminidase (NA) spike is composed of four polypeptides (tetramer). The M protein is embedded in the lipid bilayer membrane. (From Dulbecco, R., and Ginsberg, H.S.: Virology. Hagerstown, Harper and Row, 1980, p. 1126.)

Table 12–1. General Properties of Influenza Virus

Properties of nucleic acid: ss RNA; virion RNA complementary to mRNA; composed of 8 segments; total mol wt 4×10^6 daltons.
Virion: enveloped; contains hemagglutinating and neuraminidase activities on separate glycoproteins.
Capsid symmetry: helical
Size (nm): 80–120
Site of replication: cytoplasm

It is noteworthy that influenza A and B viruses readily undergo antigenic drift but influenza C virus does not. The ability of influenza viruses to undergo Ag variation stands in marked contrast to most viruses, which are for the most part antigenically stable.

Currently, the nomenclature of influenza viruses consists of the following information: type (A, B, or C), the place or species of isolation,

Table 12–2. Size of the RNA Segments and Type of Protein Encoded by Each Influenza Virus RNA Segment

RNA Segment	Size of RNA ($\times 10^6$)	Protein Encoded
1	1.10	P_3
2	1.06	P_1
3	1.05	P_2
4	0.85	HA (Peplomer)
5	0.67	NP (Capsid)
6	0.56	NA (Peplomer)
7	0.34	M
8	0.23	NS1 + NS2

P_1, P_2, P_3 = internal proteins
NP = nucleocapsid protein
NS = nonstructural proteins
M = matrix protein
HA = hemagglutinin
NA = neuraminidase

the isolate number (optional), the year of isolation, and the H and N type. For example, A/Singapore/1/57 (H2N2) is a type A virus isolated in Singapore in 1957 whose HA and NA cross-react with those of other H2N2 viruses, but not with H1N1 viruses.

Host Range and Culture. Type A influenza viruses occur naturally in swine, horses, and birds, as well as in humans, whereas types B and C appear to be parasites of man only. The restricted host range of influenza B and C viruses may account for their inability to undergo antigenic shift. Influenza viruses can be propagated in embryonated eggs and adapted to grow in the respiratory tracts of monkeys and rodents. Some strains of influenza viruses have also been adapted to grow in chicken, dog or monkey kidney cell cultures.

Replication. Attachment of influenza virus to host cell receptors occurs via the HA glycoprotein. The specificity of HA is for carbohydrates containing N-acetyl-neuraminic acid (sialic acid), a ubiquitous component of cellular, as well as serum glycoproteins. The viral neuraminidase allows the virus to elute from receptors. The ability of the virus to elute is probably important in preventing permanent attachment to irrelevant i.e., non-cellular receptors.

Entry of virus into host cells occurs via endocytosis and fusion of the viral envelope with the endocytic vesicle. Fusion is enhanced if the HA peptide is proteolytically cleaved into two disulfide-linked subunits (HA_1 and HA_2), in a fashion similar to that of the F glycoprotein of paramyxoviruses (Chapter 13). Cleavage increases viral infectivity without affecting the receptor-binding activity of HA. It can be accomplished with trypsin-like proteases such as plasmin or trypsin itself.

After virus entry all 8 RNA segments of viral (−) RNA (Chapter 3) are transcribed by the virion RNA-dependent RNA polymerase. Synthesis of virus (+) mRNA requires synthesis of new cellular messenger-type RNA. This cellular RNA appears to be used as a primer by the virion RNA polymerase, which cannot make faithful (+) mRNAs without exogenous primer. During transcription the 5′ end of the cellular mRNA including the cap and a few extra nucleotides becomes the 5′ end of the influenza virus mRNA. Transcription of (+) mRNAs stops about 20–30 nucleotides short of the 5′ end of the virion (−) RNA templates. Thus, like other (−) strand viruses, the mRNAs are shorter than genomic RNA.

Transcription of the genes for the nonstructural and matrix proteins is unusual. Two different mRNAs are made from the same genome segment and each mRNA codes for a separate protein (e.g., NS_1 and NS_2). NS_1 is coded for by almost the entire NS gene. The mRNA for NS_2 includes internal sequences that overlap those coding for NS_1. NS_1 and NS_2 are translated in different reading frames so that the two proteins have different amino acid sequences. Similar overlapping genes code for M_1 and M_2.

Replication of (+) RNA to make new virion (−) RNA occurs in replicative intermediates (RIs) that are distinct from the transcriptive complexes that make mRNA (Chapter 3). Two types of RI are necessary for replication, one containing virion (−) type RNA as a template to make full length (+) RNA and one containing the (+) strand as a template to make virion (−) RNA (Chapter 3). The (+) strand template used for replication of viral RNA differs in several respects from viral mRNA. The template (+) strand lacks a 5′ cap and 3′ poly A, and it is genome length. Regulation of synthesis for mRNA and replicative (+) RNA is probably accomplished by viral proteins. In the absence of continuous viral protein synthesis, only transcription of mRNA occurs, and replication of (+) RNA is prevented.

The mRNAs for the HA and NA glycoproteins are translated on membrane-bound polyribosomes. The proteins remain membrane-bound as they are transported through the endoplasmic reticulum and Golgi apparatus to the plasma membrane. During this process of synthesis and transport through host membranes, the viral glycoproteins acquire their carbohydrate groups. Glycosylation is mediated by host glycosyl transferases.

In contrast to the mRNAs for HA and NA, the mRNAs for the internal viral proteins are translated on free polyribosomes. Nucleocapsids are assembled in the cytoplasm of the infected cell. The nucleocapsids associate with the plasma membrane in the process of virus maturation, which occurs by budding from the plasma membrane. Little is known of the mechanism by which recognition of the budding site by the internal virion components occurs. A likely possibility is that the viral glycoproteins span the membrane lipid bilayer and interact with the internal virion components, either the matrix protein or the nucleocapsid. In any event, the viral proteins are incorporated into the envelope to the virtual exclusion of host membrane proteins. In contrast, there is little if any selectivity for which membrane lipids are incorporated into the viral envelope. Thus the lipid composition of the viral envelope generally reflects that of the host plasma membrane.

The specificity of the budding process is such that only nucleocapsids containing (−) RNA strands are enveloped even though both (+) and (−) RNA strands are encapsidated. Another dilemma, which is how 8 independent genome segments get enveloped together to form a complete influenza virus particle, has not been solved. Some recognition mechanism may exist to ensure that all 8 segments get enveloped together. Alternatively, the genome segments could be enveloped randomly. However, this mechanism would probably require that more than 8 pieces of RNA be enveloped in each particle to ensure an adequate number of virions that have at least one copy of each genome segment. The latter possibility seems unlikely since heteropolyploidy (Chapter 4) between influenza virus strains infecting the same cell is relatively rare. In addition, genome segments do not segregate randomly in recombinants. Although this could be due to selection of recombinants on a non-random basis, it is also consistent with non-random envelopment of genome segments.

REFERENCES

Krug, R.M.: Priming of influenza viral RNA transcription by capped heterologous RNAs. Curr. Top. Microbiol. Immunol., *93*:125, 1981.

Palese, P., and Young, J.F.: Variation of influenza A, B, and C viruses. Science, *215*:1468, 1982.

Ward, C.W.: Structure of the influenza virus hemagglutinin. Curr. Top. Microbiol. Immunol., *94*:1, 1981.

Webster, R.G., Laver, W.G., Air, G.M., and Schild, G.C.: Molecular mechanisms of variation in influenza viruses. Nature, *296*:115, 1982.

Chapter 13

Paramyxoviruses

A. PARAMYXOVIRIDAE FAMILY

Viruses of the Paramyxoviridae family are divided into 3 major genera (*Parainfluenza virus, Morbillivirus,* and *Pneumovirus)* based on presence or absence of hemagglutinin and neuraminidase activities on the surface of virion envelopes (Table 13–1). Human pathogens include 4 serotypes of parainfluenza virus, mumps virus, measles virus and respiratory syncytial virus. Knowledge of the replication of parainfluenza viruses was gleaned from studies with animal strains of viruses (e.g., Sendai virus, Newcastle disease virus, and simian virus 5). The typical morphology for a paramyxovirus particle is presented in Figure 13–1. The suggested arrangement of structural components is presented in Figure 13–2.

1. Parainfluenza Virus.

The first human parainfluenza virus was isolated in 1955 from infants with croup. The virus, called croup-associated virus, was recognized by the characteristic syncytial cytopathic effect it produced in cell cultures. Subsequently, 3 additional antigenically distinct viruses were recovered from infants and children with respiratory diseases. The viruses were recognized by the capacity of infected cells to absorb guinea pig red blood cells. For this reason they were called *"hemadsorption"* viruses. Subsequently, the viruses were grouped together and called parainfluenza virus types 1 to 4 because of their morphologic resemblance to influenza virus, with which they were first classified.

Structure and Antigenic Properties. Parainfluenza viruses are large

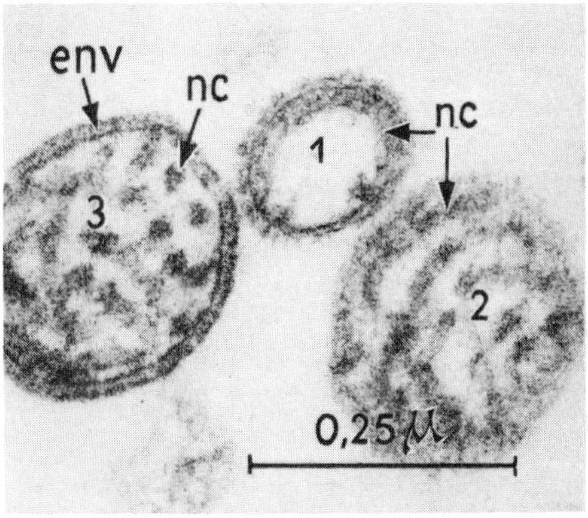

Fig. 13–1. Thin section of paramyxovirions. The envelope (env) is covered with short projections. The nucleocapsid (nc) shows various arrangements. In particle 1 the nc is parallel to the env. In particles 2 and 3 the nc is arranged more irregularly. (With permission from Berkaloff, A.: Etude au microscope electronique de la morphogenèse de la particle du virus sendai. J. Microscopy, 2:633, 1963.) (× 92,000.)

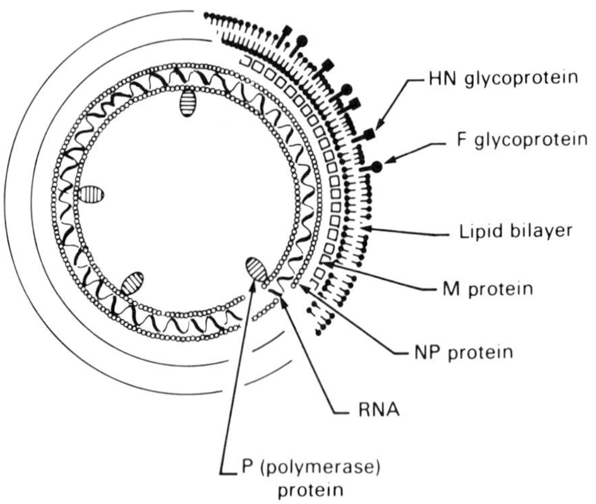

Fig. 13–2. Diagram of a paramyxovirus. (Different forms are shown for the F and HANA glycoproteins to indicate their chemical differences, although they are not distinguishable in electron micrographs.) The F and HANA glycoproteins appear to penetrate into the lipid bilayer and may traverse the entire envelope. The M protein forms the inner layer of the envelope and maintains its structure and integrity. The actual arrangement of the NP protein in the nucleocapsid is unknown. (With permission, from R. Dulbecco and H.S. Ginsburg: Virology, Hagerstown, Harper and Row, 1980, p. 1142.)

(150–300 nm in diameter) pleomorphic, spherical or filamentous particles (Fig. 13–1). The outer membrane envelope contains two types of glycoprotein surface projections (spikes) i.e., hemagglutinin and neuraminidase (HANA) and fusion (F) (Fig. 13–2). In contrast, hemagglutinin and neuraminidase activities of influenza viruses reside on separate glycoproteins (Chapter 12). The F glycoprotein is associated with 2 activities i.e., hemolytic and cell fusion. Both of these activities result from the ability of virus envelopes to fuse with cell membranes. Hemolysis may result from fusion of "leaky" virus envelopes with erythrocyte membranes; cell-to-cell fusion may result from simultaneous fusion of one viral envelope with the membranes of two different cells. Fusion is necessary for viral infectivity. The internal proteins of parainfluenza viruses include a nonglycosylated matrix (M) protein which shares many properties with the M protein of influenza viruses (Chapter 12). The nucleocapsid is a helical structure enclosing the single-stranded RNA genome. The nucleocapsid contains a major protein subunit (NP) and two proteins (P and L) that probably constitute the virion RNA-dependent RNA polymerase. The viral genome is complementary to mRNA and has a molecular weight of about 6×10^6 (Table 13–2).

Host Range and Culture. Parainfluenza viruses can be propagated in primary cultures of human or monkey kidney cells and in embryonated eggs. Most continuous cell lines (e.g., monkey kidney) will not

Table 13–1. Major Groups and Host Range of Paramyxoviridae

Virus Genus and Members	Host Range	Envelope Spikes Hemagglutinin	Neuraminidase
Parainfluenza		+	+
Parainfluenza virus (4 serotypes)	human		
Mumps virus	human		
Sendai virus	mouse		
SV5	monkey		
Newcastle disease virus	chicken, human		
Morbillivirus		+	–
measles virus	human		
canine distemper virus	dog		
rinderpest virus	cattle, sheep		
Pneumovirus		–	–
respiratory syncytial virus	human		
pneumonia virus of mice	mouse		

support multiple cycles of replication of Sendai virus because they lack an appropriate protease to cleave the viral F glycoprotein which is necessary for viral infectivity.

Replication. Attachment of parainfluenza virions to host cellular receptors is similar to that of influenza viruses (Chapter 12). Entry of paramyxoviruses into host cells occurs by fusion of the viral envelope via F glycoprotein with the host cell membrane. When parainfluenza viruses are grown in certain tissue culture cell lines, virions are released from cells with the F glycoprotein in the form of an inactive precursor (F_0). F_0 can be converted to the active form F by proteolytic cleavage of F_0 into 2 disulfide-linked subunits, F_1 and F_2. F_2 is derived from the N-terminus of F_0, while F_1 spans the viral envelope lipid bilayer. Cleavage of F_0 to F can be accomplished in vitro by treatment of tissue culture-grown virions with trypsin, which restores hemolytic and cell-fusing activities and infectivity. The structure of the F glycoprotein is analogous to the HA glycoprotein of influenza virus which also contains two disulfide-linked subunits. Following virus entry into host cells, transcription of the (–) RNA viral genome into mRNA occurs via the virion-associated RNA polymerase. The polymerase has 3 enzymatic activities: (1) RNA polymerase, (2) 5′capping activity, and (3) 3′poly A polymerase. All proteins of the transcriptive complex (NP, P, and L) must be present for any of the above enzyme activities to be expressed. Independent mRNAs are made for each viral protein. The mRNAs have a 5′cap and 3′poly A. Transcription occurs in a sequential manner from a single initiation site near the 3′ end of the genome. A short "leader" sequence is transcribed prior to transcription of the first mRNA. Since RNA synthesis is initiated at a single site, genes near the 3′ end of the genome must be transcribed in order for genes near the 5′ end to be transcribed.

A consequence of this mechanism of transcription is that the genes can be ordered, or mapped on the RNA, based on their sensitivity to inactivation by uv light. Ultraviolet irradiation causes premature termination of transcription at the site of uv reaction. Genes near the 5′ end are more sensitive to uv inactivation than those near the 3′ end, since a "hit" anywhere between the gene and the 3′ end is sufficient to prevent transcription. The order of uv sensitivity and therefore, the order of genes on the RNAs of parainfluenza viruses has been found to be 3′-NP-P-M-F_0-HANA-L-5′. Two possible mechanisms to account for the sequential transcription are (1) a large precursor RNA is made, which is subsequently cleaved by an endonuclease, or (2) the RNA polymerase stops and then re-starts polymerization at the junction between genes following an obligatory initiation at a single site.

As in the case of other (−) strand viruses (Chapter 12) replication of new virion RNA is a separate process from transcription (Chapter 3). Replication occurs in specialized structures referred to as replicative intermediates (RIs). Two types of RIs are found in infected cells, one using (−) strands as templates and the other using full-length (+) strands as templates. The 3′ ends of the (+) and (−) strands, which contain the binding site for the RNA polymerase, have similar but not identical sequences. Thus, regulation of the relative proportions of the two types of RIs (most RIs have (+) strand templates) may depend on recognition of small differences in nucleotide sequence at the RNA polymerase binding site.

As in the case of influenza viruses (Chapter 12) regulation between transcription and replication is probably accomplished by newly synthesized viral proteins. When protein synthesis is inhibited, RNA synthesis shifts from replication to transcription. It appears that all of the RNA in RIs, including the nascent RNA strands, are covered with newly synthesized NP. This immediate encapsidation may serve a useful function in preventing hybridization of newly synthesized (−) strands with viral mRNA, which would have the effect of inhibiting viral protein synthesis. The fact that all or almost all of the replicating RNA is covered with protein probably accounts for the absence of recombination by crossing-over in (−) strand viruses (Chapter 4). The mechanism of recombination probably involves hybridization of related nucleic acids from both parents. This hybridization would not be expected to occur in RIs because exogenous RNA does not have access to the replicating RNA.

Virion assembly resembles that of influenza viruses (Chapter 12).

2. Mumps Virus

Mumps is an acute contagious disease common among children and young adults (Chapter 29). The disease is characterized by nonsuppur-

Table 13–2. General Properties of Paramyxoviruses

Properties of nucleic acid: (−) single-stranded RNA; virion RNA is complementary to mRNA; 6×10^6 daltons
Virion: enveloped, contains a glycoprotein with hemagglutinating and neuraminidase activities and F-glycoproteins
Capsid symmetry: helical
Size (nm): 150–300
Site of replication: cytoplasm

ative enlargement of one or both parotid glands; complications include orchitis (infection of the testes) in young males and meningoencephalitis.

Structure and Antigenic Properties. Structurally, mumps virus is a typical paramyxovirus (Fig. 13–1) that contains hemagglutinin and neuraminidase, and hemolytic activities associated with the virus envelope. Virions are helical and range in size from 150–300 nm in diameter. There is only one serotype of mumps virus; however, mumps virus shares minor Ags with other paramyxoviruses.

Host Range and Culture. The natural host of mumps virus is man. The virus can be grown in the amnion and yolk sacs of embryonated chicken eggs and assayed by hemagglutination. Primary cultures of human or non-human primate kidney cells are used for virus isolation. Mumps virus is identified by hemadsorption-inhibition tests in infected cell cultures. Cytopathic effects of mumps-virus-infected tissue cultures include syncytial cell formation and acidophilic intracytoplasmic inclusions.

3. Respiratory Syncytial Virus

This virus was initially isolated in 1956 from a chimpanzee with an upper respiratory tract (URT) infection and subsequently from laboratory workers in contact with infected chimpanzees. Originally referred to as chimpanzee coryza agent it is now called respiratory syncytial virus because of its pronounced tendency to form syncytia in tissue cultures. The virus is recognized as being one of the most important causes of severe lower respiratory tract infections in infants.

Structure and Antigenic Properties. Respiratory syncytial virus resembles the other paramyxoviruses morphologically. However, in contrast to the parainfluenza viruses, the respiratory syncytial virion does not possess hemolytic activity, a hemagglutinin, or neuraminidase (Table 13–1) despite the presence of surface projections on the envelope. Respiratory syncytial virus is unstable at 37°C and can be stored at 4°C for only a few hours without loss of infectivity; approximately 90 percent of its infectivity is lost following slow freezing. The virus can be preserved only when it is rapidly frozen in the presence of protein, serum, or hypertonic sucrose and stored at −70°C.

The respiratory syncytial viruses can be differentiated into 4 types on the basis of minor differences in the surface V-antigens that are detectable only by sensitive neutralization tests. All respiratory syncytial viruses share a common complement-fixing (CF) Ag that does not cross-react with that of other paramyxoviruses.

Host Range and Culture. Only man and the chimpanzee appear to develop disease from respiratory syncytial virus infections. However, inapparent infections can be produced in ferrets, monkeys, and many other mammals.

Respiratory syncytial virus can be isolated and propagated best in continuous human cell lines of which HEp-2 cells are the most sensitive. Multiplication is slow and is detected by the appearance of large syncytial cells that contain dozens of nuclear and cytoplasmic inclusion bodies.

4. Measles (Rubeola) Virus

Although measles was long assumed to be of viral etiology, the causative virus was not isolated until 1954, when J.F. Enders and co-workers developed techniques for propagating the virus in human kidney tissue cultures. Following isolation of the virus, serologic procedures of basic importance in studying measles infection also became available. The basic research of Enders ultimately led to the development of the attenuated strains of virus presently used in live virus vaccines against measles.

Structure and Antigenic Properties. The measles virion shares common morphologic properties with other parainfluenza viruses (Fig. 13–1). Hemolytic activity and a hemagglutinin for monkey erythrocytes are associated with the virion (Table 13–1). The virus contains no neuraminidase, since hemagglutination is maximal at 37°C and is not followed by elution of the virus from the agglutinated cells. Measles virus is highly temperature-sensitive; it is rapidly inactivated at 37°C or by refrigeration. Specimens for virus isolation must either be inoculated into tissue cultures shortly after collection or frozen at −70°C.

Only one serotype of measles virus has been isolated. There is no antigenic relationship between measles virus and the other paramyxoviruses that infect man. However, an antigenic relationship has been demonstrated between the viruses of measles, rinderpest (which infects cattle), and canine distemper (Table 13–1).

Host Range and Culture. Man is the only natural reservoir for measles virus. Although monkeys are susceptible, they are not infected in their natural habitat. Monkeys acquire the virus only after capture and contact with human beings infected with measles virus. The virus can be adapted to the newborn mouse and hamster.

Measles virus can be cultured in a number of cell systems, either on primary isolation or after adaptation. Primary cultures of human and monkey kidney cells are the most satisfactory; human amnion cells are less sensitive. The propagation of measles virus in human and monkey mononuclear leukocytes supports the suggestion that these cells may play a role in the transport of virus during the incubation period of the disease. After initial isolation in primary cell cultures, measles virus will multiply in continuous cell lines of human or monkey origin and in primary dog kidney cells. The virus can be adapted to other cell systems such as chick embryo fibroblasts, which were used to develop the attenuated Edmonston vaccine strain. Strains adapted to human amnion cells will subsequently grow in the amniotic cavity of the chick embryo.

The cytopathic effects produced by measles virus include the formation of syncytia and multinucleate giant cells, as well as the formation of spindle-shaped fibroblast-like cells. Both types of cytopathic effects may occur in the same culture; initially, the changes are focal but, as virus spreads continuously from cell-to-cell, the entire cell culture is destroyed. Numerous acidophilic inclusion bodies can be observed in the cell cytoplasm and nuclei of stained preparations. Chromosomal breaks are regularly demonstrated by standard karyologic techniques.

REFERENCES

Choppin, P.W., and Compans, R.W.: Reproduction of paramyxoviruses. *In* Comprehensive Virology. Vol. 4. H. Fraenkel-Conrat and R.R. Wagner (eds.). New York, Plenum Press, 1975, pp. 95–178.

Dowling, P.C., Giorgi, C., Roux, L., Dethlefsen, L.A.., Galantowicz, M.E., Blumberg, B.M., and Kolakofsky, D.: Molecular cloning of the 3′-proximal third of Sendai virus genome. Proc. Natl. Acad. Sci. (U.S.A.), *80*:5213, 1983.

Huang, Y.T., and Wertz, G.W.: The genome of respiratory syncytial virus is a negative-stranded RNA that codes for at least seven mRNA species. J. Virol., *43*:150, 1982.

Leppert, M., Rittenhouse, L., Perrault, J., Summers, D.F., and Kolakofsky, D.: Plus and minus strand leader RNAs in negative strand virus-infected cells. Cell, *18*:735, 1979.

Mahy, B.W.J., and Barry, R.D. (ed.): Negative Strand Viruses and the Host Cell. New York, Academic Press, 1978.

Miller, C.A., and Raine, C.S.: Heterogeneity of virus particles in measles virus. J. Gen. Virol., *45*:441, 1979.

Chapter 14

Coronaviruses

A. Coronaviridae Family

The Coronaviridae are ubiquitous viruses which cause a broad spectrum of diseases in their natural hosts including humans. Human strains (human respiratory virus) are associated with acute upper respiratory illnesses in adults. Animal strains include avian infectious bronchitis virus, mouse hepatitis virus, and transmissible gastroenteritis virus of swine.

1. Coronavirus

a. Human Respiratory Virus. In 1965, investigators at the Common Cold Research Unit in England reported the isolation of another common cold virus. The virus was obtained from nasal washings collected from a schoolboy suffering from a typical common cold. Attempts to isolate a rhinovirus or other known respiratory agents had failed. However, colds could be induced in human volunteers by the intranasal inoculation of the original specimen. Organ cultures consisting of small bits of human embryonic trachea were inoculated with throat washings from the volunteers who developed URT infections. Culture fluids collected up to 8 or 9 days later induced colds in volunteers. Sera from the volunteers were titrated for Abs against the viruses known to cause respiratory infections, but no rise in titer was detected. Electron microscopy of infected organ cultures revealed a virus morphologically indistinguishable from the mouse hepatitis and avian bronchitis viruses. It was concluded that a new human respiratory virus, now classified as a coronavirus, had been isolated.

Structure and Antigenic Properties. The coronaviruses are pleomorphic enveloped particles 70–120 nm in diameter; the loosely wound nucleocapsid is helical (Table 14–1). Surface projections on the envelope are distinctive pedunculated or petal-shaped structures rather than the spikes typical of influenza virus. In electron micrographs, the virions resemble a crown or solar corona (Fig. 14–1), hence the name coronavirus. Several serotypes have been isolated from man. Some types cross-react antigenically with mouse hepatitis virus but not with avian infec-

Table 14–1. General Properties of Coronavirus

Properties of nucleic acid: (+) single-stranded, nonsegmented RNA; virion RNA functions as mRNA, 9×10^6 daltons
Virion: enveloped; peplomeres are petal-shaped; pleomorphic
Capsid symmetry: helical
Size (nm): 70–120
Site of replication: cytoplasm

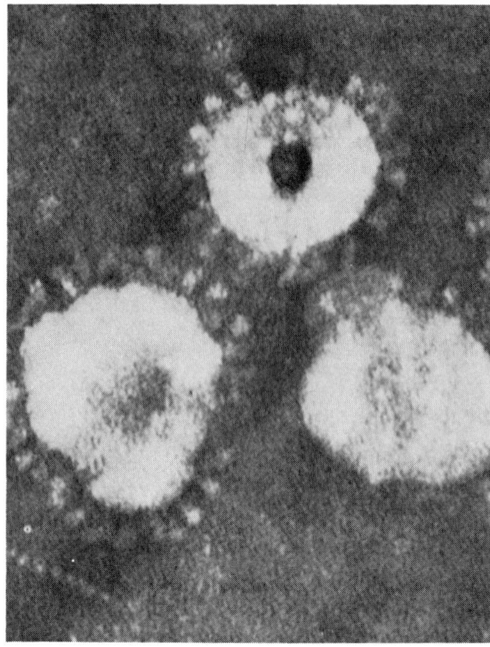

Fig. 14–1. Electron micrograph of coronavirus. Magnification × 300,000. (Reproduced with permission from Bradburne, A.F., and Tyrell, D.A.: Coronaviruses of man. Prog. Med. Virol., *13*:373, 1971.)

tious bronchitis virus. Antigenic relationships between some of the human serotypes also have been demonstrated.

Host Range and Culture. The human serotypes of coronavirus, with a few exceptions, grow only in human cells; the avian virus replicates only in avian cells, and the mouse virus in mouse cells. Although some human strains have been adapted to grow in mice, these viruses are probably species-specific in nature. Optimal temperature for growth is 33°–35°C.

Most human coronaviruses can be isolated only in fragments of human trachea, but some have been isolated in diploid strains of human embryonic fibroblasts; some also can be propagated in human transformed cell lines. The cytopathic effect develops slowly and usually does not involve the entire cell sheet. Infected cells become elongated and "stringy" in appearance; the cytoplasm becomes vesiculated, and the cells eventually degenerate.

Replication. Little is presently known concerning the replication cycle for coronaviruses. Virus attachment probably occurs via specific cell surface receptors. Penetration seems to involve viropexis and membrane fusion. The mechanism of uncoating is not known. Virion RNA has (+) polarity (same nucleotide sequence as mRNA); that is, virion

RNA functions as mRNA (Table 14–1). There appears to be four major groups of coronavirus specific RNA species: genomic RNA, replicative intermediates, discrete mRNA and defective or deleted RNA. There are nine viral-specific proteins synthesized in infected cells. One or more of the proteins functions as RNA polymerase responsible for mRNA and genomic RNA synthesis. Assembly of nucleocapsids occurs at the cisternal membranes of the endoplasmic reticulum and Golgi apparatus.

REFERENCES

Dennis, D.E., and Brian, D.A.: RNA-dependent RNA polymerase activity in coronavirus infected cells. J. Virol., *42*:153, 1982.

Hierholzer, J.C.: Purification and biophysical properties of human coronavirus 229E. Virology, *75*:155, 1976.

Lai, M.M.C., and Stohlman, S.A.: The RNA of mouse hepatitis virus. J. Virol., *26*:236, 1978.

Lomniczi, B.: Biological properties of avian coronavirus RNA. J. Gen. Virol., *36*:531, 1977.

Macnaughton, M.R., Madge, M.H., and Reed, S.E.: Two antigenic groups of human coronaviruses detected by using enzyme-linked immunosorbent assay. Infect. Immun. *33*:734, 1981.

Robb, J.A. and Bond, C.W.: Coronaviridae. *In* Comprehensive Virology. H. Fraenkel-Conrat and R. Wagner (eds.), Vol. 14. New York, Plenum Press, 1979, p. 193.

Siddell, St. H., Wege, H., and ter Meulen, V.: The structure and replication of coronaviruses. Curr. Topics Microbiol. Immunol., *99*:131, 1982.

Wege, H. Siddell, St. H., and ter Meulen, V.: The biology and pathogenesis of coronaviruses. Curr. Top. Microbiol. Immunol., *99*:165, 1982.

Chapter 15

Togaviruses

A. TOGAVIRIDAE FAMILY

This group includes well over 250 different viruses, of which about 65 have been shown to cause disease in man. These viruses characteristically require a suitable bloodsucking vector (invertebrate host) and an effective reservoir (vertebrate host) in their complex biologic cycle. The name togavirus includes many viruses previously known as arboviruses (arthropod-borne viruses) that multiply in bloodsucking insects and are transmitted to vertebrates by insect bites. Togaviruses of medical importance in the U.S.A. are currently subdivided into the genus *Alphavirus* (old arbovirus, group A), the genus *Flavivirus* (old arbovirus, group B), and the genus *Rubivirus* based on shared Ags demonstrable by neutralization, HAI or CF tests. The neutralization test is considered to be the most specific and is generally employed to differentiate the viruses within each group. Viruses classified as togaviruses are named either after a geographic location in which they occur (e.g., St. Louis encephalitis virus) or after the disease they cause (e.g., phlebotomus fever virus, rubella virus). This classification scheme has resulted in some rather unusual and exotic names, such as O'nyong-nyong, and Chikungunya viruses.

1. Alphavirus

a. Eastern Equine Encephalitis (EEE). This virus was first isolated from a horse that died from encephalitis in 1933, hence the name equine. Since then, EEE virus has been found to be responsible for disease in horses and man throughout the eastern and southeastern states, as well as in parts of Central and South America.

b. Western Equine Encephalitis (WEE). Of the equine encephalitis viruses in the U.S.A., WEE is the most widespread. Although the virus was first isolated in states west of the Mississippi river, it is now known to occur in many states east of the Mississippi river as well. Of interest is the fact that human disease has been reported to occur only in the continental U.S.A. and Brazil. In contrast to EEE, the disease WEE is rare on the Eastern coast of the U.S.A.

Structure and Antigenic Properties. All alphaviruses are spherical and contain an electron-dense core surrounded by a phospholipid bilayer envelope (Fig. 15–1). The nucleocapsid has 32 capsomeres and a diameter of 35–40 nm. The glycoprotein spikes in the viral envelope contain the virus hemagglutinin. A schematic model of an alphavirus is shown in Figure 15–2.

Host Range and Culture. Most alphaviruses multiply in embryonated chicken eggs or in tissue cultures derived from hamster, duck embryo, or other laboratory mammals. The natural vertebrate host of most alphaviruses is a wild bird or mammal; several host species are usually

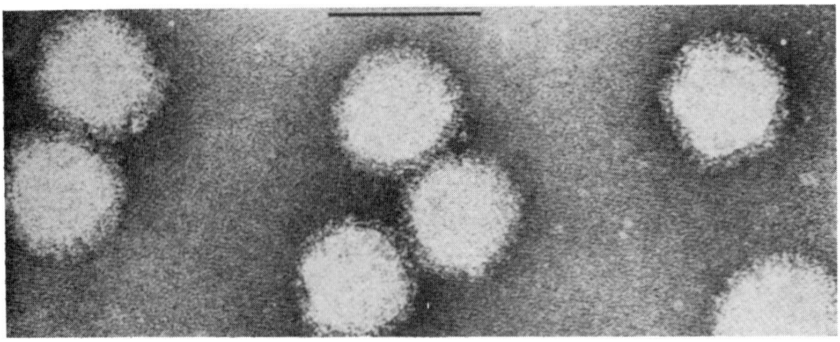

Fig. 15–1. Negative-stained virions of group A (genus Alphavirus) togaviruses. × 240,000. (Reprinted with permission from Simpson, R.W., and Hauser, R.E.: Basic structure of group A arbovirus strains Middleburg, Sindbis, and Semliki Forest examined by negative staining. Virology, *34*:358, 1968.)

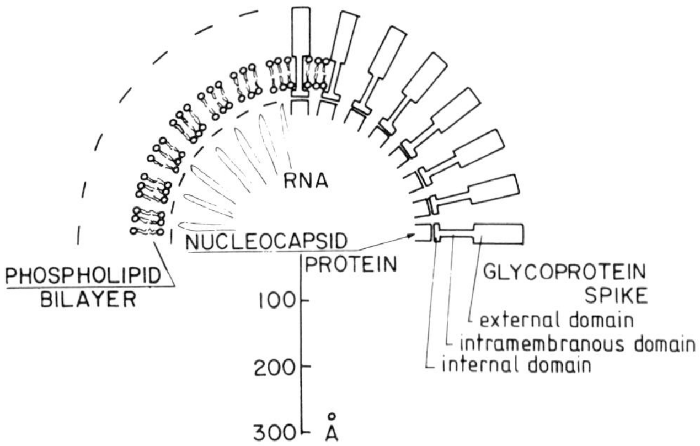

Fig. 15–2. Schematic model of an alphavirus particle. (With permission from Garoff, H., Kondor-Koch, C., and Riedel, H.: Structure and assembly of alphaviruses. Curr. Top. Microbiol. Immunol., *99*:1, 1982.)

involved in the biologic cycle of the virus. The most susceptible laboratory animal is the mouse or day-old chick. All alphaviruses replicate in the cell cytoplasm and acquire their lipoprotein envelope by budding through cell membranes (e.g., vacuolar or plasma membranes). Virus multiplication can be detected by the cytopathic effects produced, virus-specific immunofluorescence or hemadsorption tests. Alphavirus infectivity titers can be easily quantified by plaque assay in cultures of chick embryo cells as well as kidney cells of hamster or nonhuman primate origin.

2. Flavivirus

Yellow fever (YF) virus is the prototype of the genus *Flavivirus* of the family Togaviridae. Also included in this genus is Saint Louis encephalitis virus and dengue virus.

a. Yellow Fever Virus. Human isolates of YF virus exhibit viscerotropic and neurotropic characteristics. Viscerotropism is manifested by infection and injury of the kidneys, liver and heart; neurotropism signifies infection and injury of cells of the CNS. Following successive brain passages in mice, YF virus becomes adapted and exhibits more neurotropism and less viscerotropism than the original virus. Prolonged cultivation of the virus in chick embryos has produced an attenuated strain, 17D, which is widely used as a vaccine.

b. Saint Louis Encephalitis Virus. This virus was first recognized as a causative agent of disease during an epidemic of encephalitis in St. Louis, Missouri, in 1933. It has subsequently been found to occur frequently in the U.S.A. in widespread epidemics or as sporadic cases in areas where the virus is endemic. SLE virus is the major cause of disease associated with the genus *Flavivirus.*

c. Dengue Virus. This virus is responsible for a mosquito-borne infection characterized by fever, muscle and joint pain, lymphadenopathy and rash. The virus is stable at −70°C and in the lyophilized state at 5°C. Human blood may remain infectious for several weeks if kept at 5°C.

Structure and Antigenic Properties. Yellow fever virus is a small (30 to 40 nm) single-stranded, enveloped, RNA virus (Table 15–1). The virus is stable at 4°C in 50 percent glycerol and withstands lyophilization. It is rapidly inactivated by boiling water (100°C) or by 0.1% formalin. The YF virus contains at least two distinct Ags, one a hemagglutinin and the other a CF Ag. Only one antigenic type of YF virus exists in nature.

SLE virus shares properties in common with other togaviruses (Table 15–1). On the basis of serologic tests SLE virus is grouped with the genus *Flavivirus.*

Dengue virus exists in nature as 4 distinct serologic types, designated types 1–4. Although dengue virus shares structural relationships with YF and SLE viruses, there is no cross-immunity.

Table 15–1. General Properties of Togaviruses

Properties of nucleic acid: single, (+) stranded RNA; virion RNA serves as mRNA; 4.2×10^6 daltons.
Virion: enveloped; contains 2 envelope glycoproteins and one capsid protein
Capsid symmetry: icosahedral
Size (nm): 46–75
Site of replication: cytoplasm

Host Range and Culture. YF, SLE and dengue viruses grow in a variety of mammals (primates, mice) as well as mosquitoes and can be cultivated in chick embryos and cell cultures which results in characteristic cytopathic changes.

Replication. Togaviruses replicate in the cell cytoplasm. When the genome is uncoated, about 70 percent is translated into a single large polyprotein which is subsequently cleaved into four smaller proteins. Two of these are RNA polymerase subunits. The RNA polymerase transcribes the parental (+) strands into (−) strands which then serve as templates for making more (+) strands. Two types of (+) strands are synthesized: The first type are full length progeny (+) strands which can serve as mRNA; these act as templates for making more (−) strands, or become encapsidated into progeny virions. The second type of (+) strands are short pieces of RNA which code for three to four structural proteins of togavirus particles. One of these proteins, the C protein, forms the nucleocapsid and the remaining proteins make up the envelope glycoproteins or spikes. Maturation of progeny virions involves formation of nucleocapsids consisting of the (+) stranded RNA and C protein and budding through patches of cell membrane into which the glycosylated envelope proteins have been inserted.

3. Rubivirus

The first clinical description of rubella appeared in Germany in the early 1800s, hence the name "German measles." Although a viral etiology of the disease was suspected, efforts to isolate the agent were unsuccessful until 1962, when rubella virus was isolated in cell cultures. The availability of specific diagnostic tools during the 1964 epidemic led to a marked advance in knowledge of the epidemiology and control of this disease.

Structure and Antigenic Properties. The rubella virion is roughly spherical and is about 60 nm in diameter; however, pleomorphic forms have been observed. It is composed of an internal helical nucleocapsid of approximately 30 nm surrounded by a lipoprotein envelope with numerous short projections at the surface. The virus contains a hemagglutinin for red blood cells from day-old chicks, adult geese, or pigeons. The virus is relatively labile, but infectivity can be preserved by storage at −70°C.

Only one serotype of rubella virus has been recognized; there is no antigenic relationship to other known viruses.

Host Range and Culture. In nature, rubella appears to be limited to man. However, inoculation of monkeys and ferrets results in subclinical infection; a chronic infection may develop in newborn ferrets. Virus administered to pregnant rabbits may cause congenital abnormalities in the offspring. These abnormalities include cataracts and interstitial pneumonitis. Infection of the Japanese quail results in a febrile reaction and alterations in the genital system. These experimental models may be useful in delineating further the pathogenesis of congenital rubella in man.

Rubella virus can be propagated in a wide range of cell types from several animal species; however, visible cytopathic effects are not produced in all cells. One of the procedures used for detecting rubella virus depends upon the interference phenomenon, i.e., the capacity of rubella virus to interfere with the subsequent multiplication of a cytopathogenic challenge virus. Primary green monkey kidney cells are inoculated with the specimen, incubated for 7–10 days, and challenged with an enterovirus such as echovirus 11. If typical enterovirus cytopathic effects fail to develop, the specimen is presumptively considered to be positive for rubella virus; serologic tests with known positive antisera can be used to confirm the diagnosis. A number of continuous cell lines in which the virus multiplies and produces a cytopathic effect are now available; these include RK-13 (rabbit kidney), SIRC (rabbit cornea), BHK-21 (baby hamster kidney), and Vero (green monkey kidney). The cytopathic effect begins in isolated microfoci, spreads slowly, but seldom involves the entire culture. The hemadsorption technique has been used to detect virus multiplication before typical cytopathic effects appear in BHK-21 cells but apparently is unsatisfactory in many other cell systems.

REFERENCES

Garoff, H., Kondor-Koch, C., and Riedel, H.: Structure and assembly of alphaviruses. Curr. Top. Microbiol. Immunol., *99*:1, 1982.

Garoff, H., Simons, K., and Dobberstein, B.: Assembly of the Semliki Forest virus membrane glycoproteins in the membrane of the endoplasmic reticulum *in vitro*. J. Mol. Biol., *124*:587, 1978.

Helenius, A., Kartenbeck, J., Simons, K., and Fries, E.: On the entry of Semliki Forest virus into BHK-21 cells. J. Cell Biol., *84*:404, 1980.

Huang, C.H.: Studies of Japanese encephalitis in China. Adv. Virus Res., *27*:71, 1982.

Kaluza, G., and Pauli, G.: The influence of intramolecular disulfide bonds on the structure and function of Semliki Forest virus membrane glycoproteins. Virology, *102*:300, 1980.

Pfefferkorn, E.R.: Genetics of togaviruses. Comprehensive Virology, *9*:209, 1977.

Sawicki, D.L., Kaariainen, L., Lambek, C., and Gomatos, P.J.: Mechanism for control of synthesis of Semliki Forest virus 26S and 42S RNA. J. Virol., *25*:19, 1978.

Westaway, E.G.: Strategy of the flavivirus genome: Evidence for multiple internal initiation of translation of proteins specified by Kunjin virus in mammalian cells. Virology, *80*:320, 1977.

Wirth, D.F., Katz, F., Small, B., and Lodish, H.F.: How a single Sindbis virus mRNA directs the synthesis of one soluble protein and two integral membrane glycoproteins. Cell, *10*:253, 1977.

Chapter 16

Bunyaviruses

A. BUNYAVIRIDAE FAMILY

The Bunyaviridae family of arboviruses is the largest taxonomic subset of arboviruses. The other major subsets are *Alphaviruses* and *Flaviviruses* (Togaviridae family; Chapter 15) *orbiviruses* (Reoviridae family; Chapter 11) and *rhabdoviruses* (Rhabdoviridae family; Chapter 18). Classification of bunyaviruses is based on neutralization tests and HAI tests. All bunyaviruses multiply in arthropod vectors. The most important bunyaviruses associated with disease in humans belong to the genus Bunyavirus, which includes California encephalitis, Bunyamwera, and La Crosse viruses.

1. California encephalitis virus.

This virus is responsible for a febrile meningoencephalitis in humans.

Structure and Antigenic Properties. The California encephalitis virus contains a segmented, (−) single-stranded RNA genome about 6 × 10^6 daltons. Virions are helical enveloped particles about 90–100 nm in diameter. The virus envelope is covered by surface projections composed of 2 major glycoproteins (G1 and G2) that are clustered to form hollow cylindrical morphologic units arranged in an icosahedral surface lattice. A hypothetical bunyavirus particle is shown in Figure 16–1 and an electron micrograph is presented in Figure 16–2. General properties of bunyaviruses are shown in Table 16–1.

Host Range and Culture. California encephalitis virus like other bunyaviruses cause cytopathology in infected cell lines (e.g., HeLa and BHK 21). These viruses can also infect and replicate in newborn and weanling mice.

Replication. Information on the mechanism of bunyavirus repli-

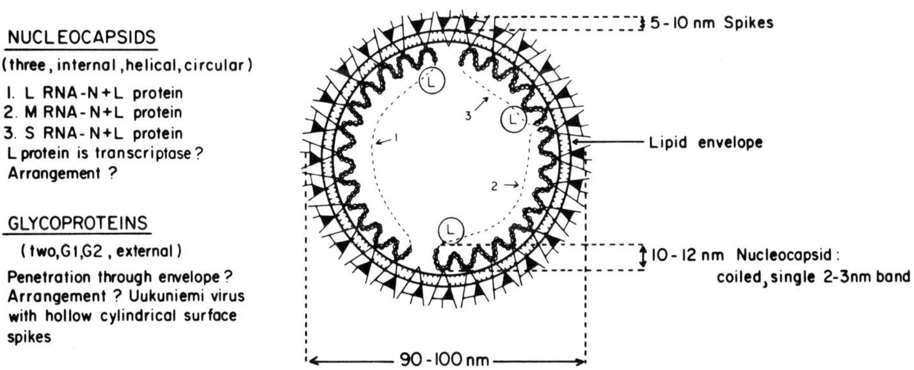

Fig. 16–1. Schematic representation of a bunyavirus particle. The particle consists of three segments of RNA (long L, medium M, and short S) and four structural proteins (N, L, G1, G2). (From Bishop, D.H.L., and Shope, R.E.: Genetic potential of bunyaviruses. Curr. Top. Microbiol. Immunol., *86*:1, 1979.)

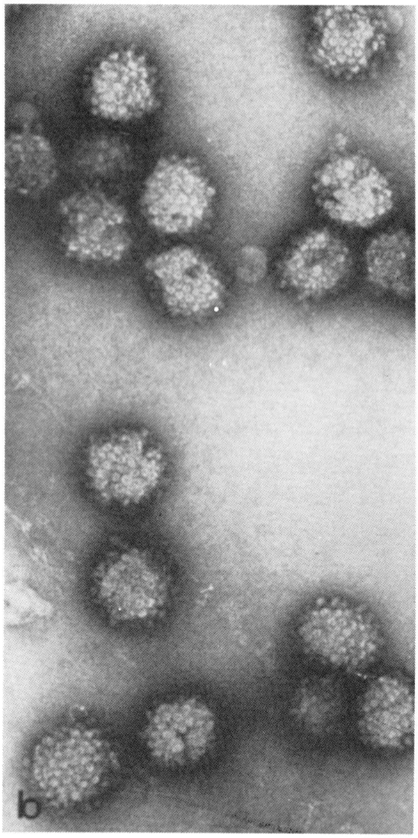

Fig. 16–2. An electron micrograph of a bunyavirus particle × 100,000. (From Bishop, D.H.L., and Shope, R.E.: Genetic potential of bunyaviruses. Curr. Top. Microbiol. Immunol., *86*:1, 1979.)

Table 16–1. General Properties of Bunyaviruses

Properties of nucleic acid:	(−) single-stranded RNA, composed of three segments; long (L), medium (M), and short (S); genome is complementary to mRNA; 6 × 10^6 daltons.
Virion:	enveloped; consists of two glycoproteins designated G1 and G2; virion-associated RNA polymerase.
Capsid symmetry:	helical.
Size (nm):	90–100
Site of replication:	cytoplasm

cation is incomplete. The negative polarity of the segmented RNA genome suggests that bunyaviruses replicate in the cell cytoplasm in a manner analogous to orthomyxoviruses (Chapter 12). The virion-associated RNA polymerase copies virion RNA into mRNA termed "primary transcription." After protein synthesis the rate of mRNA synthesis increases (secondary transcription) due to the availability of newly synthesized progeny RNA templates in the infected cells. Three virus specific proteins, G1, G2, and N, can be identified in infected cell extracts. Assembly of progeny virions occurs by a process of budding into the Golgi cisternae.

REFERENCES

Bouloy, M., Krams-Ozden, S., Horodniceanu, F., and Hannoun, C.: Three-segment RNA genome of Lumbo virus (bunyavirus). Intervirology, *2*:173, 1973–74.

Bishop, D.H.L.: Genetic potential of bunyaviruses. Curr. Top. Microbiol. Immunol., *86*:1, 1979.

Bishop, D.H.L., and Shope, R.E.: Bunyaviridae. Comprehensive Virology., *14*:1, 1979.

Clewley, J., Gentsch, J., and Bishop, D.H.L.: Three unique viral RNA species of snowshoe hare and La Crosse bunyaviruses. J. Virol., *22*:459, 1977.

Gentsch, J., Bishop, D.H.L., and Obijeski, J.F.: The virus particle nucleic acids and proteins of four bunyaviruses. J. Gen. Virol., *34*:257, 1977.

Gentsch, J.R., and Bishop, D.H.L.: Small viral RNA segment of bunyaviruses codes for viral nucleocapsid protein. J. Virol., *28*:417, 1978.

Gentsch, J.R., Rozhon, E.J., Klimas, R.A., El Said, L.H., Shope, R.E., and Bishop, D.H.L.: Evidence from recombinant bunyavirus studies that the M RNA gene products elicit neutralizing antibodies. Virology, *102*:190, 1980.

Short, N.J., Meek, A.D., and Dalgarno, L.: Seven infected-specific polypeptides in BHK cells infected with bunyamwera virus. J. Virol., *43*:840, 1982.

Vezza, A.C., Repik, P.M., Cash, P., and Bishop, D.H.L.: In vivo transcription and protein synthesis capabilities of bunyaviruses: Wild-type snowshoe hare virus and its temperature-sensitive group I, group II, and group I/II mutants. J. Virol., *31*:426, 1979.

Chapter 17

Arenaviruses

A. ARENAVIRIDAE FAMILY

1. Lassa, Lymphocytic Choriomeningitis, Junin and Machupo Viruses

The Arenaviridae family consists of three groups of viruses: *lymphocytic choriomeningitis* (LCM), *Lassa fever, Junin* and *Machupo viruses* which infect humans. With the exception of LCM, the diseases caused by these viruses represent zoonoses and occur in defined geographic areas. Although these viruses were initially classified as arboviruses, they have recently been reclassified as Arenaviridae (arena L., sandy) based on similar morphologic, immunologic and clinical characteristics. Arenaviruses do not require arthropods for transmission. Their natural hosts appear to be rodents in which they often produce chronic infections. Virus is spread to humans in excretions of naturally infected rodents. Lassa, Junin and Machupo viruses are highly contagious and can cause serious, fatal hemorrhagic fever in man; LCM virus causes sporadic, relatively benign meningoencephalitis.

Structure and Antigenic Properties. All arenaviruses appear similar when seen by electron microscopy (Fig. 17–1). The particles are usually round or oval but pleomorphic forms are sometimes observed. The (−) single-stranded RNA exists in two segments. There is no discrete nucleocapsid structures but the interior of the particles contain variable numbers (1–10) of electron dense granules which are cellular ribosomes.

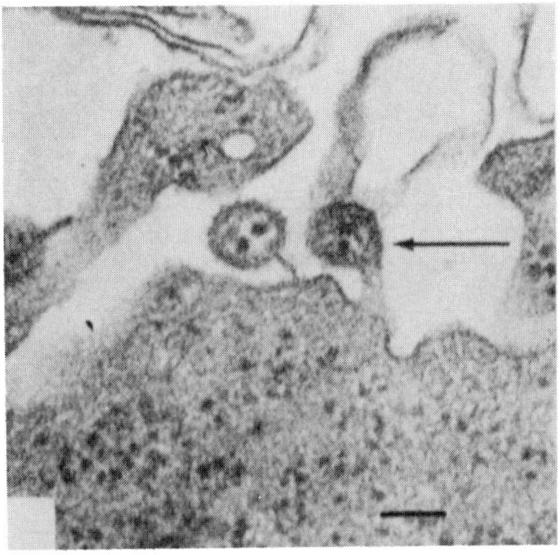

Fig. 17–1. LCM virus particle budding (arrow) from an infected mouse 3T3 cell. Peplomeres (spikes) are apparent on the surface of the other particle. × 70,000. (Reprinted with permission from Dalton, A.J., Rowe, W.P., Smith, G.H., Wilsnack, R.E. and Pugh, W.E.: Morphological and cytochemical studies on lymphocytic choriomeningitis virus. J. Virol., 2:1465, 1968.)

Table 17–1. General Properties of Arenaviruses

Properties of nucleic acid: (−) single-strand, 2 segments of RNA; virion RNA complementary to mRNA; 1.6×10^6 daltons
Virion: enveloped; contains 2 envelope glycoproteins and 2 nonglycosylated proteins; each virus particle contains 1 to 10 granules (20 nm) which are cellular ribosomes
Capsid symmetry: spherical and pleomorphic
Size (nm): 110–130
Site of replication: cytoplasm

Virions are enveloped and show spike-like surface projections (Fig. 17–1). The structural polypeptides consist of 2 glycosylated and 2 nonglycosylated proteins. One of the nonglycosylated proteins is tightly associated with the virion RNA. Antigens are broadly grouped as type-specific and type-common (cross-reacting). Envelope-associated antigens appear to be similar to those present on the surface of virus infected cells. General properties of arenaviruses are shown in Table 17–1.

Host Range and Culture. Many mammals, including man, mice, dogs, monkeys and guinea pigs, can be infected with arenaviruses. Chronically infected animals shed virus in the urine and feces and represent a hazard to animal handlers. Female mice can transmit LCM virus to their offspring *in utero;* the offspring in turn can become healthy carriers. Later in life these animals may develop a fatal debilitating disease of the CNS or chronic glomerulonephritis due to immune complexes. Arenaviruses can be propagated in mouse embryo fibroblasts (L cells) and baby hamster kidney cells (BHK 21).

Replication. The mechanism of arenavirus multiplication is largely unknown. It can be assumed that arenaviruses like LCM replicate in a manner similar to other (−) stranded RNA viruses (Chapter 3). Electron microscopic studies revealed that LCM virus maturation occurs by a process of budding at the plasma membrane (Fig. 17–1). Viral infection in tissue culture cells is not associated with extensive cell damage. Therefore, LCM virus may go through several replication cycles in a host and cause persistent or chronic infections. The production of defective virus which interferes with nondefective virus replication has also been suggested as a mechanism for maintenance of a persistent or chronic infection.

REFERENCES

Auperin, D.D., Compans, R.W., and Bishop, D.H.L.: Nucleotide sequence conservation at the 3′-termini of the virion RNA species of new world and old world arenaviruses. Virology, *121*:200, 1982.

Buchmeir, M.J. and Oldstone, M.B.A.: Identity of the viral protein reponsible for serologic cross-reactivity among the Tacaribe complex arenaviruses. In Negative Strand Viruses and the Host Cell. Mahy, B.W.J. and Barry, R.D. (eds.). New York, Academic Press, 1978, pp. 91–97.

Dutko, F.J. and Pfau, C.J.: Arenaviruses defective interfering particles mask the cell-killing potential of standard virus. J. Gen. Virol., *38*:195, 1978.
Rawls, W.E. and Leung, W.-C.: Arenaviruses. Comprehensive Virology, *14*:157, 1979.
Vezza, A.C., Clewley, J.P., Gard, G.P., Abraham, N.Z., Compans, R.W., and Bishop, D.H.L.: Virion RNA species of the arenaviruses Pichinde, Tacaribe and Tamiami. J. Virol., *26*:485, 1978.

Chapter 18

Rhabdoviruses

A. RHABDOVIRIDAE FAMILY

The Rhabdoviridae family includes at least 27 different viruses that can infect animals, insects, and plants. Man is infected only rarely by a few of the viruses which infect animals. Many viruses are classified as rhabdoviruses on a tentative basis because of their characteristic bullet-shaped morphology. Rabies virus is the most important member of this family. It is now classified in the genus *Lyssavirus* (Gr. Lyssa, rage).

The genus *Vesiculovirus* includes vesicular stomatitis virus of which there are seven distinct serotypes. This virus commonly infects cattle, pigs and horses and is transmitted by insects (arthropod-borne virus). Humans can become accidentally infected with the virus and develop a mild, febrile disease.

There are two other viruses (Marburg virus and Sudan-Zaire Hemorrhagic Fever Virus or Ebola virus) unrelated to rabies virus that have a morphology characteristic of rhabdoviruses and have been tentatively classified with the rhabdoviruses. They are responsible for outbreaks of severe hemorrhage and febrile diseases in humans.

1. Genus: *Lyssavirus*

a. Rabies and Related Viruses.

Structure and Antigenic Properties. All rhabdoviruses share the common property of being bullet-shaped (Fig. 18–1) with one flattened end. Evenly spaced projections or knobs cover the surface of the virion. Shorter bullet-shaped or cylindrical particles can be seen by electron microscopy. The nucleocapsid is helical and surrounded by an envelope.

The viral envelope which is composed of a lipid bilayer has surface

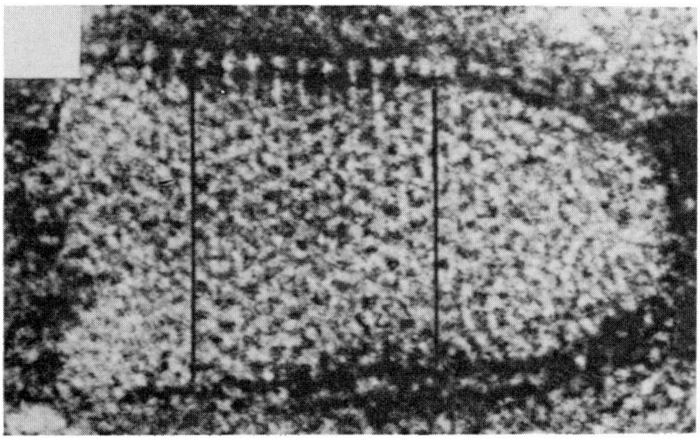

Fig. 18–1. Negative-stained preparation of a typical rabies virus; genus *Lyssavirus* particle. Note the "bullet-shaped" morphology and knoblike peplomeres arranged on the surface. × 400,000. (With permission, from Hummelar, K., Kaprowski, H. and Wiktor, T.J.: Structure and development of rabies virus in tissue culture. J. Virol., *1*:152, 1967.)

projections composed of a G glycoprotein and two matrix nonglycosylated proteins designated M1 and M2 (Table 18–1). The G protein acts as a hemagglutinin (binds to goose rbcs) and serves in the attachment of virions to host cells. Abs to the type-specific G protein mediate immunity to the virus. The nucleocapsid is composed of a (−) single-stranded RNA with virion RNA complementary to mRNA, a N protein and a RNA polymerase. The RNA polymerase is composed of a large L protein and a small NS protein (Table 18–1). The N protein of the nucleocapsid is a group-specific Ag.

The term *street* virus is used to describe rabies virus isolated in nature from domestic or wild animals, as opposed to *fixed virus,* which is considered to be an "attenuated" variant of a street virus. These terms were coined by Louis Pasteur, who chose the term "fixed virus" to describe viruses obtained by serial passage of street viruses in the brains of rabbits. Fixed viruses were used in Pasteur's first human rabies vaccine.

Host Range and Culture. Rabies virus has a wide host range that includes man and all other warm-blooded animals. In infected animals, the virus is distributed in the nervous system, saliva, urine, lymph, milk, and blood. In bats the virus is present in the salivary glands. Infected bats can transmit rabies virus in aerosolized saliva for extended periods without having apparent disease. Latent rabies virus are known to be reactivated in infected animals (e.g., skunks, foxes, dogs). The virus can be grown in chick or duck embryos and newborn mice as well as hamster kidney cell or human diploid cell cultures.

2. Genus: *Vesiculovirus*

a. Vesicular Stomatitis Virus (VSV).

Structure and Antigenic Properties. VSV shares structural properties with other rhabdoviruses (Table 18–1) having one flattened end and being bullet-shaped. The virion is composed of five structural proteins. The N, NS and L proteins are associated with the virus nucleocapsid, whereas the M and G proteins comprise the viral envelope. The NS and L proteins constitute the RNA polymerase. Antibodies directed toward the M or G proteins neutralize virus infectivity. A model depicting the structure and location of structural proteins in VSV is presented in Figure 18–2.

Table 18–1. General Properties of Rhabdoviruses

Properties of nucleic acid: (−) single-stranded RNA; virion RNA complementary to mRNA; 3.5–4.6 × 10^6 daltons
Virion: enveloped; surface glycoprotein (G protein); two matrix nonglycosylated proteins (M1 and M2); RNA polymerase composed of a large (L) protein and small (NS) protein; bullet-shaped.
Capsid symmetry: helical
Size (nm): 70 by 170
Site of replication: cytoplasm

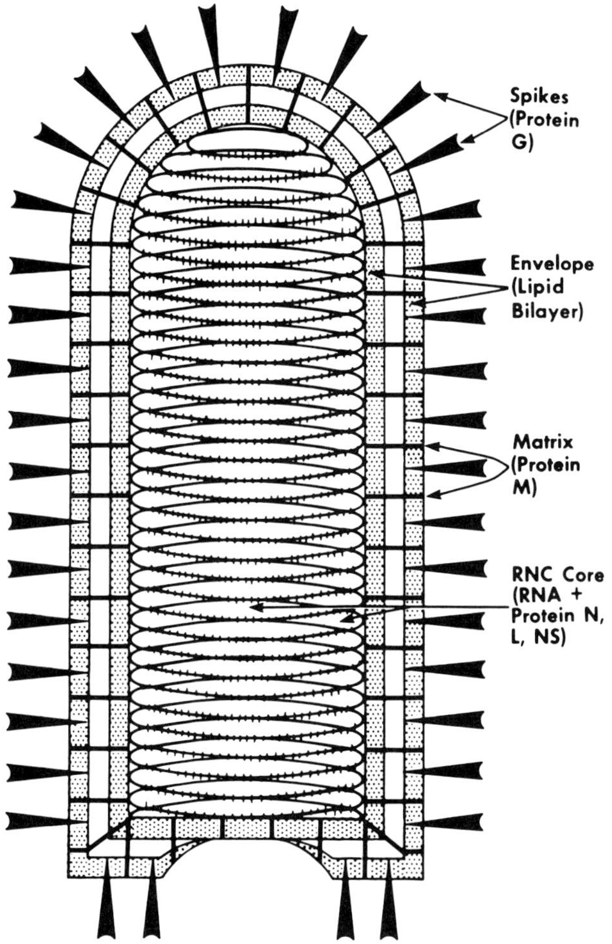

Fig. 18–2. Model of the idealized structure of VSV showing the ribonucleocapsid (RNC), the envelope, and the spikes with their associated protein constituents. (With permission from Wagner, R.R.: Reproduction of Rhabdoviruses. Comprehensive Virology, 4:1, 1975.)

Host Range and Culture. The virus grows in a variety of animal cells including rat, mouse and hamster. When VSV is passaged repeatedly at high multiplicity, the progeny often includes nondefective virus particles and defective virus particles that are capable of interfering with the homologous nondefective virus. Defective viruses (1) contain cell structural proteins, (2) contain only a part of the viral genome i.e., deletion mutants, (3) require nondefective "helper virus" in cells for replication, (4) in the absence of homologous helper virus, defective viruses can nevertheless express certain functions i.e., shut-off of host cell biosynthesis, induce synthesis of some viral proteins, and (5) interfere with replication of homologous nondefective virus (Chapter 4). Since virus cytopathology is minimal, persistent infections or chronic infections of tissue culture cells are readily established.

Replication. The mechanism of replication of rabies virus and VSV are probably the same. Virus enters host cells by a process of membrane fusion. The (−) stranded virion RNA is transcribed into (+) stranded mRNA by a virus-associated RNA polymerase (transcriptase) in the cytoplasm. Translation of mRNA produces a replicase which copies (+) stranded RNAs into virion (−) stranded RNAs. The transcriptase has a single promoter on the viral genome and sequentially copies the genome into mRNAs. The mRNAs are produced by processing the large transcript or the transcriptase restarts transcription at each junction between messages. Viral RNA replication occurs via a replicative intermediate (Chapter 3).

Helical nucleocapsids concentrate in the cell cytoplasm in "factories." These masses of nucleocapsids form inclusion bodies identified by specific immunofluorescence. Presence of intracytoplasmic inclusions (Negri bodies) is diagnostic for rabies virus. Virions are assembled by a process of budding of nucleocapsids from cytoplasmic membranes.

REFERENCES

Bishop, D.H.L. and Smith, M.: Rhabdoviruses. *In* The Molecular Biology of Animal Viruses. K.D. Maya (ed.). New York, Marcel Dekker, 1976, pp. 167–225.

Emerson, S.U.: Vesicular stomatitis virus: structure and function of virion components. Curr. Top. Microbiol. Immunol., *73*:1, 1976.

Flamand, A., Delagneau, J.F., and Bussereau, F.: An RNA polymerase activity in purified rabies virions. J. Gen. Virol., *40*:233, 1978.

Pringle, C.R.: Genetics of rhabdoviruses. Comprehensive Virology, *9*:239, 1979.

Saghi, N. and Flamand, A.: Biochemical characterization of temperature-sensitive rabies virus mutants. J. Virol., *31*:220, 1979.

Wagner, R.R.: Reproduction of rhabdoviruses. Comprehensive Virology, *4*:1, 1975.

Zakowski, J.J. and Wagner, R.R.: Localization of membrane-associated proteins in vesicular stomatitis virus by use of hydrophobic membrane probes and cross-linking reagents. J. Virol., *36*:93, 1980.

Chapter 19

Retroviruses

A. RETROVIRIDAE FAMILY

RNA tumor viruses were formerly classified as oncornaviruses, leukemia-sarcoma viruses or leukosis agents. The latest classification groups all RNA tumor viruses into the Retroviridae family (L. retro, backward) based on the presence of a unique virion-associated enzyme, reverse transcriptase, which copies (+) single-stranded virion RNA into a double-stranded DNA copy (provirus). Various retroviruses (oncoviruses; Gr. onco, tumor) induce sarcomas, leukemias, lymphomas and mammary carcinomas (Chapter 34). Retroviruses are considered weakly oncogenic or highly oncogenic (Chapter 34).

Retroviruses are unique for the following reasons: (1) A viral enzyme, reverse transcriptase, initiates infection by copying virion RNA into DNA (provirus). (2) The provirus DNA integrates into host DNA and retroviruses rarely kill their host cell. (3) Retrovirus genomes evolved with their germ lines of many animal species. (4) Some retroviruses carry genes that direct neoplastic transformation of the host cell.

Retroviruses can be divided into the following classes based on the host for virus replication. (1) *Ectropic* viruses replicate in their natural species of origin. (2) *Xenotropic* viruses replicate in other species not of their natural origin and they are nontransforming. (3) *Amphotropic* viruses share properties in common with both xenotropic and ectropic viruses, but amphotropic viruses are oncogenic. They differ from ectropic and xenotropic viruses in having unique envelope antigens.

Retroviruses are classified into various genera according to their morphology in the electron microscope. *Cisternavirus A* particles have an electronlucent center surrounded by a double shell. They are considered immature forms of type B and C particles. *Oncovirus B* particles have an eccentric core and they are considered mature forms of mammary tumor viruses. *Oncovirus* C particles are mature forms of leukemia and sarcoma viruses of avian, mammalian and reptilian origin; they possess a central core. *Oncovirus D* particles of Mason-Pfizer monkey virus are intermediate to *Oncovirus B* and C particles.

Virions are composed of two identical single-stranded RNA molecules (35S) held together by a dimer linkage structure near the 5′ end and by hydrogen bonds along several points on their lengths to form a 70S RNA genome (diploid). The 35S (haploid) RNA unit is genetically self-sufficient and can initiate virus infection if introduced in the host cell in the proviral DNA form (Table 19–1).

RNA tumor viruses are enveloped and contain three classes of structural proteins i.e., internal structural proteins (p), envelope glycoproteins (env) and reverse transcriptase (Fig. 19–1). The core of the retrovirus is an icosahedron composed of genomic RNA and several internal structural p proteins varying in molecular weight. The virus core also

Table 19–1. General Properties of Retroviruses

Properties of nucleic acid:	diploid, (+) single-stranded 70S RNA; genomic RNA same polarity as mRNA; consists of two identical 35S subunits complexed with host tRNA.
Virion:	enveloped, consists of internal group specific (gag) antigens, envelope glycoproteins and reverse transcriptase; morphologic forms are A particles with double shell and electronlucent center; B particles with eccentric core; C particles with central core or D particles with morphology intermediate to B and C particles.
Capsid symmetry:	icosahedral
Size (nm):	100
Site of replication:	replication of nucleic acid occurs via DNA intermediate (provirus) in the nucleus; virus assembly occurs in the cytoplasm and plasma membrane.

Table 19–2. Structural Composition of Retroviruses

Gene	Precursor Polyprotein	Virion Polypeptide Avian	Virion Polypeptide Murine	Structural Component	Virus Substructure
env	gp90[a]	gp85-S	gp71-S	Knob	
		|	|	Spike	Envelope
		gp35-S	p15E-S	Envelope-	
		p10	p12E	associated	
gag	p70–76	p19	p12	Inner coat[b]	Inner coat[c]
		p27	p30	Core-shell	
		p15	p15C	Core-associated	Core exterior[c]
		p12	p10	Nucleoprotein	
pol	p200	p91(β)	p70(α)	Reverse-transcriptase	Ribonucleoprotein complex[c]
		p64(α)			

[a]number indicates molecular weight; gp = glycoprotein; p = nonglycosylated protein; s-s = disulfide bond; E = envelope; C = core
[b]Represent phosphoproteins which may bind to virion RNA during maturation
[c]Substructure of the nucleocapsid

contains reverse transcriptase. The envelope contains glycosylated virus-specific proteins (Fig. 19–1; Tables 19–1 and 19–2).

Three viral genes (gag, pol, env) are necessary for replication of retroviruses (Fig. 19–2). The gag gene encodes for a large precursor polyprotein (p70–76) which is later cleaved into four interior core proteins (Table 19–2). The pol gene encodes for a reverse transcriptase. The env gene encodes for a large precursor polyprotein (gp90) which is probably inserted into the plasma membrane before it is cleaved into two glycoproteins and a nonglycosylated envelope-associated protein (Table 19–2). The envelope encloses the internal virus core during budding of virus cores from the plasma membrane. Other retroviruses contain an additional v-src gene not required for virus replication that is needed to transform the host cell into a neoplastic cell. The v-src gene is responsible for both the induction and maintenance of neoplastic transformation.

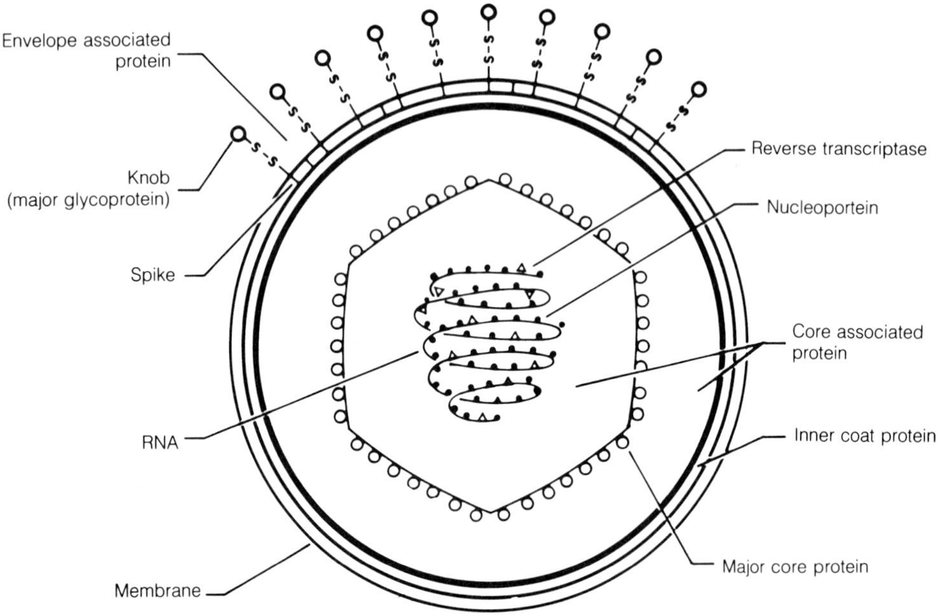

Fig. 19–1. Composition of a mature retrovirus particle. (Modified from Bolognesi, D.P., et al.: Science, *199*:183, 1978. Copyright 1978 by the American Association for the Advancement of Science.)

(A) 5′ Cap | LTR | gag | pol | env | c | LTR 3′ poly A

Leukosis Virus

(B) 5′ Cap | LTR | gag | pol | env | src | c | LTR 3′ poly A

Sarcoma Virus

(C) 5′ Cap | LTR | gag | onc | env | c | LTR 3′ poly A

Acute Leukemia Virus

Fig. 19–2. Genetic structure of a nondefective, nontransforming leukosis virus (A), a nondefective, transforming avian sarcoma virus (B), and a defective, transforming acute leukemia virus (C). The 5′ end at the left has a cap structure. LTR is the long terminal redundancy. C is a highly conserved region. The 3′ end at the right has a poly A sequence.

The product of v-src is a 60,000 dalton phosphoprotein (designated pp60src; protein kinase) which phosphorylates cellular proteins. Phosphorylation of specific cellular targets might account for transformation of the host cell by v-src.

The v-src gene serves no useful purpose in the replication of retroviruses. The origin of v-src is the host cell. Recent experiments dictate that recombination between a deletion mutant of avian sarcoma virus and a locus in the chick cell DNA reconstituted a viral oncogene indistinguishable from wild-type v-src. There is evidence for expression of retrovirus oncogenes in uninfected cells. The cellular homolog (c-src) of avian sarcoma virus encodes a 60,000 dalton phosphoprotein (protein kinase) that is indistinguishable from the viral transforming protein kinase pp60src. There is suggestive evidence the c-src gene functions during normal growth and development of cells. The oncogene designated for acute leukemia virus (Fig. 19–2) is circumstantial. Since the oncogene is usually inserted within a replicative gene and is accompanied by a deletion of a portion of a gene, the acute leukemia viruses are defective in replication and require a "helper" sarcoma virus to replicate progeny leukemia viruses.

Retroviruses infect both permissive and nonpermissive host cells. Infected permissive cells support replication of progeny virions and the cell may become neoplastic. Nonpermissive cells may be transformed by infection but progeny virions are not produced.

1. Genera

a. *Cisternavirus A.* The structure, composition and antigenic properties of cisternavirus A suggests they are immature or defective murine mammary tumor virus (MMTV) nucleocapsids. They are seen by electron microscopy in the cisternae of the endoplasmic reticulum in embryos and murine cells and are often associated with murine mammary tumor virus.

b. *Oncovirus B.*

1. Murine Mammary Tumor Virus (MMTV). This virus, also known as *milk factor* or *Bittner virus,* is the prototype exogenous virus that induces mammary tumors in mice. It is transmissible from mother to offspring through the milk (the milk agent). Exogenous MMTV are also recognized in animals of many strains that develop spontaneous mammary neoplasms. The provirus is present in all mouse strains, but not in other species.

STRUCTURE AND ANTIGENIC PROPERTIES. The MMTV resembles the leukemia murine viruses. It has an envelope with peplomeres (Fig. 19–3) and contains reverse transcriptase and a 70S RNA genome. The virus presents two morphologic types (Fig. 19–4); a small immature particle (called type A) and a larger mature particle (called type B). Particles

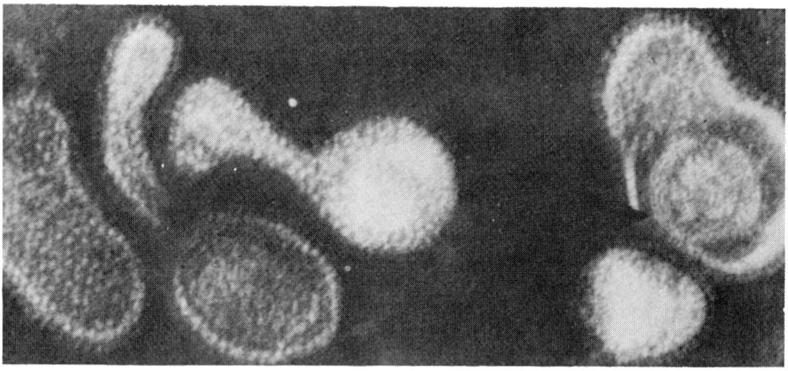

Fig. 19–3. Structure of purified murine mammary tumor viruses as observed by negative staining in the electron microscope. Note knobs (peplomeres) on the surface of the virus envelope × 200,000. (With permission from Lyons, M.J., and Moore, D.H.: Isolation of mouse mammary tumor virus: Chemical and morphological studies. J. Natl. Cancer Inst., *35*:549, 1965.)

morphologically similar to MMTV have been observed in human milk and human mammary carcinoma cells; the particles possess a 70S RNA genome and reverse transcriptase activity (Table 19–1).

The MMTV possesses type-specific envelope Ags and internal gs Ags; the latter are antigenically related to gs Ags of murine retroviruses.

HOST RANGE AND CULTURE. Murine mammary tumor viruses are naturally found in certain genetically susceptible "high cancer" strains of mice. Virus multiplies in the mammary gland and is released into the milk. Newborn or adult mice are susceptible to virus given by the oral, sc, or ip routes. Females can develop inapparent infections and transmit the virus to both male and female offspring.

c. *Oncovirus C.*

1. Avian Leukemia-Sarcoma Viruses. This virus group consists of related avian viruses that induce leukemias (lymphomatosis virus, myeloblastosis viruses, erythroblastosis viruses) or sarcomas in chickens (e.g., Rous sarcoma virus).

STRUCTURE AND ANTIGENIC PROPERTIES. The virions of this group have a mean diameter of about 100 nm and are similar in structure. They are enveloped and contain a single-stranded RNA (70S) genome as well as reverse transcriptase activity. Avian leukemia-sarcoma viruses share a common gag Ag that can be detected by using the complement-fixation avian leukosis (COFAL) test employing sera from rodents bearing Rous sarcoma virus (RSV) tumors. These viruses have been classified into 7 subgroups (A to G) based on envelope glycoprotein Ags and their ability to replicate in genetically defined avian cells. Certain oncogenic viruses can promote their own replication, whereas other "defective viruses"

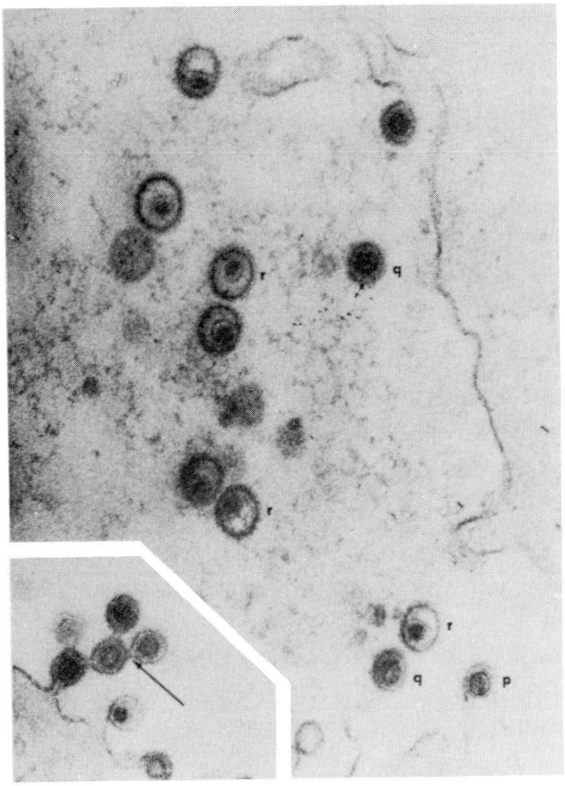

Fig. 19–4. Electron micrograph of a thin section of a mouse mammary tumor virus infected cell. q = immature type-A particles; r = mature type-B particles; p = virus particle budding through the plasma membrane. Arrow in insert indicates viruses sharing a common envelope × 100,000. (With permission from Lyons, M.J., and Moore, D.H.: Isolation of the mouse mammary tumor virus: Chemical and morphological studies. J. Natl. Cancer Inst., *35*:549, 1965.)

require the help of another virus (helper virus) to replicate. The helper activity results from phenotypic mixing of the two viruses within the same cell (Chapter 4).

HOST RANGE AND CULTURE. Most of the avian leukemia viruses replicate in chick embryo fibroblast cultures without causing any CPE or cell transformation. Virus replication is detected by immunofluorescence assay or by failure of the infected cells to transform when superinfected with RSV (interference test).

Unlike avian leukemia viruses, avian sarcoma viruses (e.g., RSV) are unique in that they can induce almost 100 percent malignant transformation of infected chick embryo cells. Infection with RSV results in production of foci of transformed cells (Fig. 19–5); virus infectivity is assayed by determining the number of focus-forming units (FFU) in a given volume of inoculum (one infectious unit of virus gives rise to one

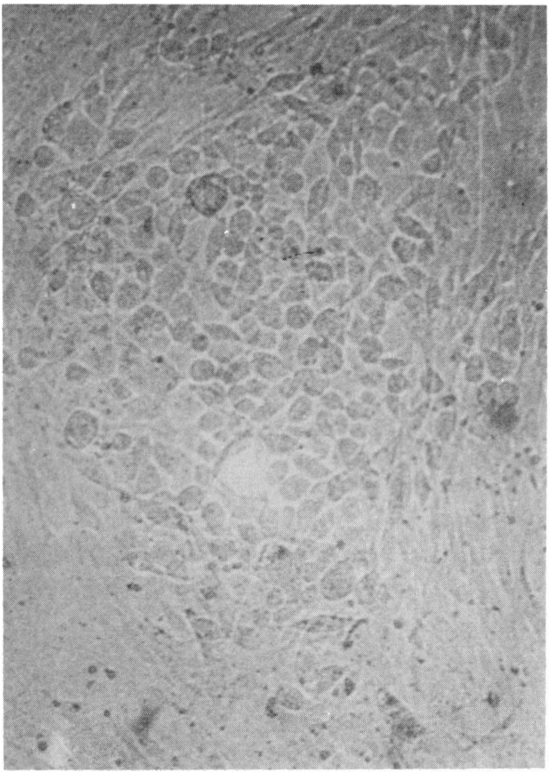

Fig. 19–5. Focus of chick embryo cells transformed by Rous sarcoma virus × 80. (With permission from Temin, H., and Rubin, H.: Characteristics of an assay for Rous sarcoma virus and Rous sarcoma cells in tissue culture. Virology, *6*:669, 1958.)

focus of transformed cells). Some strains of RSV will induce transformation of mammalian (mouse, rat, hamster, primate, human) fibroblast cells. However, assay for infectious RSV in mammalian transformed cells is accomplished by planting the transformed cells on chick embryo cell monolayers and assaying for FFU.

Many stocks of RSV contain a second virus, Rous-associated virus (RAV). The RSV is "defective" in that it can cause cell transformation but cannot replicate to form new infectious virus in the absence of RAV "helper" virus. Since cells transformed by RSV alone do not produce virus, they are called nonproducer (NP) cells. Nonproducer cells contain the gag avian leukemia-sarcoma Ags, reverse transcriptase, and the RSV-70S genome. When NP cells are superinfected with RAV "helper" virus, the cells yield new infectious RSV and RAV. Since the "helper" virus codes for RSV envelope proteins (envelope glycoproteins), RSV possesses the envelope Ags and exhibits the host range of its "helper virus."

*2. **Mammalian Leukemia-Sarcoma Viruses.*** *I. Murine.* Many different strains of murine leukemia and sarcoma viruses, named for the investigator who first described them (e.g., Gross, Friend, Maloney, Harvey, Rauscher), are associated with leukemias and sarcomas in mice.

STRUCTURE AND ANTIGENIC PROPERTIES. Purified murine leukemia viruses possess an envelope, which is essentially smooth, and a condensed central nucleoid (core) (Fig. 19–6). The viral core is surrounded by a shell containing subunits that are regularly arranged to form icosahedral symmetry (Chapter 1).

Murine leukemia viruses possess internal gag and envelope glycoprotein Ags. Neutralization tests have revealed at least two distinct gag Ags, one common to Friend, Maloney, and Rauscher viruses and the other common to Gross and Harvey viruses. These gag Ags are detectable in both transformed and nontransformed infected cells by the CF test employing sera from rats carrying a transplantable Rauscher leukemia virus-induced lymphosarcoma. Murine sarcoma viruses are antigenically related to the murine leukemia viruses.

HOST RANGE AND CULTURE. Murine leukemia viruses can be grown in mouse embryo fibroblast cells or in mice. Virus replication occurs without observable cytopathic effects (CPE); however, infected cells contain viral Ags and continue to release virus particles. Virions produced in tissue culture systems are 10,000 times less infectious than particles generated in leukemic mice. Cells infected with murine leukemia virus are resistant to subsequent transformation by murine sarcoma virus (viral interference test), similar to the phenomenon observed with avian leukemia viruses and RSV.

A plaque assay for murine leukemia viruses has been developed. Rat cells (XC cells) transformed by the Schmidt-Ruppin strain of RSV are nonproducer cells. However, for some unknown reason, XC cells fuse with murine leukemia virus-infected mouse cells to form giant syncytia (Chapter 2), which are detected as virus plaques.

Some murine leukemia viruses (e.g., Friend, Gross, Rauscher) can cause in vitro transformation of mouse, rat, and hamster embryo cells. The transformed cells contain the viral specific gag Ags, reverse transcriptase, and 70S RNA; they release infectious viruses that are oncogenic in vivo.

Newborn mice or rats (immunologically immature) are more susceptible to leukemogenic viruses than older animals; genetic factors also play an important role in the susceptibility of mice to the virus. Thymectomy reduces the incidence of acute lymphocytic leukemia (T cell malignancy) but not of myeloid leukemia; however, it has no effect on the multiplication of murine leukemia virus in other organs of the host. Large amounts of infectious viruses are released into the blood of infected animals and can be transmitted congenitally.

Murine sarcoma virus (MSV) is a "defective" virus that requires a murine leukemia "helper" virus to form infectious MSV in tissue culture

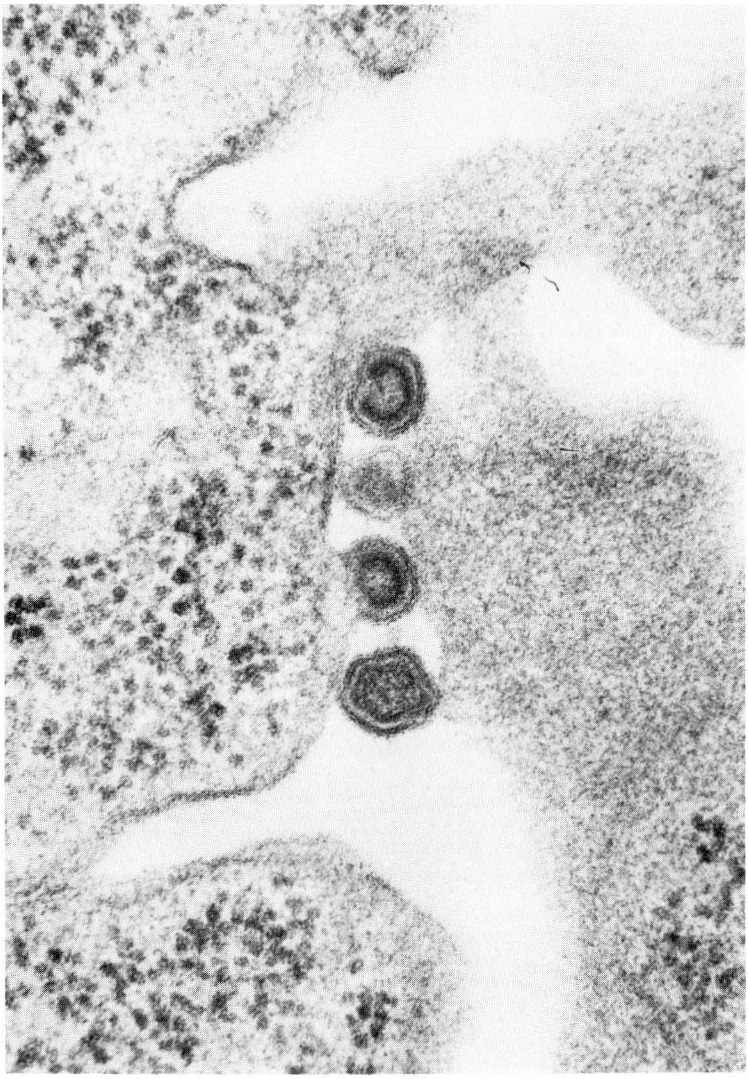

Fig. 19–6. Electron micrograph illustrating leukemia virus particles. The particles are in a leukemic spleen of an AKR/Dm strain mouse that developed spontaneous leukemia. The micrograph illustrates two stages of budding and a fully developed mature leukemia virus particle × 50,000. (With permission from Dr. Leon Dmochowski, Department of Virology, The University of Texas System Cancer Center; M.D. Anderson Hospital and Tumor Institute, Houston, Texas 77030.)

cells (derived from rat, mouse, or hamster embryos). Infection with MSV and "helper" virus results in foci of transformed cells as well as virus production that can be measured by FFU. Murine sarcoma viruses cause rhabdomyosarcomas in newborn mice, rats, and hamsters.

II. Feline Leukemia-Sarcoma Viruses. This group of viruses has the capacity to induce leukemias and sarcomas in cats, dogs, rabbits, and monkeys. They resemble C-type particles associated with murine leukemias and sarcomas. Feline leukosis viruses possess unique gag Ags and envelope glycoprotein Ags.

III. Hamster Leukemia-Sarcoma Viruses. Hamster sarcoma viruses resemble, but are antigenically distinct from murine sarcoma viruses; they are oncogenic for hamsters but not for mice. The hamster leukemia viruses have some properties in common with known murine leukemia viruses; however, they possess unrelated envelope glycoprotein Ags.

IV. Porcine, V. Bovine, VI. Primate and VII. Reptilian Leukemia-Sarcoma Viruses. Particles with the morphology of retroviruses have been detected in malignant tumors of reptiles, guinea pig, swine, cows, nonhuman primates, and reptiles. Information on the biologic properties of these viruses is incomplete. There is evidence that type C viruses related to nonhuman primates occur in human cancer cells.

d. *Oncovirus D.*

1. Mason-Pfizer Monkey Virus (MPMV). The MPMV is an exogenous virus isolated from a rhesus monkey mammary carcinoma and also from normal rhesus monkey tissues. The virus shares common glycoprotein antigens with an endogenous baboon virus. MPMV resembles oncovirus C in morphology and size of virion proteins. It resembles oncovirus B in the mechanism of virus budding. The virus can replicate in Rhesus monkey and human foreskin cells.

e. *Lentivirus E.*

1. Visna Virus, Maedi Virus. Visna virus is the cause of a slowly progressive meningoencephalitis of sheep. The agent probably represents a neutropic variant of maedi virus, the etiologic agent of progressive pneumonia of sheep. The two viruses are similar in ultrastructure and morphogenesis; they are related serologically, and visna-like virus particles can be recovered from the brains of sheep infected with the agent that causes progressive pneumonia in sheep.

STRUCTURE AND ANTIGENIC PROPERTIES

Visna virus contains a 70S single-stranded RNA genome surrounded by a lipid-containing envelope. Extracellular virus possesses a reverse transcriptase, as an integral component of the virion. The virus is pleomorphic (diameter 60–100 nm). The virion surface is covered with spikes (Fig. 19–7); a central core of 35 nm is evident in sectioned particles.

Visna and maedi viruses resemble the oncornaviruses in several ways:

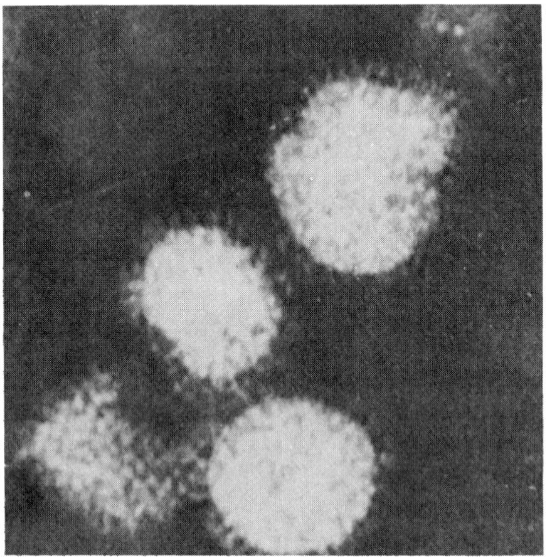

Fig. 19–7. Structure of visna virus as observed by electron microscopy. Note outer envelope covered with numerous spikes × 250,000. (With permission from Thormar, H.: The structure of visna virus studied by the negative staining technique. Virology, *25*:145, 1965.)

(1) sensitivity to physical and chemical agents, (2) ultrastructural appearance of the virions, (3) possession of a 70S RNA genome, and (4) presence of reverse transcriptase activity in virions. Visna and maedi viruses can cause transformation of embryonic mouse cells in culture and interfere with subsequent replication of murine leukemia-sarcoma viruses. However, based on molecular hybridization tests, visna and maedi viruses are genetically and antigenically unrelated to other retroviridae. Antibody made against visna virus partially neutralizes the closely related maedi virus.

HOST RANGE AND CULTURE. Visna virus can infect all organs of sheep; however, pathologic changes are confined primarily to the brain, lungs and the RES. There is a long incubation period, with virus release lasting up to 4 years after infection. Visna and maedi virus can be grown in vitro in sheep choroid plexus and bovine tracheal cells. Maximum yields of virus are obtained in cell cultures 40–72 hours after infection. Virus-induced cytopathic effects include polykaryon formation followed by cell degeneration.

REPLICATION. Retrovirus particles attach to host cell receptors via virion envelope glycoproteins. Penetration of virus occurs when the virus envelope fuses with the cellular membrane. In the cytoplasm, the viral RNA is reverse transcribed into a RNA:DNA hybrid and then into double stranded DNA using the virion associated reverse transcriptase. Initially the DNA is a linear molecule containing a continuous virus negative-strand and a discontinuous positive-strand. The discontinuous positive-strand is converted to a full continuous strand 6–9 hours after infection. The DNA moves to the nucleus and becomes cyclic. By 24 hours some of the cyclic DNA becomes integrated (provirus) into cellular chromosomes. All forms of retrovirus DNA (linear, cyclic and provirus) contain a complete genome copy and are infectious. Progeny plus RNA is made by transcription of the integrated provirus.

The 35S (+) RNA strand serves as template for synthesis of the gag polyprotein. The env polyprotein is made from a 21S (+) RNA derived from the 35S RNA by splicing of a 500-base fragment. The pol protein is made from a 35S (+) RNA by occasional read-through beyond the gag gene producing a gag-pol polyprotein. All polyproteins are later cleaved into final products. The env proteins are incorporated into the plasma membrane prior to virus budding.

Some of the 35S (+) RNA present in the cytoplasm is packaged with viral core proteins into virus particles which then mature by budding from the cell surface at sites where viral env proteins are already present in the membrane.

Infection with retroviruses is not a lytic event. Productively infected cells continue to grow and shed infectious virus into the extracellular fluids. Infected cultures can be maintained indefinitely.

REFERENCES

Bishop, J.M.: The molecular biology of RNA tumor viruses. A physician's guide. N. Engl. J. Med., *303*:675, 1980.

Fan, H.: Expression of RNA tumor viruses at translation and transcription levels. Curr. Top. Microbiol. Immunol., *79*:1, 1978.

Hayward, W.S. and Neel, B.G.: Retroviral gene expression. Curr. Top. Microbiol. Immunol., *91*:217, 1981.

Hunter, T. and Sefton, B.M.: Transforming gene product of Rous sarcoma virus phosphorylates tyrosine. Proc. Natl. Acad. Sci. (U.S.A.), *77*:1311, 1980.

Hynes, N.E. and Groner, B.: Mammary tumor formation and hormonal control of mouse mammary tumor virus expression. Curr. Top. Microbiol. Immunol., *101*:51, 1982.

Lee, W.H., Bister, K., Pawson, A., Robins, T., Moscovici, C., and Duesberg, P.H.: Fujinami sarcoma virus: An avian RNA tumor virus with a unique transforming gene. Proc. Natl. Acad. Sci. (U.S.A.), *77*:2018, 1980.

Panganiban, A.T., and Temin, H.M.: The terminal nucleotides of retrovirus DNA are required for integration but not virus production. Nature, *306*:155, 1983.

Taylor, J.M.: DNA intermediates of avian RNA tumor viruses. Curr. Top. Microbiol. Immunol., *87*:23, 1979.

Chapter 20

Hepatitis Viruses

A. UNCLASSIFIED HEPATITIS VIRUSES

Acute liver injury and jaundice may result from a variety of physical, chemical, viral, and bacterial agents. Acute viral hepatitis is commonly caused by one or the other of two specific viruses; hepatitis A virus (HAV) or infectious hepatitis (IH) virus, the causative agent of infectious hepatitis; and hepatitis B virus (HBV) or serum hepatitis (SH) virus, the causative agent of serum hepatitis and possibly hepatocellular carcinoma (Chapter 34). There is at least one additional hepatitis C virus (HCV), which is considered the major etiologic agent of hepatitis following transfusions. Yellow fever virus is responsible for an acute, febrile, arthropod-borne illness (yellow fever) that is characterized by jaundice. Other viruses (e.g. Epstein-Barr virus) may cause hepatic enlargement with jaundice as part of a more generalized viral infection (Chapter 29).

There is sufficient historic, epidemiologic, and experimental evidence to show that IH and SH are caused by antigenically distinct viruses. Investigations on the properties of these viruses are still incomplete, consequently, hepatitis viruses have not yet been grouped into one of the major virus groups.

In 1964, Blumberg, a human geneticist, discovered an Ag referred to as Australia Antigen (Au Ag) in the serum of an Australian aborigine. Subsequently, Au Ag was recognized as a specific marker of SH virus infection in patients who had received multiple blood transfusions; in addition, this Ag has been detected in the sera of patients with Down's syndrome, lepromatous leprosy, or chronic renal failure as well as in patients on immunosuppressive therapy. These findings have greatly aided our understanding of the clinical, epidemiologic, and immunologic aspects of SH. Attempts to demonstrate a specific antigenic marker in recognizing infection of patients with IH virus have been unsuccessful thus far; however, some advances in this area of research have been made, particularly with nonhuman primates.

1. Hepatitis A Virus

Structure and Antigenic Properties. Hepatitis A virus resembles the genus *Enterovirus* in the family Picornaviridae (Chapter 10) in morphology (Fig. 20–1) with an average diameter of 27 nm, icosahedral in symmetry and no envelope (Table 20–1). Virions contain single-stranded RNA, three major polypeptides and are stable to ether and pH 3.0.

IH viruses are resistant to heat (60°C for 20 hours), ultraviolet light, freezing (−20°C, for more than 20 years), and many chemical disinfectants. Formalin, activated glutaraldehyde, or preparations of hypochlorite are generally used for inactivation of hepatitis viruses. The remarkable resistance of hepatitis viruses to physical agents emphasizes the need for extra precaution in dealing with hepatitis patients and their

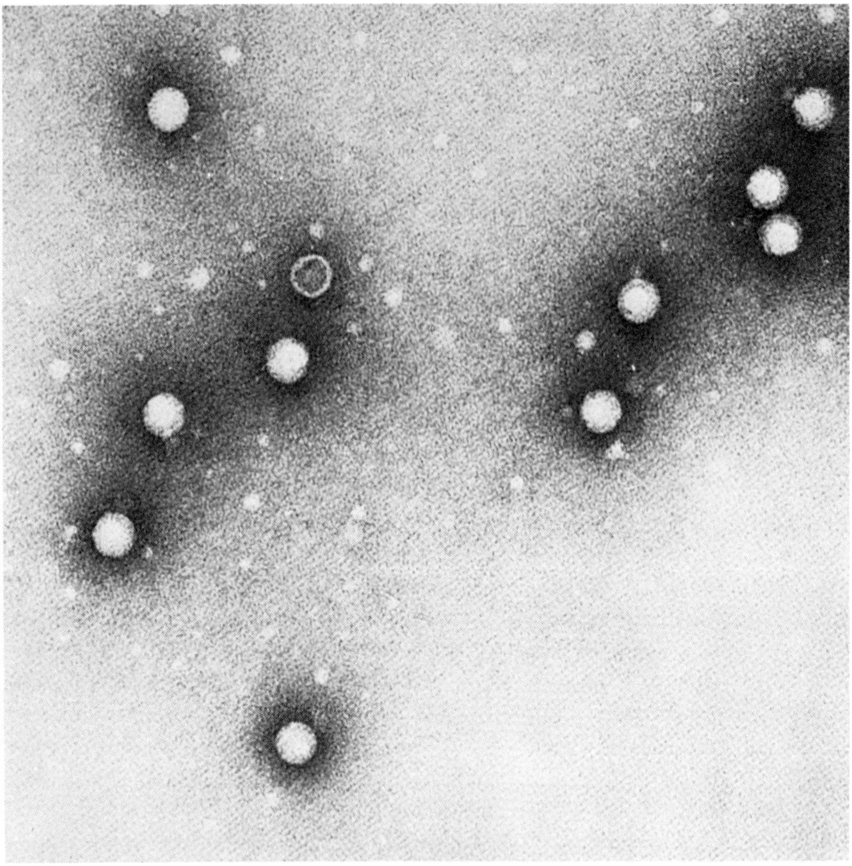

Fig. 20–1. Electron micrograph of HAV particles purified from human feces and stained with phosphotungstic acid. (With permission from Robinson, W.S.: Viruses of Human Hepatitis A and B. Comprehensive Virology, *14*:471, 1979.)

Table 20–1. General Properties of Hepatitis Viruses

	Hepatitis A Virus	Hepatitis B Virus
Properties of nucleic acid:	single-stranded RNA	partially double-stranded, circular DNA
Virion:	nonenveloped; three major polypeptides; resembles picornaviruses	enveloped; two major polypeptides HBsAg and HBcAg; associated DNA polymerase
Capsid symmetry:	icosahedral	pleomorphic spheres or filamentous
Size (nm):	27	42 (Dane particle) 28 (cores) 22 (spheres)
Site of replication:	cytoplasm	nucleus

excretion products. IH virus is antigenically distinct from SH virus. There appears to be only a single immunologic type of IH virus.

Host Range and Culture. Man and perhaps nonhuman primates are the only natural hosts of both IH and SH viruses. Dogs, mice, ducks, and turkeys are susceptible to their own specific hepatitis viruses but are resistant to hepatitis viruses of human origin. In addition to replication of HAV in nonhuman primates, the virus can multiply in a line of fetal rhesus monkey kidney cells and in hepatocytes growing in marmoset liver explant cultures. The virus multiplies in the cell cytoplasm; however, the amount of progeny virus produced is too small to allow adequate biochemical studies of virus synthesis.

2. Hepatitis B Virus

Structure and Antigenic Properties. Hepatitis B virus exists in three distinct morphologic forms (Fig. 20–2): (1) the Dane particle composed of a double-layered sphere of 42 nm in diameter with an electron-dense core 28 nm in diameter surrounded by an envelope. (2) Spherical particles with a diameter of 22 nm (average size). (3) Filamentous forms ranging in length from 22–200 nm (Fig. 20–3).

The Dane particles and isolated 28 nm core structures contain a partially double-stranded circular DNA associated with a DNA polymerase (Table 20–1; Fig. 20–2). One strand of the DNA has a gap of variable length and consequently a region which is single-stranded. The DNA polymerase of the Dane particle can polymerize the single stranded region to form a complete double-stranded DNA molecule.

The 42 nm enveloped Dane particle contains two distinct antigens, a surface Ag (HbsAg) formerly called the Australia Ag and a core Ag (HBcAg). The 22 nm spherical particle and filamentous particle (Fig. 20–3) contain only the HBsAg (Fig. 20–2). The HBsAg contains several antigenic determinants designated a, d/y and w/t. The HBcAg is a single antigenic type and is found only in the Dane particle and in free core particles. A third antigen, designated HBeAg is commonly detected in sera along with circulating Dane particles. There is a possibility that HBeAg may be associated with HBV particles.

Host Range and Culture. The host range for HBV is similar to that described above for HAV. HBV virus multiplication has recently been studied in hepatocytes from infected humans or chimpanzees and in HeLa cells exposed to cloned HBV DNA excised from a hybrid plasmid of pBR322 with a 3200-base pair of insert DNA from HBV.

Replication. After virus entry and uncoating in host cells, HBV replicates its DNA in the nucleus. The incomplete viral (−) stranded DNA is polymerized into full length (−) stranded DNA and transcribed in a (+) stranded RNA (pre-genome). After translation into viral proteins, the pre-genome is encapsulated into immature cores and reverse

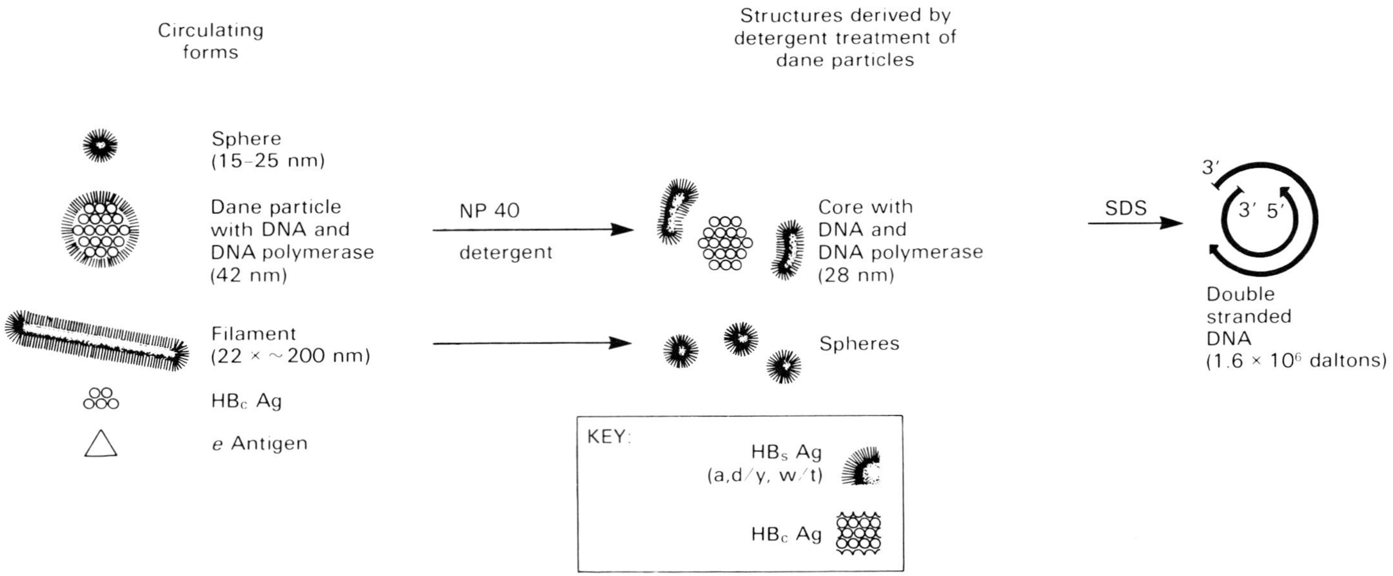

Fig. 20–2. Schematic drawing of HBV circulating forms and structures derived by detergent treatment of Dane particle. (From Robinson, W.S. and Lutwick, L.I.: N. Engl. J. Med., *295*:1232, 1976. With permission from the New England Journal of Medicine.)

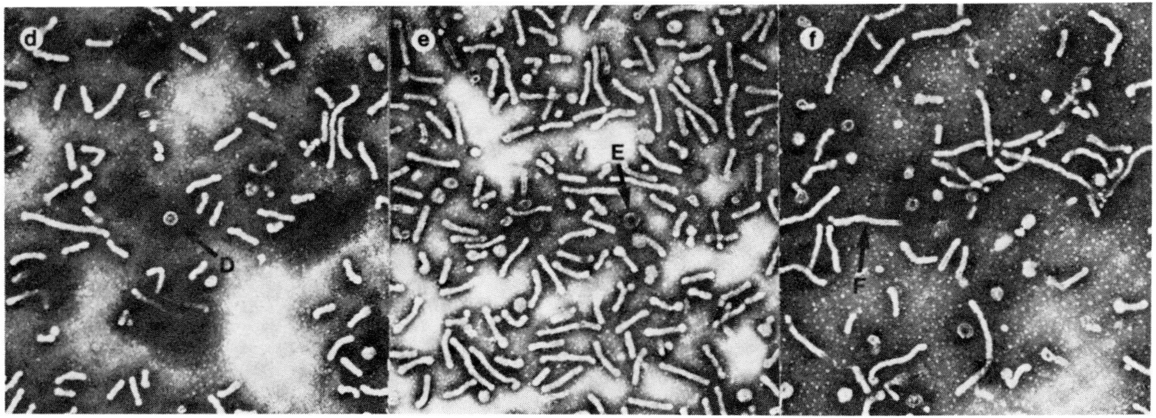

Fig. 20–3. Electron micrographs of HBV particles after sedimentation in sucrose gradients. Panel d contains Dane particles (D) with electron-dense centers. Panel e contains Dane particles with empty centers. Panels d, e, and f contain filamentous forms. (With permission, from Robinson, W.S.: Viruses of Human Hepatitis A and B. Comprehensive Virology, *14*:471, 1979.)

transcribed into (−) stranded DNA. The pre-genome is degraded and the (−) stranded DNA is partially copied into (+) stranded DNA.

3. Hepatitis C Virus

Recently, a new hepatitis virus (hepatitis C virus) has been transmitted from infected humans to chimpanzees. The virions are nonenveloped, icosahedral (average diameter = 27 nm) and resemble picornaviruses in morphology. Hepatitis C virus (also referred to as non A and non B hepatitis virus) is antigenically distinct from HAV and HBV.

4. Delta Agent

The Delta Agent is ubiquitous in nature. It is found most frequently in patients who receive massive blood transfusions, intravenous drug users, and close personal contacts of infected subjects. Infection with the Delta Agent is dependent on coinfection with HBV and synthesis of HBsAg. The Delta Agent appears to be a defective virus (Chapter 4) that can interfere with the replication of HBV.

REFERENCES

Coulepis, A.G., Locarnini, S.A., and Gust, I.D.: Iodination of hepatitis A virus reveals a fourth structural polypeptide. J. Virol., *35*:572, 1980.

Hirschman, S.Z.: The hepatitis B virus and its DNA polymerase: The prototype three-D virus. Mol. Cell Biochem., *26*:47, 1979.

Hirschman, S.Z., Price, P., Garfinkel, E., Christman, J., and Acs, B.: Expression of Clonal hepatitis B virus DNA in human cell cultures. Proc. Natl. Acad. Sci. (U.S.A.), *77*:5507, 1980.

Marx, J.L.: Is hepatitis B virus a retrovirus in disguise? Science, *217*:1021, 1982.

Maupas, P., and Melnick, J.L.: Hepatitis B virus and primary liver cancer. Prog. Med. Virol., *27*:1, 1980.

Provost, P.J. and Hilleman, M.R.: Propagation of human hepatitis A virus in cell culture in vitro. Proc. Soc. Exp. Biol. Med., *160*:213, 1979.

Robinson, W.S.: Viruses of human hepatitis A and B. Comprehensive Virology, *14*:471, 1979.

Szmuness, W.: Hepatocellular carcinoma and the hepatitis B virus: Evidence for a causal association. Prog. Med. Virol., *24*:40, 1978.

Summers, J., Smolec, J.M., and Snyder, R.: A virus similar to hepatitis B virus associated with hepatitis and hepatoma in woodchucks. Proc. Natl. Acad. Sci. (U.S.A.), *75*:4533, 1978.

Chapter 21

Viruses Causing Slow Infections

Viral infections of the CNS are usually associated with acute diseases characterized by inflammatory changes in the cerebrospinal fluid and brain (Chapter 27). Recently, it has been observed that some viruses cause subacute or chronic neurologic disease in which inflammatory changes are lacking and in which the pathogenesis involves a gradual degenerative or demyelinating process. Viruses that cause subacute or chronic diseases have long incubation periods lasting several weeks to years before the onset of clinical symptoms, hence the name "slow" viruses. Once initiated, such diseases follow a protracted course for several months that usually ends in death. Slow viruses are differentiated from latent viruses (Chapter 8), such as herpes simplex or varicella-zoster, which cause recurrent acute diseases. Similarly, slow viruses are differentiated from other viruses that cause chronic infections in which the clinical course tends to be irregular and unpredictable (e.g., hepatitis virus) (Chapter 20). Tumor viruses (Chapter 19) characteristically cause cell proliferation, whereas slow viruses cause cell dysfunction. The term "slow viruses" is a misnomer because most of the slow viruses can multiply rapidly under appropriate conditions. Some agents are considered atypical viruses or "viroids" in that they consist of a low molecular weight RNA genome without associated capsid proteins; they do not invoke a demonstrable immune reponse in the intact host.

This chapter presents available virologic and immunologic information about typical and atypical viruses associated with slowly progressive neurologic diseases in man and animals. A summary of properties pertinent to the etiologic agents associated with these diseases is presented in Table 21–1.

A. TYPICAL VIRUSES

Examples of slowly progressive neurologic diseases associated with typical viruses are presented in Table 21–2. Viruses associated with two of these diseases (subacute sclerosing panencephalitis [SSPE] and progressive multifocal leukoencephalopathy [PML] will be discussed. The etiologic agents of the six other diseases listed in Table 21–2 were discussed elsewhere (togavirus, rubella, Chapter 15; retrovirus, Chapter 19; arenavirus, Chapter 17; parvovirus, Chapter 5).

1. SSPE Virus

The etiologic agent of SSPE is considered to be a virus closely identified with the measles virus (Chapter 15). Capsid antigens and nucleocapsid morphology of SSPE virus are related to measles virus. The RNA of SSPE virus contains about 10 percent more nucleotides than measles virus. SSPE virus has been isolated in tissue culture by cocultivating infected brain tissue with susceptible monkey kidney cells.

Table 21–1. General Properties of Slow Viruses

	Subacute Sclerosing Panencephalitis Virus	JC Virus	Viroids (Prions)
Properties of nucleic acid:	(−) single-stranded RNA; virion RNA complementary to mRNA; virion RNA has 10% more nucleotides than measles virus	circular double-stranded DNA	single-stranded RNA: hairpin-like structure 110,000–127,000 daltons
Virion:	enveloped nucleocapsid; contains RNA transcriptase	nonenveloped	no structural units
Capsid symmetry:	helical	icosahedral	—
Size (nm):	100–150	45	—
Site of replication:	cytoplasm	nucleus	nucleus

Table 21–2. Slowly Progressive Neurologic Diseases Associated With Typical Viruses

Disease	Host	Target Organ	Etiologic Agent
Subacute sclerosing panencephalitis	man	brain	paramyxovirus (measles virus)
Progressive multifocal leuko-encephalopathy	man	brain	papovavirus (SV40 and JC virus)
Progressive encephalitis	man	brain	togavirus (rubella virus)
Visna	sheep	brain	retrovirus
Maedi	sheep	lung	retrovirus
Progressive pneumonia	sheep	lung	retrovirus
Lymphocytic choriomeningitis	mouse	kidney, brain, liver	arenavirus
Aleutian mink	mink	reticuloendo-thelial system	parvovirus

2. JC Virus

Two viruses have been isolated from cases of progressive multifocal leukoencephalopathy, a virus which is almost identical to SV40 virus (Chapter 6) and a virus (JC) structurally related to SV40.

Structure and Antigenic Composition. JC virus associated with PML exhibits icosahedral symmetry, a naked capsid, and a diameter of about 45 nm (Table 21–1). Although it is structurally related to SV40 (papovavirus) of primates, it appears to be antigenically distinct from papovaviruses (Chapter 6) as judged by immunofluorescent staining.

Host Range and Culture. Original isolation of the virus was achieved by inoculating a homogenate of brain tissue from a patient JC with PML onto monolayers of human fetal brain cells. Virus has also been grown in tissue culture by fusing infected brain cells to African green monkey kidney cells in vitro.

B. ATYPICAL VIRUSES

1. Viroids (Prions)

Examples of slowly progressive neurologic diseases associated with atypical viruses are presented in Table 21–3. The etiologic agents of Kuru, Creutzfeldt-Jakob disease and mink encephalopathy appear to share properties in common with a "viroid."

Structure and Antigenic Properties. The term "viroid" is used to describe the smallest known infectious agent that is pathogenic for higher plants. They consist of single-stranded RNA assuming a highly structured hairpin-like configuration in which short base-paired regions

Table 21–3. Slowly Progressive Neurologic Diseases Associated With Atypical Viruses

Disease[a]	Host
Kuru	man
Creutzfeldt-Jakob disease	man
Scrapie	sheep
Mink encephalopathy	mink

[a]For these diseases the etiologic agent is considered as a viroid (prion for scrapie) and the target organ is the brain.

appear to alternate, with short single-stranded loops. Viroids are devoid of capsid protein (Table 21–1). They are highly resistant to heat (80°C for 45 min) and formalin. In human diseases (e.g., Kuru) the viroid-like agents are nonantigenic both in natural and experimental hosts; for example, they fail to induce Abs in the nonsusceptible rabbit. The fact that nucleic acids are poor Ags may account for their lack of antigenicity. Until recently, the etiologic agent of scrapie in sheep (Chapter 32) was considered a viroid. Current information on the etiologic agent of scrapie describes it as a *prion*. Unlike viroids, prions contain a hydrophobic protein that is essential for infectivity.

Host Range and Culture. The prion of scrapie can be maintained in vitro in mouse spleen-clot cultures derived from infected animals and in primary mouse embryo cell or mouse brain cell cultures. They do not produce cytopathology in cultured cells. The agent is assayed by infectivity tests in mice, sheep, goats, monkeys, rats, or hamsters. The infected animals rub their bodies and bite their skin, hence the name scrapie for the disease (Chapter 32). They exhibit fatigue, weight loss, involuntary tremors, and bizarre behavior. The disease is progressive and invariably fatal. Prions are recoverable from lymph nodes, spleen, salivary glands, thymus, lungs, and brain of infected sheep.

REFERENCES

Diener, T.O.: Viroids: structure and function. Science, *205*:895, 1979.
Diener, T.O. and Hadidi, A.: Viroids. Comprehensive Virology, *11*:285, 1977.
Diener, T.O., McKinley, M.P., and Prusiner, S.B.: Viroids and prions. Proc. Natl. Acad. Sci. (U.S.A.), *79*:5220, 1982.
Gibbs, C.J., Jr. and Gajdusek, D.C.: Atypical viruses as the cause of sporadic, epidemic, and familial chronic diseases in man: slow viruses and human diseases. Perspectives in Virology, *10*:161, 1978.
Grill, L.K. and Semancik, J.S.: Viroid synthesis: the question of inhibition by actinomycin D. Nature, *283*:399, 1980.
Stroop, W.G. and Baringer, J.R.: Persistent, slow and latent viral infections. Prog. Med. Virol., *28*:1, 1982.
Wechsler, S.L. and Meissner, H.C.: Measles and SSPE viruses: similarities and differences. Prog. Med. Virol., *28*:65, 1982.

Chapter 22

Pathogenesis of Viral Infections

When a susceptible individual becomes infected with a virus, and a certain threshold of virus replication is reached, disease usually becomes apparent. In some cases pathologic changes may be induced by the virus in the absence of detectable virus multiplication. The summation of viral-induced events which result in the symptoms associated with a given stage of disease are determined by the biochemical, immunologic, physiologic, and cytologic changes induced by the infecting virus. The pathogenicity of the virus at the level of the multicellular host is determined by several factors: (1) susceptibility of cells, (2) availability of susceptible cells to the virus, (3) effects of toxic products produced by virus-cell interactions, and (4) the physiologic and immunologic responses of the host.

Viral infections can be subdivided into the following major types: inapparent (subclinical or latent) and apparent (acute and persistent). An *inapparent infection* is one which is not associated with clinical signs of disease. This term is used in contrast to *apparent infection* where there are obvious clinical signs of disease (e.g., cell destruction in poliomyelitis or cell proliferation in human wart virus infections). Apparent infections can be subdivided into *acute* and *persistent*. An *acute infection* usually lasts for a relatively short period of time (days to weeks), and it is generally followed by disappearance of the virus from the tissues and organs of the host (e.g., influenza). In contrast, a *persistent viral infection* is characterized by slowly progressing infection which persists for months or years (e.g., infectious hepatitis) and may eventually terminate in death (e.g., slowly progressive neurologic diseases, Kuru).

A *latent infection* is a type of inapparent infection in which production of infectious virus is not detectable, and the host and virus appear to be in equilibrium with each other. However, if this equilibrium should be disturbed and multiplication of infectious virus occurs, then apparent disease can develop. There are two general patterns of latent infections recognized in man. One pattern is exemplified by herpes simplex virus type 1 (HSV-1), the etiologic agent of coldsores, and the second pattern by varicella-zoster (V-Z) virus, the etiologic agent for chickenpox and shingles. Both viruses establish latency by infecting sensory ganglia. HSV-1 is acquired early in life, establishes a latent infection in trigeminal ganglia, and under conditions of physical, or environmental stress abandons its latent state and causes apparent *recurrent* infections (e.g., coldsores).

However, in the case of V-Z, the first attack by the virus causes chickenpox. Following recovery from chickenpox and production of lifelong immunity to this form of the disease, the virus remains latent in the cells of the dorsal root ganglia, but subject to reactivation by trauma, stress, or malignancy or some form of immunodeficiency. Reactivation of the latent V-Z virus results in a second disease called "shingles" (herpes zoster). The interval between the two diseases may vary from a few months to as long as 40 years or more.

Since viruses must become intracellular before they can produce disease, it is important that students of medicine know how viruses (1) enter tissues, (2) are disseminated in the host, (3) are excreted, and (4) are transmitted to other individuals.

A. VIRUS ENTRY, DISSEMINATION, EXCRETION, AND TRANSMISSION

Under natural conditions, infection of a susceptible host is usually initiated by a small number of virus particles. At the onset of infection, the number of primary foci of infected cells will be few in comparison with the number of cells which ultimately become infected during the height of the disease. Viruses enter the host by a variety of routes, (1) mucous membranes (respiratory tract, alimentary tract, conjunctivae), (2) skin, (3) genital tract, or (4) placenta (Table 22–1). Entry of a particular virus is governed by several factors such as source of infecting virus (direct contact, air or aerosols, water, fomites, insects), distribution of susceptible cells in the host, stability of virus to various conditions (surface integuments, body secretions, pH, temperature, humidity) and age, physiologic and immunologic status of the host.

1. Mucous Membranes

a. Respiratory Tract. The respiratory tract and alimentary tract are the principal sites of entry for the majority of viruses infecting man. Entry of viruses into the respiratory tract occurs via aerosols or droplets from the nasopharynx of infected individuals. This is an important route of entry for many common respiratory viruses of man: e.g., rhinoviruses, adenoviruses, orthomyxoviruses, paramyxoviruses, coronaviruses, and certain enteroviruses. Evidence suggests that rabies virus may also infect man by this route. The Influenza A virus, a member of the Orthomyroviridae family, is a useful and instructive model for understanding the pathogenesis of respiratory viruses. The virus enters in the form of wet or dry aggregates (droplets) and localizes in the respiratory tract. The virus then attaches to the N-acetyl neuraminic acid receptor on the cell surface of epithelial cells of the upper and (to varying degrees) the lower respiratory tract. Upon entry of the virus core (nucleocapsid), new infectious virus is produced and spread from cell to cell through intercellular and superficial fluid layers. Multiplication of the virus and its subsequent spread to other cells in the respiratory tract is restricted by (1) specific secretory IgA, (2) production of interferon (natural antiviral component) by infected cells, and (3) mucus which contains a natural glycoprotein that can neutralize the infectivity of extracellular influenza A virus. The consequences of virus multiplication, which is almost always limited to the respiratory tract, are cell necrosis and desquamation of

Table 22–1. Entry of Viruses in Man

Route	Viruses
Mucous membranes	
Respiratory tract	Rhinoviruses
	Adenoviruses
	Herpesviruses
	Orthomyxoviruses
	Paramyxoviruses
	Coronaviruses
	Reoviruses[1]
	Enteroviruses[1]
	Variola
	Rubella
	Mumps
	Rabies[1]
Alimentary tract	Enteroviruses
	Reoviruses
	Infectious hepatitis virus
	Serum hepatitis virus[1]
	Adenoviruses
Conjunctivae	Rhinoviruses
	Herpesviruses
	Adenoviruses
Skin: as the result of:	
Abrasion	Human papillomavirus
	Molluscum contagiosum
	Orf
Arthropod bite	Togaviruses (Arboviruses)
Animal bite	Rabies; B virus (simian herpes)
Hypodermic needle	Serum hepatitis virus
	Infectious hepatitis virus[1]
Genital tract	Condylomata acuminata (human papilloma)
	Herpes simplex virus
Placenta	Rubella[1]
	Cytomegalovirus[1]

[1]Enter more commonly by another route.

the epithelium of the alveoli, bronchioles, and trachea. The disease is easily spread to other individuals because large amounts of virus are shed in droplets discharged by talking, sneezing, and coughing. Since the incubation period for influenza is short (2–3 days), the disease can easily become epidemic in a few weeks. Some viruses (mumps) enter the respiratory tract via droplets, produce a primary infection of the epithelial cells, and then spread via the blood to infect other organs.

b. Alimentary Tract. Enteroviruses are the major group of viruses which infect man by way of the alimentary tract. Because enveloped viruses (e.g., herpes simplex) are inactivated by acid conditions in the stomach, the only viruses which successfully infect the alimentary tract are the non-enveloped enteroviruses (e.g., poliovirus), adenoviruses, reoviruses, and hepatitis viruses. Virus infection occurs by way of ingestion of contaminated food, drink, or fomites. In the case of poliovirus infection (Fig. 22–1), the virus initially multiplies in the pharynx and lymphoid tissues (tonsils and Peyer's patches) and then in the epithelial cells lining the small intestine. Spread of virus to the draining lymph nodes leads to viremia and dissemination of virus throughout the body. During the viremic stage of the disease, the infection may spread to the anterior horn cells of the spinal cord and motor cortex of the brain. The probability of neural damage with paralysis is influenced by several factors including age, pregnancy, and trauma. Excretion of virus occurs in the feces.

c. Conjunctivae. The conjunctivae may serve as a portal of entry for many of the viruses which cause upper respiratory tract infections including the rhinoviruses (common cold), herpesviruses, and certain adenoviruses. The source of infection can be nasopharyngeal secretions (e.g., rhinoviruses), latent infections of the trigeminal nerve root ganglia (e.g., herpesviruses), water from swimming pools or dust (e.g., adenoviruses). Also, the conjunctiva is a sensitive indicator of some systemic viral diseases such as measles and certain enteroviruses.

HSV-1 is an etiologic agent of keratoconjunctivitis, causing serious eye infection which can lead to blindness. The basic lesion is a "dendritic ulcer," a creeping tree-like ulcer of the cornea. Spread of virus to the

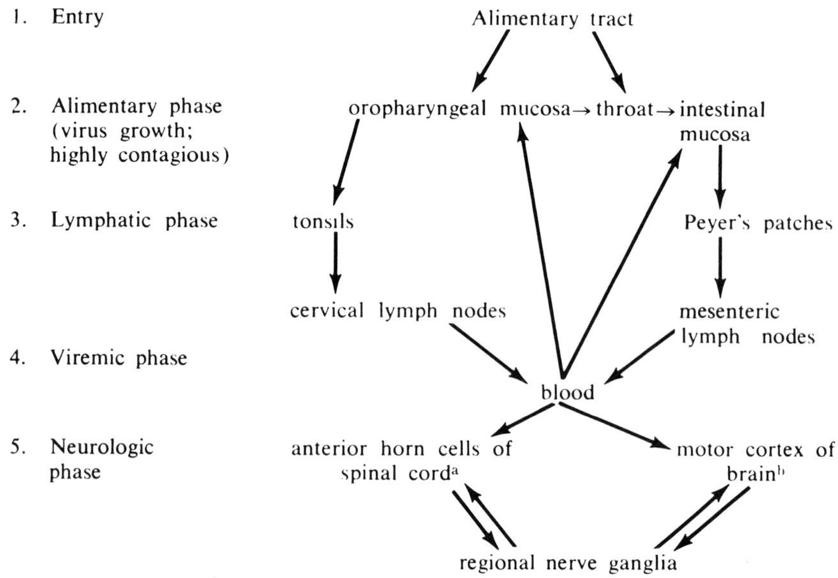

Fig. 22–1. Pathogenetic pattern of poliovirus in a nonimmune individual.

[a]Infection of anterior horn cells of spinal cord or brain stem results in bulbar polio; usually fatal due to cardiac and respiratory failure.

[b]Infection of motor cortex results in encephalitic type; only 1 to 2% infections invade CNS and give rise to paralytic clinical syndrome.

deeper layers of the cornea results in severe necrosis. Recurrent infections of the cornea are common in some patients because of reactivation of a latent herpes simplex virus in the trigeminal nerve root ganglia.

2. Skin

The normal skin acts as a natural barrier against most virus infections. Viruses may enter through the skin by way of (1) abrasions (e.g., molluscum contagiosum, orf, human papilloma virus), (2) bite of infected arthropods (e.g., togaviruses), (3) animal bite or lick on an abraded area (e.g., rabies), or (4) hypodermic needle (e.g., hepatitis viruses). Most viral skin diseases (e.g., measles, rubella, varicella, smallpox) occur following entry of virus by the respiratory route and subsequent systemic spread of the virus by way of macrophages, lymphocytes, or other leukocytes. Rabies virus spreads in man by way of dorsal root ganglion cells of peripheral nerves serving the inoculated area. The only localized viral infections of the skin in man are caused by the human papilloma virus and the proliferative inflammatory lesions of the skin produced by poxviruses (e.g., vaccinia, molluscum contagiosum and orf). More commonly, the rashes seen in the exanthemata of childhood (e.g., varicella, measles, rubella) are considered to be allergic in nature resulting from an "IgE-like" antibody on mast cells reacting with viral Ag(s) in the skin or associated with the capillary endothelium. The viral Ag(s) originates from viruses which have attached to cells of the vascular walls and produced a focus of infection.

With respect to excretion and transmission of virus reponsible for skin infections, it is notable that poxviruses (e.g., orf, smallpox) are highly resistant to drying and can persist in an infectious state in scabs and exfoliating epidermis or bedding and clothing for extended periods. V-Z virus also is transmitted by materials from lesions, whereas measles virus is excreted and transmitted by secretions from the nose, throat, or the conjunctivae.

3. Genital Tract

Two human viruses, a papilloma virus which causes genital warts (condylomata acuminata) and HSV-2, are spread by sexual intercourse. The virus of condylomata causes a marked proliferation of cells of the epidermis with extensive hyperkeratosis. Virus is found mainly in the keratinized layers of the wart. Acidophilic inclusions may be found in cell nuclei and more rarely in the cytoplasm. This is the only type of human wart that may become malignant. HSV-2 is acquired early in childhood and is usually retained in the body for life. Although rare, severe and often fatal disease can be acquired by the infant at the time of birth from virus present in the mother's vagina. The infant's skin becomes covered with vesicles which rupture to form ulcers. Systemic dissemination of the virus often occurs which can develop into hepatitis, meningitis, encephalitis, or meningoencephalitis.

4. Placenta

Entry of virus through the placenta represents a specialized route of infection. Rubella virus may cross the placenta and exert its teratogenic (production of abnormal development) action especially on the lens, heart, and brain of the fetus. Cytomegalovirus also crosses the placenta and infects the fetus which usually dies in utero or may be born with permanent neurologic sequelae attributable to microcephaly and microgyria (convolution of the brain). The virus can sometimes be isolated from the mother's urine. The fact that subsequent babies are not affected by cytomegalovirus infection indicates that infection of the mother does not persist for more than a few months. Several other viruses may infect the fetus and lead to abortion. For example, if a virus infection occurs late during pregnancy, the baby may be born with an acute viral disease caused by herpes simplex virus type 2; a similar circumstance can occur with varicella virus or poliovirus. There is speculation that some retroviruses also may be passed in utero. Evidence for this mode of transmission has been obtained with viruses causing leukemia (Chapter 34) in animals.

B. FACTORS INFLUENCING THE INCUBATION PERIOD

The *incubation period* of a virus infection is the interval between infection and the first manifestation of symptoms. It will be short (1–3 days) in diseases in which the symptoms are due to viral growth at the site of entry, as in respiratory tract infections (Table 22–2). In contrast, a moderate to long incubation period (5–30 days) is associated with generalized infection (chickenpox, measles, rubella, mumps). In these diseases virus enters the respiratory tract and then spreads in a stepwise fashion in the host before producing disease in the target organ. The length of the incubation period can be influenced by other factors; for example, the direct intravenous injection of togaviruses by an infected arthropod would result in a short (1–4 days) incubation period. On the other hand, the long incubation period which characterizes human papilloma virus infection (warts) is due to the unusually slow growth rate for these viruses. A phenomenon of great importance, which at present is poorly understood, is the unusually long incubation period of the so-called neurologic virus diseases (Chapter 32) in which virus-cell interaction is initially noncytocidal and symptoms of disease are not apparent for many months or even years after infection.

C. PATHOGENETIC PATTERNS

Viruses characteristically infect certain organs of the body and not others (i.e., display tissue and organ tropism) and may follow prescribed

Table 22–2. Periods of Incubation and Communicability of Some Common Viral Diseases in Man

Disease	Mode of Transmission	Incubation Period[a] (Days)	Period of Communicability[b]	Incidence of Subclinical Infections[c]
Influenza	Respiratory	2–3	Short	Moderate
Common cold	Respiratory	1–3	Short	Moderate
Bronchiolitis	Respiratory	3–5	Short	Moderate
Eastern equine encephalitis	Mosquito bite	3–5	Short	Low
Dengue	Mosquito bite	5–8	Short	Moderate
Herpes simplex infection	Contact	5–8	Long	Moderate
Enterovirus infection	Alimentary	6–12	Long	High
Poliomyelitis	Alimentary	5–20	Long	High
Measles	Respiratory	9–12	Moderate	Low
Smallpox	Respiratory	12–14	Moderate	Low
Chickenpox	Respiratory	13–17	Moderate	Moderate
Mumps	Respiratory	16–20	Moderate	Moderate
Rubella	Respiratory	17–20	Moderate	Moderate
Mononucleosis	Contact	30–50	? Long	High
Hepatitis (infectious)	Alimentary	15–40	Long	High
Hepatitis (serum)	Hypodermic needle	50–150	Very Long	High
Rabies	Animal bite, Respiratory	30–100	None	None
Warts	Contact	50–150	Long	Low

[a]Until first appearance of prodromal symptoms. Diagnostic signs (e.g., rash or paralysis) may not appear until 2 to 4 days later.

[b]Most viral diseases are highly transmissible for a few days before symptoms appear. Short = 1–4 days. Moderate = 5–9 days. Long = 10–30 days. Very long = months–years.

[c]High = 90 percent; low = 10 percent.

Modified from Fenner, F., and White, D.: Medical Virology, New York, Academic Press, 1970.)

pathogenetic patterns. Poliomyelitis can serve as a useful example to illustrate the pathogenetic pattern of a viral disease (Fig. 22–1). In this disease, virus enters by way of the alimentary tract and multiplies in the oropharyngeal mucosa, the tonsils, Peyer's patches, or lymph nodes which drain these tissues. Virus then begins to appear in the throat and feces and is spread via the blood to other susceptible tissues (distant lymph nodes and CNS). If virus multiplication reaches a high level in the CNS, the disease may become fatal or motor neurons are destroyed and paralysis follows. Involvement of the CNS may be prevented by poliovirus-neutralizing Abs induced by prior infection or by vaccination.

D. CONCEPTS OF VIRAL PATHOGENESIS

Two primary types of pathogenesis may follow virus-host interactions: (1) cell injury or destruction, and (2) neoplastic transformation. In discussing the mechanisms of pathogenesis of disease both viral aspects and host responses to the virus-host cell complex must be considered.

1. Virus-Related Pathogenesis

a. Cell Injury or Destruction. The majority of human viral illnesses are acute rather than chronic in nature. Cellular injury or destruction during acute disease may be caused in several ways: (1) general toxic effect produced directly by virions (e.g., penton protein of adenoviruses) or by toxic products released from the infected cells which also may be responsible for symptoms of fever, headache, and malaise which accompany many viral diseases; (2) synthesis of a virus-coded protein(s) which specifically blocks cellular biosynthesis (e.g., poliovirus infection leads to inactivation of a 5′ terminal cap binding protein needed for cellular mRNA during initiation of protein synthesis; this inhibition does not affect poliovirus protein synthesis because the virus mRNA does not require the cap-binding protein); (3) induction of chromosomal aberrations in host cells thereby inhibiting cell mitosis; or (4) production of viral inclusion bodies (i.e., intracellular masses of new viral material) which may disrupt the structure and function of host cells to cause their death.

In some cases of exanthematous disease caused by rubella, rubeola, variola, V-Z, and certain togaviruses, a severe form of pathogenesis occurs which involves hemorrhagic eruptions and disseminated intravascular coagulation. The organs most frequently involved include the kidney, brain, pituitary, lungs, liver, adrenals, and intestinal mucosa. The generalized spread of exanthematous viruses is sometimes associated with a defective immune response of the RE system, suppression by radiation, drugs, or corticosteroid therapy. Following viremia, the endothelial cells of the visceral blood vessels become infected. When infected endothelial cells are injured, platelets agglutinate and thrombi are formed at the site. The mechanism of pathogenesis may include (1) direct agglutination of platelets by the virus, (2) cell injury which may predispose to platelet agglutination and thrombi formation, or (3) replication of the virus in endothelial cells which may form Ag-Ab complexes capable of initiating intravascular coagulation. The morbidity and mortality associated with the severe forms of exanthematous diseases can be reduced by anticoagulant therapy (e.g., heparin).

b. Neoplastic Transformation. The interaction between potential oncogenic (cancer-producing) viruses and host cells can result in virus multiplication with or without transformation of the cell. For example, it is known that in animals infected with leukoviruses most organs are producing new infectious virus. However, the incidence of neoplastic transformation is low in terms of the total number of cells infected. Although cancer cells may produce infectious virus, multiplication of the virus is not required for neoplastic transformation (pathogenesis), but presence of virus genetic information is required. Neoplastic transformation, manifested by alterations in the cell plasma membrane, changes in cell morphology and orientation, presence of tumor specific Ags, and the capacity to grow at an increased rate in vivo, depends upon integration and expression of the virus genome. Neoplastic transformation of Ab-precursor cells by leukoviruses may reduce the total number of Ab-producing cells, the amount of Ab produced per cell, and the type of immunoglobulin produced. Presumably a combination of these effects accounts for the immunodepression associated with leukemia.

2. Host-Related Pathogenesis

a. Inflammatory Reactions. During the course of a virus disease infected cells become altered, thereby eliciting an inflammatory response which is generally related to release of virus or diffusion of certain by-products from the injured or destroyed cells. The nature of the inflammatory response may produce a specific clinical character to the lesions produced by a given virus (e.g., localized lesions with herpes simplex and maculopapular lesions with measles). Infection of neuronal cells in man by poliovirus causes an inflammatory response characterized by an intense perivascular infiltration of inflammatory cells which may result in accumulation of macrophages and host cell death. In contrast to acute bacterial diseases, the inflammatory cells in viral infections are predominately mononuclear cells consisting of macrophages, plasmacytes, and lymphocytes. A transitory polymorphonuclear infiltration of the infected site may occur on occasion. Leukopenia is a general characteristic of the acute phase of viral infections; the leukocyte count often decreases to 3000 per mm^3.

b. Humoral and Cell-Mediated Immunopathology. In some virus diseases immune mechanisms of the host are responsible for the pathogenesis of disease. This view has evolved from the consideration that maternal serum Ab (IgG) passively acquired by young infants does not provide protection against the most severe form of obstructive bronchiolitis associated with respiratory syncytial (RS) virus infections but instead may cause the condition. Children over 2 years of age seem to develop a less severe form of the disease because they lack specific maternal IgG Ab which is evidently responsible for injury in young infants. Injury apparently results from the lysis of infected cells due to the reaction of maternal IgG Ab with virus Ag(s) plus complement on the surface of infected cells or from immune complexes consisting of soluble viral Ag(s) and IgG Ab. Other respiratory viruses do not produce

RS virus-like disease because of their failure to cause a high incidence of infection during the first year of life when specific maternal IgG is present in the infant's serum.

The pathogenesis of some viral diseases may be directly related to immunologic alterations induced by the infecting virus. For example, it is known that viruses can profoundly affect humoral immunity (e.g., rubella, retrovirus), cellular immunity (rubeola, influenza, varicella, polio, hepatitis, and Epstein-Barr (EB) viruses) and phagocytosis by the RE system (e.g., mumps, influenza, and coxsackie viruses). Several hypotheses have been proposed to explain the mechanism of immunodepression by viruses: (1) alteration in Ag uptake, (2) depression of Ab synthesis, (3) destruction of Ab-producing cells and their precursors, and (4) competition between the infecting virus and immunizing Ag for noncommitted Ab-producing cells. Lymphocytes from individuals infected with infectious hepatitis virus are defective in their ability to undergo blast transformation and polymorphonuclear leukocytes infected with mumps, influenza, or coxsackie viruses have a reduced capacity to phagocytize bacteria.

Virus-induced tissue damage can stimulate the production of tissue-specific immunoglobulin, particularly with enveloped viruses which form their outer coat from cell membranes. In this way the new virions from damaged cells carry cell Ags which stimulate the production of anti-cell immunoglobulins. Evidently virus-induced immunologic alterations play a major role in the pathogenesis of (1) certain chronic virus infections (e.g., hepatitis), (2) disseminated intravascular coagulation in exanthematous diseases, (3) chronic viral glomerulonephritis, (4) "autoimmune" disease, and (5) malignancy.

There is strong evidence that delayed-type hypersensitivity to viral Ags may be important in the pathogenesis of some viral diseases. For example, in a small proportion of measles (rubeola) cases, postinfection encephalitis may occur 1–2 weeks following disappearance of the rash. Although measles virus cannot be isolated from the brain, lymphocytic infiltration and demyelination changes are apparent. Another example is vaccinia infection of the skin in which dermal swelling, edema, and vesicle formation in the epidermis are thought to represent a delayed-type hypersensitivity response. Arthus-type vascular lesions may also result from the complexing of circulating Ab with vaccinia virus Ags present in blood vessel walls.

REFERENCES

Bablanian, R.: Structural and functional alterations in cultured cells infected with cytocidal viruses. Prog. Med. Virol., *19*:40, 1975.

Cheville, N.F.: Cytopathology in viral diseases. Monog. Virol., *10*:1, 1975.

Howe, C., Coward, J.E., and Fenger, T.W.: Viral invasion: morphological, biochemical and biophysical aspects. Comprehensive Virology, *16*:1, 1980.

Lonberg-Holm, K., Philpson, L.: Early interaction between animal viruses and cells. Monog. Virol., *9*:1, 1974.
Mims, C.A.: The pathogenesis of infectious disease. New York, Academic Press, 1976.
Mims, C.A.: General features of persistent virus infections. Postgrad. Med. J., *54*:581, 1978.
Notkins, A.L., and Oldstone, M.B.A. (eds.): Concepts in Viral Pathogenesis. New York, Springer-Verlag, 1984.
Robb, J.A.: Virus-cell interactions: a classification for virus caused human disease. Prog. Med. Virol., *23*:51, 1977.
Rose, J.K., Trachsel, H., Leong, K., and Baltimore, D.: Inhibition of translation by poliovirus: inactivation of a specific initiation factor. Proc. Natl. Acad. Sci. (U.S.A.), *75*:2732, 1978.
Trachsel, H., Sonenberg, N., Shatkin, A.J., Rose, J.K., Leong, K., Bergmann, J.E., Gordon, J., and Baltimore, D.: Purification of a factor that restores translation of vesicular stomatitis virus mRNA in extracts from poliovirus-infected HeLa cells. Proc. Natl. Acad. Sci. (U.S.A.), *77*:770, 1980.
Wolinsky, J.S. and Johnson, R.T.: Role of viruses in chronic neurological diseases. Comprehensive Virology, *16*:257, 1980.

Chapter 23

Control of Viral Infections

A. APPROACHES TO THE CONTROL OF VIRAL INFECTIONS

Considerable success has been achieved in the treatment of bacterial diseases by the use of chemotherapeutic agents. Consequently, an intensive search for chemical substances for use in the prevention or treatment of viral diseases has emerged. There are three practical approaches to the control of viral illnesses: immunologic, chemoprophylactic, and the use of interferon (Table 23–1). Immunologic procedures have afforded significant protection against many viral diseases (Table 23–2). Recent successes with drugs that modify the course of viral infections have been encouraging. The artificial induction of interferon, a natural antiviral agent, provides a new approach for enhancing resistance against viral diseases. In addition, general supportive therapy continues to be an essential part for the treatment of all viral diseases.

B. VACCINES

1. Active Immunization

Active immunization with vaccines is a highly effective means for controlling a number of virus diseases. The concept of using viruses for human "vaccination" originated nearly two centuries ago when an English country doctor, Edward Jenner, scratched the arm of a Gloucestershire farm boy and deliberately contaminated the wound with pus from a milkmaid's cowpox sores in a trial to prevent smallpox. Since that time, routine vaccination for smallpox has been a cornerstone of public health practice. However, routine vaccination for smallpox was discontinued in 1971 due to the apparent eradication of smallpox disease in the world. Whereas routine vaccination for smallpox was discontinued, the use of live-virus vaccines is being recommended for other diseases in need of control, especially in instances in which (1) the antigenic types of the etiologic agents are few, (2) there is systemic invasion of the host, and (3) highly effective immunity follows a natural infection; e.g., measles, mumps, rubella, poliomyelitis, and yellow fever. In contrast, a disease (1) which is caused by numerous antigenic types, (2) in which the infection is superficial, and (3) in which naturally acquired immunity is

Table 23–1. Efficacy of Approaches Used For Controlling Viral Infections

Approach	Level of Effectiveness	Characteristic Antiviral Spectrum	Duration of Effect
Immunologic	Usually high	Very narrow	Relatively long
Chemoprophylactic	Moderate	Narrow	Very short
Interferon	Moderate to high	Very broad	Relatively short

not long-lasting, generally can be successfully controlled by the use of killed-virus vaccines; e.g., influenza. Viruses of certain groups, such as the rhinoviruses and certain enteroviruses of many antigenic types, cannot be controlled in a practical way with either live- or killed-virus vaccines.

Vaccines consist of either live attenuated (reduced virulence) or inactivated (dead) viruses derived from "wild-type" virus that caused natural disease. The principles of animal virus genetics have been utilized for developing attenuated poliovirus vaccines. It is possible to obtain a wide variety of temperature-sensitive (ts) mutants (i.e., having inability to multiply at 40° C) from the wild-type polioviruses or select for genetic changes that affect the viral capsid. The ts mutants retained immunogenicity but lost their neurovirulence properties. Although a single-step mutant has an appreciable leak or reversion rate back to the wild-type, this problem is circumvented by selection of multiple-step mutants (e.g., mutants with altered capsids which are unable to grow at 40°C). Alternately, the procedure used for producing the attenuated viruses currently employed in many vaccines was deliberate and repeated serial passage of the original human wild-type virus in embryonated eggs or non-human tissue cultures. This procedure resulted in the chance selection of strains of viruses which have greatly reduced virulence for man (e.g., yellow fever, measles, mumps, and rubella vaccines). Once a suitable virus strain is developed and tested for safety, and its immunogenicity is established in animals and human volunteers, a large volume of a single seed stock of virus can be stored in an unaltered state for many years and used for production of large batches of vaccines from time to time.

Inactivated vaccines are preparations of "dead" virus in which virus infectivity has been destroyed, usually by treatment with formalin. The only inactivated-virus vaccine in widespread use today is influenza vaccine. A survey of the current status of vaccines available for use against important viral infections is presented in Table 23–2. The vaccines currently available for administration together with the schedules for immunization are listed in Table 23–3. Table 23–4 lists the advantages and disadvantages of live-virus and inactivated-virus vaccines. There is a general consensus among clinical virologists that live-virus vaccines produce subclinical infections which probably lead to the same type of immune response as is seen following natural infection. Usually, a single dose of live-virus vaccine produces prolonged immunity; however, in the case of poliomyelitis and yellow fever, revaccination is recommended (Table 23–3). It is noteworthy that live-virus vaccines are not completely free of possible dangers or drawbacks such as contamination with live animal cancer viruses, interference by a preexisting virus (such as interference of poliovirus vaccine by enterovirus), and the lability of such vaccines. Inactivated-virus vaccines are free from most of the potential dangers associated with live-virus vaccines, but the disadvantages are

Table 23–2. Present Status of Vaccines Against Important Human Viral Diseases

Disease	Virus	Virus Condition: Live, attenuated	Virus Condition: Inactivated
Smallpox	Vaccinia	+	
Rabies	Rabies	+ (veterinary use)	+
Encephalitis	WEE, EEE, VEE[a]		+
Respiratory	Influenza A & B	+[b]	+
	Adenovirus[c]	+	+
Measles			
German	Rubella	+	
Rubeola	Measles	+	
Mumps	Mumps	+	+
Poliomyelitis	Polio[d]	+	+
Yellow Fever	Yellow Fever	+	
Hepatitis	Hepatitis B Virus		+ (subunit)

[a]Western, Eastern and Venezuelan equine encephalomyelitis viruses.
[b]Experimental
[c]Types 3, 4, 7 and 21.
[d]Types 1, 2 and 3.

Table 23–3. Immunization Schedules for Virus Vaccines

Vaccines	Recommendations for First Injection	Recommendations for Booster Injection(s)
Live Vaccines		
Smallpox	Not recommended[1]	Not recommended[1]
Poliomyelitis	2 to 6 months of age	One year later and then every 5 years
Yellow fever	Before travel through endemic areas	Every 10 years
Measles[2]	1 year of age	Presently not recommended
Rubella[2]	1 year of age	Presently not recommended
Mumps[2]	1 year of age	Presently not recommended
Inactivated Vaccines		
Influenza	Autumn before expected epidemic	4 weeks later and then annually if indicated
Rabies	Immediately after bite or lick by rabid animal	6 doses recommended[3]

[1]Vaccination required only when travelling to areas where smallpox cases have recently been reported.
[2]Measles, rubella, and mumps vaccines are available in a trivalent virus vaccine.
[3]The vaccine for humans is inactivated virus grown in human diploid cell culture.

Table 23–4. Advantages and Disadvantages of Live- Versus Inactivated-Virus Vaccines

Advantages	Disadvantages
Live-virus Vaccines	
Easily administered in single dose usually by mouth[1]	Possibility of live virus reverting to virulence
Administered by natural or unnatural route; produce subclinical infections and induce same immunity as natural infection	Possibility of natural spread[2]
Wide spectrum of immunoglobulins produced (IgG, IgM, IgA) and cell-mediated immunity	Cancer virus contaminants May have complications of fever, rash, malaise
Possibility of local suppression of wild-type virus infection	Viral interference by existing viral infection may prevent good immune response
Immunity is long-lasting Reduced antigenic mass required since virus multiplies in the host	Lability of virus
Inactivated-virus Vaccines	
Use of polyvalent vaccines	Booster doses needed
Stability	No development of secretory IgA
No natural spread	High concentration of virus-antigen needed Must administer vaccine by injection

[1]Booster dose may be required.
[2]Particularly important with rubella if the vaccine strain were teratogenic.

that large amounts of virus Ag must be administered and several booster injections are required. Some of these difficulties have been overcome, particularly with hepatitis B, through the use of disrupted virions as "subunit" vaccines. With the advent of many new vaccines and the expectation that others will be available in the near future (Table 23–2), efforts are being made to develop more simplified methods for vaccine administration of combined vaccines. Examples of combined vaccines include poliovirus-DPT (diphtheria, pertussis, tetanus) vaccine and a combination of measles, mumps, and rubella in a trivalent vaccine. The intranasal administration of a live "hybrid" influenza A virus vaccine is currently being tested for efficacy and safety in human volunteers.

2. Passive Immunization—Human Immune Serum and Hyperimmune Gamma Globulin

Protection against viral diseases can also be achieved by prophylactic administration of either human immune serum (normal pooled human serum having detectable Abs to some viruses) and hyperimmune gamma globulin from pooled sera derived from hyperimmunized human volunteers.

Inoculation of hyperimmune gamma globulin before infection or early in the incubation period may prevent or modify diseases with long incubation periods (greater than 12 days) such as measles, rabies, infectious hepatitis, poliomyelitis and mumps (Table 23–5).

Human immune serum has proved to be effective in the prophylaxis of infectious and serum hepatitis. However, it should be emphasized that passive immunization with human immune serum or hyperimmune gamma globulin should only be regarded as an emergency procedure for the immediate and short-term protection of unimmunized individuals or other individuals (e.g., cancer patients) at special risk. Passive immunization is effective in measles if the immunoglobulin is administered within 5 or 6 days following exposure to the disease. Passive immunization is less effective against chickenpox and is of questionable value in preventing congenital abnormalities in pregnant women exposed to rubella. In suspected rabies, combined active and passive immunization should be considered. Because of the frequent complications associated with the use of rabies immunoglobulin of equine origin, a human rabies immunoglobulin developed in human volunteers is currently being made available for administration to persons hypersensitive to equine immune serum.

3. Virus Neutralization by Specific Immunoglobulins

The mechanism of virus neutralization by specific Ab does not require saturation of the virion with Ab molecules. In fact, a single Ab molecule is usually sufficient to neutralize the infectivity of a virus particle. Neutralized virions may or may not adsorb to and penetrate host cells,

Table 23–5. Commercially Available and Uses of Human Immune Serum (H.I.S) or Hyperimmune Gamma Globulin (H.G.G.)

Disease	Type of Passive Immunity	Indications
Measles	H.I.S.	Exposed nonimmune under 3 years old; debilitated or chronically ill; pregnant persons or during institutional outbreaks
Rubella	H.I.S.	Exposed nonimmune pregnant women
Poliomyelitis	H.I.S.	Nonimmune persons, during institutional outbreaks
Hepatitis	H.I.S.	Close household contacts; medical personnel at risk; travelling to endemic areas
Rabies	H.G.G.	Prophylaxis following bite or lick by suspected rabid animal
Vaccinia	H.G.G.	Following severe reactions to vaccinia virus vaccination
Zoster	H.G.G.	Prophylaxis against varicella-zoster virus in immunosuppressed persons

depending on the number of neutralizing Ab molecules attached to the virion. However, when such Ab-neutralized virions become intracellular they are inactivated by cellular enzymes and fail to initiate an infection.

Experiments with agammaglobulinemic patients indicate that they recover from viral infections in a normal fashion despite low levels of immunoglobulin. This observation has led to a reassessment of the possible role of cell-mediated immunity and interferon activity in recovery from viral diseases. Animal experiments involving lymphocyte transfer, neonatal thymectomy, and antilymphocyte serum support the view that cell-mediated immunity plays an important role in recovery from viral diseases. However, the precise mechanisms involved are not known.

C. ANTIVIRAL PROPHYLACTIC AND THERAPEUTIC DRUGS

The discovery of antibiotics which are highly effective against bacterial diseases has stimulated an extensive search for drugs of comparable value for preventing or curing viral diseases. Modest success has been achieved, and the future looks promising.

The most encouraging results in chemoprophylaxis of viral infections have come from work with the following agents: (1) thiosemicarbazones (methyl-isatin-β-thiosemicarbazone, Methisazone) given orally for protection against smallpox, (2) amantadine (Symmetrel) for prophylaxis against influenza, and (3) metabolic inhibitors (including iododeoxyuridine, trifluorothymidine) for treating corneal infections caused by herpes simplex virus type 1, vidarabine (arabinosyl adenine, ara-A) and cytarabine (arabinosyl cytosine, ara-C) for treatment of serious herpesvirus infections including herpes simplex, cytomegalovirus and varicella-zoster. The guanosine analog acyclovir (acycloguanosine, Zovirax) is a highly selective antiherpes drug for treatment of venereal herpes simplex virus infection. The best hope for therapy of viral diseases depends on the development of drugs which can selectively block viral synthesis, reduce tissue damage, and enhance the resistance of the host. The search for useful antiviral drugs is made difficult by the very close relationship and dependence of viruses upon host cell biosynthetic processes, the fact that total virus multiplication in the host is nearly complete by the time symptoms appear, and the emergence of drug-resistant mutants. Several compounds have been found which inhibit both cellular and viral biosynthetic events. For example, puromycin and cycloheximide block both cellular and viral protein synthesis, whereas actinomycin D inhibits transcription of DNA to mRNA. Also, (α-hydroxybenzyl)benzimidazole (HBB) and guanidine inhibit enterovirus and picornavirus multiplication. Although these drugs are not suitable for human use, they are valuable tools for use in elucidating the mechanisms involved in virus replication (Chapter 3).

It is now well-accepted that there are several viral-specific sites con-

Table 23–6. Mechanism of Action of Prophylactic and Therapeutic Agents

Site and Action	Example
Neutralization of extracellular virus	Specific Ab*
Inhibition of virus penetration	Amantadine (Symmetrel)*
Inhibition of viral nucleic acid synthesis	Iododeoxyuridine (IUdR, Stoxil), trifluoromethyldeoxyuridine, arabinosyl cytosine (ara-C), arabinosyl adenine (ara-A)
Inhibition of viral transcription	actinomycin D
Inhibition of viral translation	Thiosemicarbazones, methisazone, puromycin, cycloheximide
Inhibition of virus release	Specific Ab
Inhibition of cytopathic or histopathic effects	? Anti-inflammatory drugs
Suppression of symptoms in the host	? Anti-inflammatory drugs
Enhancement of host recovery and/or immune responses	Agents that stimulate interferon*

*Mainly effective only as prophylactic agents.

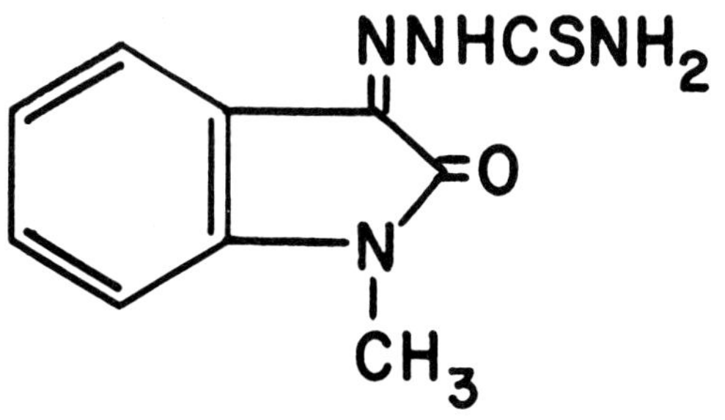

Fig. 23–1. Structural formula of N-methyl-isatin-β-thiosemicarbazone (Methisazone).

cerned with virus replication which can be attacked by antiviral drugs (Table 23–6). The ideal antiviral drug is one which would inhibit intracellular viral multiplication without harming the host cells. An alternative approach to therapy is to seek means of counteracting the effects of viral pathogenesis rather than the virus itself. Since the pathogenesis of certain viral infections (e.g., viral exanthems) is concerned in part with allergic hypersensitization and other immune phenomena (see Chapter 22), anti-inflammatory drugs have been found to be beneficial.

1. Thiosemicarbazones

The thiosemicarbazones (Fig. 23–1) are of special interest because they provide a link between bacterial and viral chemotherapeutic agents (see Chapter 26). Originally, the only known antimicrobial effects of the thiosemicarbazones was their tuberculostatic activity. It was later found that methyl-isatin-β-thiosemicarbazone (Methisazone) was active against vaccinia or smallpox viruses when given orally or subcutaneously to infected mice. Further experimental studies in human smallpox contacts indicated that methisazone is effective in reducing the incidence of smallpox by 75–95 percent of household contacts. Vaccinia reactions in eczematous patients (eczema vaccinatum) and progressive generalized vaccinia (vaccinia gangrenosa) in patients with defective cellular immunity are unaffected by hyperimmune serum but have been treated successfully with Methisazone. The drug acts by preventing the translation of viral-specific mRNA into capsid proteins and virus maturation is halted.

2. Amantadine

This compound is a synthetic primary amine that has specific prophylactic activity against certain strains of influenza A2 viruses and rubella virus. Amantadine (Symmetrel) (Fig. 23–2) appears to act in an early stage of virus infection by inhibiting uncoating of virus particles after their entrance into the host cell. It has been clearly demonstrated in several double-blind studies that Symmetrel given twice daily in 100 mg doses will decrease the clinical attack rate of natural or experimentally induced influenza in young adults by 50 percent. In addition, long-term studies have shown that treatment of older patients with similar doses of Symmetrel is associated with only minimal to moderate toxicity. Despite these favorable data, the drug is only prophylactic and, as a consequence, is not effective for treating clinical influenza or for other viral respiratory infections (e.g., rhinovirus). However, it is possible that

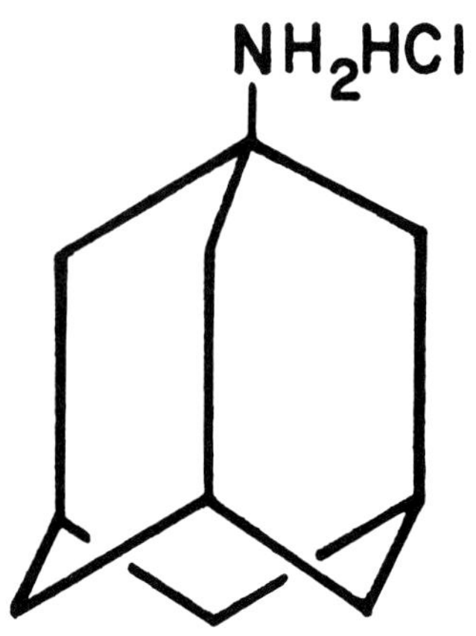

Fig. 23–2. Structural formula of amantadine hydrochloride (Symmetrel).

treatment of debilitated persons with the recommended doses of Symmetrel may reduce the incidence of pneumonia and death.

3. Iododeoxyuridine

The demonstration that iododeoxyuridine (IUdR, idoxuridine, Stoxil) inhibits vaccinia and herpesviruses in tissue culture, and the successful clinical application of this drug in the early treatment of vaccinial or herpetic eye infections in man provided the first example of a rational approach to the development of an antiviral drug. IUdR is a halogenated pyrimidine (Fig. 23–3) which is readily phosphorylated by the viral or cellular thymidine kinase. The phosphorylated derivative can become incorporated into newly synthesized DNA virus resulting in the production of abnormal proteins. Although the drug could have the same effect on cellular DNA synthesis, viral DNA synthesis in corneal cells proceeds at an accelerated rate and is more vulnerable to the action of IUdR than DNA synthesis by the slowly growing corneal cells. For this reason, IUdR should be applied only to corneal lesions caused by herpesviruses. Unsuccessful treatment with IUdR is usually due to emergence of drug-resistant viral mutants or progression of the lesion to the deep stromal layers of the cornea where infectious virus does not normally multiply. It should be pointed out that recovery from the disease depends on the natural defense mechanisms of the patient as well as on the drug. Since infection of the cornea with herpesviruses is a serious disease in man which often leads to blindness in untreated cases, IUdR is a potentially valuable drug for treating this disease. However, it is

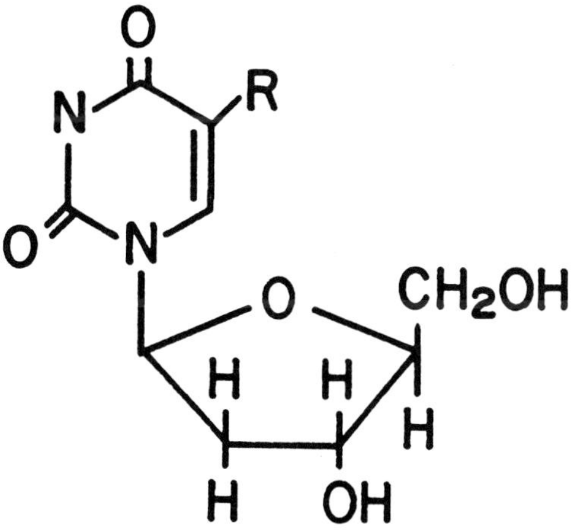

$R = CH_3$: Thymidine
I: 5-Iododeoxyuridine
F: 5-Fluorodeoxyuridine
Br: 5-Bromodeoxyuridine

Fig. 23–3. Structural formula of iododeoxyuridine (IUdR, Stoxil) and other 5-substituted deoxyuridines.

necessary to apply the drug repeatedly to the affected site and to use it only topically because systemic administration could result in neoplastic changes, infertility, or serious genetic mutation.

4. Arabinosyl Cytosine and Arabinosyl Adenine

These drugs are pyrimidine nucleosides like IUdR but, in contrast to IUdR, arabinosyl cytosine (ara-C) and arabinosyl adenine (ara-A) contain a metabolically normal base but an abnormal sugar (arabinose in place of ribose). Ara-C (Fig. 23–4) and ara-A (Fig. 23–5) both inhibit the multiplication of viruses (e.g., herpesviruses and vaccinia) by interfering with viral DNA polymerases as well as the ribonucleotide reductase which catalyzes the synthesis of deoxyribonucleotide precursors of DNA synthesis.

Adenine arabinoside is the drug of choice for clinical treatment of herpesvirus encephalitis. Following systemic administration, ara-A is rapidly deaminated to hypoxanthine arabinoside which retains significant biologic activity, unlike uracil arabinoside, the by-product of ara-C metabolism. A comparison of the antiviral activity of IUdR, ara-C, and ara-A in experimental DNA virus infections in animals indicates that ara-A exhibits superior therapeutic activity.

5. Acycloguanosine

Acycloguanosine (acyclovir, Zovirax) is one of the newest antiherpes drugs (Fig. 23–6) with high selectivity. The drug is rapidly converted to

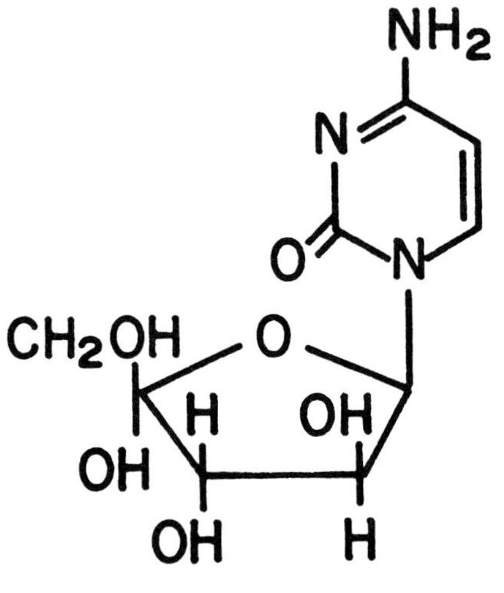

Fig. 23–4. Structural formula of arabinosyl cytosine (ara-C).

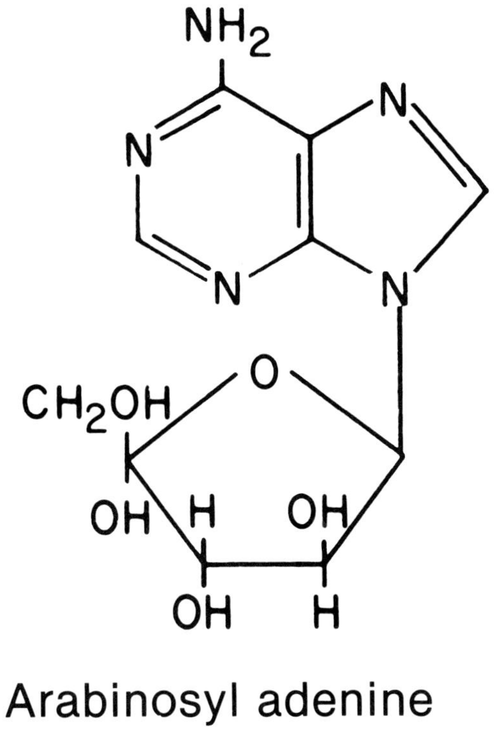

Fig. 23–5. Structural formula of arabinosyl adenine (ara-A).

an active form by herpesvirus thymidine kinase and cellular di- and triphosphate kinases. The activated acycloguanosine triphosphate markedly inhibits viral DNA polymerase activity. The drug has proved effective in the treatment of herpes keratitis and herpetic skin lesions including zoster (shingles).

D. INTERFERON

By definition, interferon (IFN) consists of a family of low molecular weight glycoproteins (MW = 12,000 or 24,000) which are stable to acid pH (pH 2.0). They are produced by living cells in response to a viral infection or other inducers such as fungal extracts containing the mycophage RNA (helenine and statalon). It is well-recognized now that nonviral agents (e.g., *Hemophilus sp.*, endotoxin, rickettsiae) and synthetic polyriboinosinic acid:polyribocytidylic acid (poly I:C) are also capable of inducing interferon in cells. It is now recognized that humans and animals can produce at least three forms of interferon (e.g., fibroblast, leukocyte and immune) which differ in antigenicity, isoelectric point and molecular weight. The three forms of human interferon have

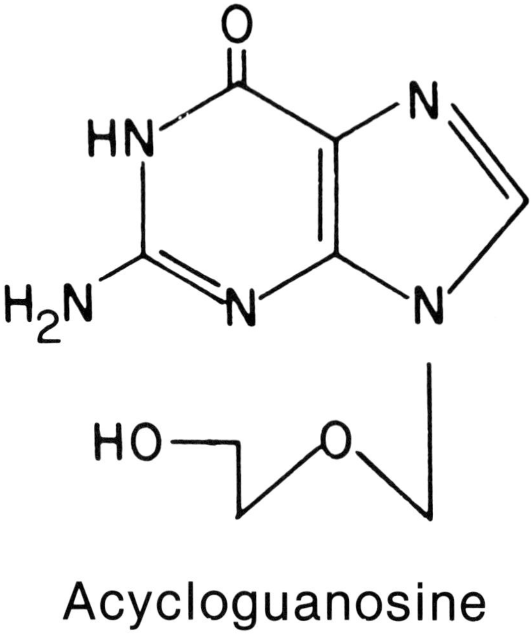

Fig. 23–6. Structural formula for acycloguanosine (acyclovir, Zovirax).

Table 23–7. Nomenclature of Human Interferons and Sensitivity to pH 2.0

Interferon, IFN	Sensitivity to pH 2.0
α, leukocyte, Le, type 1	Stable
β, fibroblast, Fe, type 1	Stable
γ, immune, type 2	Labile

recently been reclassified as IFN-α, Type 1, leukocyte (Le); IFN-β, Type 1, fibroblast (Fe) and IFN-γ, immune, type 2 (Table 23–7).

IFN's are host cell specific, not virus specific, in both their production and action. IFN types α and β act by inhibiting the multiplication of both RNA and DNA viruses, irrespective of the nature of the inducing agent. There are three important aspects in discussing the antiviral activity of interferon: (1) production, (2) nature, and (3) mechanism of action (Fig. 23–7).

1. Production

During a natural viral disease, the infecting virus nucleic acid stimulates the cell to synthesize IFN. IFN is probably the first host defense

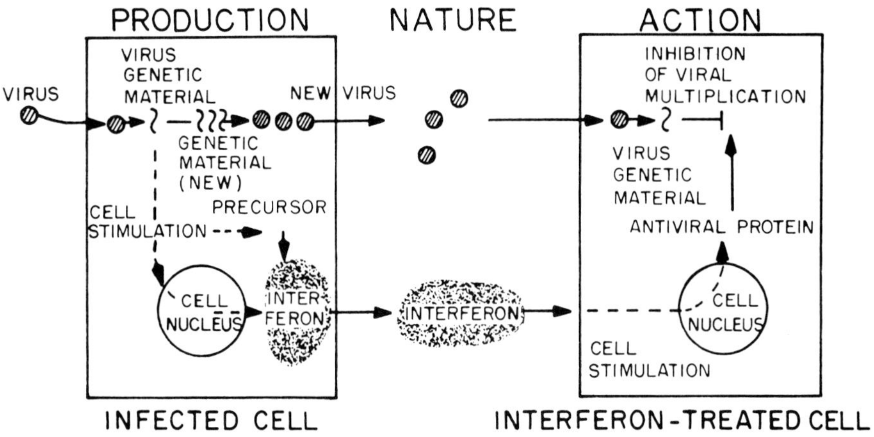

Fig. 23–7. Production, nature, and mechanism of action of interferon against viruses. During infection of a cell, virus nucleic acid either derepresses a cellular gene(s) to produce interferon or induces the release of an interferon precursor molecule. Interferon is then released into the extracellular spaces and rapidly attaches to membrane receptors of surrounding host cells where it derepresses a cellular gene(s) to produce a second component (AVP) which inhibits viral protein synthesis.

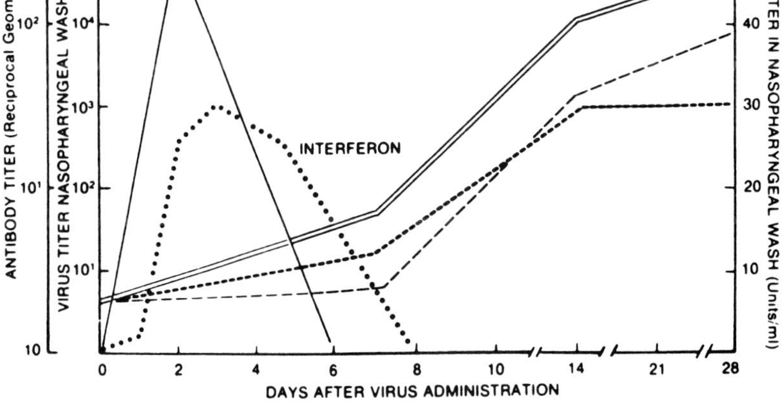

Fig. 23–8. Interferon is the earliest defense the host has against viruses. (From a study by Murphy, B., et al., National Institutes of Health.)

mechanism induced by a viral infection (Fig. 23–8). The early appearance of IFN is important in the ultimate outcome of virus infection. The genetic information for IFN synthesis resides in the host cell and not in the infecting virus. For example, in cells infected with an RNA virus like parainfluenza, actinomycin D (inhibitor of DNA-dependent RNA synthesis) blocks the production of IFN but does not prevent parainfluenza virus multiplication. IFN does not directly inactivate viruses; therefore, infected cells may continue to produce more virus and IFN simultaneously. However, the IFN they produce is released like a hormone, is rapidly taken up by uninfected cells where it induces resistance to virus replication, and thus promotes recovery from viral diseases (Fig. 23–7). In the absence of virus infection or a nonviral inducer, IFN production normally remains repressed in the cell. Experiments to measure the IFN yield in cell cultures derived from various organs and exposed to different strains of influenza virus showed that both viruses and cells differ in their capacity to induce or produce IFN.

2. Nature

Some general properties of IFN are listed in Table 23–7. Different viruses apparently induce the same class of IFN in human cells. IFN induced by one virus (e.g., influenza virus) will protect against many different DNA and RNA viruses (e.g., herpesviruses, polioviruses). In contrast to this lack of virus specificity, IFN is considered host-species specific; e.g., chick interferon will protect chickens but not humans from viral disease.

3. Mechanism of Action

It has been postulated that several cellular proteins are involved in interferon action: (1) a receptor site protein reponsible for recognizing an interferon-inducing molecule, (2) interferon itself, (3) a repressor of interferon synthesis, (4) an endonuclease, and (5) a protein kinase. The mechanism of action of IFN requires the production of new antiviral proteins responsible for antiviral activity since inhibitors of mRNA (actinomycin D) or protein (cycloheximide) synthesis completely block IFN induction of antiviral activity. IFN by itself has absolutely no effect on viruses. Once IFN binds to specific cell surface receptors, it induces a signal that is transmitted to the cell nucleus. Specific cellular genes are derepressed that code for endonuclease and protein kinase. The action of interferon-induced endonuclease and protein kinase activities are inhibition of protein synthesis (Fig. 23–9). Interestingly, viruses do not acquire resistance to IFN. IFN also possesses nonantiviral activities, e.g., inhibition of tumor cell growth, enhancement of antibody secretion and cytotoxicity, inhibition of T- and B-lymphoid cell proliferation and enhancement of phagocytic macrophage activity.

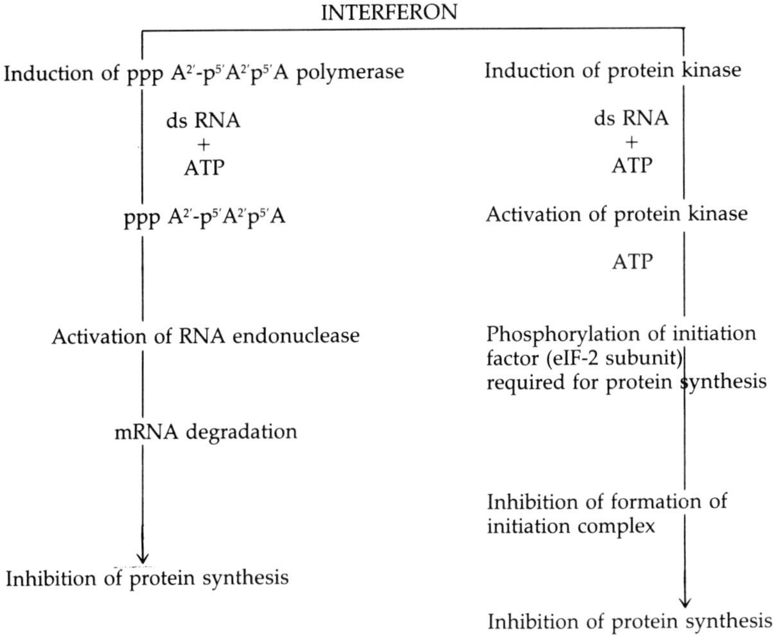

Fig. 23–9. Two suggested mechanisms of action of IFN on viral protein synthesis. One mechanism involves induction of an enzyme that is activated by ds RNA and ATP. The activated enzyme synthesizes an unusual adenine trinucleotide with a 2′-5′ phosphodiester linkage. The trinucleotide activates an RNA nuclease. The other mechanism involves induction of a protein kinase that inactivates an initiation factor eIF-2 by phosphorylating one of the subunits. Both mechanisms inhibit protein synthesis.

E. CLINICAL APPLICATION OF INTERFERON

Interferon is synthesized during most virus infections. It has been detected in the serum of patients with mumps, yellow fever, chickenpox, and influenza and in nonimmune individuals after vaccination with an attenuated strain of measles virus. Interferon has several advantages for clinical use: (1) it is weakly antigenic; (2) repeated doses can be administered; (3) it has little or no toxicity; (4) it is a highly active biologic material. A major disadvantage is that animal IFN are inactive or at most only slightly active in humans; therefore, all IFN used in man must be made from human or possibly primate cells. The production of large quantities of IFN for human use is being approached using recombinant DNA technology.

REFERENCES

Ankel, H., Krishnamurti, C., Bescericon, F., Stefanos, S. and Falcoff, E.: Mouse fibroblast (type I) and immune (type II) interferons: Pronounced differences in affinity for gangliosides and in antiviral and antigrowth effects on mouse leukemia L-1210R cells. Proc. Natl. Acad. Sci. (U.S.A.), *77*:2528, 1980.

Becker, Y. and Hadar, J.: Antivirals 1980—An update. Prog. Med. Virol., *26*:1, 1980.

DeMaeyer, E. and DeMaeyer, Guignard, J.: Interferons. Comprehensive Virology, *15*:205, 1979.

Koplan, J.P., and Axnick, N.W.: Benefits, risks and costs of viral vaccines. Prog. Med. Virol., *28*:180, 1982.

Krim, M.: Towards tumor therapy with interferons, Part I. Interferons: Production and Properties. Blood, *55*:711, 1980.

Miyamoto, N.G. and Samuel, C.E.: Mechanism of interferon action. Interferon-mediated inhibition of reovirus mRNA translation in the absence of detectable mRNA degradation but in the presence of protein phosphorylation. Virology, *107*:461, 1980.

Rubin, B.Y. and Gupta, S.L.: Interferon-induced proteins in human fibroblasts and development of the antiviral state. J. Virol., *34*:446, 1980.

Scott, G.M. and Tyrrell, D.A.J.: Interferon: therapeutic fact or fiction for the '80s? Br. Med. J., *280*:1558, 1980.

Stewart, W.E., II, and Havell, E.A.: Characterization of a subspecies of mouse interferon cross-reactive on human cells and antigenically related to human leukocyte interferon. Virology, *101*:315, 1980.

Taniguchi, T., Fujii-Kuriyama, Y. and Muramatsu, M.: Molecular cloning of human interferon cDNA. Proc. Natl. Acad. Sci. (U.S.A.), *77*:4003, 1980.

White, D.D.: Antiviral chemotherapy, interferons and vaccines. *In* Monographs in Virology. Vol. 16. J.L. Melnick (ed.). New York, S. Karger, 1984, pp. 1–112.

Chapter 24

Principles of Laboratory Diagnosis

A. INDICATIONS FOR LABORATORY DIAGNOSIS IN CLINICAL VIROLOGY

In recent years the laboratory diagnosis of viral infections has gained increasing importance and value in clinical medicine. Formerly, specific viral diagnosis was based almost exclusively upon recovery and identification of the etiologic agents in living host systems or by demonstration of a rise in viral antibody levels over the course of the patient's illness, both of which are expensive and relatively slow to yield results. Since viral diagnosis was usually retrospective, and since specific methods for treating viral diseases were not available, laboratory diagnosis was often considered to be largely an academic exercise and of limited assistance in guiding physicians in patient management.

However, newer methods now permit direct demonstration of a variety of viruses in clinical materials, and more rapid identification of viruses recovered in laboratory hosts. Thus, laboratory assistance in viral diagnosis is being increasingly sought to aid in patient management. Treatment and prophylaxis of viral infections with antiviral agents and management of pregnant women exposed to rubella or of high-risk, immunocompromised individuals exposed to herpesvirus infections are examples where rapid and specific viral diagnosis is critical. Specific identification of a viral infection of the central nervous system, eye, or genital tract may prevent indiscriminate use of antibiotics. Also, specific virus identification may be of prognostic value, as in distinguishing between meningoencephalitis due to herpes simplex virus or that due to an echovirus. In the long run, specific viral diagnosis educates physicians to associate particular viruses with certain clinical syndromes. Diagnostic virology has done much over the years to clarify the etiology of various central nervous system, respiratory, ocular, gastrointestinal, hepatic and perinatal diseases, as well as certain exanthemata and syndromes occurring in immunosuppressed individuals.

From a public health standpoint, viral diagnosis is essential for the surveillance and control of epidemic diseases such as influenza, poliomyelitis, viral hepatitis, certain togavirus (arbovirus) infections, and highly dangerous diseases such as Lassa fever. Rapid diagnosis in index cases may initiate immunization programs or suitable containment measures. Antibody surveys are also important in determining immunity status, efficiency of immunization programs, and prevalence of particular virus infections in various population groups.

B. APPROACHES TO LABORATORY DIAGNOSIS OF VIRAL INFECTIONS

There are three basic approaches to the laboratory diagnosis of viral infections (Table 24–1). These are (1) direct examination of clinical ma-

Table 24–1. Approaches to the Laboratory Diagnosis of Viral Infections

1. Direct demonstration of virus or viral Ag in clinical specimens—Rapid viral diagnosis.
   a. Electron microscopy
   b. Immunologic methods
      1) Untagged Ab
         Immune electron microscopy (IEM)
         Immunodiffusion (ID)
         Counterimmunoelectrophoresis (CIEP)
      2) Tagged Ab
         Immunofluorescence (IF)
         Enzyme immunoassay (EIA)
         Radioimmunoassay (RIA)
2. Isolation and identification of virus
   a. Cell culture host systems
      1) Rapid identification by IF, EIA, CIEP, RIA
      2) Conventional identification by neutralization, hemagglutination-inhibition (HI), hemadsorption-inhibition (HAd-I) or complement fixation (CF)
   b. Suckling or adult mice
      1) Rapid identification by IF, EIA, CIEP, RIA
      2) Conventional identification by neutralization, HI, CF
   c. Embryonated eggs
      Identification by HI, CF, neutralization
3. Serologic procedures
   a. Applicable to routine or large-scale use
      CF, HI, EIA, RIA, single radial diffusion, Hemolysin-in-gel, passive hemagglutination.
   b. Special or small-scale use
      Neutralization, indirect IF or IP, HAd-I, IEM

terials for virus or viral Ag, (2) isolation of virus in living host systems and subsequent identification by immunologic methods, and (3) serologic diagnosis based upon the demonstration of a significant increase in viral Ab over the course of the patient's illness. Methods included under the first approach are those which permit rapid viral diagnosis without the need for virus propagation, and this is the area in which there has been the most interest in recent years. Direct methods permit detection and diagnosis of viral infections caused by agents which cannot be propagated in standard laboratory host systems, namely, the agents of hepatitis A and hepatitis B, the rotaviruses and parvo-like viruses which cause non-bacterial gastroenteritis, and human papilloma (wart) virus.

1. Direct Examination

Although direct demonstration of virus in clinical specimens is the most rapid and economical approach to viral diagnosis, it is limited by the small size of viruses, their close association with host material, and the relatively low concentration of virus at sites of replication accessible for specimen collection.

For successful detection of virus by electron microscopy (EM), the agent must be present at concentration greater than 10^6 particles per ml, and the virus must have a distinctive morphology which is readily distinguishable from debris. EM with negative staining is most applicable to detection of herpesviruses and poxviruses in lesion material and to detection of rotavirus gastroenteritis agents in fecal specimens, as well as to detection of hepatitis B Ag in serum. This approach can identify major groups of viruses on the basis of their morphologies but cannot, for example, differentiate between different herpesviruses such as herpes simplex and varicella-zoster virus.

Direct identification methods for which the specimen is treated with specific viral Ab permit detection and identification of the viral agent in a single step. Sensitivity is increased over that of direct EM because virus is amplified by the specific Ab, either by aggregation or through a "tag," or label, on the Ab. This approach requires high-titered specific antisera, and one must have some idea of the agent involved, since it is feasible to test a specimen against only a few appropriate antisera.

For immune electron microscopy, the specimen is mixed and incubated with viral Ab, and with a known negative serum for a control, negatively stained with phosphotungstic acid, and then observed by EM for the presence of virus-Ab complexes. Virus-Ab complex formation aids in concentration, detection, as well as identification of the virus. This approach permitted the initial detection of hepatitis A virus and the Norwalk gastroenteritis virus, which cannot be isolated in laboratory host systems.

Direct examination of clinical specimens for viral Ags which react with viral antisera to produce immunoprecipitates in gels is of limited value because of the high concentration of Ag required to produce a visible precipitate in immunodiffusion or counterimmunoelectrophoresis systems, but these methods are applicable to detection of hepatitis B surface Ag in serum, where Ag may attain a concentration of greater than 10^{12} particles per ml.

Viral Abs used for direct examination may be tagged with a fluorochrome, an enzyme or a radioisotope to aid in the detection of virus-Ab complexes.

2. Immunofluorescence Techniques

Immunofluorescence (IF) staining is based upon tagging Ab with fluorescein isothiocyanate and demonstrating virus-Ab complexes in the specimen by microscopic examination with ultraviolet illumination. Either the direct or indirect method may be employed. For the direct procedure the specimen is treated with labeled specific viral Ab; for the indirect method the specimen is reacted with unlabeled viral Ab which in turn is detected through the use of fluorescein-labeled immune globulins directed against Ig of the species of antiserum used in the initial reaction. The direct method is generally more specific, since fewer reagents are involved, whereas the indirect method is more sensitive, since the intermediate antiserum increases the surface area available for attachment of the labeled immunoglobulins. Another IF technique, with high sensitivity for detection of certain herpesvirus Ags, is the anticomplement immunofluorescence (ACIF) method, for which virus-Ab complexes are detected through the addition of C, followed by fluorescein-labeled Abs to C. IF staining has been the most widely applied method for rapid viral diagnosis, and it is suitable for demonstration of a variety of respiratory viruses, measles virus and rubella virus in cellular material from nasopharyngeal exudates, for detection of herpesviruses and poxviruses in cells from vesicular lesions, and for demonstration of viral Ags such as rabies, herpes simplex, and measles directly in brain tissue.

3. Enzyme Immunoassays

Enzyme immunoassays (EIA) employ Abs labeled with an enzyme rather than fluorescein, and labeled Abs complexed with virus are detected by the addition of a substrate upon which the enzyme acts to produce a colored product. Theoretically EIAs are more sensitive than IF staining, since the enzyme label has a continuous action on the substrate, producing additional reaction product and thus amplifying the initial reaction.

If the EIA system utilizes a substrate which produces an insoluble reaction product, the colored product at the site of the Ag-Ab reaction is detected microscopically, using an ordinary light microscope. The so-called immunoperoxidase (IP) staining method employing horseradish peroxidase as an enzyme label and various substrates yielding insoluble colored reaction products, has been applied to rapid viral diagnosis in the same manner as IF staining, using either direct or, more commonly, indirect procedures. One problem has been the endogenous peroxidase activity of certain types of host tissue, which causes nonspecific reactivity. However, methods have been developed for destroying this activity and thus permitting valid examination by IP staining.

The so-called enzyme linked immunosorbent assay (ELISA) system employs enzyme-labeled Abs (either peroxidase or alkaline phosphatase) and substrates which give a soluble colored reaction product. The reaction product was observed visually. A more sensitive method is to measure the quantity of product by spectrophotometry. For virus detection, specific viral Ab is coated onto a solid phase (plastic cup or bead), the test material is added, and virus present is bound to the "capture" Ab. Bound virus is then detected through the use of a second viral antiserum ("detector" Ab) labeled with an enzyme, or by unlabeled viral Ab from a different species than the "capture" Ab, followed by labeled Abs directed against Ig of the species of the detector Ab. ELISA systems have been particularly useful for detection of viruses which cannot be cultivated in laboratory host systems, such as hepatitis A and B viruses and human rotaviruses. They have also been applied to direct detection of herpes simplex and respiratory syncytial virus.

Antibodies labeled with a radioisotope, usually ^{125}I, can be used in the same way as enzyme-labeled Abs, in either direct or indirect systems, to detect viral Ags in clinical materials. Binding of detector Ab to virus is demonstrated by counting the radioactivity of the specimen in a gamma counter after completion of the reaction, and demonstrating a significant increase in radioactivity over that of negative controls. Again, radioimmunoassay (RIA) has had its widest application in detection of hepatitis A and B viruses and certain gastroenteritis agents. It is also applicable to detection of herpes simplex and measles viruses directly in brain tissue.

For those agents which can be isolated in laboratory host systems, direct methods are generally not as sensitive as virus isolation procedures for virus detection. Small amounts of virus which might be missed by direct examination are amplified by replication in a susceptible host. In some instances, however, particularly late in the course of infection, positive results may be obtained by direct examination but not by virus isolation attempts, possibly because of complexing of virus with specific Ab, and possibly because viral infectivity is more labile than are viral particles or Ags.

4. Neutralization of Virus Infectivity

Neutralization tests are based upon the ability of specific Ab to combine with virus and render it noninfectious. The isolate is mixed with known antisera, incubated, and then the mixtures are inoculated into a host system in which the virus produces an observable effect. Identification is based upon suppression of the effect (CPE, hemadsorption, death, etc.) by a specific viral antiserum, but not by a control serum. The system requires the use of living host systems, and a relatively long incubation period to allow for the virus to exert its effect in the host. Neutralization tests to identify viruses belonging to large groups are facilitated by the use of viral antisera combined into pool schemes in which a type-specific antiserum is incorporated into one, two or three pools, and identification is made by demonstrating neutralization in the pool or pools sharing a common type-specific antiserum.

5. Hemagglutination and Hemadsorption

Certain in vitro tests are more rapid and economical than neutralization tests and can be applied to identification of viruses which agglutinate red blood cells. If sufficiently high levels of viral hemagglutinins are released from host cells into cell culture medium or egg fluids, the fluids can be used as Ag preparations for typing the isolate by hemagglutination-inhibition (see below) against known viral antisera; this approach is useful for strain-specific identification of influenza viruses. If levels of hemagglutinin in the culture fluids are not high enough to permit typing by hemagglutination-inhibition (HI), but if the virus adsorbs red cells to the infected host cells (hemadsorbs), it can be identified by pretreatment of the infected cells with different viral antisera, and demonstrating inhibition of the hemadsorption reaction by a given viral antiserum, but not by control or heterologous sera; this phenomenon is called hemadsorption-inhibition (HAd-I), and is used primarily for identification of parainfluenza viruses and low-titered influenza virus isolates.

6. Complement Fixation

Virus identification is sometimes done by preparing a complement-fixing (CF) Ag from the isolate and testing it against known viral antisera. However, this is relatively insensitive, and may require considerable effort to produce an adequately potent Ag. Identification generally can be accomplished much more readily by IF or enzyme immunoassay.

C. VIRUS ISOLATION AND IDENTIFICATION

1. Specimen Collection and Processing

The success of viral diagnosis depends to a large extent upon the quality of the specimens examined. The pathogenesis of the disease and the target organs of a particular virus determine the specimens which are most likely to yield virus. Chapters dealing with individual viruses should be consulted to determine the most appropriate test specimens. Specimens should be collected aseptically and placed in sterile containers to ensure that they are not contaminated with viruses from other sources or with microbial agents which might hamper viral examination. Specimens should always be accompanied with specific information on the date of onset of illness, type of disease suspected, major clinical findings, any unusual travel or exposure, and immunizations; this guides the laboratory in the selection of appropriate tests and in the interpretation of results.

One of the most important factors influencing the success of viral diagnosis is collection of specimens as soon as possible after onset of illness, when virus is being excreted at high levels and has not yet been bound by Ab. Most viruses are recoverable from the nasopharynx and throat, or from vesicular lesions, for only 3–4 days after onset of illness. Virus excretion in the intestinal tract may be more prolonged in some instances, but diminishes greatly after the first week of illness. Since serological diagnosis of viral infections is based upon demonstration of an increase in Ab titer over the course of the patient's illness, it is essential to collect an acute-phase blood specimen at onset of illness to examine with a convalescent-phase serum collected 14–21 days later. If the acute phase specimen is not collected early enough, a substantial Ab response may already have occurred, making it impossible to demonstrate a significant Ab titer increase (4-fold or greater) between the acute and convalescent phase specimens.

If swabs or lesion materials are collected into a holding medium, the medium should contain protein (known to be free from viral Abs or inhibitors) to protect against loss of viral infectivity. Virus recovery is most successful if specimens are inoculated into suitable host systems as soon as possible after collection. It is well recognized that freezing and thawing specimens reduces virus isolation rates. Specimens should be held at 4°C if they can be inoculated within 24 hours; for longer periods they should be kept at −70°C.

2. Isolation of Virus

Most virus isolation work is done in cell culture systems. However, laboratories studying togavirus (arbovirus) and certain group A coxsackievirus infections require suckling mice. Laboratories attempting to isolate new strains of influenza viruses need embryonated chicken eggs.

No single cell culture system is sensitive enough for recovery of all viruses likely to be encountered in clinical specimens. The combined use of primary monkey kidney cells and a human fetal diploid cell line is satisfactory for recovery of the most common human viruses.

Most types of specimens are treated with antibiotics before they are inoculated into laboratory host systems in order to suppress microbial contaminants which might also grow in the host system.

Growth of virus in inoculated cell culture systems is evidenced by degeneration of the cell cultures, in some cases by viral cytopathic effect (CPE), by hemadsorption of red cells to the infected cell monolayers for certain orthomyxoviruses and paramyxoviruses which do not produce a clear-cut viral CPE, or by interference with a cytopathic challenge virus in the case of rubella virus. Viral isolates are recognized by production of illness and death in suckling mice. In embryonated chicken eggs viral isolates may cause death of the embryo, lesions on the egg membranes, or simply presence of viral Ags in egg membranes or fluids.

If emphasis is being placed upon the recovery of only a few types of viruses, e.g., respiratory viruses in the orthomyxovirus or paramyxovirus groups, it is feasible to use IF staining to detect virus in inoculated cell cultures earlier than it can be demonstrated by CPE or hemadsorption.

The characteristics of the viral CPE, a positive hemadsorption reaction, or characteristics of the viral effect in animals or embryonated eggs provides a clue to the major group to which the isolate may belong, and guides in the selection of the most appropriate tests and immune reagents to be used for specific identification. Identification is based upon demonstrating an immunologic reaction between the isolate and known viral Abs. Determining the physical and chemical properties of a virus, such as size, ether sensitivity and nucleic acid content is generally not a useful early step in virus identification. These procedures require considerable time and effort and give little significant information beyond that obtained from the viral CPE or hemadsorption reaction.

3. Identification of Virus Isolate

Once a virus is isolated in a laboratory host system, its identification can be expedited greatly through the use of immunologic methods such as IF, EIA, hemagglutination-inhibition (HI), or hemadsorption-inhibition (HAd-I) with specific viral antisera. These methods are more sensitive and specific for identification of virus than direct examination since the virus concentration is increased over that in the original specimen by replication in susceptible host cells. Also, virus is freed from host material or Ab which might interfere with virus detection in the original specimen. Identification by these methods can be accomplished within a few hours after one has a clue as to the possible identity of the virus by its effect on the host system. Unfortunately, these methods are not feasible for identification of viruses in the large groups such as enteroviruses or rhinoviruses which consist of many distinct immunotypes. Laboratory personnel must resort to more time consuming and expensive neutralization tests for typing these isolates.

Table 24–2 outlines steps in the identification of the most common human viruses. It should be emphasized that the source of the specimen also plays a large role in determining the viruses to be considered. For example, viruses with essential lipids (orthomyxoviruses and herpesviruses) or which are sensitive to low pH (rhinoviruses) are rarely encountered in specimens from the enteric tract, and relatively few kinds of viruses would be expected in vesicular lesion specimens.

D. PROBLEMS IN VIRUS ISOLATION AND IDENTIFICATION

Problems which may be encountered in virus isolation and identification include possible toxicity of the specimen for the host system, which may mask the specific effect of viruses. This usually can be overcome with a subpassage of the material. Simian virus contaminants in monkey kidney cells may be mistaken for isolates from the specimen, as may mycoplasma contaminants in host cell cultures. These possibil-

Table 24–2. Detection and Identification of Common Viral Isolates

Host System	Evidence of Infection	Viruses to be Considered	Specific Identification Procedure
Cell cultures, Primary monkey kidney (MK) or human fetal diploid strain	CPE typical of picornaviruses	Enterovirus,	Neutralization
		Rhinovirus	Demonstration of lability at low pH, neutralization
	CPE typical of adenoviruses	Adenovirus	IF[a] for group identification, neutralization or HI for typing
	CPE typical of herpesviruses	Herpes simplex	IF or neutralization
		Varicella-Zoster	IF
		Cytomegalovirus	IF (ACIF Method)[b]
	CPE typical of reoviruses	Reovirus	Neutralization, HI
	"Syncytial" CPE with hemadsorption of guinea pig red cells	Parainfluenza	IF, HAd-I, HI, neutralization
		Mumps	IF, HAd-I, HI, neutralization
		Influenza	IF, HAd-I, HI, neutralization
	"Syncytial" CPE, neg. hemadsorption	Respiratory syncytial	IF, neutralization
		Measles	IF, neutralization
	Little or no CPE, hemadsorption of guinea pig red cells	Parainfluenza	IF, HAd-I, HI, neutralization
		Mumps	IF, HAd-I, HI, neutralization
		Influenza	IF, HAd-I, HI, neutralization
	No CPE, rubella suspected	Rubella	IF, interference in MK
Mice <1 day of age	Paralysis, death	Coxsackievirus	Neutralization
		Herpes simplex	IF, neutralization
		Togavirus (Arbovirus)	IF, neutralization, HI, CF
Embryonated eggs	Death or lesions on membranes, no hemagglutinins	Herpes simplex	IF, neutralization
		Vaccinia	IF, neutralization
	Hemagglutinins in fluids	Influenza	HI, neutralization
		Mumps	HI, neutralization
		Newcastle disease	HI, neutralization

[a]Where IF is indicated, EIA (immunoperoxidase staining) and RIA would also be appropriate.
[b]ACIF: Anticomplement immunofluorescence method.

ities should be recognized, and problem isolates should be tested against antisera to the most likely viral contaminants and cultured for mycoplasma. A portion of the original specimen should always be retained for retesting if necessary. The human origin and association of the isolate with the patient's illness can be strengthened by testing paired sera and demonstrating a 4-fold or greater rise in Ab titer to the isolate over the course of illness.

E. SEROLOGIC DIAGNOSIS EMPLOYING PATIENTS' SERA

In order to demonstrate a significant rise in Ab titer (4-fold or greater) to a given viral Ag over the course of the illness, it is essential to examine the acute and convalescent phase sera in the same test. Test-to-test variation inherent in most viral Ab assay systems makes this essential for valid interpretation of differences in Ab titer. Because of the delay in obtaining the convalescent-phase specimens, serologic diagnosis is retrospective, which limits its usefulness in some clinical situations. However, serologic procedures have the advantage of being more economical than virus isolation, and they are particularly useful for diagnosis of infections from which virus rarely can be isolated, e.g., nonfatal togavirus infections and certain respiratory infections in which virus is excreted for only a short period of time. It is anticipated that markedly more sensitive enzyme immunoassays may permit earlier detection of significant rises in Ab titer, within the first few days of illness.

The conventional methods for serologic diagnosis of viral infections are the neutralization, HI and CF tests. Major limitations of the first method are its expense and time required to obtain results; sensitivity and specificity are advantages. HI tests are based upon the fact that many viruses have sites on their surfaces (hemagglutinins) which attach to erythrocytes and agglutinate them (hemagglutination); however, combination of virus with specific Ab prevents this reaction, and thus Ab can be assayed by demonstrating the ability of the test serum to inhibit hemagglutination by a standard dose of virus. The major drawback of this test is the fact that sera also contain non-Ab substances which inhibit hemagglutination, and these must be removed before a valid Ab assay can be performed. Also, a number of viruses do not have the property of hemagglutination. The CF test is most widely used for viral serodiagnosis because of its versatility, broad reactivity and ability to demonstrate significant Ab titer rises; however, it is relatively insensitive, particularly for detecting Ab of long duration, and reliable CF Ags are not available for all important viruses.

1. Newer Techniques

Various newer techniques for detection of viral Abs have been described in recent years which offer possible advantages over the above conventional methods. Tests of the future should have improved sensitivity, specificity, speed and economy.

An alternative to the standard CF test system is the immune adherence hemagglutination (IAHA) test. This is based upon the fact that binding of C to a virus-Ab complex activates the C3 component so that it attaches to receptors on human erythrocytes and agglutinates them. The test gives higher Ab and Ag titers than those obtained by CF, and it is equally simple and versatile. IAHA has been particularly useful for detection of Abs to hepatitis A virus and certain gastroenteritis viruses.

Among the most promising new techniques for viral Ab assays are indirect enzyme immunoassays (ELISA methods) for which viral Ag is adsorbed to a solid phase consisting of a plastic cup or bead, test serum is added, and virus-Ab reactions are detected through the addition of enzyme-labeled Abs directed against the species of the test serum, followed by a substrate giving a soluble, colored reaction product. The system is relatively simple, more sensitive than most conventional assays, and non-Ab inhibitors do not present a problem. Nonspecific reactivity at lower serum dilutions has been encountered, but this can be reduced by the use of more highly purified Ags and suitable diluents to prevent nonspecific adsorption of reagents to the solid phase. These tests are applicable to a wide variety of viral agents, to detection of viral Abs in different subclasses, and the tests can be automated and attached to a computer for automatic print-outs.

Indirect RIA procedures based upon the same principle as the ELISA method, but using a radiolabeled anti-species globulin to detect Ab bound to viral Ag in the solid phase, are highly sensitive and specific, but they are most suitable for large-scale studies rather than for routine serology, because of the instability of the labeled reagent and the special conditions and equipment required for handling radioisotopes.

Simple viral Ab asays in which Ab levels can be estimated from tests on a single serum concentration are also receiving increasing attention, particularly for immunity studies and other Ab surveys. The single radial immunodiffusion system utilizes viral Ag immobilized in a gel and serum placed in wells cut in the gel. Antibody present in the test serum diffuses radially and produces an opalescent halo around the well, due to the increased light-scattering by the virus-Ab complexes. Antibody can be quantitated by measuring the reaction zone. The test is simple and useful in the field, since the reagents are stable. However, it requires high concentrations of purified or partially purified Ag in the gel. The hemolysis-in-gel system is similar in principle, but utilizes viral Ag coated onto erythrocytes immobilized in a gel. In the presence of C, Ab diffusing from wells in the gel lyses the Ag-coated erythrocytes and produces a zone of hemolysis around the well, with a diameter proportional to the content of Ab in the serum. The test is more sensitive than radial immunodiffusion, and it requires less Ag. However, the Ag-containing gel is less stable.

REFERENCES

Benjamin, D.R.: Use of immunoperoxidase for rapid viral diagnosis. *In* Microbiology 1975. Washington, D.C., American Society for Microbiology, 1975.

Emmons, R.W., and Riggs, J.L.: Application of immunofluorescence to diagnosis of viral infections. *In* Methods of Virology. Vol. 6. K. Maramorosch and H. Koprowski (eds.). New York, Academic Press, 1977.

Forghani, B.: Radioimmunoassay. *In* Diagnostic Procedures for Viral, Rickettsial and Chlamydial Infections. Vol. 5. E.H. Lennette and N.J. Schmidt (eds.). Washington, D.C., American Public Health Assoc., 1979.

Gardner, P.S., McQuillin, J., and Grandien, M.: Rapid Virus Diagnois. Application of Immunofluorescence. 2nd Ed. London, Butterworth, 1980.

Hawkes, R.A.: General principles underlying laboratory diagnosis of viral infections. *In* Diagnostic Procedures for Viral, Rickettsial and Chlamydial Infections. Vol. 5. E.H. Lennette and N.J. Schmidt (eds.). Washington, D.C., American Public Health Association, 1979.

Hsiung, G.-D., Fong, C.K.Y., and August, M.J.: The use of electron microscopy for diagnosis of viral infections. An overview. Prog. Med. Virol., *25*:133, 1979.

Kurstak, E., and Morisset, R. (eds.): Viral Immunodiagnosis. New York, Academic Press, 1974.

McIntosh, K.: Sense and nonsense in viral diagnosis—past, present and future. Curr. Top. Microbiol. Immunol., *104*:1, 1983.

Schmidt, N.J.: Laboratory diagnosis of viral infections. *In* Antiviral Agents and Viral Diseases of Man, G.J. Galasso, T.C. Merigan, and R.A. Buchanan (eds.). New York, Raven Press, 1979.

Chapter 25

Viral Respiratory Diseases

Viral respiratory infections can be inapparent or can present symptoms ranging from mild, self-limiting infections represented by the common cold to severe and even fatal pneumonias. Viruses associated with respiratory infections have a predilection for a particular level of the respiratory tract (e.g., nasopharynx, trachea, bronchi, bronchioles, or lungs). The level of the respiratory tract susceptible to infection by a given virus can be modified by the age of the individual, immune response, or immunosuppression. The relative frequencies of the different clinical syndromes caused by the major respiratory viruses are presented in Table 25–1.

A. COMMON COLD

The common cold is a syndrome caused by viral infection of the mucous membranes of the upper respiratory tract (URT). More specifically, the term refers to afebrile, acute coryza, or rhinitis of viral etiology. The main differences between the common cold and other viral or bacterial respiratory tract infections is the absence of fever and the relative mildness of systemic symptoms. In many cases, severe systemic viral infections begin with symptoms similar to the common cold. The common cold syndrome alone accounts for a loss of more than 200 million man days of work and school in a single year in the U.S.A.

1. Etiologic Agents

Over 113 antigenically different viruses can cause the common cold syndrome. *Rhinoviruses* account for 20–40 percent of acute coryza in young adults and 8–10 percent in children. *Coronaviruses* also appear to cause a substantial proportion of minor URT diseases, since up to 98 percent of a given population tested had Abs to one or more of the human respiratory viruses. The *parainfluenza* and *respiratory syncytial viruses,* which cause severe respiratory infections in children, usually produce only common colds in adults. Although the *adenoviruses* and *influenza viruses* may be associated with more severe infections, some individuals infected with these agents develop only acute coryza or rhinitis. *Coxsackievirus A* 21, 24 and *B* 4, 5, as well as *echovirus* types 11, 20, and possibly others, also cause the common cold. These mild illnesses probably represent the most common clinical manifestation of coxsackievirus and echovirus infections in infants and children and are frequently described as "summer grippe." These same agents cause less frequently other more severe illnesses, such as aseptic meningitis, epidemic myalgia, herpangina, myocarditis neonatorum, and various exanthems. *Reoviruses* have been isolated from children with mild respiratory infections as well as from some children with pneumonia and intestinal, hepatic, and central nervous system disease. However, the relationship of these viruses to clinical illness remains uncertain. During

Table 25–1. Clinical Syndrome

Virus	Coryza	Pharyngitis (Conjunctivitis; Otitis)	Croup	Bronchitis	Other
Influenza A, B	+ +[a]			+ +	+ + + +
Influenza C	+ +			+	
Parainfluenza 1, 3	+ + +	+ +	+ + +	+	+ + +
Parainfluenza 2	+		+ + + +		
Parainfluenza 4	+ + + +				
Respiratory syncytial	+ + +		+ +	+ + + +	+
Reovirus 1	+				
Adenovirus 1, 2, 5	+	+ +			+
Adenovirus 3, 4, 7, 14, 21	+ +	+ + + +		+	+
Coxsackie A 2, 3, 4, 6, 8, 10	+	+ + + +	+		+ +
A 21, 24	+ + + +	+			
Coxsackie B 2, 3, 4, 5	+	+	+	+	+ + + +
ECHO 4, 5, 7, 8, 9, 11, 19, 20, 25	+ +	+ +	+		+ +
Rhinovirus	+ + + +	+		+	
Coronavirus	+ + + +	+	+		

[a]The relative frequency of the different clinical syndromes resulting from infection with some of the different viruses responsible for the common respiratory infections is indicated by + to + + + +. Modified from Jackson, G.G.: Nonbacterial pharyngitis. *In* Cecil-Loeb Textook of Medicine. 13th ed. P.B. Beeson and W. McDermott (eds.). Philadelphia, W.B. Saunders Co., 1971, pp. 357–363.

human volunteer studies, most of the subjects inoculated intranasally with reovirus showed no signs of illness; some developed variable symptoms, and no clear-cut clinical picture of the infections could be established.

2. Clinical Symptoms

The portal of infection for these respiratory viruses is the nasopharynx. The incubation period is short, usually 1–3 days, during which time the virus proliferates in the nasal mucosa.

The major clinical symptoms (Table 25–2) of the common cold differ appreciably from person to person, but for any given individual they tend to be similar. Sneezing, headache, and malaise are the initial signs of infection and are followed by a chilly sensation, sore throat, and nasal congestion. Fever of any significant degree is absent, and the constitutional symptoms, if present, usually last only 1 to 2 days. As the cold progresses, the nasal discharge, which at the onset is clear and watery, may become mucopurulent. These symptoms usually run their course within 5–7 days. However, a cough may appear as a prominent symptom during the acute illness and may persist for several weeks. As the symptoms recede, virus excretion ceases.

A significant number of individuals can serve as asymptomatic carriers of viruses causing acute coryza.

3. Pathogenesis

Virus appears to be confined to the upper respiratory tract (nasopharynx). The pathologic changes in the mucous membranes of the respiratory passages include edema, hyperemia, transudation, and ex-

Table 25–2. The Syndrome of the Common Cold

Symptoms		Frequency %
Severe		
	Nasal discharge	100
	Nasal obstruction	99
Moderate		
	Sore or dry throat	96
	Malaise	81
	Postnasal discharge	79
	Headache	78
	Cough	76
Mild		
	Sneezing	97
	Feverishness	49
	Chilliness	43
	Burning eyes and mucous membranes	28
	Muscle aching	22

From: Jackson, G.G. The common cold. *In* Cecil-Loeb Textbook of Medicine. 13th Ed. P.B. Beeson and W. McDermott (eds.). Philadelphia, W.B. Saunders Co., 1971, pp. 358–361.

udation. During the acute phase of infection, immunoglobulins, especially IgA become more abundant in nasal secretions, and other substances of cellular origin accumulate. Picornaviruses (rhinoviruses, coxsackieviruses) cause more metaplasia and degeneration of cells from the nasal turbinates than the other viral causes (e.g., coronaviruses, adenoviruses, influenza, parainfluenza, respiratory syncytial viruses) of the common cold. However, tissue damage is rapidly repaired.

4. Immunity

Specific immunity develops following infection with any of the etiologic agents of the common cold. In human volunteers, resistance against reinfection with the same strain of rhinovirus has been shown to persist for a month to several years. Even if specific Ab does not prevent reinfection, it usually results in an attenuated illness. However, because of the large number of viruses that can cause the common cold and the highly specific protection provided by Ab, infection with a different virus or even different strains of the same virus is likely to occur.

5. Diagnosis

Diagnosis of the common cold usually is made on the basis of the characteristic clinical symptoms. It is important to recognize or exclude the possibility that the patient may have a more severe disease. Prodromal symptoms of other infections, as well as some allergic, vascular, and neoplastic diseases, can mimic the common cold.

Virus isolation provides the only satisfactory method for establishing a definitive diagnosis. Rhinoviruses are clearly distinguished from other picornaviruses by their sensitivity to inactivation at acid pH and stability at 50°C. Rhinoviruses which grow only in human cells (H strains) are more heat stable than those that multiply in monkey cells (M strains). Serologic tests employing the patients' paired sera are not practical because of the large number of virus types that can cause the common cold. In most instances, etiologic diagnoses are made only for epidemiologic purposes.

6. Epidemiology

The common cold is worldwide in its distribution, and during a given outbreak, the same virus often can be demonstrated even in isolated areas. Colds are most frequent in infants and children; the incidence of infection in adults declines significantly after middle age.

Although the common cold is spread by respiratory droplets and possibly by fomites contaminated with infected secretions, other factors also appear to be involved in disease transmission. There is support for the idea that colds increase following marked changes in temperature, humidity, or air pollution. However, the popular belief that cold weather, wet feet, and chilling increase susceptibility to common colds is not supported by the results from human volunteer experiments. The importance of person-to-person contact is indicated by epidemiologic studies performed in isolated areas; the incidence of URT infections remains low until contact is made with "the outside world," at which time there is a dramatic increase in the incidence of the common cold.

In the U.S.A., waves of common colds usually occur each year; one appears in the fall, another in midwinter, and a third in the spring. It has been established that asymptomatic carriers shed virus and can participate in the spread and perpetuation of infection.

7. Treatment and Prevention

Therapy is confined to nonspecific remedies such as aspirin and decongestants. At present, there are no drugs that are effective against the viruses that cause the common cold. Antibiotics effective against bacteria are useless and should not be given unless there are secondary bacterial complications.

B. ACUTE FEBRILE PHARYNGITIS, PHARYNGOCONJUNCTIVAL FEVER, AND ACUTE RESPIRATORY DISEASE

Acute febrile pharyngitis is an illness of only a few days duration that is characterized by a mild sore throat, fever, and cough, and sometimes by mild cervical lymphadenitis.

Pharyngoconjunctival fever resembles the foregoing illness but is accompanied by conjunctivitis, which may be unilateral; gastrointestinal symptoms occasionally develop, particularly in children.

Acute respiratory disease (ARD or recruit fever) is an influenza-like illness that lasts about a week and is characterized by fever, malaise, mild sore throat, cough, hoarseness, and rhinitis. The disease, which seldom occurs in civilian populations, appears almost exclusively in young adults shortly after their arrival in recruit camps.

1. Etiologic Agents

Acute febrile nonbacterial pharyngitis can be caused by almost any of the respiratory viruses (e.g., *adeno, coxsackie, influenza, parainfluenza, respiratory syncytial, echo,* and occasionally *rhinoviruses*).

Pharyngoconjunctival fever is caused by *adenovirus* types 3 and 7, and, less frequently, by types 4, 14, and 21.

2. Clinical Symptoms

The incubation period for adenovirus infections that result in pharyngoconjunctival fever, ARD, and febrile pharyngitis is 5–10 days. In the case of respiratory syncytial virus and parainfluenza virus infections, the incubation period is 4–6 days, and for the coxsackieviruses, echoviruses, and rhinoviruses, 2–5 days.

Febrile pharyngitis and *pharyngoconjunctival fever* often are preceded by symptoms of the common cold. Fever has a gradual onset and reaches a maximum of 103°–104°F on the second or third day; fever may be high for 5–6 days. Nontender submandibular lymphadenopathy is commonly present. Conjunctivitis, when present, is mild to moderate and may persist longer than respiratory symptoms. It is an acute, nonpurulent, follicular conjunctivitis with marked erythema, suffusion, and narrowing of the palpebral fissure. There is usually no involvement of the cornea or uveal tract.

ARD is a more severe illness than pharyngitis or pharyngoconjunctival fever. Fever reaches a maximum of 103°–104°F on the second or third day. Pharyngitis, which is the most prominent localized manifestation of the disease, develops 4–5 days after infection, increases in severity for 2 or 3 days, and then gradually decreases. Malaise and headache are constant features of the disease, and bronchitis and laryngitis are frequent. Virus persists in the nose, throat, and feces for at least a month after recovery.

3. Pathogenesis

Little information is available on the pathogenesis or the pathologic changes produced as a result of these viral infections of the URT mainly because of their relative mildness. The age and immunologic status of the host determine to a considerable degree the response to infection.

Depending upon the route of infection, virus multiplies initially in the pharynx, conjunctivae or small intestine. Examination of tissues from rare fatal cases of infantile pneumonia reveals massive necrosis of the bronchial and tracheal epithelium. The nuclei of infected cells contain basophilic inclusion bodies similar to those produced by adenovirus in cell cultures. Adenoviruses have been isolated from fragments of mesenteric lymph nodes as well as from tonsils and adenoids (lymphoid tissues). It is apparent that these viruses frequently become latent in lymphoid tissue following a primary infection and may persist for long periods of time.

Although certain adenoviruses can induce tumors in animals, there is no evidence that these viruses are oncogenic in humans.

4. Immunity

Neutralizing and complement-fixing (CF) Abs are detectable about a week after infection. Of particular importance is the development of a local IgA response.

In contrast to the short duration of immunity to most respiratory viruses, neutralizing (i.e., protective) Ab against adenoviruses can be detected at least 8–10 years after infection, perhaps as a result of viral persistence in the body. As a consequence, second attacks of illness due to the same type are rare. There is evidence that infection with some adenovirus types provides protection against other types within the same immunologic group.

5. Diagnosis

Pharyngitis caused by adenoviruses must be differentiated from similar illnesses caused by bacteria for which specific therapy is available. For example, it is essential to diagnose and treat β-hemolytic streptococcal infections promptly to prevent the development of rheumatic fever or glomerulonephritis.

When conjunctivitis is prominent, the differential diagnosis of *pharyngoconjunctival fever* includes influenza, measles, herpangina (caused by coxsackievirus A), as well as leptospirosis, inclusion conjunctivitis, and ocular trauma.

The differential diagnosis of *ARD* should rule out influenza and other viral respiratory diseases, URT infections caused by *Mycoplasma pneumoniae,* and purulent sinusitis.

Virus isolation and serologic studies are not performed routinely in the case of URT infections. If a specific diagnosis is required, nasal washings or throat swabs, anal swabs, or fecal specimens for virus isolation, as well as acute and convalescent serum samples for serologic studies should be submitted to the laboratory. Serologic tests, including CF and neutralization tests, are useful for determining the group to which the virus belongs. Since primary infection with any of the possible viruses involved in these illnesses most likely occurs during childhood, a 4-fold or greater rise in Ab titer must be demonstrated.

6. Epidemiology

Epidemics and sporadic cases of pharyngitis and pharyngoconjunctival fever occur throughout the world. There are at least 2 routes by which naturally occurring respiratory illness caused by adenoviruses may be transmitted. When the conjunctival sac is exposed to adenovirus, conjunctivitis is the most common symptom; however, there may also be respiratory involvement. Thus, during outbreaks of pharyngoconjunctival fever, infections may occur after ocular irritation commonly experienced in swimming pools. Volunteers who inhale a virus aerosol develop ARD or pneumonia. Ingestion of adenovirus does not usually initiate disease, but the virus can multiply in the gastrointestinal tract, and specific immunity develops.

Epidemic ARD is confined primarily to recruit centers. Under the crowded conditions in recruit camps, aerosols containing large amounts of infectious virus are generated by the sneezing and coughing of infected recruits. Inhalation of these virus-rich aerosols could account for the high incidence of severe adenovirus infections observed in these populations. In contrast, the occurrence of serious adenovirus infections in children most likely reflects a lack of specific resistance to the virus rather than an overwhelming inoculation.

7. Treatment and Prevention

Specific treatment is not available for adenovirus infections; aspirin and cough syrup may help alleviate the symptoms.

Swimming pools should be avoided during outbreaks of pharyngoconjunctival fever, and care should be exercised to prevent person-to-person spread by way of discharges from infected eyes.

Immunization with vaccines consisting of enteric-coated live adenovirus types 4 and 7 has been shown to be safe and highly effective in preventing respiratory illness. Because of the low incidence and sporadic nature of adenovirus infections in civilian populations, the use of these vaccines is limited to military recruits. Based on the results of the field trials, it was established that vaccination of recruits with both adenovirus types 4 and 7 was required for effective protection against ARD. It was estimated that 26,979 ARD hospitalizations were prevented, with a saving of 7.53 million dollars.

Another potentially fruitful approach to adenovirus vaccines that is under investigation is the use of immunogenic subunit vaccines containing only viral capsid proteins. Such a vaccine would eliminate the problem of the potential oncogenicity of adenovirus DNA.

C. HERPANGINA

Herpangina is a mild, infectious disease caused by a group A coxsackievirus. It is characterized by fever, malaise, and small papular, vesicular, and ulcerative lesions in the palate and the faucal areas. The disease is one of the most frequent causes of summer illness in early childhood but may go unrecognized unless it occurs in epidemic form. Although the disease occurs most often in children, it is also observed in young adults.

1. Etiologic Agents

The *group A coxsackieviruses* were suggested as etiologic agents of herpangina because this disease is one of the most common manifestations of certain types of *group A coxsackievirus* infection. Although serologically distinct from one another, group A coxsackievirus types 1 to 6, 8 and 10 commonly cause herpangina. In addition, herpangina-like illnesses may occasionally be caused by *echoviruses* and certain *group B coxsackieviruses*.

2. Clinical Symptoms

The incubation period for herpangina is about 4 to 6 days. Children between the ages of 1 and 7 years are most commonly afflicted. The disease is characterized by a sudden elevation in temperature

(102°–105°F), severe sore throat, nausea, and vomiting. In infants, convulsions may occur, whereas older children usually develop only a sore throat. The fever reaches a peak during the first 24–48 hours after symptoms appear. Minute papules or petechiae develop on the soft palate and in the tonsillar pillars; 12–24 hours later, superficial ulcers with grayish bases surrounded by red areolae are present at these sites (Fig. 25–1). The lesions increase in size and number for 2–3 days and heal within 4–5 days. The systemic symptoms also begin to subside within 4–5 days, and total recovery occurs within a week.

3. Pathogenesis

The pathogenesis of herpangina in man is not known. Since the illness is not fatal, necropsy tissues are not available for examination.

4. Immunity

Immunity to each infecting coxsackievirus appears to be long-lasting. However, typical herpangina caused by serologically distinct virus types

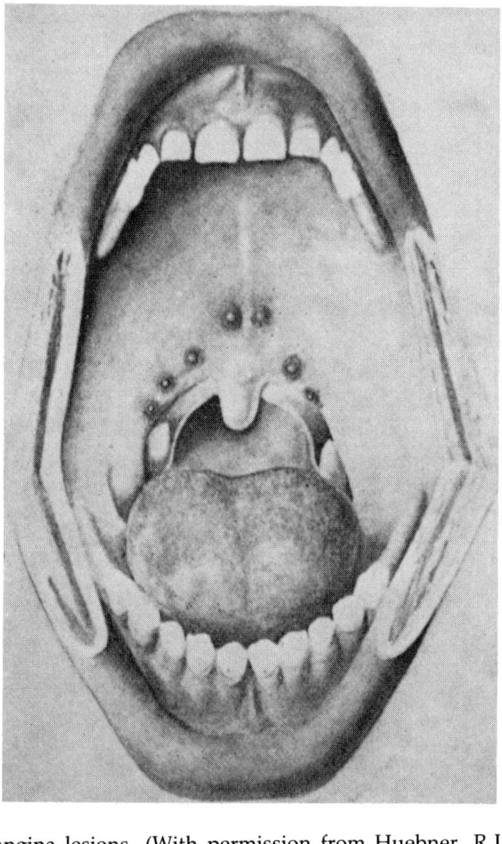

Fig. 25–1. Herpangina lesions. (With permission from Huebner, R.J.: Herpangina. *In* Cecil-Loeb Textbook of Medicine. 12th Ed. P.B. Beeson and W. McDermott (eds.). Philadelphia, W.B. Saunders Co., 1967, pp. 75–76.)

can occur in the same or a subsequent season. Most adults living in urban areas possess Abs to more than one type of group A coxsackievirus; the spectrum of type-specific immunity increases with age.

5. Diagnosis

A clinical diagnosis of herpangina often can be made clinically by the typical appearance of pharyngeal lesions. However, not all infected individuals develop the ulcers characteristic of this disease.

The differential diagnosis must include primary infection with herpes simplex virus, which usually produces larger and more painful ulcers than are seen in herpangina; ulcers due to herpes simplex are primarily in the anterior part of the mouth rather than in the pharynx. Bacterial pharyngitis and the oropharyngeal lesions of certain viral exanthems, such as measles or chickenpox, may be confused initially with herpangina until typical lesions appear. It is also difficult, if not impossible, to separate clinically herpangina without pharyngeal lesions from abortive or nonparalytic poliomyelitis.

The laboratory diagnosis can be established by recovering a group A coxsackievirus from vesicle fluid, pharyngeal washings, or stool in combination with a 4-fold or greater rise in serum Ab titer.

6. Epidemiology

Herpangina occurs most frequently during the summer months. Most illnesses occur in early childhood; up to 10 percent of children examined at random in pediatric clinics in the U.S.A. have been found to harbor herpangina strains of coxsackievirus during July through September.

Within a household or other closed populations, nearly all susceptible contacts are rapidly infected; however, only about a third of the infected individuals manifest the typical pharyngeal lesions. A variable number of family contacts develop mild febrile illness without throat lesions. In addition, a large proportion of group A coxsackievirus infections are asymptomatic and are detected only by epidemiologic studies involving serologic examinations and virus isolation.

7. Treatment and Prevention

Treatment is confined to topical symptomatic measures. The use of anesthetics such as Benadryl elixir or butacaine for gargling will soothe the painful pharyngeal lesions. A fluid or soft food diet is recommended.

D. BRONCHITIS, CROUP, AND BRONCHIOLITIS

Acute bronchitis is an inflammation of the bronchial membranes. It is most often caused by viruses but can be caused by certain bacteria or external irritants. *Croup (laryngotracheobronchitis)* is an inflammation of the larynx or trachea that is characterized by difficult and noisy respiration and a hoarse cough. *Bronchiolitis* is an inflammation of the bronchioles.

1. Etiologic Agents

Acute bronchitis usually is caused by *respiratory syncytial, parainfluenza, influenza,* or *adenoviruses*. Bronchitis may be a complication of the common cold, especially in infants.

Parainfluenza viruses account for one-third to one-half of all cases of severe *croup* in children. *Respiratory syncytial virus* causes most of the remaining cases, but group A and group B *coxsackieviruses,* as well as some *echoviruses,* also have been isolated from patients with croup. During influenza epidemics, *influenza virus* is the main cause of severe croup.

Bronchiolitis in infants is most often caused by *respiratory syncytial virus. Influenza* (during epidemics) and *parainfluenza viruses* also can cause bronchiolitis. Although croup is the most distinctive syndrome caused by *parainfluenza viruses,* these agents cause bronchitis, bronchiolitis, and pneumonia almost as frequently as they cause croup.

2. Clinical Symptoms

Acute bronchitis is usually preceded or accompanied by symptoms of URT infection. The patient has malaise, general aching, and frequently a headache; occasionally there is a mild sore throat and a nasal discharge. A dry, unremittent cough is the most important sign. After 1–2 days, it becomes productive and looser and is accompanied by mucopurulent sputum. The cough may be associated with substernal soreness. Fever is usually present but is moderate, rarely rising above 101°F.

In the absence of complications, acute bronchitis is a self-limited disease and seldom lasts more than 5–6 days. However, cough and expectoration may persist for 1–2 weeks or longer if the patient has preexisting chronic respiratory disease. Patients with chronic bronchitis or emphysema may have severe dyspnea, hypoxia, and increased carbon dioxide retention during episodes of acute bronchitis. An episode of bronchial obstruction may be precipitated in individuals with bronchial asthma; some allergic patients have attacks of asthma only in association with acute bronchitis.

Croup causes an alarming clinical picture in children. Initially, the mucous membranes of the nose and throat are involved, causing symptoms of URT infection for 2–3 days. The inflammatory process descends into the lower respiratory tract and produces bronchial changes accompanied by laryngeal and tracheal involvement. The onset is heralded by an abrupt rise in temperature; fever is variable but may reach 101°–102°F. The first attack often occurs upon waking. There may be an arrest of respiration and the child becomes cyanotic as it struggles for breath. When the laryngeal spasm is suddenly released, air is drawn into the lungs with a high-pitched "whooping" sound. In young children, the attack may be accompanied by convulsions or tetany. Patients usually recover even when the initial symptoms are severe. Only a small proportion of cases proceed to bronchiolitis or pneumonia.

Acute bronchiolitis is observed mainly in infants 6 months of age or less. The onset is marked by a sharp rise in temperature after 2–3 days of URT infection. The temperature is variable and may be moderate or reach 102°–104°F. Bronchiolitis is characterized by a dry, persistent cough and progressive dyspnea. Respiration is difficult and shortness of breath and cyanosis may be evident from the onset of symptoms; convulsions may occur in the young child. The chest is distended and breathing is rapid and shallow; expiration is often noisy. In the absence of complications, recovery is uneventful. The fever may drop rapidly within 48 hours or gradually during 4–5 days. The mortality is variable and can be as low as 1 percent or as high as 25 percent during different epidemics.

Complications include convulsions, heart failure, bacterial pneumonia, otitis, and dehydration. The more serious complications may be fatal, and the course of the illness may be quite rapid. It has been suggested that some cases of "crib death" may be caused by viral respiratory disease.

3. Pathogenesis

In bronchitis, croup, and bronchiolitis, the mucous membranes of the nose and throat are involved initially; paranasal and eustachian tube obstruction also may occur. In bronchitis, there is an excessive secretion of thick and tenacious mucus that contains a mixture of mononuclear cells, desquamated bronchial cells, and some neutrophils. Bronchial biopsy shows edema of the submucosa accompanied by capillary dilatation and mononuclear infiltrates. The paucity of neutrophils is an aid in distinguishing bronchitis of viral etiology from bacterial bronchitis.

If more extensive changes occur in the lower respiratory tract, the larynx may become inflamed and edematous, resulting in the croup syndrome; the accumulation of mucus causes additional obstruction of the airways. In young infants, bronchiolar obstruction can result in focal areas of atelectasis. Suppurative changes resulting from secondary bacterial infection may alter the pattern of the illness and complicate the effects produced by the virus alone.

The pathogenesis of respiratory infections is influenced to a considerable extent by the age and physical condition of the host. Bronchiolitis occurs most often in infections in early infancy and is less likely to be seen with increasing age. Immunologically compromised patients and individuals with preexisting allergies are most likely to develop serious infections.

4. Immunity

Immunity to parainfluenza, respiratory syncytial, and influenza viruses is relatively weak and depends mainly upon the production of IgA Abs in the respiratory tract. Reinfection with the same virus may occur, but the disease is usually mild and limited to the URT. Second attacks of illness caused by adenoviruses of the same type are rare.

5. Diagnosis

The diagnosis of acute viral bronchitis, croup, and bronchiolitis usually can be made on the basis of clinical symptoms. The main problem is to determine whether pneumonia is present. The chest x-ray, which is normal in uncomplicated bronchitis, is useful in detecting pneumonia. As with most other viral respiratory infections, the same virus can cause different clinical pictures. The specific agent involved in each case can be determined only by the diagnostic laboratory. However, except for epidemiologic studies and persistent infections, an etiologic diagnosis is seldom requested.

Since respiratory syncytial virus and, to a lesser extent, parainfluenza virus may be inactivated by freezing, respiratory secretions, or nasopharyngeal washings for virus isolation should be submitted to the laboratory as soon as possible without freezing. If the specimen must be frozen for transport, it should be frozen quickly in a stabilizing medium. Virus isolation provides the only definitive diagnosis for infections with parainfluenza virus because the serologic responses are not always type-specific. In contrast, a 4-fold or greater rise in the serum-neutralizing or CF Ab titer to respiratory syncytial virus, influenza, or adenovirus is regarded as diagnostic.

6. Epidemiology

Infections caused by parainfluenza virus types 1 and 3 are endemic and occur throughout the year. These viruses also cause epidemics during the winter, particularly in nurseries, orphanages, and other similar institutions. Parainfluenza virus type 2 is less prevalent and more episodic in its occurrence than types 1 and 3; infections with type 4 are quite sporadic. Respiratory syncytial virus epidemics occur at intervals of 8–16 months and appear simultaneously in geographically separated locations. Usually there are sharp, well-defined limits to each epidemic. Although respiratory syncytial virus is detected less frequently during interepidemic periods, it has been isolated from respiratory illness during every month of the year. The adenoviruses play a minor role in respiratory infections among civilians; the epidemiology of these infections is discussed in section B. Influenza virus will be discussed in section E.

7. Treatment and Prevention

No specific treatment is available. Close observation and careful assessment of respiratory competence are essential in infants and young children, since a significant proportion of these patients may develop more extensive lower respiratory tract involvement. Bronchiolitis and the croup syndrome may cause respiratory insufficiency requiring urgent and decisive management; tracheotomy and mechanical assistance with ventilation may be necessary in occasional patients. A warm draft-free environment and vapor therapy are useful in reducing laryngeal stridor and bronchial and bronchiolar obstruction. Vaporization is indispensable after tracheotomy and also helps to relieve cough and dyspnea. Since oxygen in high concentrations dries the mucosa, it should not be administered routinely but only as required to reduce cyanosis. Perfusion may be necessary for rehydration. Antibiotics should not be administered in the absence of bacterial complications.

No commercial vaccines are presently available.

E. INFLUENZA

Influenza is an acute respiratory infection of specific viral etiology; it is characterized by sudden onset of fever and prostration and is ordinarily self-limited.

1. Etiologic Agents

Influenza is caused by 3 antigenically distinct groups of *influenza virus,* designated A, B, and C. The terms influenza and "flu" should be restricted to those illnesses with laboratory or bona fide epidemiologic evidence of infection with one of the influenza viruses.

2. Clinical Symptoms

Infection with influenza virus may be asymptomatic, may cause only a slight fever, or may result in the typical prostrating illness that occurs during epidemics of influenza. Asymptomatic infection is the most common form of the disease. Clinical differentiation of infections caused by influenza A and B between epidemics is not possible; influenza caused by influenza C is particularly hard to recognize clinically because of its relative mildness.

The following description depicts a typical, uncomplicated influenza A virus infection. The incubation period of influenza is usually 2–3 days. Mild prodromal symptoms of malaise, chilliness, and cough are sometimes present, but more often the subject suddenly feels acutely ill. The most common initial symptom is severe generalized and frontal headache. A rapid rise in temperature (101°–104°F) is accompanied by diffuse muscular aches, intense fatigue, and stabbing retroorbital pain. At this stage, respiratory symptoms may be entirely absent; they are most prominent when the systemic manifestations and fever begin to subside. Patients almost invariably develop a cough that is brief and spasmodic and usually nonproductive. Sneezing and a watery nasal discharge or a stuffy nose occur in most cases. Conjunctival burning and itching watery eyes are also frequently observed.

Uncomplicated influenza, although acute and prostrating, is brief. Recovery is usually complete in 2–3 days, and it is uncommon for any symptoms to persist beyond 7–10 days. However, convalescence may be prolonged by postinfection asthenia and depression.

The most frequent complications of influenza are secondary bacterial infections of the paranasal sinuses, middle ear, bronchi, and lungs. At present, the most serious complication is staphylococcal pneumonia, which tends to run a fulminating and often fatal course, particularly in geriatric patients.

The rapidly fatal *primary influenza pneumonia*, described during the 1958 epidemic, has been rare in more recent epidemics. These pneumonias were characterized by severe dyspnea, cyanosis, leukopenia, and scanty, grossly bloody sputum. Although staphylococci were isolated from the sputum of some patients, appropriate antibacterial therapy was ineffective. Influenza virus was cultured from the lungs, suggesting that the virus, rather than the bacteria, played the major role in producing the lung damage. Most patients who develop primary influenzal pneumonia have valvular heart disease, preexisting lung disease or are pregnant.

Another serious, sometimes fatal, complication of influenza is influenzal encephalopathy or postinfluenzal encephalitis. Even relatively mild influenza infections also can provoke cardiac involvement including pericarditis. Young children with influenza may develop laryngotracheitis and have difficulty in clearing the respiratory tract of secretions.

3. Pathogenesis

The primary lesion of influenza involves necrosis of the ciliated epithelium of the respiratory tract. In uncomplicated cases, epithelial damage is confined to the upper and middle portion of the tract. Studies have shown that the laryngeal, tracheal, and bronchial mucosae present an acute inflammatory reaction, with desquamation of the ciliated cells and a subepithelial exudate of mononuclear cells; the basal layer remains intact. On about the fifth day of illness, regeneration begins in the basal layer with the development of undifferentiated transitional epithelium; after two weeks, ciliated cells are again present. There are no residual lesions in uncomplicated influenza.

Fatal primary influenza virus pneumonia, presents a characteristic picture. The findings at necropsy include the following: (1) pulmonary hemorrhages, (2) marked edema of the alveolar septa and spaces, (3) hyaline membranes lining the alveolar ducts and alveoli, (4) bloody fluid in the trachea and bronchi, (5) hyperemic tracheal and bronchial mucosae, and (6) absence of ciliated epithelium in trachea, bronchi, and bronchioles. Although viremia is a transient and inconsistent feature of influenza, the virus has been isolated from heart, kidney, and other extrapulmonary tissues. This observation suggests that virus and virus products may enter the circulation and account for some of the systemic manifestations of the disease.

When influenza is complicated by bacterial pneumonia, the pathologic sequelae vary, depending upon the nature of the secondary invader.

4. Immunity

Two Ags are important in immunity, hemagglutinin (H) and neuraminidase (N). Three subtypes of hemagglutinin Ags (H1–H3) and two subtypes of neuraminidase Ag (N1, N2) are recognized among influenza viruses causing disease in humans. Both Ags exist as external glycoprotein spikes (peplomeres) on the surface of virions. They also appear on the surface of the cytoplasmic membrane of infected cells.

Specific immunity to influenza virus develops rapidly after infection but is effective for only 1–2 years. Resistance to reinfection reflects the levels of secretory IgA Ab in the respiratory tract rather than neutralizing Ab in serum. Recurrence of infection with the same antigenic type of virus may occur, especially under conditions of heavy exposure, such as those found in military barracks and dormitories. Resistance to one strain, however, does not protect against newly emerging variants. An individual's immune response to a second infection with an influenza virus is influenced by previous infection to influenza virus. New variants usually share enough antigenic determinants with earlier strains to elicit an anamnestic reponse to Ags present in previously encountered strains. After repeated infections with successive variants exhibiting minor antigenic differences, the dominant Abs present are usually those against the strain that was first encountered. This phenomenon has been called "the doctrine of original antigenic sin."

5. Diagnosis

In the context of an epidemic, influenza is easily recognized; however, the clinical diagnosis of cases at the beginning of an interepidemic outbreak may be difficult.

Radiologic findings of the lungs in uncomplicated influenza are most often normal, but pleural effusion is revealed in 10–20 percent of the cases. Accentuated bronchovascular markings, basilar streaking, small areas of patchy infiltration, atelectasis, and, less frequently, nodular densities also may be observed. The blood leukocyte count may be normal, or there may be a leukopenia 2–4 days after the onset of illness.

The differential diagnosis of influenza includes other viral as well as bacterial respiratory infections that can mimic the onset of influenza. Some adenovirus and respiratory syncytial virus infections are especially difficult to distinguish from influenza.

The definitive diagnosis of influenza depends upon isolation of the virus from throat washings or sputum or the demonstration of at least a 4-fold increase in specific Abs.

Virus isolation is best accomplished by the intraamniotic inoculation of chick embryos with throat washings obtained during the first to the third day of illness. Tissue cultures of primary human or monkey cell cultures also may be used but are less satisfactory. Fluorescent Ab staining of exfoliated nasal epithelial cells has been used recently to establish a specific and rapid diagnosis.

Serologic diagnosis can be made most reliably by CF or hemagglutination-inhibition (HI) tests using acute and convalescent serum samples. Differentiation of Abs against individual strains within each serotype can be accomplished by the HI test, the strain-specific CF test, or the neutralization test.

6. Epidemiology

Influenza A viruses cause epidemics every 2–4 years, whereas influenza B and C viruses are associated chiefly with localized outbreaks or sporadic epidemics. There is no evidence that pandemics (worldwide epidemics) of influenza B or C have occurred.

The factors involved in the periodicity of influenza A epidemics are most likely due to the decline in immunity during interepidemic periods and the periodic emergence of new strains of virus. A *major antigenic* change (antigenic shift) in the virus has occurred 3 or 4 times in this century. These major changes are probably due to genetic recombination between animal and human influenza virus strains. The antigenic variation may be so extreme that the existing immunity is inadequate to prevent infection with the new virus. Under these conditions, the new strain may become dominant and may replace the older virus in the population. People of all ages throughout the world would be susceptible to such a strain, and a pandemic could occur. A *minor antigenic* change (antigenic drift) occurs every 1–2 years due to random mutations and selection of more virulent strains of influenza viruses. A minor antigenic change is usually responsible for localized epidemics.

Because of the short incubation period of influenza, epidemics start abruptly once progressive spread begins, reach a peak in 2–3 months, and then rapidly subside. The attack rate is variable but can exceed 50 percent of urban populations. Crowding seems to be a major factor predisposing to epidemics. Illness usually is observed first among school children, then young adults, and finally in the elderly, less exposed members of the population. Since infection of the aged is more frequently followed by bacterial pneumonia than infection in younger persons, a second wave of increased mortality may coincide with infections in this age group when influenza is no longer apparent in the community at large.

Minor outbreaks of influenza occur almost every winter when new young susceptibles and individuals who escaped the previous outbreak encounter the virus. Following a pandemic, the frequency and extent of outbreaks is reduced as a result of widespread immunity to the new virus. As more members of the population become immune, inapparent or mild infections become more frequent and probably serve to maintain the virus in the community.

7. Treatment and Prevention

Administration of amantadine (Symmetrel) prophylactically may be beneficial in preventing influenza particularly in institutionalized individuals or geriatric patients. Ribovirin has been shown to prevent experimental influenza in tissue culture, animal models and in clinical trials. Codeine sulfate affords relief from incapacitating cough and irritability and is more effective than aspirin for symptomatic treatment of headache and myalgia. Aspirin often increases discomfort by causing excessive sweating and chills. Fluid should be given in abundance; high-fat foods are not well tolerated and should be avoided. Bed rest and gradual return to full activity are recommended. Treatment of primary influenza virus pneumonia usually is unsatisfactory; oxygen therapy with positive-pressure breathing devices may be useful.

Antibiotics do not affect the course of uncomplicated influenza, and routine antibacterial prophylaxis is unnecessary and inadvisable. It has been demonstrated that selective suppression of the normal microbial flora of the respiratory tract as a consequence of antibiotic therapy favors bacterial superinfection. Specific antibiotic therapy should be reserved for diagnosed secondary bacterial infections.

Inactivated *influenza virus vaccines* have been employed for more than 25 years. Polyvalent vaccines, containing a mixture of formalinized egg-grown influenza A and B viruses, are the standard preparations presently used in the U.S.A.

Hybrid influenza virus strains for possible use in living attenuated virus vaccines have been obtained by genetic recombination; however, the stability of the recombinants remains to be established. To ensure the safety of the vaccine, the attenuated parent virus should contain multiple detectable mutations, so that the attenuated vaccine strains can be identified after recombinations with a new variant. A live attenuated influenza virus vaccine administered intranasally is already in use in the Soviet Union and is reportedly effective in preventing the disease.

The use of immunogenic viral subunits, (hemagglutinin and neuraminidase—either alone or with adjuvant), also is being investigated. The results of these studies indicate that the subunit vaccine provides significant protection against infection with fewer side effects than vaccine containing intact virus.

The protection provided by the commercially available influenza vaccines is limited by the following factors: (1) inactivated vaccine given parenterally stimulates mainly serum Abs, which are much less effective in preventing infection than local secretory IgA Abs; (2) the resistance that is induced is transient and persists for only a few months to a year. The U.S. Public Health Service strongly recommends routine yearly immunization for high-risk groups, including those with chronic cardiac or pulmonary disease, diabetes mellitus, or Addison's disease, as well as pregnant women and persons over 65 years of age. The vaccine also may be useful in reducing morbidity in individuals essential for community well-being, such as physicians and policemen, and in closed populations in military camps and institutions.

The influenza virus vaccine is administered subcutaneously; however, the effectiveness of intranasal administration is being investigated. Although the seroconversion rate is low after intranasal spray, the secretory IgA Ab response is marked; the protection seems comparable to that attained after parenteral vaccination, and there are fewer side effects.

During the fall of 1976 over 35 million persons received the inactivated swine influenza ($H_{sw}N1$) virus vaccine in a widespread immunization program. For some yet unexplained reason, there was an excess frequency of Guillain-Barré syndrome (GBS) among the vaccinees. The risk was found to be 10 cases of GBS per 1 million vaccinees—an incidence 5–6 times higher than in unvaccinated persons. This complication caused much criticism and discouragement not only in the future use of influenza vaccination but of other mass immunization programs.

Although the presently available influenza vaccine contains purified virus and is generally safe, it should not be given to individuals allergic to egg proteins and should be administered only advisedly and in small doses to infants and young children.

F. VIRAL PNEUMONIAS

Viral pneumonias are acute diseases with involvement of the lungs, marked by a high fever and cough but with relatively few physical signs. For almost 20 years the descriptive term "primary atypical pneumonia" was applied to all illnesses possessing these characteristics. It was not until 1963 that the filterable agent most frequently involved in these illnesses was found to be *Mycoplasma pneumoniae* rather than a virus.

True viral pneumonias are not associated with a rise in "cold hemagglutinins" or streptococcal MG agglutinins but are essentially indistinguishable clinically from mycoplasma pneumonia. Other nonfilterable agents, including *Coxiella burnetti* and species of *Bedsonia,* also have been isolated from infections diagnosed as primary atypical pneumonia. It has been suggested that the term be dropped; however, because of the difficulties in establishing a rapid and precise etiologic diagnosis, practicing physicians most likely will continue to employ the term "atypical pneumonia" as a provisional diagnosis in many cases.

1. Etiologic Agents

Primary bronchopneumonia may be caused by a number of viruses, including *influenza virus* A, B, and C; several types of *adenovirus* or, in infants and children, *respiratory syncytial, parainfluenza* and occasionally *rhinoviruses.* Less frequently, pneumonia may result from infection with *coxsackie* (particularly group B), various types of *echo,* and even more rarely, *reoviruses.* In addition, pneumonia may develop in adults experiencing a primary *varicella-zoster* infection, in immunologically compromised children infected with *measles virus,* and occasionally in children infected with *measles virus,* and in children and adults infected with *mumps virus.* In recent years, there has been an increase in the incidence of fatal *cytomegalovirus* pneumonia in patients receiving immunosuppressive therapy.

There are still some cases of pneumonia in which an etiologic diagnosis cannot be made, suggesting that other causes of pneumonia remain to be discovered.

2. Clinical Symptoms

The incubation period for the viral pneumonias is variable and may last from 5–15 days, depending on the virus involved. At the onset, the illness is characterized by malaise and chills, fever of 100°–104°F, and occasionally pharyngeal redness and coryza. The most outstanding feature is a dry, persistent cough, which may become productive and cause thoracic pain. However, physical signs may be absent or moderate. Chest x-ray is required for a clinical diagnosis and often reveals abnormalities (Fig. 25–2) from the onset. The illness usually is mild but prolonged. The systemic symptoms often disappear within 3–10 days but may persist longer; however, the x-ray abnormalities last a month or more in 20 percent of patients.

Age of the patient is often an important factor in determining the clinical aspects of viral pneumonia. Although viral pneumonia in the newborn child is rare, the neonate may develop a diffuse involvement of the respiratory tract following a nosocomial infection. Involvement of the lower respiratory tract is characterized by a sudden deterioration of the general physical condition of the infant. The complexion becomes grayish, and there is an acceleration and inversion of respiratory rhythm

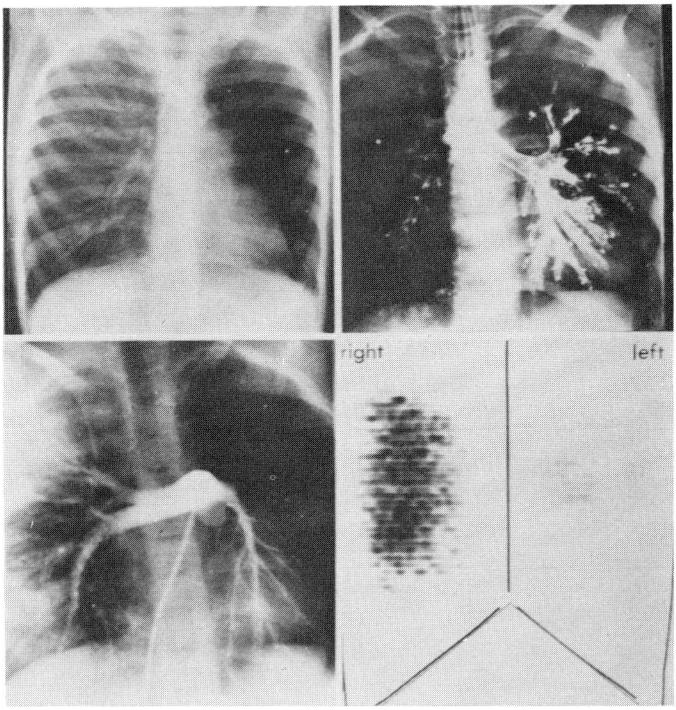

Fig. 25–2. Four-year-old female who developed a unilateral hyperlucent lung following an adenoviral infection at age 8 months. Plain chest film (upper left) shows a hyperlucent lung. Bronchogram (upper right) reveals a typical "pruned tree" appearance of bronchiolitis obliterans. Pulmonary angiogram (lower left) and lung scan (lower right) confirm that there is a marked decrease in perfusion of the left lung. (With permission from Chernick, V. and Macpherson, R.I.: Respiratory syncytial and adenovirus infections of the lower respiratory tract in infancy. Clin. Notes Respir. Dis., *10*:3, 1971.)

together with the appearance of diffuse bilateral rales. In contrast to the symptoms observed in older children and adults, cough may be infrequent and the temperature may be either normal or subnormal. Although recovery usually occurs by 1–2 weeks, the outcome may be fatal, particularly in premature infants.

3. Pathogenesis

During the acute phase of severe cases of viral pneumonia, the lesions observed include bronchial and bronchiolar inflammation, interstitial pneumopathy, and sometimes edematous or hemorrhagic alveolitis. In fatal pneumonias, alternating zones of atelectasis and emphysema have been observed. In approximately half of such cases, a hyaline membrane was present within the alveoli.

The lesions that develop in the lower respiratory tract may reduce the oxygenation of the blood substantially, resulting in shortness of breath, tachypnea, and ultimately, cyanosis. Respiratory alkalosis may develop with reactionary hyperventilation. The hyperventilation may cause some dehydration and hemoconcentration. In serious cases respiratory obstruction and alveolar blockage induce hypoventilation and acidosis. Respiratory alkalosis and acidosis is a grave sign that is observed more frequently in young children than adults. Death can occur from either vascular collapse or heart failure.

There appears to be a relationship between respiratory allergy and viral respiratory infection. It has been observed that asthmatic attacks often occur during a respiratory infection and that allergic subjects are more likely to develop respiratory obstruction than normal individuals. It has not been established whether the viral disease sensitizes the respiratory mucosae.

Some chronic respiratory diseases in later life may result from severe pulmonary infections in childhood. For example, chronic disease following adenovirus infection of the lower respiratory tract in infancy may result in persistent airway obstruction with hyperinflation. There may be recurrent attacks of pulmonary infection, particularly during the first 3–4 years of life.

Pneumonia resulting from a second infection with any given virus is rare, since specific humoral and local neutralizing Abs are produced following the initial infection. Although specific immunity to most of the viruses involved may not prevent reinfection, previous immunization undoubtedly accounts for the fact that respiratory symptoms in adults are usually moderate, and subclinical infections are common. These facts also explain, in part, the relative frequency of viral pneumonias during the first years of life and their rarity in adults.

4. Diagnosis

As in the case of influenzal pneumonia (section E), the other viral pneumonias are distinct from most bacterial pneumonias but are difficult to distinguish from those caused by *Mycoplasma pneumoniae* and certain of the *Bedsonia* and rickettsiae. The diagnostic dilemma is compounded by the fact that many different viruses may cause identical patterns of pulmonary disease and the same agent can produce pneumonitis of varying severity.

Viral pneumonias are generally less severe than bacterial pneumonias and are associated with pulmonary infiltration of lesser density. Physical signs are absent or moderate. The blood count most often is normal, although a leukocytosis may occur in serious cases. The absence of an increase in neutrophils suggests a viral rather than bacterial etiology. The x-ray picture is variable, depending on the severity of disease. Most often, there is a nonhomogeneous, hazy opacity in an inferior lobe; it is bilateral in 10–50 percent of the cases. The opacities characteristically are progressive and become increasingly apparent over a period of 10 days–2 weeks. Pleural effusion is rare. It is impossible to establish a relation between a specific virus and any particular radiologic picture.

To establish an etiologic diagnosis, the virus may be isolated from freshly collected throat washings, sputum, or feces, depending on the agent involved. A significant increase (greater than 4-fold) in specific Ab to the virus isolate between acute phase and convalescent phase sera confirms the diagnosis.

5. Epidemiology

Pneumonia due to adenovirus infection is rare in civilian adults but is observed in military recruits, usually as an extension of ARD. Sporadic and epidemic cases of highly fatal adenovirus pneumonia occur in infants. Pneumopathy caused by respiratory syncytial virus occurs most commonly in infants during epidemics in the late winter and early spring. Parainfluenza virus pneumonia, which is observed mainly in children occurs throughout the year; epidemics occur primarily in hospitals and densely populated institutions. Primary influenza pneumonia is observed only during epidemics. Fatal cases of pneumonia caused by coxsackieviruses and echoviruses are rare but have been observed in infants and young children.

6. Treatment and Prevention

In the absence of specific antiviral therapy, treatment for the viral pneumonias is symptomatic and is designed to (1) relieve the patient's discomfort, (2) maintain good hydration, (3) reduce pulmonary obstruction, (4) prevent or correct hypoxia and acidosis, and (5) prevent or control complications, including secondary infections and sequelae.

Codeine derivatives are used to suppress severe coughs and relieve thoracic pain. No attempt is made to control fever unless it is high; large doses of aspirin favor acidosis and should be avoided. Humidification of room air and vapor therapy are widely used to reduce bronchial and bronchiolar obstruction by decreasing the density and viscosity of secretions. Intravenous perfusion is sometimes required to maintain hydration, especially in children. Oxygen therapy is indicated only if cyanosis develops. Antibiotics should not be administered unless a bacterial infection is indicated.

No vaccines are commercially available. Vaccines containing living adenovirus types 4 and 7 are used to prevent acute respiratory infections in military recruits.

G. EPIDEMIC MYALGIA (BORNHOLM DISEASE, EPIDEMIC PLEURODYNIA, DEVIL'S GRIP)

Epidemic myalgia is an acute viral disease characterized by sudden onset of severe paroxysmal pain in the upper part of the abdomen or lower thorax. The pain is aggravated by breathing and movement.

The disease was first described by Daae and Homann in Norway in 1872. Following extensive observations during an epidemic of the disease on the island of Bornholm, Sylvest published a monograph in 1933 and applied the name Bornholm disease to the syndrome. The diagnostic confusion with pleurisy led to the pseudonym epidemic pleurodynia; the excruciating pain associated with the disease led to the synonym devil's grip. Epidemic myalgia describes the illness more precisely and is the preferred term.

1. Etiologic Agents

A viral cause of epidemic myalgia was postulated for many years, but all efforts to demonstrate such an agent failed until 1948, when it was discovered that a *group B coxackievirus* can cause epidemic myalgia; this disease is the most typical manifestation of infections by these viruses. Other enteroviruses, including various types of *group A coxsackie* and *echoviruses,* occasionally cause an identical clinical picture.

2. Clinical Symptoms

The incubation period for epidemic myalgia is 2–4 days. The onset is sudden and usually begins with a violent, sharp pain in the thorax; less frequently, the pain is progressive. In about 25 percent of the cases, the acute illness is preceded by 1–10 days of fever, anorexia, malaise, stiffness, and URT infection. Fever is a constant finding during the initial attack and usually fluctuates from 100°–103°F but may occasionally exceed 104°F. The most characteristic manifestation of the disease is intense pain in the lower portion of the thorax. Breathing is painful and, as a result, is rapid and shallow. Pain and spasm of anterior abdominal muscles occur in about half the cases, usually in combination with the chest pain. There is often a pronounced headache and pain in the neck, shoulders, and scapula. Nausea and vomiting, diarrhea, shivering, and pharyngeal pain may occur. After the initial attack, the patient may suddenly feel well; however, several successive recurrences of severe pain may occur at intervals of 2–8 days. The acute phase of the illness usually lasts 3–7 days, but additional relapses may occur. Weakness, discomfort, and tenderness may persist for several weeks. Despite the severity of the attacks, virtually no deaths have been attributed to uncomplicated epidemic myalgia. Other clinical manifestations of infection by the virus may accompany epidemic myalgia or may occur as complications of the disease. The most frequently reported complication is an orchitis lasting 3–7 days that occurs in 2 to 3 percent of adult males with epidemic myalgia. No after effects or secondary atrophy of the testicles have been reported. Aseptic meningitis also may precede or accompany epidemic myalgia.

3. Pathogenesis

The specific pathologic process of epidemic myalgia in man is unknown, since the disease is rarely fatal and no necropsy examinations have been reported. The presence of virus in stools, muscle biopsy material, and throat washings suggests a generalized infection.

4. Immunity

Acquired immunity to group B coxsackieviruses is type-specific and long-lasting. However, since all antigenic types of the virus can cause epidemic myalgia, recurrences theoretically are possible.

5. Diagnosis

The diagnosis of epidemic myalgia usually can be made on the basis of the characteristic clinical symptoms once the existence of an epidemic is apparent. However, during the initial stages of an epidemic or when sporadic cases occur, the disease may be confused with serious and even life-threatening conditions. Depending upon the age of the patient and the site of the pain, epidemic myalgia must be differentiated from pulmonary infarction, acute pericarditis, myocardial infarction, angina, pancreatitis, pleurisy, spontaneous pneumothorax, pneumonia, and appendicitis.

Laboratory confirmation of epidemic myalgia is essential to confirm the clinical diagnosis. The virologic diagnosis is made by demonstrating group B coxsackievirus in throat washings taken during the acute illness or in feces for up to a month after infection. The diagnosis is confirmed by demonstrating a 4-fold increase in specific serum Ab.

6. Epidemiology

Outbreaks of epidemic myalgia have been described at all latitudes and in all races. The illness occurs in all age groups but is most common in young adults and children. In temperate zones, it occurs during the summer and early autumn, usually as minor epidemics limited to several households or a community.

Epidemics of Bornholm disease frequently are associated with epidemics of other clinical manifestations of group B coxsackievirus infections, such as acute lymphocytic meningitis and, less frequently, acute infantile myocarditis. Acute myocarditis due to group B coxsackievirus is confined almost exclusively to newborns. Many group B coxsackievirus infections are asymptomatic.

The spread of coxsackieviruses depends directly or indirectly on a human source. The viruses can be found in the throats and feces of patients or healthy carriers. Infection, therefore, may occur by way of respiratory droplet transmission or by the fecal-oral route. In the community or family, children appear to be the principal agents for spreading infection. No natural reservoir other than man has been discovered for the group B coxsackieviruses.

7. Treatment and Prevention

Treatment is aimed at relieving pain. Aspirin has little effect and may cause excessive sweating; stronger analgesics may be used in severe forms of the disease. Cortisone treatment is contraindicated since it aggravates experimental coxsackievirus infections. No vaccines are available.

REFERENCES

Braude, A.I., Davis, C.E. and Fierer, J. (eds.): Medical Microbiology and Infectious Diseases. Philadelphia, W.B. Saunders Co., 1981.

Evans, A.S. (ed.): Viral Infections of Humans. Epidemiology and Control. 2nd Ed. New York, Plenum Publishing Corp., 1982.

McIntosh, K. and Fishaut, J.M.: Immunopathologic mechanisms in lower respiratory tract disease of infants due to a respiratory syncytial virus. Prog. Med. Virol., *26*:94, 1980.

McLean, D.M. (ed.): Virology in Health Care. Baltimore, Williams & Wilkins Co., 1980.

Palese, P., Brand, C., Yound, J.F., Baez, M., Siy, H.R. and Kasel, J.A.: Molecular Epidemiology of Influenza Viruses. *In* Perspectives in Virology. M. Pollard (ed.). New York, Alan R. Liss, Inc., 1981.

Robb, J.A. and Bond, C.W.: Coronaviridae. Comprehensive Virology, *14*:193, 1979.

Sweet, C., Macartney, J.C., Bird, R.A., Cavanash, D., Collie, M.H., Husseini, R.H., and Smith, H.: Differential distribution of virus and histological damage in the lower respiratory tract of ferrets infected with influenza viruses of differing virulence. J. Gen. Virol., *54*:103, 1981.

Young, F., Elliott, R.M., Berkowitz, E.M. and Palese, P.: Mechanisms of genetic variation in human influenza viruses. Ann. N.Y. Acad. Sci., *354*:135, 1979.

Chapter 26

Viral Skin Diseases

A. MEASLES (RUBEOLA)

Measles is an acute, highly contagious febrile illness characterized by a maculopapular rash, ocular symptoms, and catarrhal inflammation of the respiratory tract.

1. Etiologic Agent

Measles is caused by the *measles virus,* which is antigenically distinct from rubella virus, the etiologic agent of rubella (German measles). Measles virus is classified with the Paramyxoviridae family.

2. Clinical Symptoms

The incubation period for measles is 9–14 days. Natural infection invariably results in recognizable illness; inapparent infections rarely, if ever, occur. The early prodromal manifestations of disease are high fever (103°–105°F), malaise, myalgia, headache, conjunctivitis, excessive lacrimation, and photophobia. These symptoms are accompanied or followed by upper respiratory tract (URT) symptoms, including sneezing, hacking cough, and nasal discharge. One to two days before the onset of the rash, Koplik's spots appear on the inside of the cheek (Fig. 26–1). The lesions originally described by Koplik are small red macules or ulcers with a bluish-white center; they constitute a valuable diagnostic sign.

The rash first appears as a blotchy erythema behind the ears or on the face, spreads downward over the trunk and finally involves the extremities; the feet and hands are often spared (Fig. 26–2). The eruption consists of discrete reddish-brown macules that become slightly elevated and tend to coalesce (Fig. 26–3). After 3 to 4 days, the lesions fade in

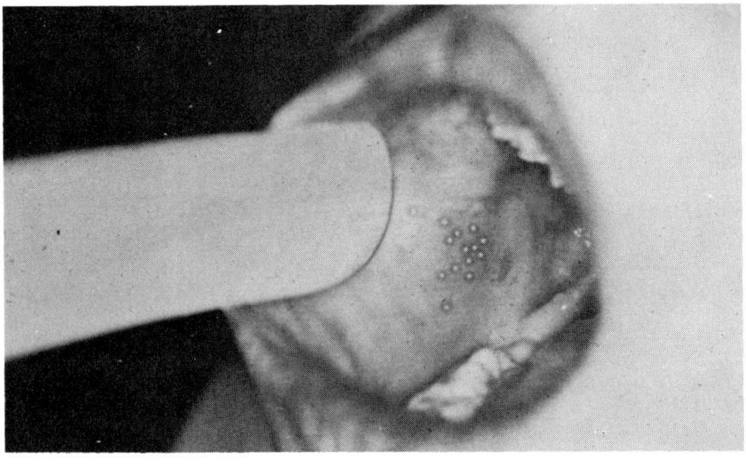

Fig. 26–1. Measles. Koplik's spots on the buccal mucosa duing the prodromal phase of measles. (With permission from Eli Lilly and Co., Physicians Bulletin, 1959; Series of slides distributed by Lily entitled "Current Advances and Concepts in Virology.")

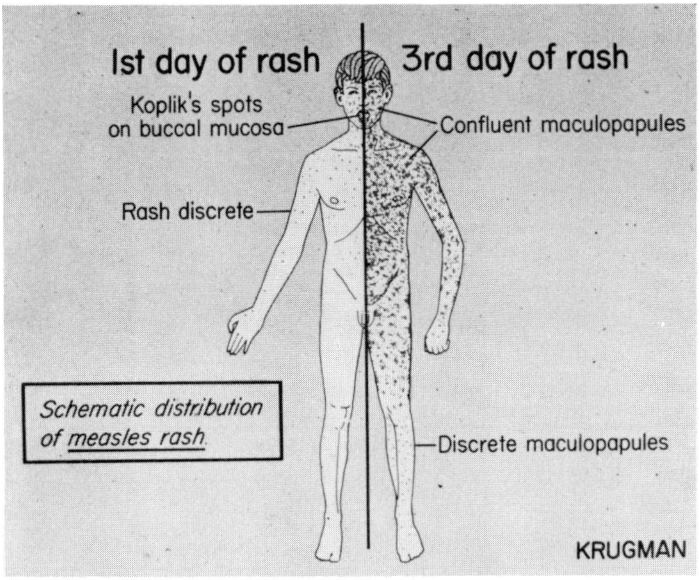

Fig. 26–2. Schematic distribution of measles rash. (With permission from Krugman, S., and Ward, R. (eds.): Infectious Diseases of Children. St. Louis, C.V. Mosby Co., 1968.)

the same order in which they appeared. The skin becomes brownish, and there is fine, powdery desquamation of granular skin. Fever and malaise usually persist until the rash reaches a maximum and then subside. In adults, the fever may follow rather than precede respiratory signs; in addition, the rash tends to be more prominent, and complications are more frequent.

Measles is usually a benign, self-limited disease; however, viral involvement of the respiratory tract may lead to croup, bronchitis, or bronchiolitis. Children suffering from leukemia may develop an interstitial pneumonia characterized by the presence of giant cells containing intranuclear and intracytoplasmic inclusion bodies. Giant-cell pneumonia may occur in the absence of rash, in which case measles may not be suspected; this form of the disease is usually fatal. However, most of the severe or fatal cases of measles in normal individuals result from superimposed bacterial infections that can cause otitis media and pneumonia.

Serious complications directly caused by the measles virus are rare. The most common complications of measles are otitis media, bronchitis, and pneumonia due to secondary bacterial infections. Complications due to measles virus include croup, bronchitis and pneumonia, demyelinating encephalomyelitis appears in about one of every thousand cases

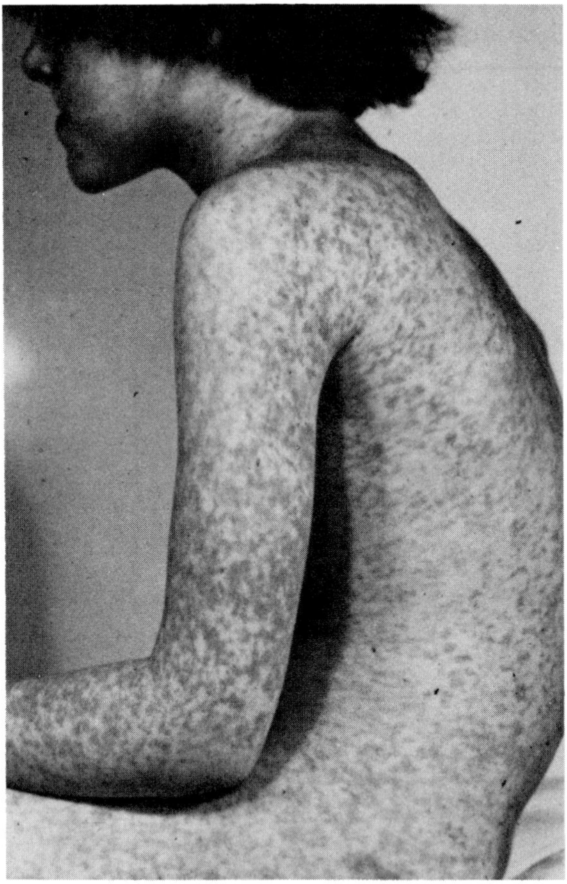

Fig. 26–3. Measles. Full-blown, coalescing skin eruptions in measles. (With permission from Eli Lilly and Co., Physicians Bulletin, 1959; Series of slides distributed by Lilly entitled, "Current Advances and Concepts in Virology.")

of measles and is fatal in approximately 10 percent of the patients; about half of these who survive suffer permanent residual CNS damage, including mental changes, epilepsy, and paralysis. Measles infection of pregnant women results in fetal death in about 20 percent of the cases; however, no teratogenic effects, such as those caused by rubella virus, have been demonstrated. Less serious complications include myocarditis, which occurs in about 20 percent of patients, and results in transient changes in the electrocardiogram but rarely causes irreversible cardiac dysfunction.

3. Pathogenesis

Measles virus enters by the respiratory route, multiplies in the epithelium of the respiratory tract and is disseminated by way of the blood to distant sites. During the prodromal period and for 1–2 days after the rash appears, the disease is highly contagious; virus is present in conjunctival and respiratory secretions, and in the blood and leukocytes as well as in lymphoid tissue. Virus may persist in the urine up to 4 days after the onset of rash.

The skin and mucous membrane lesions may be caused by virus infection of these areas or by the action of immune complexes consisting of virus and Ab. Koplik's spots arise from inflammatory mononuclear cell infiltration of submucous glands and focal necrosis of vesicular lesions of the mucosa. Rash results from the proliferation of capillary endothelium in the corium and exudation of serum and occasionally erythrocytes into the epidermis; epithelial cells become vacuolated and necrotic, and vesicles are formed. Large multinucleate giant cells are characteristic of measles virus infection. They are present in hyperplastic lymphoid tissues, skin lesions, and Koplik's spots as well as in the respiratory tract. Necrosis and sloughing of respiratory epithelium may lead to secondary bacterial infection. Changes in the brain of patients with encephalitis include diffuse focal hemorrhage, lymphocytic infiltration, and demyelinization.

There is mounting evidence that measles virus may persist in the body for many years and may be of etiologic significance in some of the chronic "slow virus" diseases (Chapter 32). Measles virus appears to be the cause of subacute sclerosing panencephalitis, a rare chronic degenerative brain disorder that may develop over a period of years following recovery from a typical uncomplicated case of measles. Elevated levels of measles virus Ab have been detected in the serum and spinal fluid of patients with multiple sclerosis. Although many viruses have been suggested as the etiologic agent of this disease, only measles virus Ab titers have consistently been found to be higher in patients with multiple sclerosis than in normal individuals. In addition, measles virus Ag has been detected in the endothelium of glomeruli and peritubular capillaries in biopsy tissue and kidneys removed from patients with systemic lupus erythematosus; the significance of the virus in the pathogenesis of this disease remains to be established.

4. Immunity

There is only one antigenic type of measles virus, and an unmodified attack of measles results in lifelong immunity to exogenous reinfection. Antibody can be detected in the blood about 2 weeks after infection, and IgG persists in a relatively high titer even in older persons. Although cases of recurrent measles in otherwise normal individuals have been reported, it appears likely that either the primary or recurring illness was incorrectly diagnosed. However, some immunologically impaired patients may be subject to repeated infections with measles virus, as well as with other systemic viruses that usually confer lasting immunity.

5. Diagnosis

Uncomplicated measles during childhood can be diagnosed on the basis of the characteristic clinical picture; in adults, the disease may be more severe and difficult to diagnose. Even before the rash appears, the diagnosis is suggested by the appearance of Koplik's spots on the buccal mucosa and pronounced catarrhal symptoms with a higher fever (Fig. 26–4). Typical giant cells with inclusion bodies are often present in stained preparations of nasal secretions or sputum, particularly in giant-cell pneumonia.

An etiologic diagnosis may be required in the absence of Koplik's spots or in modified or atypical cases of measles. Virus can be isolated, with some difficulty, by inoculating appropriate cell cultures with nasopharyngeal washings, conjunctival secretions, blood, or urine obtained during the prodromal stages or 1–2 days after the rash appears. Attempts to isolate virus thereafter are usually futile. Immunofluorescence techniques have been used successfully to identify specific virus Ag within cells obtained from nasal or conjunctival swabs. Increases in serum Ab can be detected by neutralization, complement-fixation (CF), or hemagglutination-inhibition (HI) tests.

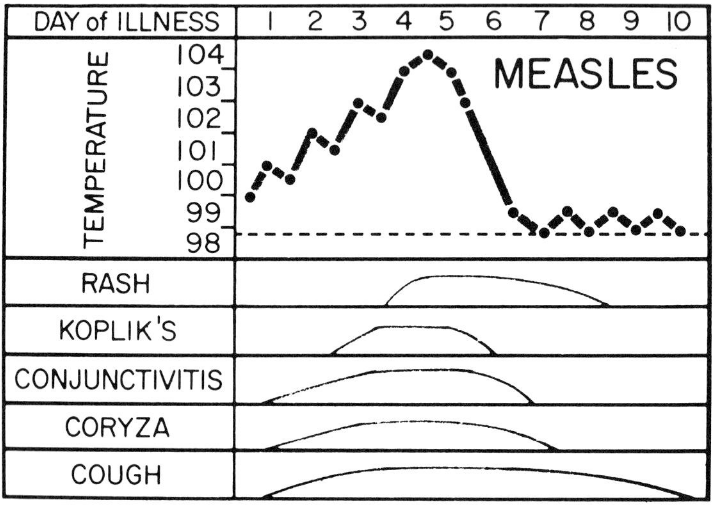

KRUGMAN

Fig. 26–4. Clinical manifestations of measles. (With permission from Krugman, S., and Ward, R. (eds.): Infectious Diseases of Children. St. Louis, C.V. Mosby Co., 1968.)

6. Epidemiology

Measles is endemic throughout the world except in isolated populations. It may occur at any time of the year but is most common in the late winter or early spring. Prior to the development of an effective vaccine, the disease occurred in epidemic cycles every 2–3 years, presumably because new susceptible populations were emerging continuously. There is no evidence that subclinical infections, infectious carrier states or animal reservoirs maintain the virus between epidemics. Thus it appears likely that virus is periodically introduced into susceptible populations by infected individuals entering the community from other areas. It is possible that widespread immunization against measles could ultimately eradicate the disease.

Measles is primarily a disease of childhood. Before active immunization became widespread, 90 percent or more of the population had neutralizing Ab against the virus by 10 years of age. Since the disease is highly contagious, most nonimmunized individuals become infected upon primary exposure to the virus. In isolated areas where measles is not endemic, introduction of the virus results in infection of nonimmune adults as well as children. The morbidity and mortality are high during these epidemics because of the high susceptibility of affected adults to secondary bacterial infection and other complications.

7. Treatment and Prevention

There is no specific therapy for measles. In the absence of complications, treatment is symptomatic. Aspirin may reduce fever, headache, and myalgia; codeine sulfate is effective when the cough is severe. Bright light does not present an ocular hazard, but a darkened room may reduce the symptoms of photophobia. Bacterial superinfections should be vigorously treated with appropriate antibiotics. However, the incidence of serious bacterial infections is not sufficient to justify routine prophylactic use of antibiotics.

The administration of pooled human gamma globulin (passive immunization) within 5 days of exposure will prevent or attenuate measles, depending upon the amount given. Passive immunization may be considered for children under 3 years of age, adults over 60, chronically ill patients, pregnant women, and individuals with impaired immune mechanisms.

Highly effective live *attenuated measles vaccines* prepared in chick embryo fibroblasts are available. The measles vaccine may be given alone or in combination with rubella and mumps vaccines. Immunization appears to protect about 98 percent of those vaccinated and is effective for at least 4–5 years. The maximum duration of immunity remains to be determined. Vaccination is recommended for all healthy children over 9 months of age who have not had the disease. Maternal Ab passively protects infants during the first 6–9 months of life; and it suppresses the effectiveness of attenuated virus vaccines given during this period; the vaccine is also relatively ineffective in patients who have received gamma globulin during the preceding 4–6 weeks. Vaccination is contraindicated for pregnant women, individuals with leukemia or other widespread malignant processes, patients being treated with immunosuppressive drugs or irradiation, and persons with allergies to eggs or other products in the vaccines.

Vaccination with killed measles virus is no longer recommended. Some subjects who received killed measles vaccine in past years developed severe, atypical measles upon subsequent infection with wild-type measles virus; other individuals experienced intense localized hypersensitivity reactions when revaccinated with attenuated virus.

B. RUBELLA (GERMAN MEASLES)

Rubella is a benign exanthematous disease of children and young adults. It was first reported to be a distinct disease entity early in the 19th century and has subsequently been recognized as one of the most common infectious diseases of childhood. The disease attracted little attention until 1941, when Gregg, an Australian ophthalmologist, reported an epidemic of congenital cataracts in infants whose mothers had rubella during early pregnancy. It is now known that contraction of rubella during the first trimester of pregnancy may lead to infection of the fetus and cause congenital abnormalities in virtually any organ.

1. Etiologic Agent

The etiologic agent of rubella is the *rubella virus,* a member of the Togaviridae family.

2. Clinical Symptoms

Postnatal Rubella. The clinical patterns of rubella present a spectrum ranging from inapparent infection to a characteristic clinical picture of lymphadenopathy, rash, and low-grade fever (Fig. 26–5). The incubation period is 17–25 days. In adolescents and adults, mild prodromal symptoms may accompany the onset of adenopathy; these include malaise, headache, low-grade fever, mild sore throat, and mild coryza. These symptoms may precede the rash by 1–5 days. In young children, the prodrome is usually absent, and the rash may be the first sign of disease. In some cases, there may be lymph node involvement without skin lesions.

The rubella rash is variable and has no characteristic features. It may be only a transient blush, but usually the lesions persist for 2–3 days. Initially, small pink macules and papules appear on the forehead and spread within 24 hours to the neck, trunk, arms, and finally the legs (Fig. 26–6). The lesions are usually discrete but may coalesce and sim-

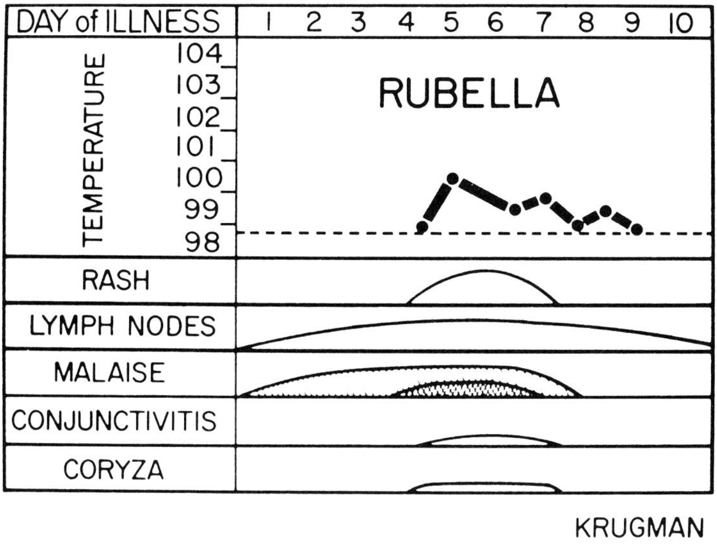

Fig. 26–5. Clinical manifestations of rubella. (With permission from Krugman, S., and Ward, R. (eds.): Infectious Diseases of Children. St. Louis, C.V. Mosby Co., 1968.)

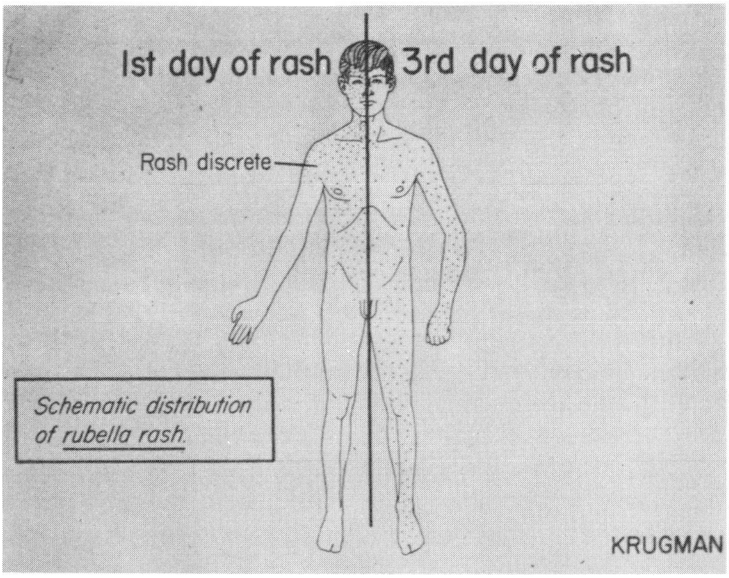

Fig. 26–6. Schematic distribution of rubella rash. (With permission from Krugman, S., and Ward, R. (eds.): Infectious Diseases of Children. St. Louis, C.V. Mosby Co., 1968.)

ulate measles or scarlet fever. The rash on the face often clears by the time a full-blown rash appears on the legs. Desquamation and discoloration of the skin are rare. Fever is usually absent or low grade in children and seldom exceeds 102°F. On rare occasions, rubella in adults may simulate measles. Such patients may experience a fever of 105°F, conjunctivitis, cough, photophobia, and generalized debilitation. Recovery is almost always prompt and uneventful; secondary bacterial infections rarely occur.

Polyarthralgia and polyarthritis may occur as complications of rubella in adult females; they occur less often in males and are rare in children. Pain and swelling, usually in the small joints, are most pronounced during the exanthematous period. Severe symptoms usually disappear in a few days to 2 weeks but may persist several months. Paresthesia, usually numbness and tingling, often accompanies and may outlast the joint symptoms.

Postinfectious encephalitis and *thrombocytopenic purpura* are rare but serious complications of rubella. Encephalitis occurs in approximately 1 out of 5,000 patients and is fatal in about 20 percent of those afflicted. In contrast to other viral encephalitides, rubella encephalopathy is not associated with demyelinization. The fact that the onset of symptoms coincides with the appearance of rubella-neutralizing Ab has resulted in speculation that this complication may have an immunologic component. The prognosis for thrombocytopenia is generally excellent, but fatalities due to CNS hemorrhage may occur on rare occasion.

Congenital Rubella. About 10–15 percent of living infants born to mothers with apparent or inapparent rubella during the first trimester of pregnancy present evidence of infection recognizable at birth or during the first year of life. The consequences of in utero infection are varied and unpredictable. The spectrum includes spontaneous abortion, stillbirth, and live birth with moderate to severe abnormalities, as well as completely normal infants. Virtually every organ system may be affected either transiently, progressively, or permanently.

The congenital rubella syndrome was originally considered to consist only of neurologic and developmental defects, cardiovascular defects, hearing loss, and eye lesions, including pearly cataracts (Fig. 26–7), glaucoma, chorioretinitis, microphthalmia, and corneal clouding. Following the 1964 epidemic of rubella, additional abnormalities were encountered in association with previously recognized manifestations. The expanded congenital rubella syndrome includes thrombocytopenic purpura, hepatosplenomegaly, metaphyseal bone lesions, interstitial pneumonia, anemia, and intrauterine growth retardation. Any combination of anomalies may occur in an individual infant. Many infants lack evidence of disease at birth, but symptoms may become apparent within a few weeks to a year.

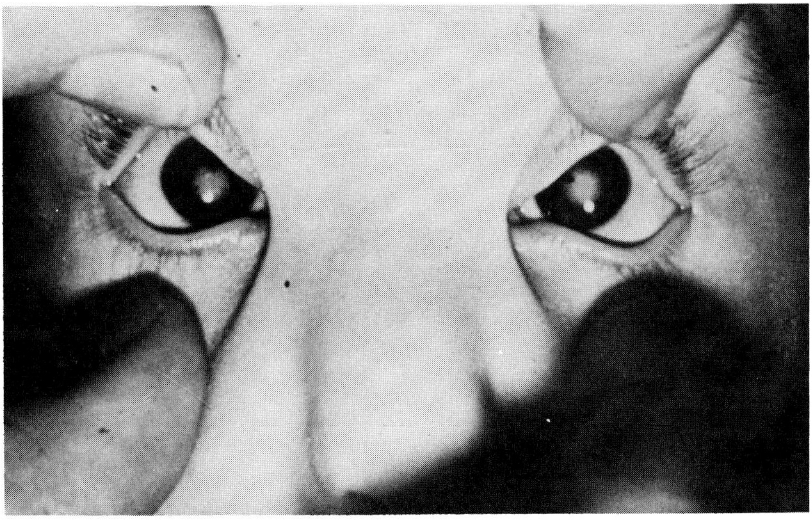

Fig. 26–7. Congenital rubella. Typical pearly, bilateral rubella cataracts. (Courtesy of Dr. Louis Z. Cooper.)

3. Pathogenesis

Postnatal rubella infection is transmitted by the respiratory route. The virus multiplies in the URT and spreads to the blood by way of the cervical lymph nodes. Virus is present in throat washings, blood, and feces for several days before the rash appears. Virus may persist in the pharynx for 1 to 2 weeks but is rarely found in the blood after the first day of the exanthem. There is suggestive evidence that the rash is immunologically mediated and is not associated with direct viral action. Since the appearance of the rash coincides with the development of detectable humoral Ab and a decrease in circulating virus, it is possible that immune complexes are involved.

Although lymph nodes show edema, hyperplasia, and loss of follicles, there are no histologic changes pathognomonic for rubella. The onset of disease is attended by a leukopenia that results from decreased levels of both lymphocytes and neutrophils; there is a marked lymphocytosis 4 to 5 days later.

Congenital rubella results from the transplacental transmission of virus during the viremic stage of maternal infection. Greatest risk of fetal damage follows infection in the first 12 weeks of pregnancy, but there is risk up to the 16th week. Organogenesis occurs during the second through the sixth week after conception, and infection is a maximum hazard to the heart and eyes during this period. Deafness may be the only overt manifestation when the infection occurs after the first 8 weeks of pregnancy. Recent studies indicate that rubella infection during the first trimester results in fetal abnormalities, abortion, or stillbirth in approximately 30–35 percent of the cases.

The mechanism of rubella embryopathy may be explained in part by the chronic nature of the infection and the inhibition of mitosis that occurs. Results from studies on tissues obtained at autopsy reveal that many body organs are underdeveloped and have fewer cells than normal. Although only a small percentage of cells are infected in utero, the infected cells are characterized by a slower growth rate when cultured in vitro. Accordingly, mitotic inhibition in vivo would undoubtedly result in derangement of growth and differentiation. Examination of lymphocyte cultures from children with the rubella syndrome also reveals an increased number of chromosomal breaks. Rubella virus is relatively noncytocidal, and cells are chronically infected rather than destroyed. Chronic infection may contribute to the acute illness seen in the newborn and also to the progressive effects occasionally observed during infancy.

4. Immunity

Immunity to rubella is lifelong following either apparent or inapparent postnatal infection. Antibody is first detected at about the time the rash appears and reaches a peak level 21–28 days later. The presence of any demonstrable level of naturally acquired Ab in the serum seems to protect against disease. However, there is some evidence that symptomatic mucosal reinfection may occur in some individuals in the presence of Ab.

The sera of newborn infants with congenital rubella contain passively acquired antirubella IgG as well as IgM produced by the fetus. Both IgM and IgG are synthesized after birth, and high levels of Abs of both immunoglobulin types can be detected at 4–6 months of age. In contrast, normal infants possess only passively acquired IgG at birth, and antirubella titers gradually decline as maternal Ab is lost. Most infants with congenital rubella have normal immunoglobulin levels by 1 year of age; however, some children may present persistent dysglobulinemia characterized by low levels of IgG with or without an elevation of IgM. Impaired cell-mediated immunity has also been observed in some cases of congenital rubella. These children fail to respond to various skin test Ags, and their lymphocytes do not undergo blast transformation when exposed to mitogens such as phytohemagglutinin.

5. Diagnosis

Except during epidemics, it is almost impossible to diagnose rubella on the basis of clinical symptoms. Many viruses, including echoviruses, coxsackieviruses, adenoviruses, paramyxoviruses, and reoviruses, may cause rubelliform rashes. Other fevers accompanied by a rash, such as exanthema subitum, as well as toxic rashes, may also be confused with rubella. It may even be difficult to differentiate rubella from mild or attenuated cases of measles; conversely, severe rubella may simulate measles or scarlet fever.

A diagnosis of rubella can be confirmed only by virus isolation or serologic procedures. Nasopharyngeal swabs collected within 4 days following appearance of the rash and inoculated directly into cell cultures yield virus in 85–90 percent of cases. In practice, feces are also cultured for virus in the event that the disease is caused by an enterovirus or other enteric agent. The final identification of rubella virus may take 2 weeks or longer, but often a presumptive diagnosis can be made within a week.

Serologic procedures used in the diagnosis of rubella include HI, neutralization, CF, and immunofluorescence tests. In patients with clinical rubella, neutralizing and HI Abs are detectable within 24–48 hours after the onset of rash and peak titers are reached in 6–12 days. In rubella without a rash, these Abs are detectable about 14–21 days after exposure. The promptness of these responses makes the serodiagnosis of rubella difficult unless acute phase blood is obtained within a few days after the rash. However, the range of retrospective diagnosis may be extended by examining the CF Ab response, which develops somewhat more slowly than the HI Ab response. It may be possible to demonstrate a rise in the titer of CF Ab in sera obtained too late to show an increase in HI Ab. In some patients, the HI Ab titer drops significantly within 6 weeks after infection. Therefore, a 4-fold or greater *decrease* in Ab in paired sera may also be considered highly suggestive of recent rubella. In addition, the serum can be examined for rubella IgM HI Ab; a 4-fold or greater decrease in rubella HI Ab after treating the serum with 2-mercaptoethanol provides evidence of a primary Ab response.

Much effort has been expended to develop rapid procedures for the diagnosis of suspected cases of rubella in pregnant women and in persons with whom they have been in contact. Direct immunofluorescent staining of infected cells from throat swabs has been found to be as sensitive as virus isolation in diagnosing rubella in children; however, the procedure is less reliable in adults. Indirect fluorescent Ab techniques have been used to detect rubella-specific IgM in serum. This procedure is more sensitive for diagnosis than the demonstration of virus Ag in infected cells and is recommended as the method of choice when early diagnosis is urgent.

Resistance or susceptibility to rubella is usually determined by testing a single serum sample for HI Ab; any detectable Ab indicates resistance.

An enzyme-linked immunosorbent assay (ELISA) test is commercially available for measuring the immune status of an individual. The test is not useful for diagnosis of an active rubella infection.

6. Epidemiology

Rubella occurs throughout the world and has its highest incidence in the springtime. The seasonal character of the disease is useful in diagnosis since many of the skin diseases with rubelliform rashes are caused by enterovirus infections which occur most frequently in the summer and fall. Rubella is endemic in many areas but also occurs in explosive epidemics at irregular intervals. The disease is most common in children of elementary and high school age; however, a small but significant number of persons reach adulthood without contracting the infection. In the U.S.A., approximately 20 to 30 percent of women of childbearing age have no detectable rubella Ab.

Individuals with inapparent as well as apparent infections are a source or rubella virus. Patients are contagious for 7 days before, and up to 7 days after the rash appears. Virus is shed from the pharynx and is transmitted by close person to person contact. A chronic carrier state does not occur following recovery from rubella acquired postnatally. In contrast, congenital rubella is characterized by chronic shedding of virus for months after birth. These infants are a particular hazard for susceptible hospital personnel. It is, therefore, important to consider only seropositive or postmenopausal nurses for assignments to care for congenital rubella patients. Similarly, the serologic testing of young women working in a rubella diagnostic laboratory is essential.

7. Treatment and Prevention

No effective treatment for either postnatal or congenital rubella has been developed. Early diagnosis of auditory and ocular abnormalities should be followed by attempts to correct the defects. Every effort should be made to prevent the shunting of infants into institutions for the retarded without adequate trials and testing in an education-rehabilitation program.

Gamma globulin has been administered to pregnant women exposed to rubella during the first trimester of pregnancy, but there is no convincing evidence that it is effective for preventing fetal infection. Passive immunization may only mask the symptoms of disease in the mother and confound the decision about the need for a therapeutic abortion.

Several *live attenuated rubella virus vaccines* produced in human diploid fibroblasts have been licensed for use in the U.S.A. Although vaccine-induced immunity appears to be effective in preventing disease, the degree of protection is much lower than that produced by natural infection with wild-type virus. In most individuals studied, Ab has persisted for 3 years or more after vaccination, but maintenance of adequate protection has sometimes been proved to result from asymptomatic infection by wild-type virus. In some cases, Ab titers in vaccinated subjects may decrease slowly to the point where they are inadequate to protect against infection. Immunized persons with inapparent infections constitute a danger to susceptible individuals who, after contact, may develop clinical disease. Spread of wild-type virus from an immunized child to its pregnant nonimmune mother is probably not common but is an ever-present danger.

Some adults, especially women, develop a rash, malaise, arthralgia, and arthritis following immunization with derivatives of the rubella high passage virus (HPV-77) vaccine. Vaccination with the Cendehill strain rubella vaccine also has caused joint manifestations, but they have been less frequent and generally milder than those associated with the HPV-77 strain. The incidence of complications is directly related to age; in children, vaccination seldom causes any adverse reactions.

The HPV-77DK12 vaccine, which is produced in dog kidney cells, should not be given to individuals allergic to dog dander and HPV-77DE5 produced in duck embryo cells is contraindicated for patients allergic to eggs. No living vaccine should be administered to persons with impaired immunologic functions.

In anticipation of a rubella epidemic in the early 1970s, a nationwide immunization program was begun in 1969 and was directed towards the vaccination of all children between the age of 1 year and puberty. The program was based primarily on the concept of "herd immunity." It was anticipated that immunizing one "herd," i.e., prepubertal children, would reduce the spread of rubella and protect a second "herd," namely, suspectible pregnant women. However, it has become apparent that the concept of immunizing one segment of a population to prevent infection in another segment may not be valid for rubella. During a rubella outbreak in 1971, involving 1,000 cases, rubella spread among nonimmune individuals despite a fairly high level of immunity in the 1- to 12-year-old group. Most of the cases occurred among nonimmunized adolescents 12–18 years old, but some pregnant women became infected.

Since the principle of herd immunity does not seem to function in a population in which some of the members have not received rubella vaccine, alternatives may be considered. The only means of eliminating rubella would involve vaccination of all infants, followed by appropriate booster injection until the total population is immune. However, this is not a feasible program because of the impossibility of reaching and maintaining contact with the entire population. Therefore, a more practical approach would be to identify and vaccinate all susceptible females approaching the childbearing age.

The difficulty with a program aimed at immunizing postpubertal women is that the vaccine itself could represent a definite threat to the fetus if inadvertently administered during pregnancy. The potential fetal hazard presented by vaccination was evaluated by giving the vaccine to 35 women already certified for legal abortions. Rubella virus was subsequently recovered from the placenta in 6 cases and from the fetus in 1 case. Virus was also isolated from 13 of 22 uterine cervical swabs taken 9 to 25 days after vaccination of seronegative mothers; no virus was found in comparable specimens obtained from women with preexisting rubella virus Ab. These results indicate the need to observe strict precautions when postpubertal females are vaccinated. It has been suggested that all premarital and pregnancy laboratory studies include a test (e.g. ELISA test) for rubella Ab. All seronegative patients who are not pregnant should be vaccinated and warned that it is imperative not to become pregnant for at least 3 months. Seronegative patients who are pregnant should be given the vaccine immediately after delivery.

C. HERPES SIMPLEX

Primary herpes simplex is usually inapparent but may be a serious and even fatal systemic disease. The most common clinical form of the disease is characterized by localized clusters of vesicles on the mucous membranes or skin. *Recurrent herpes simplex* results from the activation of latent virus that persists in the tissues after recovery from a primary infection.

1. Etiologic Agent

Herpes simplex virus type 1 (HSV-1) is associated with most nongenital infections and type 2 with the majority of genital and neonatal herpetic infections. In addition, HSV type 2 has been found in association with carcinoma of the cervix (Chapter 34). Infections with HSV-1 appear to be much more common than HSV-2 infections on the basis of results from serologic studies.

2. Clinical Symptoms

a. Primary Herpes Simplex. Primary herpetic infections may be inapparent, benign, or severe. Approximately 90 percent of the cases are subclinical, and the only evidence of infection is the appearance of specific Abs. Since most infections are inapparent, the incubation period has not been well defined; however, during several institutional outbreaks, it was observed to vary from 2–12 days, averaging about a week. In patients who develop mild or severe illness, fever and malaise are often prominent symptoms. Vesicles may develop at single or multiple mucosal or cutaneous sites.

Herpetic gingivostomatitis is the most common clinical form of primary herpetic infection. It is observed most frequently in children 1–4 years of age and less often in adults. The onset is usually gradual and is marked by fever (101°–104°F), sore throat and mouth, malaise, and irritability. White plaques and edematous vesicles develop on the buccal mucous membranes, tongue, and oropharynx. As the vesicles ulcerate or erode, the mucosal surface becomes denuded, which results in intense pain and difficulty in swallowing. The gums are inflamed, particularly at the gingival margin, and bleed easily. The regional lymph nodes are enlarged and tender. Fever and pain usually persist for 6–8 days, and the ulcers gradually heal during the following week.

Genital herpes (e.g., *herpetic vulvovaginitis*), is usually caused by HSV-2 and gives rise to extensive and painful lesions similar to those of gingivostomatitis. The vesicular stage may be followed by confluent ulcerations extending to the whole area surrounding the vulvar orifice and occasionally as far as the anus; the ulcerations are often covered with a yellowish-gray pseudomembrane. The inguinal nodes are consistently enlarged and tender. In contrast to genital infections in women, the lesions in men are usually few in number and discrete. The vesicles, most often on the penis and the prepuce, open quickly and exude a clear serous fluid, i.e., "weep." Genital herpes in both sexes may be transmitted venereally. Genital herpes infection is a disease of increasing incidence and growing importance due to morbidity of the illness, occasional complications such as aseptic meningitis, high rate of occurrence, devastating effects on the neonate and epidemiologic association with cervical carcinoma. Recurrences of genital herpes infections are most often associated with primary HSV-2 infections, more likely to occur in men than in women and are directly related to the increased titer of neutralizing Ab to HSV-2 in convalescent-phase serum.

Herpetic keratoconjunctivitis is characterized by edema and inflammation of the conjunctiva and cornea. It may occur alone or may accompany herpetic lesions on the eyelid or elsewhere. Usually, only one eye is involved; preauricular lymphadenopathy is a constant finding on the affected side. Although the primary corneal opacities may be superficial, scar formation from repeated attacks may result in permanent visual impairment. In addition to the cornea, the lens, retina, and choroid may be involved, with the subsequent development of cataracts and pigmentary alterations in the peripheral retina.

Traumatic herpes simplex is characterized by the appearance of large vesicles and pustules around a cutaneous abrasion or burn. The lesions sometimes assume a radicular distribution similar to that observed in zoster (Section D).

Eczema herpeticum (Kaposi's varicelliform eruption) is a rare but sometimes fatal form of primary herpes simplex infection that occurs in individuals with eczema or neurodermatitis. A generalized pustular eruption develops over a period of days. Groups of vesicles appear in "crops," so that lesions in all stages of development are present simultaneously on a single area of the body; in this respect, the disease is similar to varicella (Section D). The disease may be mistakenly diagnosed as eczema vaccinatum, which occurs under similar conditions following vaccination for smallpox (Section E). The eponym "Kaposi's varicelliform eruption" has unfortunately been applied to both conditions.

Herpetic meningoencephalitis is a serious life-threatening manifestation of primary herpes virus infection. The neurologic symptoms often indicate localized brain lesions in the temporoparietal region of the cerebral cortex. The mortality rate is high; patients who survive often have permanent psychic, psychomotor, purely motor, or epileptic sequelae. Occasionally, meningitis develops with no involvement of the brain, and, more rarely, encephalitis occurs in the absence of meningitis. It has been estimated that 5 to 7 percent of cases of aseptic meningitis may be caused by herpesvirus.

Neonatal herpes may develop in infants infected with HSV-2 by contact with herpetic lesions during passage through the birth canal or by transplacental infection; the illness can vary from mild to severe disseminated disease. Generally, herpetic disease of the newborn occurs only in the absence of maternally acquired IgG. Premature infants, who are less well immunized by maternal Ig, are more likely to develop the severe generalized form of the disease (visceral herpes) than full-term infants. *Visceral herpes* of the newborn is a fulminating infection, with fever, viremia, and necrotic lesions in the liver, spleen, lungs, adrenals, kidneys, and brain. The few infants who survive almost invariably have permanent brain damage.

b. Recurrent Herpes Simplex. Although specific Abs develop following either apparent or inapparent primary herpes infections, herpesvirus may persist in a latent form and be reactivated later by various nonspecific stimuli. In some individuals, the recurrent attacks appear after excessive exposure to sunlight or heat; others may be affected following febrile episodes, respiratory or gastrointestinal disturbances, trauma, or exertion. Menstruation, emotional disturbances, and pregnancy also may predispose to recurrent herpes.

The lesions of recurrent herpes are usually benign but painful and tend to occur repeatedly. The recurrences often involve the same site as the primary symptomatic or, presumably, subclinical infection. In most cases, the lesions resulting from reactivation of latent virus take the form of *herpes labialis,* i.e., fever blisters or cold sores (Fig. 26–8). The affected area(s) becomes erythematous and begins to burn, itch, or tingle several hours before the typical fragile, clear vesicles develop. The vesicles rupture soon after they appear and exude a serous fluid that forms a yellow crust as it dries. The eruptions heal without scarring within several days to a week unless they become secondarily infected with bacteria.

Recurrent generalized herpes and *recurrent meningoencephalitis* are rare but may occur in individuals who recover from eczema herpeticum or primary herpetic encephalitis. A more frequent form of severe recurrent infection occurs following primary herpetic *keratoconjunctivitis;* the most common complication is permanent stromal scarring with loss of visual acuity.

3. Pathogenesis

Herpes simplex virus can enter the host through the mucous membranes of the nasopharynx, conjunctivae, and genitalia or through abraded or traumatized skin. The virus multiplies at the portal of entry and is spread by either the hematogenous or neurogenic routes. Present evidence suggests that meningitis and extraneural visceral involvement

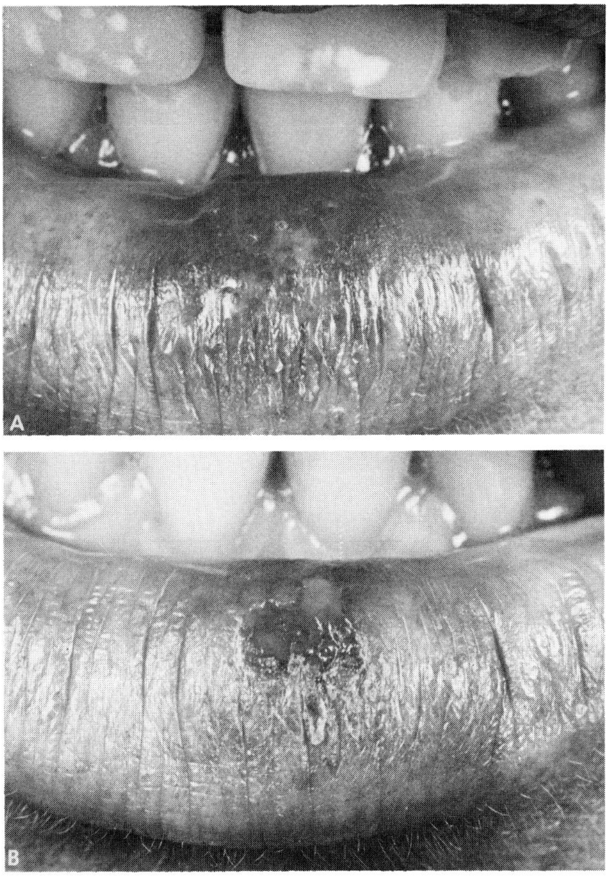

Fig. 26–8. Recurrent herpes labialis. *A,* Six hours after initial symptoms. *B,* Third day.

result from hematogenous dissemination; isolated cases of encephalitis and some cases of ocular herpes appear to be the consequence of neural spread. After resolution of the primary infection, the virus becomes latent. During the latent period, virus apparently persists intracellularly, because sufficient neutralizing Ab is present to inactivate extracellular virus. Recent evidence indicates that virus may persist within spinal ganglia between recurrent episodes. The virus has been isolated from trigeminal ganglia removed during routine autopsy and cultured in vitro. These results indicate that the virus probably is present in a high proportion of human trigeminal ganglia in the absence of overt herpes infections. The common denominator among the factors that precipitate recurrence may be an alteration in host cell physiology that permits virus multiplication; however, the precise mechanism of reactivation is not known.

In recurrent herpes simplex, humoral Ab prevents dissemination of virus but does not inhibit the development of local lesions. The virus can spread directly from cell-to-cell by the fusion of infected cells with adjacent uninfected cells. The lesions of primary and recurrent herpes are indistinguishable microscopically. Skin biopsies taken during the early vesicular stage show dermal congestion with swelling and ballooning degeneration of prickle cells of the epidermis. Vesicles or ulcerated areas contain degenerated epithelial cells as well as multinucleated giant cells. The nuclei of infected cells are enlarged and filled with homogeneous inclusion bodies that first stain blue and later red with hematoxylin and eosin; each of the many nuclei in giant cells often contains one or more inclusion bodies. The intraepidermal vesicles do not extend below the basement membrane and hence do not cause permanent scarring. During healing, the vesicle and corium are densely infiltrated with inflammatory cells.

In fatal cases of herpes encephalitis, there is perivascular infiltration and nerve cell destruction; intranuclear inclusion bodies may be observed in glial cells and, less frequently, in nerve cells.

Visceral herpes of the newborn is characterized by multiple areas of focal necrosis with mononuclear infiltration. All organs of the body, especially the liver and adrenals, may be involved.

The branchlike *dendritic lesion* is the classic sign of ocular herpes simplex infection. An acute florid conjunctivitis, often follicular, may precede the onset of herpetic keratitis by several days. Corticosteroid therapy appears to predispose to the enlargement and ulceration of primary dendritic lesions. At later stages, an epithelial ulcer may persist for weeks or months in the absence of detectable virus multiplication. The failure of the epithelium to regenerate over the corneal stroma is thought to result from a damaged basement membrane. The stroma underlying the epithelial ulcer develops pathologic changes that often spread, even if the epithelial lesion heals. The entire thickness of the corneal stroma is invaded by inflammatory cells, with development of edema, decompensation of the endothelium and stromal scarring. This chronic form of ocular herpetic infection, referred to as *disciform keratitis,* is thought to be a hypersensitivity reaction. In contrast, involvement of the iris in stromal keratitis may be due to direct invasion of the virus or to a toxic effect.

4. Immunity

Maternal IgG usually affords protection against HSV-1 infection during the first 6 months of life. Although the majority of adults possess Ab to HSV-1, the incidence of HSV-2 infection is considerably lower. As a consequence, passively acquired immunity to HSV-2 is not common.

The primary immune response following either apparent or inapparent infection is characterized by the appearance of neutralizing Abs, which can be detected only in the presence of fresh complement; shortly thereafter, these Abs are replaced by complement-independent neutralizing Abs. The presence of serum Abs probably accounts for the attenuated clinical symptoms and absence of viremia during recurrent herpes. However, since reactivation of latent virus occurs in the presence of high levels of circulating Abs, it appears that cell-mediated immunity may play the dominant role in resistance to and recovery from recurrent infections. It has been suggested that an exaggerated hypersensitivity to viral Ags may contribute to pathogenesis.

5. Diagnosis

Virus isolation and serologic studies are used mainly to confirm the clinical diagnosis of primary herpes simplex. The history and clinical appearance of recurrent skin and eye infections are usually sufficient to establish the diagnosis. In instances of severe disease, a laboratory diagnosis may be required.

For virus isolation, specimens should be collected within the first 4–5 days of illness and inoculated into tissue cultures. Embryonated eggs and laboratory animals are now seldom used for primary isolation of the HSV but are useful for differentiating it from varicella-zoster (V-Z) virus, which replicates only in tissue cultures. HSV usually can be isolated from vesicle fluid, scrapings of ulcerations, and throat swabs or saliva. Direct examination of stained cells from the lesions or infected tissue culture cells from the lesions or infected tissue culture cells reveals eosinophilic intranuclear inclusion bodies characteristic of the herpesvirus group; vaccinia and variola, with which some herpes infections may be confused, cause cytoplasmic inclusion bodies. HSV must be specifically identified by CF, neutralization, or immunofluoresence techniques to differentiate it from V-Z virus, which causes similar lesions.

Serologic studies using patients' paired sera are of value in the diagnosis of primary herpes infections only in instances in which a rise in specific Ab titer can be demonstrated. In recurrent herpes, high Ab levels may be present, but there is no consistent rise in titer. It is important to note that patients with a recent history of V-Z virus infection may respond to HSV infection with an increase in CF Abs to both viruses. In addition to infections with V-Z virus, vaccinia, and variola viruses, the differential diagnosis of herpes simplex includes herpangina, Vincent's angina (which responds dramatically to penicillin), and secondary bacterial infection of eczema. Herpetic meningoencephalitis must be differentiated from bacterial and viral encephalitides and postinfectious encephalitis in order to institute the proper therapy as soon as possible.

6. Epidemiology

Man is the only known natural reservoir of HSV. It has been estimated that 70–90 percent of adults possess neutralizing Abs against HSV-1; this agent is probably the most ubiquitous virus in man.

Primary infection with HSV-1 usually occurs before the age of 5 years but is rare in infants under 6 months of age, who are passively protected by maternal Abs. There is a high infection rate during infancy in lower socioeconomic groups where crowded living conditions and poor sanitation exists. Herpes simplex is an endemic disease but may occur in minor epidemics within family groups, hospitals, and institutions. Patients with overt disease as well as convalescents and healthy carriers may shed the virus in vesicle fluid, saliva, genital secretions, and stools. The virus is believed to be spread from person-to-person by direct contact or indirectly from eating or drinking utensils contaminated with saliva.

Both HSV-1 and HSV-2 can be transmitted venereally. With the exception of neonatal infections acquired at or before birth, genital herpes seldom is observed before adolescence, i.e., children are infrequently infected. The incidence of HSV-2 infection is lower than HSV-1 infection and correlates with venereal exposure. For example, the incidence among prostitutes is high, but it is negligible among nuns.

7. Treatment and Prevention

HSV-1 infections were among the first of the viral infections to be successfully treated with chemotherapeutic agents. The majority of patients with primary herpetic keratitis, cutaneous infections, and even encephalitis respond dramatically when treated with iododeoxyuridine (IUdR) or adenine arabinoside (Ara-A) early in the course of the infection; many individuals are asymptomatic within a week. Topical application of IUdR every 2 hours is particularly effective in preventing the sequelae associated with dendritic keratitis. Chronic infections that involve the corneal stroma do not respond to IUdR but may be reduced by judicious topical application of corticosteroids; steroid therapy is contraindicated with dendritic keratitis since it may increase the chance of ulceration and perforation. Combined IUdR and steroid therapy should be attempted only by an experienced ophthalmologist. IUdR is toxic when administered by perfusion. Iododeoxyuridine is relatively ineffective for gingivostomatitis because it is rapidly washed away by secretions. Although the drug suppresses viral multiplication, it neither eliminates the virus (i.e., it is nonviricidal) nor prevents relapses; recurrent infections occur as frequently in IUdR-treated patients as in untreated patients. Some strains of HSV-1 and most strains of HSV-2 are resistant to IUdR. The emergence of drug-resistant or drug-dependent mutants may account for the decreased effectiveness of the drug during recurrent infections.

Results with Ara-A suggest that these adenine analogues are less toxic than IUdR and may be equally effective in treating herpetic infections. It appears to be particularly promising for use when IUdR-resistant mutants emerge and in cases of allergic or toxic reactions.

A new antiviral drug, acycloguanosine (acyclovir or Zovirax) is highly selective against herpes simplex viruses and varicella-zoster virus and shows great promise in treating skin and venereal infections associated with these viruses.

The prophylactic use of gamma globulin derived from hyperimmune serum is recommended when a newborn has been in contact with an infected individual; however, it is not effective for treating congenital infections.

No vaccines for HSV are commercially available. Some success in preventing recurrent infections has been reported following repeated intradermal and s.c. injections of an experimental inactivated vaccine. The mechanism involved in the resistance that develops is not known, but it appears to be at the cellular level, since there is no consistent increase in circulating Abs.

Results from controlled studies indicate that the formerly popular practice of administering smallpox vaccine to prevent recurrent herpes labialis has no scientific basis and is no more effective than placebos.

D. VARICELLA (CHICKENPOX) AND ZOSTER (SHINGLES)

Varicella is a highly contagious disease that occurs mainly in children. It is characterized by a disseminated vesicular eruption that develops in brief successive crops: when the disease occurs in adults, neonates, or immunologically compromised patients, the symptoms may be severe.

Zoster is a sporadic infectious disease characterized by unilateral inflammation of dorsal root ganglia or extramedullary cranial nerve ganglia. A unilateral vesicular skin eruption usually develops along the pathway of the involved nerves. The disease occurs mainly in adults and probably results from reactivation of virus that persists in latent form following recovery from chickenpox.

1. Etiologic Agent

Chickenpox and zoster are caused by a single serotype of virus, *varicella-zoster* (V-Z), a member of the Herpesviridae family. The virus is distinct from HSV, although the two share minor antigenic determinants.

2. Clinical Symptoms

VARICELLA. The incubation period from the time of exposure to the appearance of rash usually ranges from 13–17 days. There may be fever and malaise for 1–2 days before the rash appears, but more often these symptoms and the rash develop concurrently. The lesions, which appear in crops over a period of 1–5 days, are first observed on the trunk and later on the extremities. The typical chickenpox eruption begins as a maculopapule that evolves within a few hours into a fragile, teardrop-shaped vesicle surrounded by a red border. Within a day, the erythema diminishes and the vesicle collapses in the center, forming an umbilicated pustule. The lesion becomes crusted, and after several days the scab falls off. There is no scarring in the absence of secondary bacterial contamination. However, because of intense itching, children usually scratch, and vesicles may become infected. New maculopapules continue to erupt and go through a similar evolution. Because of the appearance of new crops, lesions in various stages of development (maculopapule, vesicle, pustule, and scabs) are present simultaneously in a single area (Fig. 26–9). The characteristics listed in Table 26–1 are useful for establishing a presumptive diagnosis for varicella.

Although chickenpox is usually a benign disease in childhood, it can be severe and sometimes fatal in adults, neonates, children with leukemia, and individuals on immunosuppressive therapy. In these patients, the rash often persists longer than usual and may become hemorrhagic. Primary varicella pneumonia develops in about 15 percent of adults with chickenpox. The fatal disseminated form of the disease is seen most frequently in children being treated with steroids for leukemia or other diseases of the hematopoietic system. Postvaricella encephalitis

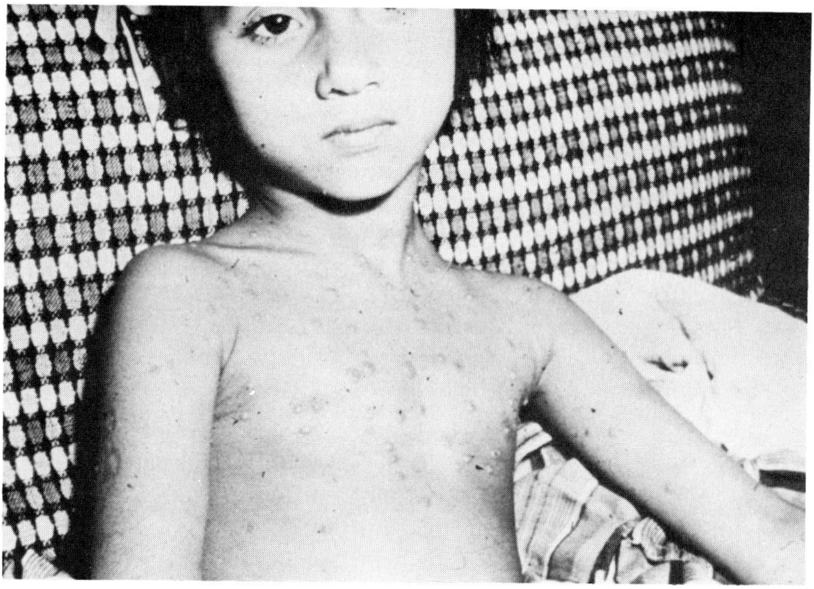

Fig. 26–9. Varicella. Lesions in various stages of development on trunk. Minimal involvement of face and arms. (Courtesy of Center for Disease Control.)

Table 26–1. Characteristics of Varicella Lesions

1. Lesions in all stages of development on a given area of skin.
2. Lesions first appear on the trunk.
3. Palms not involved.
4. Nuclear inclusions observed in skin scrapings from lesions.

is a rare but grave complication that can follow either mild or severe disease.

Zoster. The preeruptive stage of zoster may be preceded by 1–4 days of fever with pain and paresthesia over the area of the involved nerve. Subsequently, an erythematous rash appears, the individual lesions evolving as in chickenpox. The eruption is confined, for the most part, to the area of distribution of one or more spinal nerves or the sensory division of a cranial nerve. Infection of the ophthalmic division of the trigeminal nerve may result in damage to the cornea, as well as in skin eruptions on the forehead, eyelids, and nose. The corneal changes range from lesions with no stromal alteration to lesions with epithelial and stromal ulceration and opacity. The localized distribution of the lesions in zoster gives the eruption its characteristic unilateral band-like pattern (Fig. 26–10). Regional lymph node involvement is a constant finding. Some patients develop a generalized vesicular rash resembling varicella shortly after the localized lesions appear.

The course of zoster from onset to complete recovery is about 2–3 weeks. Adults over 40 years of age may not recover completely for 4–5 weeks. Postherpetic neuralgia is a serious and painful complication of zoster that may occur in elderly patients with arteriosclerosis. Frequently, there is an interval between the acute phase of the disease and the development of characteristic severe pain; pain may persist for weeks to months.

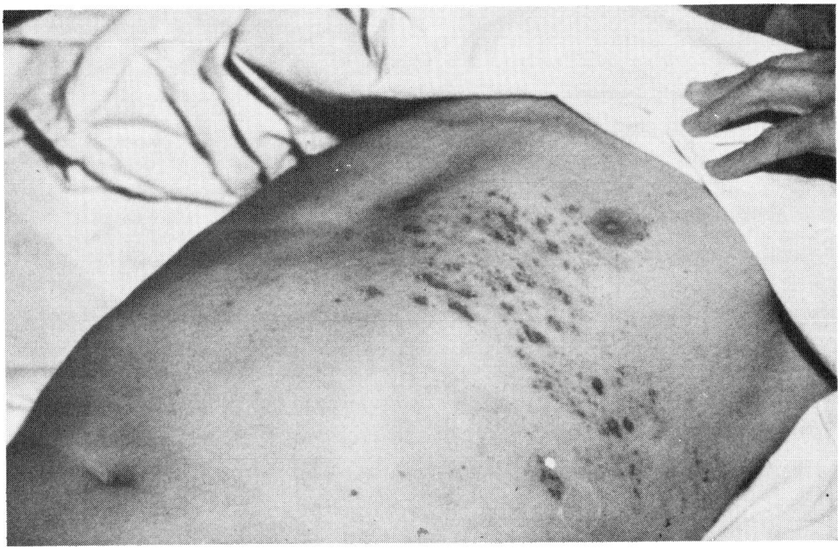

Fig. 26–10. Zoster skin lesions. The unilaterally located vesicles have ruptured. (With permission from Eli Lilly and Co., Physicians Bulletin, 1959; Series of slides distributed by Lilly entitled "Current Advances and Concepts in Virology.")

3. Pathogenesis

VARICELLA. Varicella is probably transmitted by the respiratory route. The appearance of successive crops of widely distributed skin lesions suggests that intermittent viremia may occur. Focal viral infection of blood vessels in the corium results in fluid accumulation and ballooning degeneration of cells in the basal and prickle layers of the epithelium. The resulting vesicles contain serum, epithelial and inflammatory cells, and multinucleate giant cells. Eosinophilic inclusion bodies identical with these seen in herpes simplex may be observed in the nuclei of infected cells. Virus is present in vesicle fluid for several days after eruption but is seldom detected in crusts or scales. In varicella pneumonia, the tracheobronchial mucosa, the alveolar septa, and the interstitium of the lung become edematous and are infiltrated with monocytic inflammatory cells and giant cells. The CNS changes observed in patients who develop encephalomyelitis include diffuse focal hemorrhage, lymphocytic infiltration, and demyelinization.

ZOSTER. The pathogenesis of zoster is not clear. The disease occurs in individuals with a history of varicella and apparently results from the reactivation of latent virus. The trigger mechanism is not known in some cases of zoster; in others, the disease develops following trauma, exhaustion, sunburn, injection of drugs, or immunosuppressive therapy, or is concomitant with diseases such as tuberculosis or malignancy. The cutaneous lesions of zoster are histopathologically identical with those of varicella. In addition, there is an acute inflammatory reaction of the dorsal nerve roots and ganglia. Often only a single ganglion is involved. Zoster most commonly involves areas of the skin innervated by the thoracic ganglia or, less frequently, the cervical ganglia or the ophthalmic branch of the gasserian ganglia. The affected spinal ganglion is infiltrated with mononuclear cells and presents scattered hemorrhagic areas. Intranuclear inclusions have been demonstrated in ganglia and in satellite cells. The inflammatory response may extend to the posterior horns or, less often, to the anterior horns of the spinal cord.

Virus is abundant in vesicle fluid and occasionally may be recovered from spinal fluid; V-Z virus has been detected in trigeminal nerves and ganglia by immunofluorescence and electron microscopy but it has not been isolated from these tissues.

4. Immunity

Immunity to exogenous reinfection with V-Z virus is long lasting. However, the neutralizing Abs produced do not protect against reactivation of latent virus. There is some evidence to indicate that varicella convalescent Abs may differ qualitatively from zoster Abs. Two subclasses of IgG have been separated from sera of patients with V-Z virus infection. The two classes, "slow" and "fast" IgG, differ in their neutralizing activity, depending upon the clinical manifestation of the infection. In varicella infections, neutralizing activity was demonstrable in the "slow" IgG fraction, whereas in zoster, the neutralizing Ab was in the "fast" IgG fraction. The "slow" IgG fraction following chickenpox may not prevent the subsequent development of zoster. However, a second generalized exposure to V-Z virus in the form of zoster results in significant levels of efficient neutralizing Abs. These Abs may account for the fact that second attacks of zoster are rare.

In established V-Z virus infections, the virus most likely spreads by direct cell-to-cell transfer, and cell-mediated immunity probably plays an important role in recovery from infection.

5. Diagnosis

Varicella. Varicella can usually be diagnosed by the character of the rash and a history of recent exposure. A rapid presumptive diagnosis can be made by examining stained scrapings of early vesicles or biopsy tissue for the presence of characteristic multinucleated giant cells containing intranuclear inclusion bodies. Complement-fixation and agar-gel diffusion tests are useful for demonstrating specific Ags in vesicle fluids. Atypical varicella occasionally must be differentiated from other generalized vesicular eruptions, including complicated herpes simplex, rickettsialpox, and some coxsackievirus infections. When a specific etiologic diagnosis is required, virus may be isolated in tissue cultures inoculated with fluid obtained from vesicles in an early stage of development. V-Z virus, in contrast to HSV, does not multiply in either embryonated eggs or laboratory animals. Serologic tests using patients' paired sera may be used to confirm a diagnosis of varicella when virus can no longer be isolated.

Zoster. The diagnosis of zoster is difficult before the appearance of the characteristic unilaterally distributed vesicles. During the preeruptive stage, the disease often is confused with more common causes of intense pain, such as pleurisy, a collapsed intervertebral disc, or appendicitis. If the eruption is atypical, zoster may be clinically indistinguishable from recurrent herpes simplex, which may also follow radicular lines. Since second attacks of zoster are rare, most reported cases of recurrent zoster are probably examples of HSV infection. A specific diagnosis can be established by isolating and identifying the etiologic agent.

6. Epidemiology

Varicella is one of the most common contagious diseases of childhood. The attack rate is 70 percent or more among susceptible individuals exposed to the virus; inapparent infections are rare. Although the disease is endemic in the U.S.A., epidemics occur during the winter or spring every 2–5 years. Children 5–8 years of age are most commonly affected, but newborns, younger children, and adults who escaped infection during childhood also may develop the disease. Chickenpox presumably is spread by droplet transmission as well as by direct or indirect contact with skin lesions. The infectious period extends from 1–2 days before the rash up to about a week after lesions appear.

In contrast to chickenpox, zoster is a sporadic disease without seasonal prevalence and occurs more commonly in adults than in children. If a zoster patient transmits the virus to a susceptible child, the child would develop typical chickenpox and possibly initiate an epidemic. However, the incidence of zoster does not increase during epidemics of chickenpox. The increasing use of immunosuppressive drugs for tumor therapy and organ transplantation has led to a rise in the incidence of zoster, especially in the life-threatening disseminated forms of zoster.

7. Treatment and Prevention

Uncomplicated varicella and zoster are self-limiting diseases. Therapy is symptomatic and consists primarily of ointments, such as calamine lotion, to relieve the itching. In the acute phase of zoster, aspirin and codeine usually control the pain; no treatment is completely satisfactory for relieving severe postherpetic neuralgia. Treatment of ocular lesions includes atropine and cortisone, which should be administered by an ophthalmologist. Secondary bacterial infections should be treated with appropriate antibiotics.

Controlled studies are in progress to evaluate the reported effectiveness of acyclovir (Zovirax) for treating severe cases of chickenpox and zoster. Iododeoxyuridine appears to be ineffective against zoster when given in the usual doses; however, continuous local application may reduce the duration of pain. Although steroids are contraindicated during the incubation period of chickenpox, it has been demonstrated that pain following zoster may be diminished in duration by early therapy with steroids.

Passive immunization with pooled gamma globulin is not effective for preventing varicella or for treating generalized zoster or severe varicella. However, it has been demonstrated that varicella can be prevented or attenuated by giving zoster (but not chickenpox) hyperimmune globulin to susceptible children within 72 hours after exposure. Immune zoster-globulin is in short supply and should be used only to protect high-risk patients (e.g., newborns, children with leukemia or immune deficiency diseases, and individuals on immunosuppressive drugs).

No vaccines are available for V-Z virus infections.

E. SMALLPOX

Although the World Health Organization (WHO) declared that smallpox has been eradicated from the world, due to confinement and vigorous programs of immunization, the disease has great historic significance. Smallpox occurred in epidemic form in China and Africa as early as the 12th century, B.C. and was introduced into the U.S.A. by Europeans in the 15th and 16th centuries. In 1798, Edward Jenner showed that protection from smallpox was possible by inoculation with material from cowpox lesions. This demonstration began the era of vaccination (vacca = cow). The last known case of smallpox occurred as a result of a laboratory accident in England during 1978. Because of the many important contributions that investigators have made during the course of studying smallpox, a discussion of the disease is important to students of medicine and allied sciences.

Smallpox (variola major) is a contagious febrile disease characterized by vesicular and pustular lesions. *Alastrim* (variola minor) is a form of smallpox that is clinically indistinguishable from mild cases of variola major but consistently has a lower mortality rate. *Vaccinia* is a disease of the skin induced by inoculating vaccinia virus for the prevention of smallpox.

1. Etiologic Agents

Smallpox is caused by *variola virus*. Alastrim is caused by a stable attenuated variant of variola virus that is antigenically indistinguishable from the virulent virus; the origin of the attenuated virus is unknown. Vaccinia is caused by *vaccinia virus*, which was originally obtained from cowpox lesions.

2. Clinical Symptoms

Smallpox may assume many different clinical forms, ranging from a minor febrile illness with no rash *(variola sine eruptione)* to a rapidly fulminating disease.

A typical moderate case of smallpox is diphasic; it can be divided into a preeruptive (prodrome) and an eruptive phase. After an incubation period of approximately 12 days, the illness begins with a vague syndrome consisting of fever (102°–106°F), headache, abdominal pain, vomiting, backache, limb pains, and prostration. In a few patients, the prodromal symptoms include a transient erythematous or petechial rash. After 3–4 days, the fever subsides and there is marked clinical improvement.

When the eruptive phase begins, the patient is usually afebrile. Early manifestations include painful ulcers on the buccal mucosa and macules that appear first on the face and forearms and rapidly develop into papules. The papules increase in number and spread from the face and distal extremities to involve the trunk. A characteristic feature of smallpox is the fact that lesions in any one area are all in the same stage of development. Lesions are generally found on the palms of the hands, which are seldom involved in chickenpox (Fig. 26–11). Within 2 or 3 days after the appearance of the focal rash, the papules develop into

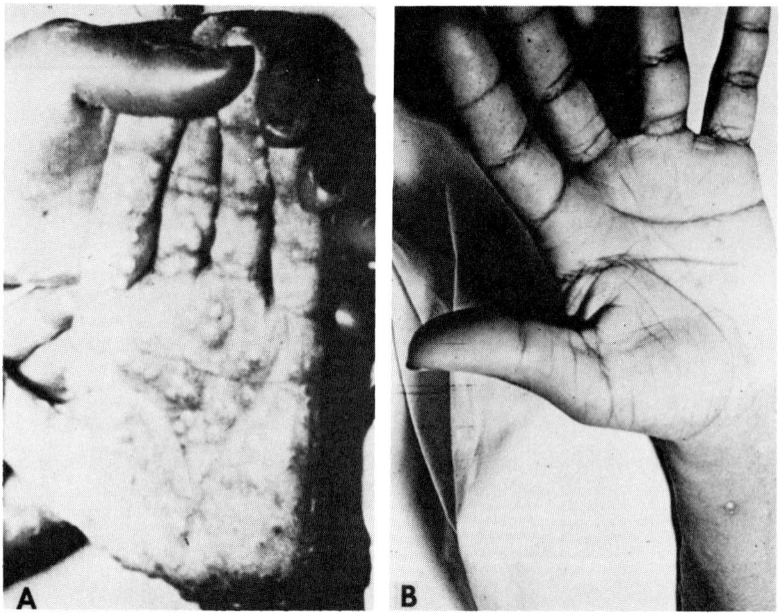

Fig. 26–11. *A*, Variola major; extensive eruptions on palm. *B*, Varicella; minimal involvement of palm. (Courtesy of Center for Disease Control.)

vesicles that contain clear fluid. Shortly thereafter, the vesicles become cloudy and pustular as a result of infiltration of pus cells and desquamated epithelial cells. Concomitant with pustule formation, there is a secondary rise in temperature proportional to the severity of the disease. The fever, which probably results from absorption of toxic products released by cell necrosis, persists until healing begins. About 8 to 9 days after onset of the eruption, the pustules umbilicate and begin to form crusts. The scabs drop off within 3 to 4 weeks after the beginning of the disease. Pitted scarring of the skin is more pronounced on the face and distal parts of the arms and legs, where the eruption is characteristically more severe (Fig. 26–12).

In naturally immunized or vaccinated individuals, the focal eruption may be absent or scant. Following the usual incubation period and prodrome, a rash resembling chickenpox may develop. Regardless of how mild an index case of smallpox may be, susceptible contacts may suffer severe disease. Incorrectly diagnosed or undiagnosed cases frequently have been the cause of severe epidemics.

The case mortality rate for smallpox ranges from 2 percent in patients with a scant, discrete rash to 25 percent if the rash is semiconfluent to 50 percent if the rash is confluent. A case of severe variola major is illustrated in Figure 26–13. The fulminating hemorrhagic type of smallpox results from an overwhelming infection and is almost always fatal.

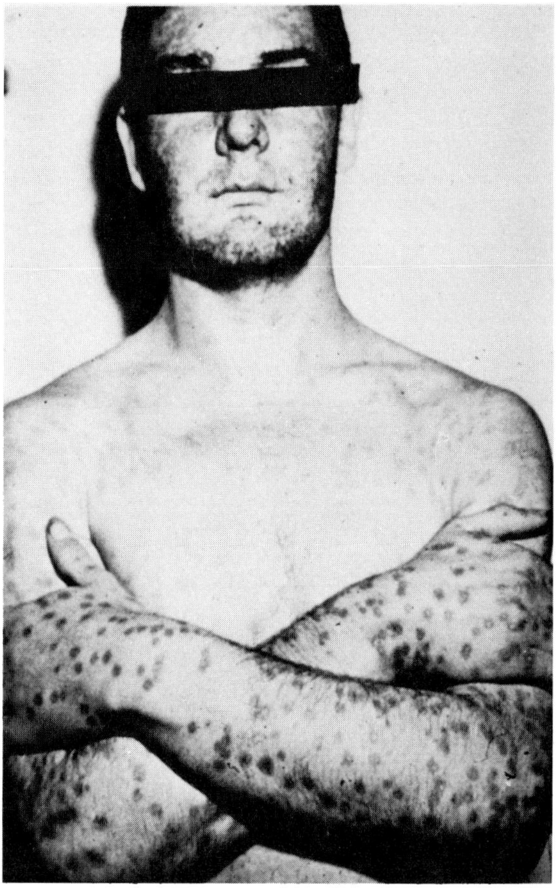

Fig. 26–12. Variola major. Late disease, early scarring. (Courtesy of Armed Forces Institute of Pathology.)

After the usual incubation period, the patient develops a severe prodromal illness characterized by prostration, high fever, bone marrow depression, hemorrhagic skin lesions, bleeding from any or all orifices of the body, shock, coma, and death. The disease may progress from onset to death within 3 to 4 days without evidence of the typical focal eruption.

Alastrim resembles the mild to moderate forms of smallpox in that it has the same incubation period and a similar, but less severe, prodromal illness. The focal eruption is less extensive and is of shorter duration than the focal eruption in smallpox (Fig. 26–14); the case mortality of alastrim is less than 1%. The mildness of the illness makes it difficult to distinguish from chickenpox and benign forms of variola major.

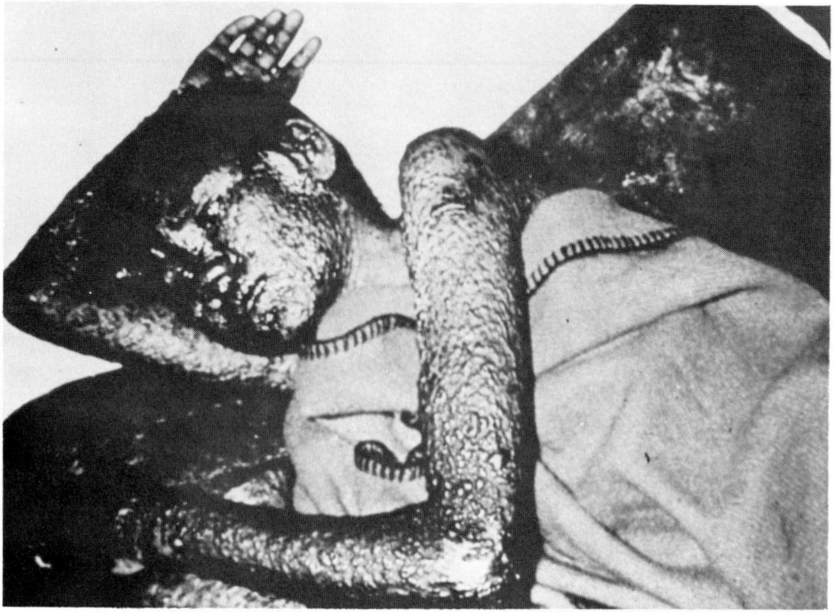

Fig. 26–13. Variola major. Severe infection, 10th to 12th day. (Courtesy of Center for Disease Control.)

3. Pathogenesis

The portal of entry for variola virus is probably the URT. During the incubation period, the virus multiplies at an unknown site, most likely in regional lymphoid tissue. The fact that patients are not infectious during the incubation period suggests that there are no open lesions in the respiratory mucosa. A primary viremia occurs at the onset of fever and persists for the first 2 or 3 days of the prodrome. Virus localizes in the cells of the RES and undergoes a second phase of multiplication. Release of virus from these tissues causes an intense secondary viremia. From the blood stream, virus is distributed to the mucous membranes and skin where it produces the typical focal lesions.

The first pathologic changes leading to skin lesions include capillary dilatation, plasma cell infiltration of the corium, and proliferation of the prickle-cell layer. Infected epithelial cells become swollen and vacuolated and undergo ballooning degeneration. Intercellular and intracellular edema lead to the formation of vesicles where the epithelial cells have been destroyed. The accumulation of epithelial cell debris and the infiltration of neutrophils into the vesicle fluid converts the lesion into a pustule. Infected epithelial cells contain acidophilic intracytoplasmic inclusion bodies, called *Guarnieri bodies,* which are surrounded by a clear unstained halo.

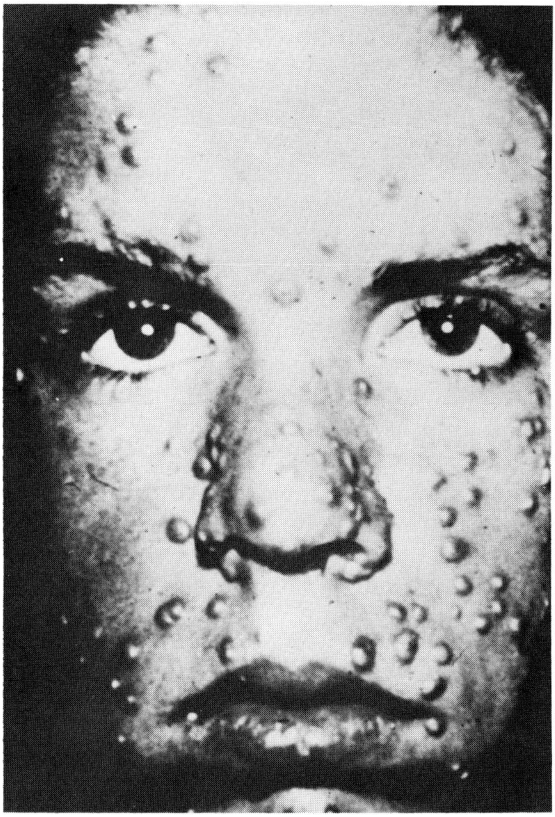

Fig. 26–14. Alastrim (variola minor). Second week. (Courtesy of Center for Disease Control.)

4. Immunity

Active immunity following recovery from smallpox persists for many years and is probably lifelong. The immunity that develops following vaccination is of shorter duration but appears to be effective for at least 3–5 years.

Antibodies can be detected in the blood of smallpox patients as early as the fourth day of the disease, but they do not prevent progression of the local lesions. The role of cell-mediated immunity has not been clearly defined, but it is probably important in suppressing cell-to-cell spread of virus within the lesions. Patients with hypogammaglobulinemia usually respond normally to vaccination and develop immunity, whereas those with defects in both cell-mediated immunity and humoral Ab production develop progressive, often fatal, disease.

5. Diagnosis

Typical cases of smallpox are easily diagnosed on the basis of clinical symptoms, particularly in endemic areas and during epidemics. In countries where smallpox was not endemic, the first case may have been difficult to diagnose. This is particularly true if the patient has either a very mild form of the disease, as a result of previous vaccination, or the fulminating hemorrhagic type in which death may occur before the disease reaches the eruptive phase. Unless there is a history of exposure, a clinical diagnosis cannot be made during the prodromal phase. In the eruptive phase, the most useful clinical features in establishing the diagnosis are (1) the centrifugal distribution of the lesions; (2) the progressive appearance of lesions that spread from the face and arms to the legs and finally the trunk; and (3) the fact that all lesions in a given area of the body are in the same stage of development.

Laboratory procedures usually are required to establish or confirm the diagnosis of the first case of smallpox that appears in the community. A rapid presumptive diagnosis of smallpox can be made by either demonstrating intracytoplasmic inclusion bodies in stained smears of cells from papules and vesicles or by electron microscopic demonstration of poxvirus particles in vesicle fluid. These two procedures are not satisfactory after the vesicular stage of the disease has passed. Viral Ag in vesicles, pustules, and scabs may be demonstrated by fluorescent microscopy, agar gel diffusion or CF tests. Serologic procedures, such as CF, neutralization, and HI tests, may be used to detect an increase in specific Abs during the eruptive phase.

Virus isolation is more time-consuming but is the most reliable way to establish a specific diagnosis. Isolation procedures are particularly important in differentiating smallpox from generalized vaccinia, which may occur following vaccination during an epidemic. Virus may be isolated from the blood during the preeruptive phase of severe disease, particularly the hemorrhagic type in milder cases the viremia may last only a few hours, and often virus cannot be isolated. The focal lesions invariably contain virus at any stage of the disease. The specimens are inoculated onto the chorioallantoic membrane of the chick embryo or susceptible cell cultures. Virus isolates are identified by appropriate serologic tests.

6. Epidemiology

Man is the only natural host and reservoir of variola virus. Although smallpox affects monkeys, they do not appear to be an important source of virus in the absence of human cases.

Smallpox patients are not infectious during the incubation period or the prodrome but become highly infectious in the early eruptive phase. The virus may be spread by either direct or indirect contact. The buccal lesions, which appear before the skin eruptions, ulcerate and cause gross contamination of saliva and respiratory secretions. During this period, virus is spread by droplet transmission. Vesicles and pustules contain large amounts of virus and, when ruptured, provide a major source for dissemination. Virus is also present in exfoliating scabs where it may persist in infectious form for long periods of time. Bed linens and clothes contaminated with virus from open lesions or secretions may serve as an indirect source of infection; in this case, the virus is probably acquired by the inhalation of infected dust.

7. Treatment and Prevention

Treatment. There is no effective treatment for smallpox. Beta-thiosemicarbazone, which is effective prophylactically, is of no value in the treatment of established cases.

Vaccination. The commercially available *live vaccinia virus vaccine* is highly effective in preventing smallpox. This fact has been demonstrated beyond question during the 175 years since it was first popularized by Jenner.

Smallpox vaccine is prepared from vaccinia virus propagated in the skin of calves or sheep. The material is treated with 1 percent phenol to reduce bacterial contamination and is either lyophilized or stabilized with glycerol and stored at −10°C. At −10°C, the vaccine retains its infectivity for about a year. The lyophilized vaccine, however, can be preserved at 0°C or lower for about 5 years; in temperate zones, it retains its potency for approximately a year without refrigeration, but only for a month in the tropics. Once the vaccine has been reconstituted, it can be kept for only a few days in the refrigerator. Recently, a vaccine has been developed that eliminates these inherent disadvantages of the currently available vaccines. It consists of a trifurcated needle device precoated with freeze-dried vaccinia virus. The vaccine appears to be stable for long periods without freezing or refrigeration and can be administered percutaneously without rehydration.

The preferred site for vaccination is the outer aspect of the upper arm over the insertion of the deltoid muscle. Reactions are less likely to be severe in this area than on other parts of the body. Vaccine may be introduced by a variety of techniques, including multiple-pressure, multiple-puncture, and jet injector. The scratch method gives satisfactory results in persons being vaccinated for the first time but may not be effective for revaccination.

Following successful *primary vaccination*, a vesicle develops after 3 to 5 days; subsequently, the lesion becomes pustular and achieves its maximal size after 8 or 9 days (Fig. 26–15A). A scab is then formed, which separates at 14–21 days, leaving a typical vaccination scar.

Revaccination is considered to have been successful if, on examination after 6–8 days, there is either a pustular lesion or an area of definite

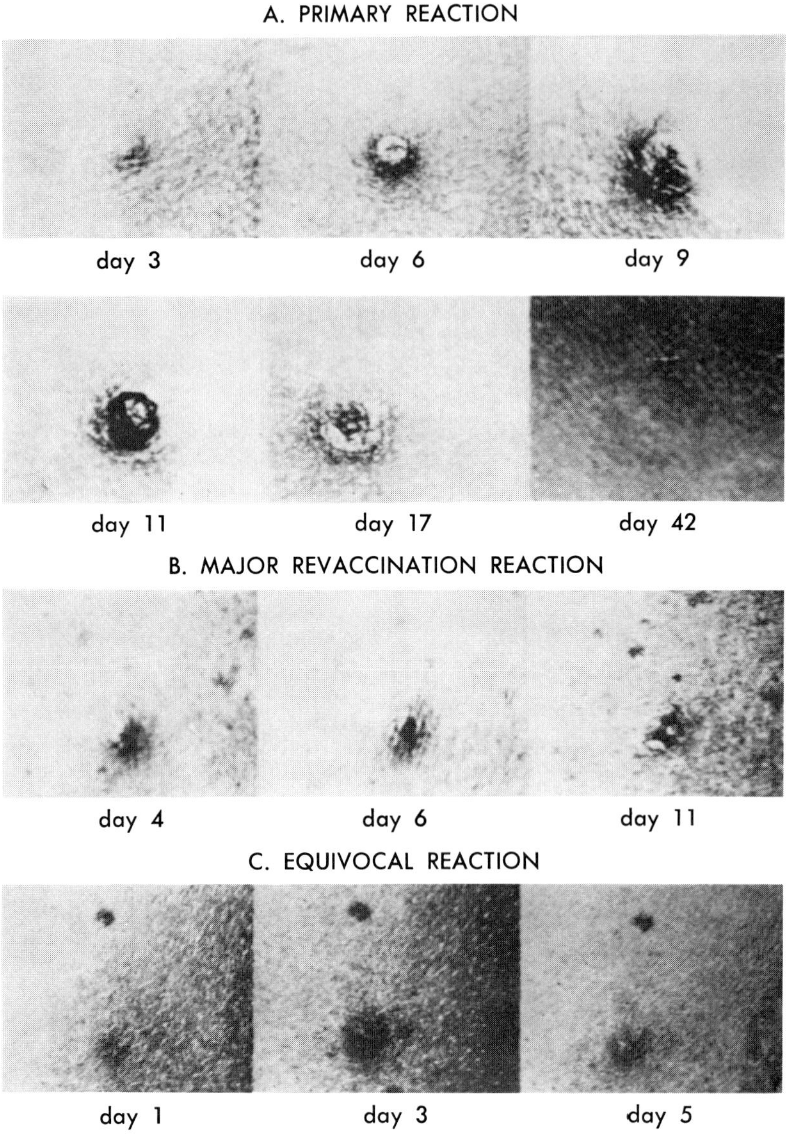

Fig. 26–15. Typical responses to smallpox vaccination.

A, Primary reaction: day 3, papule with early vesiculation; day 6, vesicle; day 9, vesicle with early scab formation; day 11, scab; day 17, scab shed, leaving area of desquamation and discoloration; day 42, remaining scar.

B, Major revaccination reaction: day 4, papule with central scab; day 6, papule covered by scab; day 11, scab shed, leaving small area of desquamation.

C, Equivocal reaction: day 1, erythema and papule; day 3, papule larger, surrounding erythema fading; day 5, papule fading.

(Modified from Elisberg, B.L., McCown, J.M., and Smadel, J.E.: Vaccination against smallpox II. Jet injection of chorioallantoic membrane vaccine. J. Immunol., *77*:340, 1956.)

induration or congestion surrounding an ulcer of a scab. This is termed a "major" reaction (Fig. 26–15B); all other responses to revaccination are termed "equivocal reactions" (Fig. 26–15C). Equivocal reactions, previously classified as accelerated (vaccinoid) or immediate (immune) responses, may be exhibited by persons with varying degrees of immunity. However, an equivocal reaction may also develop in individuals who are allergic to the vaccine or who have been vaccinated with an inactive vaccine. When equivocal reactions are observed in isolated vaccinees who are likely to be exposed to smallpox, the procedure should be repeated at a different site with fresh vaccine.

Complications Resulting from Vaccination. A small but definite risk of serious complications is associated with smallpox vaccination. The risk of death from all complications is about 1 per million primary vaccinees and 1 per 100,000 revaccinees; the risk of death for infants under 1 year of age is approximately 5 per million primary vaccinations. Vaccination is contraindicated for individuals with immune deficiency diseases or conditions requiring the use of immunosuppressive drugs, steroids, or radiation therapy. In addition, pregnant women and individuals with eczema or persons in contact with eczematous patients should not be vaccinated.

Eczema vaccinatum may result when vaccinia virus is transferred from the inoculation site to preexisting skin lesions of either the vaccinated person or a contact (Fig. 26–16). Although many cases of eczema vaccinatum resolve satisfactorily, some progress and may prove fatal.

Progressive vaccinia (vaccinia necrosum) is a rare and serious complication occurring in persons with impaired cell-mediated or humoral immunity. The initial lesion fails to regress and becomes progressively necrotic (Fig. 26–17). Foci may develop in other parts of the body, including the bones and internal organs. The disease is usually fatal.

Postvaccinal encephalitis is a rare complication that occurs most often during the second week after primary vaccination; it is extremely infrequent after revaccination. There are no known predisposing factors. Although the etiology is not clear, the timing suggests that the encephalitis may be caused by Ag-Ab complexes. Whereas most persons affected with postvaccinal encephalitis recover completely, a few die, and a few of those who recover have residual neurologic sequelae.

Autoinoculation or *accidental infection* of contacts may result when vaccinia virus is transferred from the vaccination site to mucous membranes or to abraded skin surfaces. Localized lesions develop at the infection sites in the absence of predisposing factors. Except in the rare instances in which ocular infection is followed by permanent scarring of the cornea, this complication has no serious consequences.

Generalized vaccinia may develop following a transient viremia. Vaccinial lesions begin to appear 5–10 days after vaccination and may be widely distributed over the body. There are no known predisposing

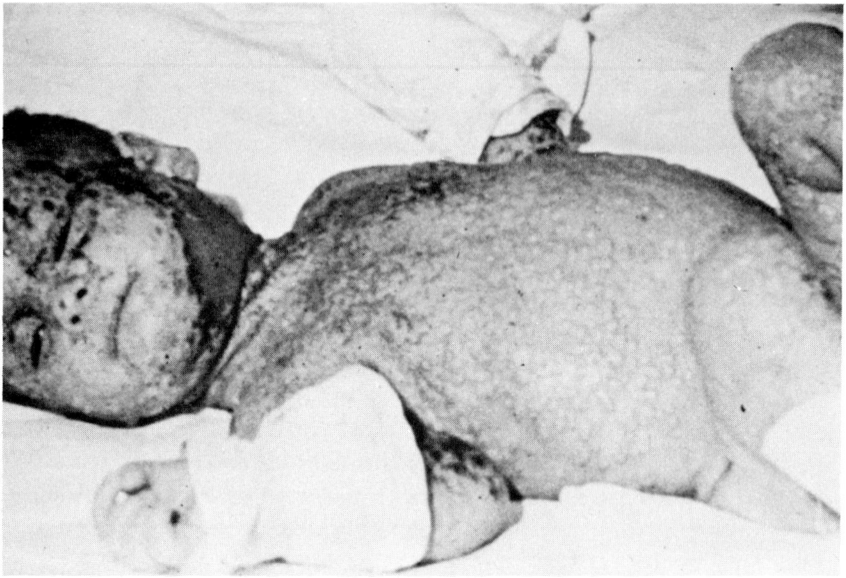

Fig. 26–16. Eczema vaccinatum. Eight-month-old boy with eczema vaccinatum who acquired vaccinia from a recently vaccinated sibling. (Courtesy of Center for Disease Control.)

factors; generalized vaccinia may occur in individuals with no detectable immunologic deficiencies and in the absence of preexisting skin lesions. This complication is not progressive and is never fatal.

Hypersensitivity reactions in the form of urticarial, morbilliform, and erythema multiform eruptions are sometimes seen and may be confused with generalized vaccinia.

Fetal vaccinia may follow maternal vaccination in the 3rd to the 24th weeks of gestation. Most fetal infections result in stillbirths. Infants who survive may have extensive hemorrhagic lesions (Fig. 26–18). Depending upon the severity of the infection, the disease can be fatal or the infant may recover with no effects other than scarring.

Patients with eczema vaccinatum, progressive vaccinia, generalized vaccinia, or ocular vaccinia should be treated promptly with vaccinia immune globulin. The use of immune globulin has reduced the fatality rate of eczema vaccinatum from 40 percent to about 1 percent. Although globulin administered at the time of vaccination reduces the incidence of encephalitis, it is not effective in the treatment of encephalitis.

F. CONTAGIOUS PUSTULAR DERMATITIS (ORF) AND MILKERS' NODULE

Contagious pustular dermatitis is a disease of sheep that is occasionally transmitted to man. Milkers' nodule is a benign infectious disease that

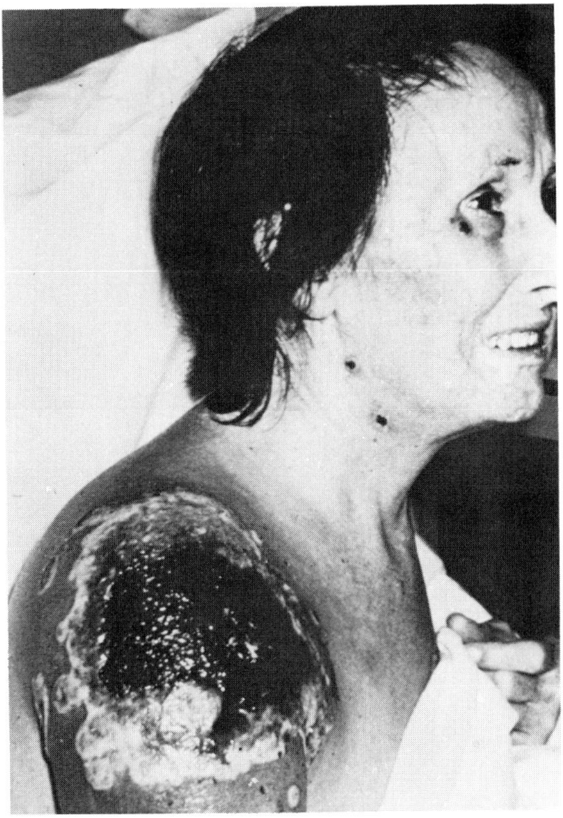

Fig. 26–17. Progressive vaccinia. Sixty-two-year-old patient with chronic lymphocytic leukemia who was vaccinated as a therapeutic measure for herpes simplex. (Courtesy of Center for Disease Control.)

man contracts from cows. Both diseases are characterized by painless dark red papules that become enlarged and may ulcerate.

1. Etiologic Agents

Contagious pustular dermatitis and *Milkers' nodule viruses* are morphologically identical poxviruses that are distinguished mainly on the basis of their natural hosts. The relationship between the viruses is obscure. It has been shown that inoculation of sheep with bovine virus from milkers' nodule causes lesions identical to those of Orf.

2. Clinical Symptoms

In man, the incubation period for both diseases is 4–11 days. The eruption begins as single or multiple lesions, usually on the hands or face. The initial lesions are small, reddish-blue macules or papules that

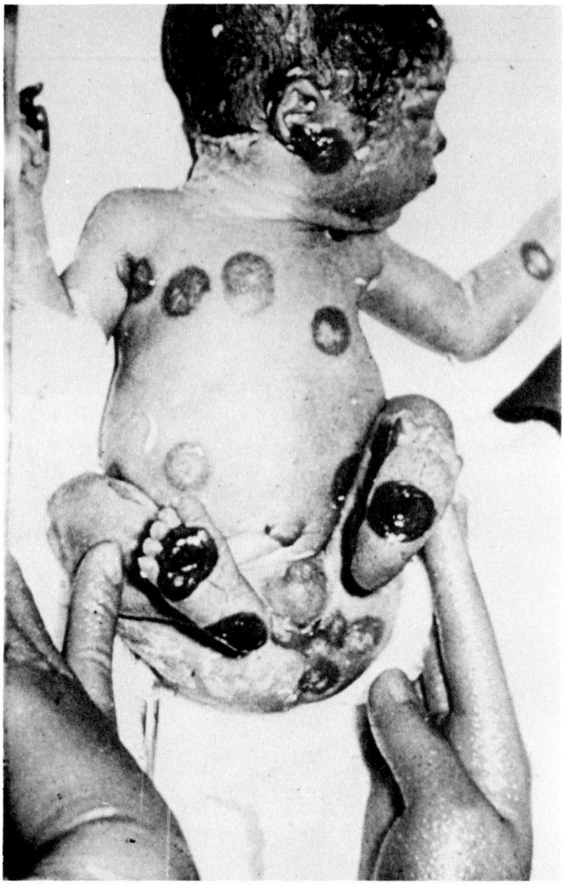

Fig. 26–18. Fetal vaccinia in a child born in the 28th week of gestation. (Courtesy of Center for Disease Control.)

enlarge to form vascularized bullae from which little fluid can be expressed. The surface gradually becomes white and sodden, a crust forms, and a granulomatous reaction may develop. The mature lesions are elevated or tumor-like and well demarcated from normal skin. Systemic manifestations are rare, except for occasional low-grade fever and mild regional lymph node involvement. Healing occurs spontaneously in 4–6 weeks with little or no scar formation.

3. Pathogenesis

These infections result in proliferation, ballooning, and reticular degeneration of epidermal cells. There is multilocular vesiculation and hyperplasia that can mimic malignancy of the epithelium. A granulomatous reaction is present in the dermis.

4. Immunity

Neutralizing and CF Abs have been demonstrated in the blood of recovered animals and man. A single attack seems to confer lasting immunity to reinfection.

5. Diagnosis

Diagnosis usually is on the basis of the characteristic lesions in individuals with a history of contact with sheep or cows. The diagnosis can be confirmed by virus isolation in tissue culture and by demonstrating a rise in titer of serum neutralizing Ab against the specific agent.

6. Epidemiology

Orf is a worldwide disease of sheep and goats that causes crusty, warty, or granulomatous lesions on the lips, nose, or eyelids. Milkers' nodule is a natural infection of cows and causes chronic small papulovesicular lesions on the teats. The diseases may be transmitted to sheepherders, dairy farmers, butchers, and veterinarians by direct contact with infected animals or animal products or virus-contaminated dried crusts shed from lesions into grazing pastures. Both viruses are resistant to drying and remain infectious for months to years in dried crusts.

7. Treatment and Prevention

Treatment of both diseases is symptomatic and is directed toward prevention or cure of secondary bacterial infection. Prevention of infection with these viruses in man involves control of the diseases in animals. Vaccination with live vaccines and quarantine of infected pastures are effective in eliminating contagious pustular dermatitis in sheep. Flocks of sheep and herds of cattle in which the disease is widespread should be destroyed.

G. MOLLUSCUM CONTAGIOSUM

Molluscum contagiosum is a benign infectious disease of the skin characterized by the presence of small pearly nodules.

1. Etiologic Agents

The *molluscum contagiosum virus* has not been grown in the laboratory but has been classified morphologically with the poxviruses.

2. Clinical Symptoms

The incubation period of molluscum contagiosum is not known; on the basis of human volunteer studies, different investigators have reported it to range from 2 weeks–2 months. The disease usually occurs in children or young adults. It is characterized by the formation of painless, pearly white, wart-like nodules limited to the epidermal layer of the skin and mucous membranes. The lesions appear most frequently on the face, trunk, and anogenital areas; the conjunctiva, lips, and buccal mucosa are rarely affected, and no lesions develop on the palms of the hands or soles of the feet. The lesions may vary in size and number, ranging from 1 mm–2 cm in diameter. The nodules are umbilicated and contain a white core of curdlike material that contains numerous virus particles. The lesions usually persist for 6 months to a year or more, but spontaneous regression eventually occurs without scarring.

3. Pathogenesis

The nodules of molluscum contagiosum are characterized by epithelial cell proliferation, hyperplasia, skin thickening, and degeneration. Light microscopy reveals that infected cells are greatly enlarged and contain large intracytoplasmic, eosinophilic inclusion bodies that displace the nucleus to one site. These "molluscum bodies" consist of masses of virus particles embedded in a spongy protein matrix that may be divided into cavities.

4. Immunity

Convalescent human serum contains Abs against Ag extracted from skin lesions. Some sera possess CF Abs against a soluble, heat-labile Ag obtained from suspensions of nodules. The role of these Abs in immunity is not known, and the immune mechanisms that mediate regression remain to be elucidated.

5. Diagnosis

Molluscum contagiosum is diagnosed on the basis of the characteristic appearance of the lesions and histologic examination of the core which is composed of clusters of cells containing the diagnostic giant eosinophilic inclusion bodies.

The virus has not been propagated in tissue cultures or experimental animals, and no satisfactory serologic tests are available for diagnostic use.

6. Epidemiology

Man is the only known natural host of molluscum contagiosum virus. The disease has a worldwide distribution and has occurred in epidemics within orphanages and family groups. The exact mode of transmission is not known, but the virus appears to be spread by direct contact or by contaminated fomites. Autoinoculation from suppurating lesions appears to be common and may account for spread of the virus from one part of the skin to another; outbreaks have occurred among wrestlers.

7. Treatment and Prevention

There is no effective antiviral treatment. Lesions can be removed surgically.

H. WARTS (VERRUCAE)

Warts are epithelial tumors of the skin and adjoining mucous membranes.

1. Etiologic Agent

In man, warts are caused by human *papillomavirus*. One virus, or closely related strain of a single virus appears to cause several different types of warts, depending upon the site of infection and host reaction.

2. Clinical Symptoms

The incubation period, determined by inoculation of volunteers, varies between 1 and 20 months and averages about 4 months. Single or multiple warts can occur at any site on the skin or mucous membrane adjacent to the skin. Warts may persist and spread, presumably by autoinoculation, for several years. However, most lesions eventually regress spontaneously without scarring, but they may recur even after treatment. The lesions are classified according to morphology and location.

Sessile or *common warts (verruca vulgaris)* are raised, gray or brown lesions with a rough surface. They are usually seen on the hands and around or under the fingernails; they also may occur on the feet, legs, face, and neck. They vary in size from 1 mm–2 cm; clusters of lesions may become confluent.

Filiform warts are horny finger-like projections that are a few millimeters in diameter and several millimeters long. They usually occur on the bearded area of the face, the neck and the scalp. The lips and eyelids may also be involved.

Plantar warts are flat and have a horny surface; they are demarcated from normal skin by a hyperkeratotic ring. The mass of the lesion is beneath the skin surface and is conical, with the pointed end projected inward. Plantar warts occur principally beneath pressure points on the soles of the feet and are often quite painful. Some plantar warts are not visible from the surface and are recognized only by acute pain at a calloused pressure point. In contrast, "mosaic plantar warts," produced when a number of lesions coalesce, are relatively painless keratotic lesions that may not be recognized.

Flat warts (verruca planae) are only slightly elevated, smooth, skin-colored papules that are 1–5 mm in diameter. They usually occur in multiples on the face, neck, chest, and back of the hands.

Genital warts (condylomata acuminata) are moist, soft, pink or white lesions that may be clustered together to produce cauliflower-like growths. They appear most frequently on the moist skin of the external genitalia and the perianal region and on the vulvar and vaginal or anal mucosae. However, they may also occur at other moist sites, such as the conjunctiva, margins of the mouth, and between the toes and the larynx. Because of their location, moist warts may become macerated and malodorous. Condylomata acuminata is the only type of human wart that may become malignant.

3. Pathogenesis

The principal pathologic changes observed in warts occur in the epidermis. Depending upon the type and location of the lesion, varying degrees of hyperplasia of the prickle cell, granular, or horny layers may occur. Characteristic large vacuolated cells appear in the upper prickle cell and granular layers; the nuclei of these cells may be large and variable in size. Nuclear inclusion bodies are frequently observed in plantar warts but only occasionally in common warts.

4. Immunity

Low titers of specific IgM Abs against human papilloma virus can be detected in patients carrying warts; as the warts regress, serum IgG Abs develop. The role of the immune response in spontaneous regression and resistance is by no means clear. Following treatment of one wart, untreated lesions sometimes disappear. The increased incidence of warts in patients receiving immunosuppressive drugs after kidney transplantation suggests that the immune response is important in resistance.

5. Diagnosis

Most warts can be recognized by their appearance. The diagnosis may be confirmed by histologic examination. A persistent single moist wart on the genitalia should be biopsied to rule out a malignancy such as squamous cell carcinoma. In addition, some lesions of secondary syphilis may be confused with virus-induced genital warts. Mosaic plantar warts may not be apparent until the epidermal layers are successively removed to reveal the infected rounded white mass.

6. Epidemiology

Warts are quite common; their distribution is worldwide, and both sexes and all ages are affected. They are contagious, but the sources of infection and modes of transmission are frequently unknown. Minor abrasions may be important in establishing the infection. Autoinoculation apparently accounts for the appearance of satellite lesions around traumatized older warts.

7. Treatment and Prevention

Treatment of warts is empirical and depends upon the number, location, and type of lesion. Since most warts regress spontaneously, treatment should be conservative to avoid excessive local irritation and scarring. In general, therapy involves the physical destruction or removal of the lesions. If deemed necessary, most sessile, filiform, or flat warts can be eliminated by electrodesiccation or freezing with liquid nitrogen or dry ice. Weekly application of podophyllin in tincture of benzoin is recommended for moist warts. Plantar warts, which are especially hard to eliminate, may require daily application of salicylic acid or other caustic agents.

There are no known methods for preventing warts.

REFERENCES

Arita, I.: Virological evidence for the success of the smallpox eradication program. Nature, *279*:293, 1979.

Arvin, A.M., Kushner, J.H., Feldman, S., Baehner, R.L., Hammond, D., and Merigan, T.G.: Human leukocyte interferon for the treatment of varicella in children with cancer. N. Engl. J. Med., *306*:761, 1982.

Corey, L., Nahmias, A.J., Guinan, M.E., Benedetti, J.K., Critchlow, C.W., and Holmes, K.K.: A trial of topical acyclovir in genital herpes simplex virus infections. N. Engl. J. Med., *306*:1313, 1982.

Froser, K.B. and Martin, S.J.: Measles virus and its biology. New York, Academic Press, 1978.

Morgan, E.M., and Rapp, F.: Measles virus and its associated diseases. Bacteriol. Rev., *41*:636, 1977.

Oldstone, M.B.A. and Tishon, A.: Immunologic injury in measles virus infection. IV. Antigenic modulation and abrogation of lymphocyte lysis of virus-infected cells. Clin. Immunol. Immunopathol., *9*:55, 1978.

Saral, R., Burns, W.H., Laskin, O.L., Santos, G.W., and Lietman, P.S.: Acyclovir prophylaxis of herpes simplex virus infections: A randomized, double-blind, controlled trial in bone marrow transplant recipients. N. Engl. J. Med., *305*:63, 1982.

Weibel, R.E., Buynak, E.B., McLean, A.A., and Hilleman, M.R.: Persistence of antibody after administration of monovalent and combined live attenuated measles and rubella virus vaccines. Pediatrics, *61*:5, 1978.

Wollensak, J.: Herpes simplex of the eye and possibilities of therapeutic control. Adv. Ophthalmol., *38*:99, 1979.

Yancey, R.B. and Smith, J.G., Jr.: Interferon: status in treatment of skin disease. J. Am. Acad. Dermatol., *3*:585, 1980.

Chapter 27

Viral Neurotropic Diseases

Viruses are responsible for most cases of meningitis and essentially all cases of encephalitis in man. *Viral encephalitis* by definition is an inflammatory disease of the cerebrospinal axis produced by the direct effect of virus on neurologic tissues. The inflammation may spread to the meningeal spaces, resulting in *meningoencephalitis,* or to the spinal cord, resulting in *meningomyeloencephalitis.* In some cases of encephalitis, disease is not due to direct action of virus. For example, *postinfectious encephalitis* following recovery from measles is considered to involve an allergic reaction to viral Ag. The etiology and pertinent characteristics of viral diseases involving the central nervous system (CNS) will be discussed in this chapter.

A. POLIOMYELITIS

Until the present century, poliomyelitis was primarily a disease of infants, hence the name *infantile paralysis.* However, as a result of improved sanitation over the past 50 years, the age distribution of the disease changed to include young adults. During a peak incidence of poliomyelitis in the U.S.A. in 1953, there occurred about 1,500 deaths and 7,000 cases with residual paralysis. The unfortunate crippling of many survivors of polio, including Franklin D. Roosevelt, caused great concern about the disease. Although it was one of the most feared diseases only some 3 decades ago, some medical students of today may never have occasion to diagnose and treat a clinical case of poliomyelitis. This remarkable change in the incidence of poliomyelitis was due to research, supported in large part by the March of Dimes Fund. Enders, Weller, and Robbins (1949) were the first to report the successful growth of poliovirus in tissue cultures of nonneural cells. This work, which earned for them the Nobel Prize in Medicine, provided great impetus to modern research in virology and led to the subsequent production of the Salk and Sabin vaccines. Widespread use of these vaccines has almost completely eradicated poliomyelitis from developed countries.

For example, in the U.S.A., the annual number of cases of paralytic disease decreased from more than 18,000 in 1954 to an average annual number of less than 13 in 1973–1980. Small outbreaks have occurred in 1970, 1972 and 1979 as a result of introduction of wild-type virus into susceptible populations in communities with low immunization levels. Inapparent infection with wild type strains of virus no longer contributes significantly to establishing and maintaining immunity, making universal vaccination of infants and children extremely important.

The history of the research that led to a successful poliomyelitis vaccine is especially deserving of comment because, like many other important areas in science and medicine, it illustrates admirably how formidable a problem can be before needed concepts and techniques are developed, and how simple the solution seems in retrospect. The first breakthrough in poliomyelitis came in 1947 when strong arguments and data were presented to support the concept that the virus grows in nonneural as well as neural tissue, a view that opposed the long held dogma that the organism was obligatorily neurotropic. The second breakthrough, the cultivation of the virus in cultures of nonneural cells, followed rapidly (1949); the development of the first vaccine (Salk) was finally accomplished in 1953. Thus, within the short span of a few years after the acceptance of a new concept, an effective method for controlling this important disease, so terrifying to parents, was at hand.

Because of the great historic importance of poliomyelitis and the knowledge that has accumulated about the virus and the disease, together with the contributions that poliomyelitis research has made to the study and development of other important virus vaccines, a discussion of poliomyelitis and poliovirus is instructive and important to students of medicine and related sciences.

Etiologic Agents. Three antigenic types of poliovirus (types 1, 2 and 3) exist that are distinguishable by neutralization tests using specific antisera. *Poliovirus type 1 is the most common cause of paralytic polio.* Avirulent mutants of poliovirus have become widely distributed in nature because of the widespread use of attenuated virus vaccines (Sabin).

Clinical Symptoms. When a nonimmune individual is infected with poliovirus, one or more of the following responses may occur: (1) inapparent infection (occurs most often); (2) mild infection characterized by fever, malaise, and sore throat with or without nausea and headache; (3) aseptic meningitis; or (4) paralytic poliomyelitis. *Only about 1 percent of persons with apparent poliovirus infection develop paralysis.* After an incubation period of 5–20 days, the disease may follow a *diphasic course:* The initial mild infection may be followed by a few symptom-free days prior to the development of muscle stiffness and pain commencing 1–3 days later. Paralysis develops rapidly after infection of the motor cortex of the brain. Infection of the anterior horn cells or brain stem results in bulbar polio, which may be fatal because of cardiac and respiratory failure. Some degree of recovery of motor function may occur during the first few months but paralysis remaining at the end of this period is permanent.

Pathogenesis. Virus enters the alimentary tract following ingestion of contaminated food or drink. The virus first multiplies in the tonsils, lymph nodes of the neck, Peyer's patches, and the mucosa of the small intestine. It appears in the throat and feces before symptoms develop. One week after the onset of symptoms, virus is found in the blood; virus is excreted in the feces for several weeks despite the development of high levels of circulating Abs. Occasionally, the CNS is invaded during the viremic phase of the disease. Virus can spread from the blood or along axons of peripheral nerves to the anterior horn cells of the spinal cord. In severe cases, the intermediate gray ganglia as well as the posterior horn and dorsal route ganglia are involved. Intracellular multiplication of the virus in neurons leads to complete destruction of nerve cells. Inflammation occurs secondary to infection of neurons, with infiltration of lymphocytes, sometimes neutrophils, plasma cells and microglial cells. Cells injured as the direct result of virus infection or by the inflammatory response of the host may recover completely during convalescence. The probability of CNS involvement with residual paralysis is enhanced in adults and in circumstances such as pregnancy, tonsillectomy, trauma, and fatigue. In particular, tonsillectomy increases the incidence of bulbar poliomyelitis. Irritation resulting from injection of various materials or trauma predisposes to paralysis in the affected limb. Nerves in areas of irritation or trauma are highly susceptible to invasion by virus. Virus spreads from these areas along peripheral nerves to reach the corresponding segment of the spinal cord or brain stem.

In contrast to the thousands of cases of paralysis caused by polioviruses in the past, there have been relatively few cases of pseudopoliomyelitis caused by other viruses (echoviruses and coxsackieviruses A and B). These enteroviruses were discovered by accident in the course of investigations on poliomyelitis. The differential diagnosis of enterovirus infections is difficult clinically, and virologic diagnosis is imperative. Polioviruses have a high affinity for the nervous system, whereas other enteroviruses have a weak affinity for this organ system. Although rare, a poliovirus infection can coexist with infection due to either echovirus or coxsackievirus.

Immunity. Passive immunity against polioviruses is transferred from mother to offspring during pregnancy. It results from maternally derived Abs and gradually decreases during the first 6–9 months of life. In contrast, passive immunity conveyed by the administration of hyperimmune serum lasts only 3–5 weeks. Naturally acquired active immunity is permanent and is usually *type-specific (homotypic),* which explains why second attacks of poliomyelitis can occur. Occasionally, a low degree of *cross-resistance (heterotypic immunity)* between types 1 and 2 polioviruses can develop.

Diagnosis. During the early stages of disease, virus can be cultured from pharyngeal secretions by inoculating monkey kidney cells. After paralysis has become apparent, virus is readily isolated from the intestinal tract, but is not easily recovered from the cerebrospinal fluid (CSF). Virus multiplies rapidly (1–3 days), and the isolate can be identified by tissue culture neutralization tests using hyperimmune serum. The immune status of patients can be established by determining their serum titers of neutralizing Ab against the 3 poliovirus serotypes.

Epidemiology. The only known reservoir of poliovirus is man. The virus is maintained in the population by carriers and is usually spread by direct contact. Poliovirus can be recovered from the pharynx and feces of patients for several weeks after the acute phase of the disease, as well as from healthy carriers, and is moderately stable in the external environment. These facts explain how, under crowded conditions and poor sanitation, the virus can easily spread from infected to noninfected individuals.

When conditions of hygiene and sanitation are poor, almost all infants are heavily exposed to the virus while they are still under the protection of maternally-derived Ab and thus acquire active immunity early in life without appreciable danger of developing paralytic disease.

Poliovirus infections are endemic throughout the world. *In isolated populations,* such as the Eskimos, poliomyelitis occurs in all age groups. *In crowded primitive areas with poor sanitation,* clinical poliomyelitis is infrequent because essentially all children and older individuals have developed active immunity from heavy exposure to the virus. In these areas, the susceptible individuals are primarily those few infants who do not receive sufficient maternal Ab to protect them. *Epidemics in temperate zones* are most likely to occur during the summer and fall. Spread of infection takes place most readily among nonimmunized school children and family contacts. Poliovirus type 1 is the most common serotype in Western countries. *In many developed countries,* the greatest age incidence of the disease has shifted to older, unvaccinated individuals. This change has been brought about by strong "herd immunity" induced in the younger age groups by the widespread use of poliovirus vaccine. The practices of immunization and improved hygiene have essentially eliminated the wild-type virus from many communities, and in consequence, have reduced the booster value of frequent exposure to poliovirus for maintaining immunity in later years. Although waning immunity in an individual residing in a virus-free population does not ordinarily pose appreciable risk to him, he may be at high risk if he travels to an area where virulent virus is abundant in the population.

Compared with most diseases, paralytic poliomyelitis in unvaccinated populations is a paradox, the lowest incidence of paralytic disease being favored when crowding is greatest and sanitation is the poorest.

Treatment and Prevention. Both live (Sabin) and killed (Salk) vaccines are available for general use (Chapter 23). *Immunization with the Salk or Sabin vaccine is of prophylactic but not therapeutic value.* The Sabin vaccine is administered orally; it contains live, attenuated virus grown in cultured monkey kidney cells. Viral contaminants may be present in some cultured monkey kidney cells; hence, the human diploid cell strain (WI-38), free of detectable microbial contamination, has recently been licensed for vaccine production. The poliovirus contained in the oral vaccine invades the intestinal epithelium and multiplies; as a consequence, it immunizes the recipient by inducing the production of humoral Abs (mainly IgA Abs) in the wall of the small intestine. Recipients are protected not only against spread of natural polioviruses through the bloodstream to the spinal cord, but also against initial multiplication of such viruses in the small intestine. Repeated vaccinations are recommended to establish permanent immunity. The protection rate induced by Sabin vaccine is 80–90 percent; however, due to the high "herd immunity" and reduction in carriage rate that results from vaccination, the overall effectiveness of vaccination approaches 100 percent following 2 doses of vaccine. The live poliovirus vaccine is the vaccine of choice for primary immunization of children when the benefits and risks of the entire population are considered.

The killed (Salk) poliovirus vaccine induces humoral Abs and protects the CNS from invasion of wild-type virus. *In contrast to the Sabin vaccine, the Salk vaccine does not prevent intestinal infection by poliovirus.* An improved killed poliovirus vaccine has been developed in Europe. Studies with the vaccine revealed essentially 100 percent seroconversion following 2 doses.

Hyperimmune gamma globulin affords limited (3–5 weeks) prophylactic protection against the paralytic disease; it has no value after symptoms appear.

During an epidemic, which is defined as 2 or more local cases caused by the same serotype of poliovirus in any 4-week period, bed rest is recommended for children with fever. Nose and throat operations should be avoided. Quarantine of patients or household contacts is ineffective for controlling spread of the disease. Food and human excreta should be protected from flies. All pharyngeal discharges are considered to be infectious and should be disposed of properly. Once the serotype of poliovirus responsible for the epidemic has been determined, type-specific monovalent Sabin vaccine should be administered to all susceptible individuals in the population.

B. ASEPTIC MENINGITIS

Most cases of viral meningitis are secondary to infection elsewhere in the body: most are of the aseptic type. The term "aseptic" refers to the type of meningitis in which the spinal fluid is clear, in contrast to the purulent spinal fluid often seen in bacterial meningitis. Before the advent of antibiotic therapy, bacteria, rather than viruses, were the most frequent cause of meningitis. *Mumps virus*, certain *coxsackieviruses*, and *echoviruses* are among the common causative agents. Less common etiologic agents of aseptic meningitis include *enteroviruses* (e.g., *polioviruses*), *herpes simplex*, *mengo*, and *lymphocytic choriomeningitis (LCM) viruses* (Table 27–1).

Untreated cases of viral meningitis carry an excellent prognosis; this is in sharp contrast to bacterial meningitis. Even though viral meningitis occurs in epidemic proportions, mortality generally is low. Lumbar punc-

Table 27–1. Etiology of Viral Aseptic Meningitis

Viruses	
Common	*Less Common*
Mumps	Other Coxsackie and ECHO Serotypes
Coxsackie A 7, 9, 23	Polio types 1, 2, 3
ECHO 4, 6, 9, 16, 30	Herpes simplex
Coxsackie B 1–6	Mengo
	Lymphocytic choriomeningitis
	Hepatitis

ture is necessary in order to distinguish between viral and bacterial meningitis.

1. Mumps Virus Meningitis

Meningitis may appear before, during, or after mumps parotitis. In some cases, meningitis may be the only apparent manifestation of mumps virus infection. It often occurs in a crowded environment (e.g., schools and military camps). *In comparison to other viral meningitides, mumps meningitis is the most common, the most prolonged and often the most violent.*

ETIOLOGIC AGENTS. A single serotype of *mumps virus* is responsible for disease in man; structurally it resembles other paramyxoviruses (Chapter 13).

Clinical Symptoms. The incubation period ranges from 16–20 days. One out of every 3 cases of mumps virus infection is inapparent. A prodromal period of malaise and anorexia may or may not be followed by enlargement of the parotid glands before invasion of the CNS. Mumps virus is responsible for about 10–15 percent of the cases of aseptic meningitis in the U.S.A., and is more common in males than females. The disease usually develops 5–7 days after the prodromal period. The CSF contains an average of 200–600 leukocytes per cm^3, most of which are lymphocytes. Other rare complications of mumps include polyarthritis, pancreatitis, nephritis, and thyroiditis.

PATHOGENESIS. Virus enters through the respiratory tract. Primary replication of virus is thought to occur in the parotid gland or epithelium of the respiratory tract. Virus is then carried by the blood to the CNS and other organs of the body.

IMMUNITY. Acquired immunity as a result of apparent or inapparent infection is considered to be lifelong.

DIAGNOSIS. In cases of mumps aseptic meningitis, without parotitis, laboratory studies are necessary to establish the etiologic diagnosis. Virus is generally recovered from saliva, blood, and CSF for the first few days of illness and later from urine. The specimen is inoculated directly into primary human or monkey kidney tissue cultures or chick embryos. After 3–5 days incubation, the cell cultures are monitored by hemadsorption (HAd), or the amniotic fluid from inoculated chick embryos is assayed for viral hemagglutinin.

Complement fixation (CF) or hemagglutination-inhibition (HI) tests using acute and convalescent sera may be used to diagnose mumps virus meningitis. The CF test is recommended for sensitivity, specificity, and accuracy. A 4-fold increase in Ab titer in convalescent serum is considered to be diagnostic of a preceding mumps virus infection.

Skin test Ag for detecting susceptible persons who have escaped mumps virus infection is available commercially; however, the test is of no value in diagnosis. Delayed hypersensitivity to mumps virus Ag develops during convalescence; sensitivity usually appears within 3–4 weeks after the onset of symptoms. A positive skin test indicates previous exposure to mumps virus; about 80 percent of the adult population is skin-test-positive.

Differential diagnosis should include meningitis caused by *Mycobacterium tuberculosis, Leptospira sp., Cryptococcus neoformans,* and other common bacteria (e.g., *Hemophilus sp., staphylococci,* and *pneumococci).*

EPIDEMIOLOGY. Mumps virus is endemic throughout the year, with peak periods during the winter and spring. Aseptic mumps virus meningitis, which carries a low mortality, reaches its highest incidence in children 5–15 years of age; however, epidemics also occur in army camps.

Mumps virus is spread by direct contact, droplets, and fomites, and is excreted in the urine of infected individuals. The disease is communicable from about 4 days before to about 7 days after symptoms develop. More intimate contact is needed for transmission of mumps than for measles or varicella-zoster viruses. In about 30–40 percent of cases, invasion by mumps virus results in inapparent infections; these individuals can carry and transmit the virus to susceptible individuals. Protective maternal Abs specific for mumps virus cross the placenta and gradually diminish during the first 6–9 months of life.

TREATMENT AND PREVENTION. Prevention of mumps virus infection does not carry the same priority as prevention of measles, polio, diphtheria, or tetanus. However, vaccination with live attenuated mumps virus vaccine is recommended at puberty if there is no clinical history of mumps. Adults who show a negative skin test or who lack humoral Abs to mumps virus should also receive the vaccine. A single dose of vaccine induces detectable Abs in 95 percent of vaccines; however, the duration of immunity following vaccination is not known. There is no specific treatment for mumps virus meningitis.

2. Echovirus Meningitis

The echoviruses form the most important and best-defined group of viruses causing aseptic meningitis; they cause many summer epidemics of the disease.

Etiologic Agents. Echovirus meningitis is most commonly due to virus types 4, 6, 9, 16, and 30.

Clinical Symptoms. The incubation period is short (3 to 5 days), followed by a rapid onset with headache, vomiting, sore throat, muscle pains, and fever. In 35–60 percent of cases, the first symptoms disappear briefly before a second phase of clinical disease, characterized by fever and meningeal signs begins. A maculopapular rash, which may appear at the onset of the second phase, usually persists during the febrile period and greatly facilitates diagnosis during epidemics. The non-itching maculopapular rash, which is found on the face, neck, thorax, trunk, and extremities, presents a pale red lesion, about 2–4 mm in diameter. Occasionally, discrete grayish-white lesions 1–2 mm in diameter are present on the tonsils, buccal mucosa, and sometimes the tongue.

Pathogenesis. The pathogenesis of echovirus meningitis is similar to that of meningitis caused by other enteroviruses. Virus multiplication occurs in the alimentary tract (throat and small intestine) before viremia and invasion of the CNS occurs. Infection with two or more enteroviruses may occur simultaneously.

Immunity. Both neutralizing and CF Abs are detectable following recovery.

Diagnosis. The procedure of choice is to isolate the virus from feces, a throat swab, or the CSF by propagation in human or monkey kidney cell culture. If possible, the isolate should be identified by the virus neutralization test or the HI test.

Epidemiology. The epidemiology of echovirus infections is similar to the epidemiology of other enterovirus infections. The echoviruses are distributed worldwide, they produce only transitory infections, primarily in young individuals. Infections occur chiefly during the summer and autumn and are most prevalent in children from lower socioeconomic levels. Most infections are inapparent or subclinical. Echoviruses are rapidly disseminated among household contacts.

Treatment and Prevention. No specific measures for controlling echovirus meningitis are available. Very young infants should be separated from young children exhibiting an acute febrile illness with a maculopapular rash. Medical housestaff should be screened for possible carriage of echoviruses, particularly during outbreaks of aseptic meningitis.

3. Coxsackievirus Meningitis

This syndrome occurs in large epidemics, usually in late summer or early fall, and may involve individuals of every age group, including young infants.

Etiologic Agents. Coxsackievirus meningitis is caused by all serotypes of group B and 3 serotypes of group A *coxsackieviruses* viruses (7, 9, and 23).

Clinical Symptoms. The incubation period ranges from 2–9 days.

In cases of meningitis caused by coxsackie A serotypes, characteristic discrete vesicular lesions of "herpangina" may be seen on the anterior pillars of the fauces and sometimes on the palate, uvula, tonsils, and tongue. Coxsackievirus B meningitis is generally accompanied by painful myalgia. Late meningitis may occur 5–9 days after the onset of headache, vomiting, fever, neck or back stiffness, and sometimes convulsions, herpangina, or pleurodynia. A maculopapular rash may accompany coxsackievirus meningitis. Family contacts may not suffer from meningitis but instead from an inapparent infection, a mild undifferentiated upper respiratory tract infection, or perhaps from epidemic myalgia, myocarditis (coxsackievirus B), herpangina (coxsackievirus A), or exanthem (coxsackievirus A or B). Some paralysis may occur, but it is reversible.

Pathogenesis. The pathogenesis of coxsackievirus is similar to that of other enterovirus infections (e.g., poliomyelitis). Mortality is highest in infants; it is attributable to acute interstitial myocarditis and lesions that may be present in the CNS and liver as a result of a generalized systemic disease.

Immunity. Immunity against coxsackieviruses is type-specific. Passive transfer of neutralizing and CF Abs from mother to offspring occurs. In contrast to children, who often lack Abs, most adults possess Abs against several types of coxsackieviruses, suggesting that multiple exposure to these viruses occurs with increasing age.

Diagnosis. Virus can be recovered from throat swabs or blood early during infection and from rectal swabs during the first few weeks following apparent infection. Virus can also be isolated from the CSF. The laboratory procedures for the isolation and identification of coxsackieviruses are similar to those used for enteroviruses in general (e.g., poliovirus) and should include inoculation into suckling mice. The virus is identified by immunologic tests and by the characteristic lesions it produces in mice (Chapter 24).

Epidemiology. Coxsackieviruses are widely distributed in the human population. Once the virus is introduced into a household, all nonimmune persons usually become infected; however not all develop apparent disease. The coxsackieviruses share many epidemiologic characteristics with other enteroviruses. Consequently, enteroviruses often occur together in nature (e.g., in sewage or flies) and in the same human host.

Treatment and Prevention. No specific control measures for controlling coxsackieviruses are known. Infants and young children should be separated from persons exhibiting acute febrile illness, especially those with a rash.

C. ENCEPHALITIS

Encephalitis is the most serious viral disease in man; fortunately, it is rare. It is usually a complication of an inapparent infection. When apparent

Table 27–2. Etiology of Viral Encephalitides in the United States

Viruses	
Common	*Less Common*
Togaviruses (EEE, WEE, SLE, California groups)[a]	Herpes simplex
Measles[b]	Vaccinia[b]
Varicella-zoster[b]	Rubella[b]
Mumps[b]	Rabies
	Polio
	Measles (SSPE)[c]
	Mengo[b]
	Lymphocytic choriomeningitis (LCM)[d]

[a]EEE, eastern equine encephalitis; WEE, western equine encephalitis; SLE, St. Louis encephalitis.
[b]Postinfectious encephalitis is not common, but when it occurs the pathogenesis is probably due to a hypersensitivity reaction. Virus has not been recovered from the brain.
[c]Subacute sclerosing panencephalitis. Virus recovery from the infected cerebral tissue has been reported.
[d]Virus of rodents which may be transmitted to man.

disease occurs, residual effects such as mental retardation, epilepsy, paralysis, deafness, or blindness are common in those who recover. Most cases of viral encephalitis are caused by *togaviruses (arboviruses)* that cause outbreaks or epidemics in restricted areas of the world; this is in sharp contrast to *rabies virus,* which is endemic and essentially worldwide. In special cases, encephalitis due to delayed hypersensitivity may follow natural recovery from *measles, varicella-zoster, mumps,* and more rarely, *rubella* (Table 27–2). Viral encephalitis may also follow rabies or active immunization with vaccinia. Less common causes of viral encephalitis include *herpes simplex, herpes B* and *mengo viruses* and the unusual conditions called subacute sclerosing panencephalitis, which sometimes follows recovery from measles (Table 27–2).

The following discussion will encompass viral encephalitis associated with togaviruses and rabies virus and the subject of postinfectious encephalitis. Subacute sclerosing panencephalitis will be included in a discussion (Chapter 32) of the chronic progressive neurologic diseases of probable viral etiology that affects the CNS (e.g., kuru and progressive multifocal leukoencephalopathy).

1. Togavirus Encephalitis

Most cases of viral encephalitis, especially among children, are caused by viruses in the *Togaviridae* family (arboviruses or bunyaviruses) (Table 27–3). The Togaviridae family includes over 250 different viruses, of which about 65 have been isolated from man.

Etiologic Agents. The principal members of the Togaviridae family of importance in the U.S.A. include the genus *Alphavirus (Eastern equine encephalitis* (EEE) *virus, Western equine encephalitis* (WEE) *virus),* and

Table 27–3. Summary of Four Viral Encephalitides Encountered in the Americas

Virus Family	Viral Species	Clinical Disease	Pathology	Vector(s)	Distribution of Disease
Togaviridae genus: Alphavirus	EEE; Eastern equine encephalitis	Encephalitis; severe; high mortality; residual neurologic damage; mortality 80 percent	Lesions in white and gray matter; prominent in brain stem and basal ganglia; spinal cord shows milder changes	*Culex* and *Aedes* species	Eastern and SE USA, Canada, Cuba, Central and South America
Togaviridae genus: Alphavirus	WEE; Western equine encephalitis	Encephalitis; recovery usually complete; incidence of disease and permanent neurologic damage highest in children; can have inapparent disease or fever and headache; mortality 5–15 percent	Chiefly affects brain; produces lymphocyte infiltration of meninges and parenchyma; lesions consist of necrosis of neurons; glial infiltration and perivascular cuffing	*Culex tarsalis*	Western U.S.A., Canada, Mexico, Argentina
Togaviridae genus: Flavivirus	SLE; St. Louis encephalitis	Encephalitis; mild lesions; little neurologic damage; mortality 20 percent	Same as for genus Alphavirus but always less severe	*Culex tarsalis; Culex pipiens*	Texas, New Jersey, St. Louis, MO; major Group B infection in U.S.A.; most important in U.S.A.
Bunyaviridae	CEV; California encephalitis virus	Mild, undifferentiated illness, ranging from influenza-like syndrome to acute CNS disease; major sequelae uncommon; mortality low	Brain edema; neuronal degeneration; inflammation and perivascular edema	*Aedes* and *Culex* species; *Dermacentor andersoni*	California, North Carolina, Minnesota, Indiana, New York, Florida

the genus *Flavivirus (St. Louis encephalitis* (SLE) *virus).* The principal bunyavirus is *California encephalitis virus* (CEV).

Clinical Symptoms. Clinical disease caused by togaviruses may vary from an inapparent infection to an acute febrile syndrome with hemorrhagic fever and involvement of the CNS. The pathogenesis of Alphaviruses and Flaviviruses is summarized in Fig. 27–1. After a bite by an infected arthropod vector, acute encephalitis (infection of the CNS) begins with a rapid onset of convulsions and abnormalities in consciousness. Headache and fever may precede neurologic symptoms. After a few days, death or extensions of abnormalities in consciousness may occur, including coma, and neurologic and psychologic signs (delirium, confusion and excitement). In the early stages of disease, the CSF is usually abnormal, with moderate lymphocytosis and elevation of protein. The level of sugar remains normal. Prognosis in encephalitis caused by togaviruses is variable. Although recovery is common and usually takes place after about 10 days of severe illness, death may occur after a few days of illness (Table 27–3).

The disease caused by EEE virus is always severe in horses and in human beings, particularly children. It tends to run a more acute course and presents more severe symptoms than are seen in cases of encephalitis due to other togaviruses. The case mortality rate is high (80 percent), and surviving children are left with permanent physical and mental sequelae.

Most persons infected with WEE have inapparent illness or at least no involvement of the CNS. The severity is inversely proportional to age. The incidence of infection is highest among children, who experience the most severe illness and the highest death rate (5–15 percent). In children, the disease is associated with convulsions, vomiting, and excitement. Children who recover from the disease usually are left with serious physical and mental handicaps. Symptoms in adults include lethargy, neck and back stiffness, and often mental confusion and temporary coma.

Viral encephalitis caused by SLE virus is the most important togavirus disease in the U.S.A. In 1966, the worst epidemics of SLE occurred in Texas; they involved more than 700 cases and 32 deaths. It is likely that many cases of SLE virus infection escape diagnosis. *As with most togaviruses, the majority of infections are inapparent.* Estimates during large epidemics suggest that, whereas 60–70 percent of the population become infected, only a few individuals develop encephalitis. Benign forms of the disease are limited to a short febrile, influenza-like illness of a few days duration, with headache and aching stiffness of muscles. Severe clinical disease has an abrupt onset, with headache, fever, and vomiting followed rapidly by clinical evidence of encephalitis or meningoencephalitis. Neurologic symptoms are usually more severe in aged persons. The mortality rate is usually between 10 and 20 percent, and most deaths occur in persons over 50 years of age. Death may result from the primary encephalitis or from complications, including bacterial infection, pulmonary embolism, or hemorrhage in various organs of the body. Recovery may be followed by the late development of neurologic abnormalities (defects in sensory perception) or psychologic changes (irritability and instability).

Encephalitis caused by CEV has a good prognosis and complete recovery occurring in 7–14 days (Table 27–3). An incubation period of 4–6 days is followed by an acute nonexanthematous febrile stage, with headache and severe pain in muscles of the back and legs.

Pathogenesis. Man is infected by the bite of an infected mosquito (Fig. 27–1). The infected arthropod introduces its proboscis into the capillary bed beneath the skin and injects its virus-laden saliva; virus multiplication takes place in the vascular endothelium and RE cells in lymph nodes, liver, spleen, and other organs. Virus released from these cells sets up a viremia that initiates the "systemic phase" of the disease characterized by chills, fever and aches. Progress of the disease depends on the presence of neutralizing Abs in the blood and on the amount of virus present. Other organs and systems may become infected, such as joints and muscles (arthritis and myositis), skin (rash), liver and kidneys (hepatitis and nephritis), or the brain (encephalitis that may be fatal). The skin rash, if present, is characterized by swelling of the capillary endothelium, perivascular edema, and infiltration of mononuclear cells. Lesions in the skin possess eosinophilic intranuclear inclusions, *"Torres bodies"* and larger eosinophilic masses of hyaline material resulting from fusion of necrotic cells *"Councilman bodies."* Fatty degeneration is prominent in kidney tubules and is seen to a lesser extent in spleen, heart,

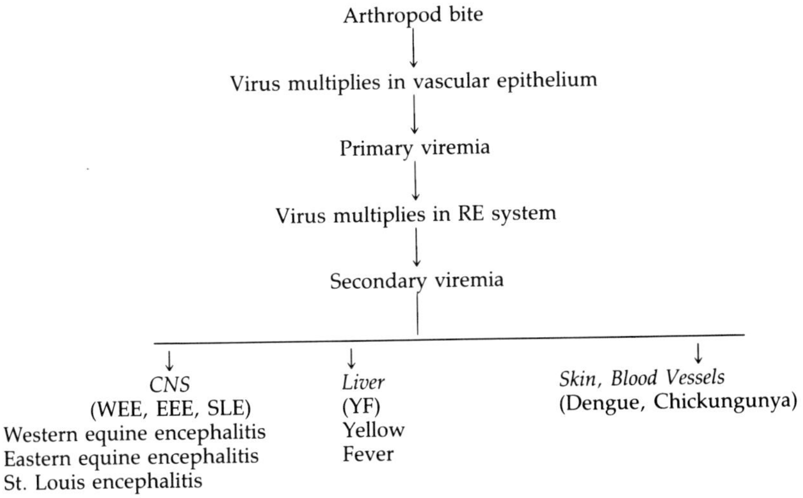

Fig. 27–1. Pathogenesis of Alphavirus and Flavivirus infections.

lymph nodes, and brain. In cases of encephalitis, inflammatory foci in the brain show necrosis of neurons and glial cells, perivascular cuffing, capillary thrombosis, infiltration of lymphocytes, and a variable degree of meningitis.

Immunity. Type-specific acquired immunity to the virus is believed to be permanent. This may be due to the strong stimulus provided by intimate contact of virus Ag(s) with the spleen and lymph nodes during the viremic stage of the disease and the booster effect of repeated exposure to subclinical infections with related viruses of the same serologic group. For example, repeated infections with one or more of the agents from the genus *Flavivirus* (e.g., SLE) broadens immunity against other *Flaviviruses* (e.g., yellow fever).

Diagnosis. Virus can be isolated (with difficulty) from the blood during the incubation period and before symptoms appear. It can usually be recovered from the brains of fatal cases, provided the specimen is taken soon after death and inoculated i.c., s.c., or i.p. into suckling mice. Tentative identification of the virus isolate can be made by the hemagglutination-inhibition or the CF test and confirmed by neutralization of virus infectivity using known antisera. More commonly, a laboratory diagnosis is made by detecting a rise in titer of specific Abs by tests on acute and convalescent sera.

Epidemiology. Persistence of togaviruses and bunyaviruses in nature rests upon a suitable bloodsucking vector (invertebrate host) and a vertebrate host (Fig. 27–2). Only the female mosquito feeds on blood and can feed and transmit the virus more than once. The growth cycle of several togaviruses in the infected mosquito has been determined. The primary site for virus multiplication involves the cells of the midgut, followed by viremia and subsequent infection of the salivary glands and nerve tissue, where secondary virus multiplication occurs. Infected mosquitoes do not become ill; although they carry the virus for a lifetime; however, their lifespan is short (several weeks to months). Infected vertebrates (e.g., rodents) rapidly recover, eliminate the virus and develop long-lasting immunity. Man, an unnatural and terminal host for most togaviruses, becomes accidentally infected when he is bitten by an infected bloodsucking arthropod carrying togaviruses (e.g., EEE and WEE). Togaviruses can persist where arthropods coexist with birds and mammals. The events concerned with survival of togaviruses in temperate zones are similar to those which take place in tropical zones during the summer months but "overwintering" is not completely understood. There is evidence which suggests that some arthropod-borne viruses (e.g., EEE and WEE) produce latent infections in hibernating mammals (e.g., bats and snakes and other reptiles). Following hibernation, a viremia occurs, and the virus is transmitted to a new host of mosquitoes

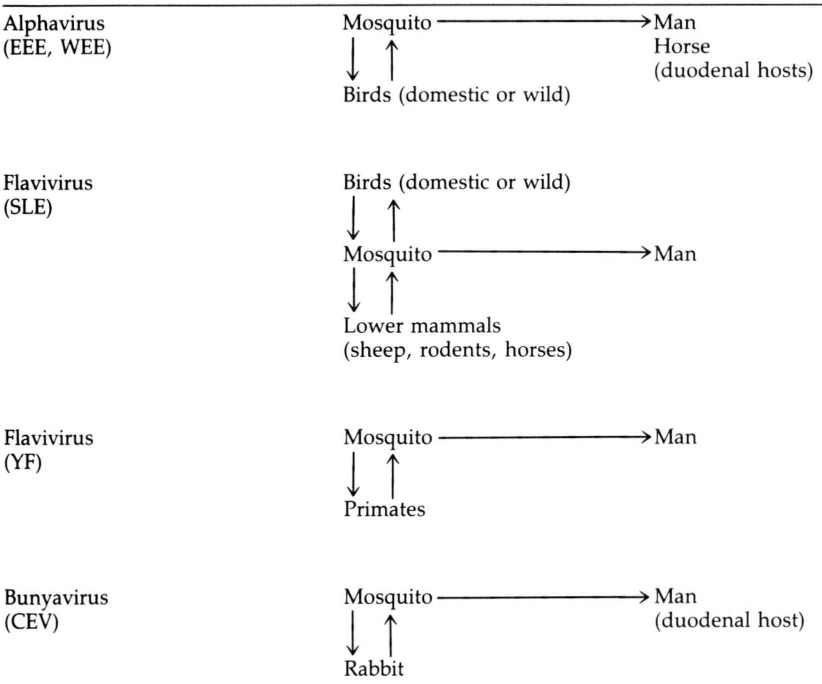

Fig. 27–2. Epidemiology of Togaviridae and Bunyaviridae.

that bite the awakened animal. Another possibility is that the virus survives the winter in mosquito vectors that can hibernate *(Culex tarsalis)*. Other arthropod-borne viruses (CEV) are maintained in ticks by transovarial transmission through successive generations of hosts.

Treatment and Prevention. There are no effective vaccines and no specific treatments for encephalitis caused by togaviruses and bunyaviruses. Prevention of infection is accomplished by (1) avoidance of exposure in arthropod-infested areas and (2) eradication of vectors.

2. Herpesvirus Encephalitis

Encephalitis caused by herpes simplex virus is probably due to a primary infection rather than a reactivated latent infection. The virus probably reaches the brain by the olfactory nerve. Once established in the brain, infectious virus is passed cell-to-cell which leads to extensive neuronal necrosis. The patient's immunologic responses contribute to the pathogenesis of disease.

Diagnosis of herpesvirus encephalitis is difficult from laboratory evidence and almost impossible on clinical grounds. Examination of brain biopsy material is often the only means of establishing an exact diagnosis. Early diagnosis for herpesvirus encephalitis is extremely important since the disease can be treated with the antiviral drug, vidarabine (adenine arabinoside, Chapter 23).

3. Rabies Virus Encephalitis

Rabies is the most lethal of all infectious diseases. Man has lived in fear of the rabid (*rabidus* = mad) dog throughout the centuries. In 1884, Louis Pasteur developed the first man-made viral vaccine. His demonstration of its efficacy in saving the life of a peasant boy bitten by a rabid dog was a milestone in microbiology.

ETIOLOGIC AGENTS. *Rabies virus* is classified in the *Rhabdoviridae* family. The most important member of the *Rhabdoviridae* family (rabies virus) belongs to the new genus, *Lyssavirus* (lyssa = mad). Only one major antigenic type exists in nature. However, 4 distinct serotypes of rabies virus have been recently isolated from infected animals in remote areas of the world.

CLINICAL SYMPTOMS. The incubation period (2–16 weeks) depends on the distance of the nerve path from the point of virus entry to the brain. Accordingly, the shortest incubation periods occur following bites about the face, neck, and arms.

In man, two prodromal signs are pathognomonic: (1) the earliest but most inconsistent sign is a tingling sensation at the site of the bite, and (2) the most consistent but later sign is a state of anxiety and laryngopharyngeal spasm during the act of swallowing (nervous hydrophobia). This phase is generally followed by convulsive seizures and usually death within 3–5 days after onset.

In dogs, the incubation period for rabies is variable, but it may be as short as 10 days. Three phases of disease have been noted: (1) prodromal (fever, change in temperament, slow corneal reflexes); (2) excitative (irritability, nervousness, exaggerated response to light and sound stimuli), and (3) paralytic (convulsive seizures, coma, and death due to asphyxia). Unlike man, hydrophobia does not occur in dogs. During the terminal stages of the disease, the dog's saliva contains a high titer of infectious virus, which makes the animal dangerous, particularly if rabies is not suspected.

PATHOGENESIS. Rabies virus multiplies in muscle and connective tissue and passes along sensory nerves to the CNS. The virus multiplies in the CNS and may spread through peripheral nerve trunks to the salivary glands. Pathogenesis involves extensive nerve cell damage in the cerebrum, cerebellum, cortex, midbrain, basal ganglia, pons, and medulla. Demyelination occurs in the white matter, and there is extensive degeneration of axons and of the myelin sheath. If the bite wound is in an arm or leg, the corresponding posterior horns in the spinal cord will be destroyed by the virus. Cellular infiltrations are usually rich in mononuclear leukocytes and tend to be extensive when the disease is prolonged.

IMMUNITY. In the past, rabies was considered to be uniformly fatal in nonimmunized persons; however, in 1970, a case of the disease was treated and the patient recovered. Animals immunized before exposure to rabies virus are in general resistant to the infection if virus-neutralizing Abs are present in their serum. The same is probably true for man.

DIAGNOSIS. Any dead or dying animal suspected of having rabies should be subjected to laboratory diagnostic tests. The technique involves sending the head of the suspected animal to the laboratory in an ice-filled container or submerged in a container filled with glycerin. The diagnosis of rabies is based on finding intracytoplasmic inclusions (Negri bodies) in nerve cells (Fig. 27–3). Negri bodies consist of an acidophilic matrix containing basophilic granules. If Negri bodies are not detectable by direct examination of fluorescent-Ab stained impression-smears of the brain, a suspension of the brain (hippocampus) or submaxillary salivary gland should be inoculated intracranially into suckling animals (mice, hamsters, rabbits). Positive diagnosis in the inoculated animals is made on the basis of clinical symptoms (e.g., convulsions) and demonstration of Negri bodies in brain sections. Laboratory diagnosis in man is of only retrospective interest but is necessary in order to establish the clinical diagnosis with certainty.

Dogs suspected of rabies are routinely quarantined for 14 days. If the

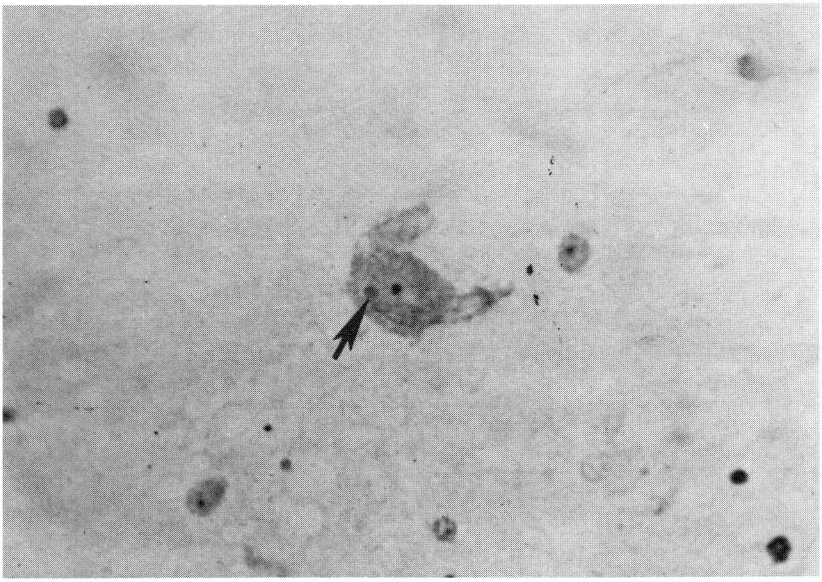

Fig. 27–3. Infected nerve cells showing intracytoplasmic inclusion bodies (Negri bodies). Presence of Negri bodies is diagnostic for rabies virus infection.

dog remains asymptomatic for this period, the animal is presumed to be healthy. However, if symptoms of rabies develop, the dog should be held in isolation for several more days before tests are made in order to permit Negri bodies to develop in the brain cells.

EPIDEMIOLOGY. Rabies is found worldwide, except in Australia and the United Kingdom. Man is an unnatural host of the rabies virus and does not serve as a reservoir for natural transmission of the disease. The natural hosts reponsible for perpetuating the virus in nature have not been defined; they may be members of the family *Mustelidae* (weasel and skunk) and the family *Viverridae* (civet and ferret). Recent evidence indicates that vampire and insectivorous bats are important natural hosts of the rabies virus. Although the virus tends to remain latent in the bat, it multiplies in the olfactory mucosa and is shed by aerosols that can transmit the disease to bats and probably to other animals and man by the respiratory route. Bat bites may also serve to transmit the disease. In consequence of these circumstances, speleologists run a high risk of contracting rabies. Dogs and cats serve as the most important sources of human infection because of their close association with man. However, many wild animals (e.g., skunks, raccoons and foxes) are common carriers of the virus and sometimes serve as sources of human infection. Spread of rabies depends on growth of the virus in salivary glands of infected animals. About 50% of rabid dogs shed virus in their saliva.

TREATMENT AND PREVENTION. Prophylaxis against rabies includes destruction of wild animal reservoirs, quarantine of animals suspected of being rabid, and routine vaccination of domestic dogs and cats. For their own protection, most dogs and cats are immunized with a live attenuated rabies virus adapted to grow in chick embryos (Flury vaccine). This vaccine is not suitable for man because of the small antigenic mass of the vaccine. Two types of vaccines are available for man in the United States. A duck embryo vaccine (DEV) and a human diploid cell rabies vaccine (HDCV). The DEV vaccine is prepared in duck embryos infected with the original Pasteur rabbit brain "fixed virus;" the virus was "fixed" by serial laboratory passage and then killed by β-propiolactone before human use. The HDCV is grown in human diploid cells (WI-38 or MRC-5) in tissue culture infected with fixed virus and inactivated with β-propiolactone. Usually 23 doses of the DEV are given (1 per day) subcutaneously in the abdomen, lower back or lateral aspect of the thigh. HDCV requires only 5 doses for post-exposure rabies prophylaxis, is less reactogenic, and is a better stimulator of antibody production. In addition, there is a human rabies immune globulin (HRIG) available that is recommended for administration with the first dose of HDCV. HRIG is the globulin fraction of serum from human volunteers previously immunized against rabies.

In the United States the number of rabies cases reported annually in man is low (1 to 3); however, 30,000 persons receive postexposure pro-

Table 27–4. Rabies Postexposure Prophylaxis Guide

Animal Species	Condition of Animal at Time of Attack	Treatment of Exposed Person*
Dog and cat	healthy and available for 10 days of observation	none, unless animal develops rabies[a]
	rabid or suspected rabid;	RIG[b] and HDCV[c]
	unknown (escaped)	consult public health officials. If treatment is indicated, give RIG[b] and HDCV[c]
Skunk, bat, fox, coyote, raccoon, bobcat, and other carnivores	regard as rabid unless proven negative by laboratory tests	RIG[b] and HDCV[c]
Livestock, rodents, and lagomorphs (rabbits and hares)	Consider individually. Local and state public officials should be consulted on questions about the need for rabies prophylaxis. Bites of squirrels, hamsters, guinea pigs, gerbils, chipmunks, rats, mice, other rodents, rabbits, and hares almost never call for antirabies prophylaxis.	

*All bites and wounds should immediately be thoroughly cleansed with soap and water. If antirabies treatment is indicated, both rabies immune globulin (RIG) and human diploid cell rabies vaccine (HDCV) should be given as soon as possible, regardless of the interval from exposure.

[a]During the usual holding period of 10 days, begin treatment with RIG and vaccine (preferably with HDCV) at first sign of rabies in a dog or cat that has bitten someone. The symptomatic animal should be killed and tested for Negri bodies in brain cells.

[b]If RIG is not available, use equine antirabies serum (ARS). Do not use more than the recommended dosage.

[c]If HDCV is not available, use duck embryo vaccine (DEV). Local reactions to vaccines are common and do not contraindicate continuing treatment. Discontinue vaccine if fluorescent-antibody (FA) tests of the suspected rabid animal are negative.

phylaxis. A guide for postexposure antirabies vaccination is given in Table 27–4. It should be noted that all persons bitten by bats should receive immediate antirabies vaccine, since 50 percent of infected bats have no detectable Negri bodies in brain smears even though they shed virus in their salivary glands and nasal secretions. Prophylactic rabies vaccination with either DEV or HDCV is recommended for persons exposed to high risk (e.g., veterinarians and speleologists).

4. Postinfectious Encephalitis

Postinfectious encephalitis is an acute or subacute disease of the CNS characterized by focal demyelinative lesions that are prominent in white matter and meninges. The disease occurs as a "postinfectious" (i.e., following recovery from an acute viral disease) complication of many common viral diseases that do not usually affect the CNS. It may also occur as a result of viral vaccination and sometimes without an apparent cause. The disease may present as acute aseptic meningitis, disseminated encephalomyelitis, optic neuritis, or polyneuritis.

Etiologic Agents. Postinfectious encephalitis has been observed following recovery from *smallpox, measles, varicella-zoster, herpes simplex, rubella, poliomyelitis, mengo, lymphocytic choriomeningitis, mumps* and *Epstein-Barr virus* infections. Postinfectious encephalitis may also develop following vaccination for smallpox or rabies.

Clinical Symptoms. The latent period is usually 2–6 days after measles, rubella, or varicella-zoster exanthems. After smallpox vaccination, neurologic symptoms appear on the 10th to the 12th day. About 14–21 days after rabies vaccination, the disease may start abruptly with convulsions, fever, headache, vomiting, and behavioral disturbances. Occasionally, the onset is more gradual. Postinfectious encephalitis following smallpox or rabies vaccination carries a case mortality of about 50 percent. Following measles, varicella-zoster, rubella, infectious mononucleosis, and mumps, the case mortality ranges from 10–20 percent. Recovery may be complete in 2–3 weeks after onset or may take several months. Permanent neurologic sequelae may follow recovery from measles encephalitis.

Pathogenesis. The consensus is that the lesions of postinfectious encephalitis result from an allergic reaction, due in most instances to an antigenic component of virus persisting in the nervous system. Evidence supporting this view is provided by the following example: In postvaccinal encephalitis, vaccinia virus can be isolated from the brain and CSF during the first 5 days of illness. Encephalitis following rabies vaccination (i.e., DEV vaccine), in which repeated doses of rabies virus is administered, is an exception that probably represents an autoallergic (autoimmune) disease. Between 1958 and 1975, 21 neuroparalytic reactions were reported among an estimated 512,000 recipients of DEV (1/24,000) including 5 cases of transverse myelitis, 7 cases of cranial or peripheral neuropathy and 9 cases of encephalopathy (2 fatal).

Immunity. Techniques are being developed to detect and measure the immune responses to allergic lesions in the CNS. These investigations should lead to a better understanding of the causative Ag(s) and the allergic mechanism(s) involved in postinfectious encephalitis.

Diagnosis. Diagnosis of postinfectious encephalitis depends on either direct virus isolation from CSF or postmortem examination. In cases of diseases caused by smallpox, vaccinia, or measles viruses, the agent can be isolated from the brain or CSF during the first few days of illness. In the case of mumps virus, diagnosis may be established by isolation of the virus from saliva of patients with accompanying parotitis as well as by isolating the virus from the CSF, by detecting a rise in Ab titer, or by noting the appearance of specific skin reactivity (delayed hypersensitivity) to mumps virus skin test antigen.

Epidemiology. Patients with exanthems and accompanying encephalitis should be separated from young children and nonimmune persons. In the case of mumps virus encephalitis, the agent is shed in the urine, which may serve as a source of infection for others.

TREATMENT AND PREVENTION. In the case of rabies prophylaxis, the newer vaccine made in human diploid cells (HDCV) should be used if possible. Immunization against measles and rubella is recommended for children and nonimmune adults to induce active immunity to these diseases; mumps virus vaccination is recommended at puberty in cases where there is no clinical history of mumps.

In one study, corticosteroid therapy started within 48 hours of onset of postinfectious encephalitis was successful in up to 50 percent of the patients treated. Hyperimmune gamma globulin is without therapeutic effect.

5. Marburg Virus Encephalitis

Marburg virus was first recognized as an etiologic agent of disease in 1967 in Marburg, Germany, when laboratory workers became infected with a virus contaminant of Green monkey kidney cells. The natural host of the virus is unknown. The disease is associated with severe hemorrhage that is systemic and has a 30 percent mortality rate. Following an incubation period of 4–7 days, there is a rapid onset of fever, headache, myalgia, vomiting and diarrhea; the second phase of disease is associated with exanthema and hemorrhage in the GI tract and nose, photophobia, encephalitis, meningitis, kidney damage, and coma. Laboratory diagnosis is made by isolation of virus in tissue culture cells and serologic tests. Importance of the virus in the United States is due to an import of 12,000 to 20,000 Green monkeys per year for medical research.

D. ENCEPHALITIDES OF PROBABLE VIRAL ORIGIN

1. Von Economo's Disease (Lethargic Encephalitis).

The recorded history of von Economo's disease began in 1915, when a worldwide outbreak occurred. The pandemic lasted for 10 years and then mysteriously disappeared. Small epidemics (60–70 cases) are still reported occasionally, particularly in the United Kingdom and West Germany. During the pandemic, von Economo, a physician, reported that essentially every practitioner in most of Europe, America, and Australia observed at least several patients and identified the disease with certainty. The disease is characterized by progressive onset of weariness, headache, weight loss, bouts of hiccup, and fever followed by a number of nervous disorders (e.g., mental, ocular, paralytic), including parkinsonian syndrome. The acute phase of the disease is dominated by drowsiness. The prognosis of the disease is unfavorable, the mortality being 30–60 percent. Sequelae may consist of attacks of headache, sleep disturbances, oculomotor and mental disorders, and complete parkinsonian syndrome.

2. Acute Necrotizing Encephalitis

This condition develops rapidly and ends in death 1–2 weeks after symptoms develop. In all cases, the disease begins with an "influenza-like" fever, followed by severe disturbances in consciousness, psychiatric changes, abnormal movements, temporal epilepsy, and meningoencephalitis. Although some reports suggest that acute necrotizing encephalitis may be due to herpesviruses, other etiologic agents may be involved.

REFERENCES

Bishop, H.L.: Genetic potential of Bunyaviruses. Curr. Top. Microbiol. Immunol., *86*:1, 1979.
Bishop, D.H.L. and Shope, R.E.: Bunyaviridae. Comprehensive Virology, *14*:1, 1979.
Buckley, S.M. and Casals, J.: Pathobiology of Lassa fever. Int. Rev. Exp. Pathol., *18*:97, 1978.
Garoff, H., Kondor-Koch, C., and Riedel, H.: Structure and assembly of Alphaviruses. Curr. Top. Microbiol. Immunol., *99*:1, 1982.
Grist, N.R., Bell, E.J., and Assaad, F.: Enteroviruses in human diseases. Prog. Med. Virol., *24*:114, 1978.
Morens, D.M.: Enteroviral disease in early infancy. J. Pediatr., *92*:374, 1978.
Schlesinger, R.W.: Dengue viruses. Virology Monographs, *16*:1, 1977.
Schlesinger, R.W. (ed.): The Togaviruses. New York, Academic Press, 1980.
Wilfert, C.M., Lauer, B.A., Cohen, M.J., Costenbader, M.L. and Meyers, E.: An epidemic of echovirus 18 meningitis. J. Infect. Dis., *131*:75, 1975.
Wolinsky, J.S. and Johnson, R.T.: Role of viruses in chronic neurological diseases. Comprehensive Virology, *16*:257, 1980.

Chapter 28

Viral Diseases of the Liver

Acute viral hepatitis is one of the most important of the communicable diseases present in developed countries. The most common etiologic agents are hepatitis A (infectious hepatitis) virus, hepatitis B (serum hepatitis) virus and Non A Non B hepatitis virus. Several other viruses may infect the liver as part of a more general disseminated disease (Table 28–1). Differential diagnosis is facilitated by certain characteristic features of various infections in which hepatitis occurs. For example, lymphadenopathy and pharyngitis are characteristic of infectious mononucleosis (Epstein-Barr virus etiology), together with a positive Paul-Bunnel test (heterophil Ab) and a significant titer of serum Abs specific for Epstein-Barr virus (EBV). Hepatitis caused by yellow fever virus should be considered in known endemic areas.

Neonatal hepatitis caused by cytomegalovirus or other viruses presents special diagnostic problems. In each of these cases, hepatitis is but one of the clinical signs of a generalized viral infection. Complete physical examination will always reveal other abnormalities that are useful in deciding the etiology of the hepatitis (see Table 28–1 and chapters dealing with the individual viruses concerned). *The diagnosis of viral hepatitis is considered to be a presumptive one when based on epidemiologic, clinical, biochemical, and histologic evidence.* The student of medicine must bear in mind that jaundice can also be caused by bacteria, protozoa, and noninfectious agents, such as hepatotoxic chemicals, and by physical obstruction of the biliary system.

A. HEPATITIS TYPE A (INFECTIOUS HEPATITIS) AND HEPATITIS TYPE B (SERUM HEPATITIS)

Viral hepatitis type A (epidemic or infectious hepatitis; IH) and viral hepatitis type B (serum hepatitis; SH) are distinct diseases. During the past decade, there has been a steady rise in the incidence of SH particularly among males in the 15- to 29-year-old age group. Narcotic addiction is now widespread and may account for the nationwide changing pattern of viral hepatitis in all socioeconomic levels. There is increasing evidence associating SH virus with hepatocellular carcinoma in humans.

Table 28–1. Causative Viruses of Acute Hepatic Injury

Common	Uncommon
Infectious hepatitis	Yellow fever
Serum hepatitis	Epstein-Barr
Non A Non B hepatitis	Cytomegalovirus[a]
	Rubella[a]
	Herpes simplex[b]
	Coxsackievirus[b]

[a]Etiologic agent of prenatal hepatitis acquired in utero.
[b]Etiologic agents of postnatal hepatitis usually acquired during or after parturition.

Serum hepatitis is a disease with a more insidious impact, longer incubation period, and higher mortality than IH. Also, the discovery of a virus-like Ag referred to as Australia antigen (AuAg) or hepatitis B surface antigen (HBsAg) and its persistence in the blood for prolonged periods in patients recovered from SH showed that SH virus has a worldwide distribution similar to that previously attributed only to IH virus. In recent years, another virus(es) referred to as Non A Non B hepatitis has been established as a separate virus by excluding SH and IH viruses in clinical disease. It is estimated that Non A Non B hepatitis virus accounts for about 90 percent of post-transfusion hepatitis in the U.S.A.

Approximately 200,000 cases of viral hepatitis occur annually in the U.S.A., primarily in young adults. The number of deaths associated with this disease ranges between 2,000 and 9,000 annually. About 6 percent of infected individuals become carriers resulting in a pool of 400,000–800,000 infected persons. With respect to hepatitis patients in the U.S.A., it is estimated that the direct costs for hospitalization and physician's services and the indirect costs due to loss of productivity exceed one billion dollars per year.

1. Etiologic Agents

It is reasonably certain that SH and the virus-like HBsAg are associated with spherical, cylindrical and Dane virus particles present in patients' sera (Table 28–2). Studies in human volunteers demonstrated that IH also is associated with a transmissible virus that shares some properties with picornaviruses (Table 28–2). The viruses of SH and IH are epidemiologically and immunologically distinct.

Table 28–2. Properties of Virus Particles Associated With Hepatitis A and Hepatitis B

Hepatitis A	Hepatitis B
1. Icosahedral particles 27 nm in diameter; resemble picornavirus; nonenveloped; ssRNA genome; contains three major structural polypeptides	1. Spherical particles 20–22 nm in diameter; no nucleic acid HBsAg found in cytoplasm of hepatocyte
	2. Cylindrical particles 20–22 nm in diameter; no nucleic acid HBsAg present
	3. Dane particle 42 nm in diameter; 27 nm core; core contains several antigens (HBcAg, HBeAg), incomplete circular 2X DNA, DNA polymerase; surface contains HBsAg; considered complete infectious particle
	Ratio of Spheres:Cylinders:Dane 1700:120:1

The Non A Non B virus class appears to comprise at least 2 viruses. A Non A Non B virus particle 27 nm has been transmitted to chimpanzees; the Non A Non B virus particle is distinct from the 20 to 27 nm particles associated with SH and IH.

2. Clinical Symptoms

Infectious hepatitis and SH have similar as well as distinctive clinical features (Table 28–3). Jaundice may or may not be present during the acute phase of infection. Factors that influence the severity of viral hepatitis include age and associated conditions (e.g., pregnancy, diabetes, abdominal surgery, and cancer).

Some cases of hepatitis may develop into chronic persisting or recurrent hepatitis that follows a period of recovery from typical acute hepatitis. Whether persistence or recurrence of hepatitis is the result of active viral multiplication or immunologic injury to the liver is not clearly understood.

3. Pathogenesis

Viremia during the incubation period, results from multiplication of virus in the liver and gastrointestinal (GI) tract. Serum and infectious types of hepatitis cause identical lesions in the liver. Lesions may extend beyond the liver to the upper GI tract and the kidney. The hepatic lesions are characterized by parenchymal cell degeneration and necrosis, proliferation of Kupffer cells, inflammation, and cell regeneration. Macrophages accumulate at sites of necrosis, as do lymphocytes, plasma cells and neutrophils. These changes disappear with full recovery from the disease.

4. Immunity

There are indications that several serotypes of these etiologic agents exist. There is no cross-immunity between different serotypes, and second attacks of both diseases occur.

5. Diagnosis

Assessment of abnormal liver function by various means, including measurement of serum glutamic oxaloacetic acid transaminase (SGOT), serum transaminase, and bilirubin in the urine and blood, thymol turbidity and bromosulfothalein tests, and histopathologic study of liver biopsy specimens, aid in the clinical diagnosis of hepatitis, particularly in cases with jaundice.

Counter immunoelectrophoresis determinations, complement-fixation (CF) tests, and radio-immunoassays are useful for detecting HbsAg in the serum or plasma of patients with SH. About 80 percent of patients with clinically diagnosed SH show positive tests for HBsAg. HBsAg appears during the preicteric period; it frequently disappears during the icteric period and may reappear. Detection of virus neutralizing Abs has not proved to be of diagnostic value; however, detection of Abs against HBsAg is of diagnostic significance. Routine isolation of hepatitis virus is not feasible.

6. Epidemiology

There are major differences in the epidemiologic features of IH and SH (Table 28–3). Infectious hepatitis is generally transmitted by person-to-person contact or ingestion of contaminated food (e.g., shellfish). However, it is possible to transmit IH by contaminated blood products or contaminated hypodermic needles. Subclinical cases of IH are common and may serve as a possible source of spread. Small clusters of cases of viral hepatitis in human beings in close contact with nonhuman primates have been reported. In these episodes, it could not be determined whether the virus was of human or nonhuman origin. Apparently, the incidence of IH depends on opportunities for ready distribution of infected material and the susceptibility of the exposed population. Poor sanitation, overcrowding, exposure of a nonimmune population to contaminated material, and a short incubation period contribute to epidemics of IH. Explosive epidemics of IH have occurred following fecal contamination of drinking water and after consumption of shellfish from polluted water.

Since serum hepatitis virus is usually transmitted by the use of contaminated needles, syringes, and blood products, patients requiring renal dialysis, and heart-lung machines are at risk. The incidence of SH in recipients of blood transfusions is 0.3–4 percent; the incidence is significantly increased following multiple transfusions. The present widely employed technique for detecting HBsAg in blood are estimated to be capable of reducing the incidence of post-transfusion hepatitis by 30 percent. Many persons have become infected with SH virus as the result of vaccination with improperly sterilized instruments; tattooing also has increased risk. Serum hepatitis virus is also transmitted on occasion by direct personal contact. Groups at highest risk of getting SH are male homosexuals, drug addicts, patients and staff at institutions for the mentally retarded, health care workers, and patients on kidney dialysis. It is noteworthy that the estimated ratio of anicteric to icteric infections is reported to be greater than 100 to 1.

On the basis of long-term studies with SH, two definitions for the chronic carrier states have emerged: *chronic persistent hepatitis* and *chronic aggressive hepatitis*. Chronic persistent hepatitis is a mild, benign disease that is not always preceded by recognizable acute illness; malaise, hepatomegaly, and minor abnormalities of liver function are the clinical features, there is no progression to cirrhosis, and the prognosis is good.

Table 28–3. Characteristics of Hepatic Viral Diseases of Man Due to IH, SH and YF Viruses

Feature	Infectious Hepatitis (IH)	Serum Hepatitis (SH)	Yellow Fever (YF)
Transmission to human subjects	Oral, parenteral routes via nasopharyngeal secretion, fecal material in water, food, and fomites or direct contact	Oral, parenteral injection, direct contact	Bite from infected mosquito
Antigenicity	No cross-immunity to SH virus	No cross-immunity to IH virus	Cross-reacts with other group B togaviruses (arboviruses), best with HAI, intermediate with CF, least with neutralization titration
Natural communicability	Contagious during illness	Can be transmitted nonparenterally	During viremic state
Incubation period	Short, 15–40 days	Long, 5–150 days	3–6 days
Onset	Acute, sudden rise in serum transaminases and abrupt fall 19 days later; same for IgM	Insidious prolonged rise in serum transaminases; IgM only slightly modified	Rapid with fever, headache, vomiting
Symptoms preceding jaundice	Fever, malaise, anorexia, nausea, diarrhea, abdominal discomfort	Fever, malaise, anorexia, nausea, diarrhea, abdominal discomfort	Bloodshot eyes, wine-colored face, fever, headache, vomiting
Preicteric phase duration	2–21 days	21–81 days	3–4 days, followed by remission
Icteric phase	Abrupt, high fever, jaundice	Insidious, low fever, jaundice	Jaundice, fever, vomiting, gastric hemorrhage, degeneration of myocardium, meninges, kidneys
Seasonal incidence	Autumn and winter	All year round	During mosquito season
Age incidence	Common in children and young adults	Rare in children, common in adults	All ages
Host range	Man, nonhuman primates susceptible	Man, nonhuman primates susceptible	Man, nonhuman primates
Severity	Mortality low (1 percent)	Mortality high (50 percent)	Mortality about 10 percent; can also occur as an inapparent infection followed by immunity

Diagnostic tests	Transaminase levels, liver biopsy	Transaminase levels, liver biopsy	Isolation of virus in newborn mice
HBsAg in blood	Not present	Present during incubation period and acute phase, may persist for months to years	Not present
Virus present in			
feces	During incubation period and acute phase	Depends on site of virus entry; not very common	Not present
blood	During incubation period and acute phase	During incubation period and acute phase	During incubation period and acute phase
Duration of carrier state			
feces	Weeks to months	Not demonstrated	Not demonstrated
blood	Unknown (maybe 8 mo)	Several years	During viremic stage
Prophylatic value of pooled human γ-globulin	Very helpful	Experimentally successful if given shortly after transfusion and one month later	Not used
Preventive measures	Good hygiene; detection of carriers; use of disposable equipment; contaminated equipment should be autoclaved, heated at 180°C for 1 hr, incinerated, or treated with formalin, hypochlorite, or activated glutaraldehyde; subjects having history of jaundice should not donate blood	Good hygiene; use of disposable needles and syringes, screening blood for HBsAg; no way to inactivate virus in blood or plasma; addition of β-propiolactone shows promise experimentally; contaminated equipment should be autoclaved, heated at 180°C for 1 hr, incinerated, or treated with formalin, hypochlorite, or activated glutaraldehyde; subjects having history of jaundice should not donate blood	Vaccination, eradicate mosquito population
Carrier rate	Widespread	2–3 percent	Primates serve as reservoir
Vaccine	In development	Formalin-inactivated HBsAg particles free of viral DNA	17D (attenuated live vaccine)

Unfortunately, only a small proportion of these patients demonstrate HBsAg. In contrast, *chronic aggressive hepatitis* is usually characterized by parenchymal necrosis, inflammatory cell infiltration, and varying degrees of hepatic cirrhosis. Young females are most often affected. Autoantibodies are usually present in the serum, and immunoglobulin levels are elevated. HbsAg has been detected in 4–60 percent of patients with chronic aggressive hepatitis.

Fortunately, purified human gamma-globulin or albumin is free of hepatitis viruses because the virus is inactivated by ethanol during the fractionation of plasma. Patients needing massive blood transfusions (e.g., patients undergoing open heart surgery and patients subjected to renal dialysis) stand the greatest risk of becoming infected. Medical personnel attending such patients are also at considerable risk.

7. Treatment and Prevention

The control of IH is most effectively accomplished through preventive measures, including sanitation and aseptic techniques that break the chain of transmission and through passive immunization. Pooled human gamma-globulin affords some protection if given intramuscularly during the incubation period or up to 6 days before the onset of symptoms; passive immunity lasts about 3–6 months. Gamma globulin is especially recommended for halting epidemics in military camps or institutions. Prophylactic immunization has been practiced among military personnel entering Vietnam as well as among missionaries, Peace Corps volunteers, and travelers entering endemic areas. Every effort should be made to maintain approved sanitation procedures that prevent fecal contamination. It is recommended that food handlers be screened by determining SGOT levels. All needles, syringes, and lancets that have come in contact with blood or blood products should be autoclaved (15 psi, 121°C, 15 min), heated to 180°C for 1 hr or incinerated.

The principal measure for controlling SH is to routinely screen blood plasma and other blood products for HBsAg. Persons who have a history of hepatitis should not be used as blood donors. Because the duration of carriage of SH virus is not known, estimates of the incidence of hepatitis carriers vary from 0.1–10 percent. As little as 0.004 ml of plasma can transmit the disease; therefore, the use of pooled plasma should be avoided whenever possible. If pools must be used, it is recommended that they be made from fewer than 6 donors to reduce the possibility of transmitting SH to patients. The use of disposable needles and syringes is recommended to assure sterility and to reduce the chances that a single syringe or needle will be used on more than one person. Physicians, nurses, and dentists should be made aware of the importance of decontaminating instruments that may have come in contact with blood or feces. Dry heat, autoclaving or incineration are the most satisfactory methods for decontamination. The choice of chemical disinfectants includes formalin, activated glutaraldehyde, or preparations of hypochlorite. Disposable dialyzers should, if possible, be used for infected patients. A record of patients using dialysis monitors is desirable to trace possible cross infections. A formalin-inactivated vaccine consisting of 20–22 nm HBsAg particles free of viral DNA is available for immunization against SH.

B. YELLOW FEVER

Yellow fever (YF) is an acute viral disease characterized by sudden onset, moderately high fever, prostration, and a relatively slow pulse rate in terms of the degree of fever. Severe cases are sometimes associated with vomiting of altered blood, albuminuria, jaundice, and formation of emboli that may lead to death (Table 28–3). The disease is endemic in tropical areas of Africa and South America. The virus is transmitted from man to man by the domestic mosquito, *Aedes aegypti (urban yellow fever)*. When the virus is transmitted from monkey to man or monkey to monkey by jungle mosquitoes, the disease is called *sylvan* or *jungle yellow fever.*

1. Etiologic Agents

The causative agent of YF is a small single-stranded RNA virus that is classified in the genus *Flavivirus* (group B) of the *Togaviridae* (arbovirus) family. Fresh human isolates of YF virus exhibit viscerotropism for liver, kidneys, and heart, and neurotropism for cells of the CNS.

2. Clinical Symptoms

Most attacks of YF are mild, with fever and headache of short duration. Epidemiologic features are therefore important in diagnosis. The incubation period is from 3–6 days; the onset of disease is sudden, with chills, headache, and vomiting. The acute stage of the disease lasts about 3 days and is referred to as the *period of infection*. The symptoms preceding jaundice may also include bloodshot eyes and a wine-colored face due to hemorrhages. After a short remission of about 4 days, the *period of intoxication* begins. During this period, jaundice may develop, with associated hemorrhage into the gums, nose, and gastrointestinal (GI) tract (Table 28–3). The absence of jaundice is much more frequent than its presence. Fatal cases often exhibit degenerative changes in the myocardium, meninges, and kidneys. Patients with severe disease may recover by the 7th or 8th day following onset of the period of intoxication. Convalescence progresses rapidly. Unlike IH or SH, relapses do not occur, and there are no permanent sequelae following recovery from YF; immunity following convalescence is lifelong.

3. Pathogenesis

Knowledge of the pathogenesis of YF has been gained largely from experiments with nonhuman primates. When the YF virus is introduced naturally into the skin of susceptible primates by an infected mosquito, the virus reaches the local lymph nodes and multiplies; it then enters the bloodstream and becomes localized in the liver, spleen, kidneys, bone marrow, and lymph glands. Animals inoculated with a highly virulent strain of YF virus show the highest concentration of virus in the liver, whereas in animals inoculated with the attenuated 17D strain (vaccine strain), the virus is limited to the spleen, lymph nodes, and bone marrow.

Injury due to YF results from replication of the virus in target organs. Death is due to destruction of parenchymatous cells of the liver and to hemorrhage; the most frequent site of hemorrhage is the mucosa at the pyloric end of the stomach.

Histologically, there may be irregular masses (inclusions) of hyaline material in the cell cytoplasm; these are called *Councilman bodies.* Inclusion bodies may also be present in the nucleus and are of diagnostic value. In patients who recover, the damaged hepatic cells are replaced, and normal liver functions are restored.

4. Immunity

Immunity is transmitted passively from mother to infant and is acquired actively by exposure to natural infection or by vaccination with the 17D strain of YF virus.

5. Diagnosis

The possibilities of isolating virus from serum is highest during the first 5 days of disease. Adult or baby mice injected intracranially with serum from a suspected case of YF develop encephalitis if virus is present. Identification of the virus is made by specific neutralization tests in mice. Neutralizing Abs may be detected in patients' sera as early as the 5th day of disease. The assay involves mixing the serum with various dilutions of YF virus and inoculating the mixtures intracranially in mice. It is imperative that samples of the original specimen be retained so that subsequent successful isolation can be confirmed by reisolation at the same laboratory and at a WHO regional reference center. Histopathologic examination of liver biopsy material for Councilman bodies from human cases is also useful for establishing a diagnosis of YF.

6. Epidemiology

Yellow fever occurs in 2 distinct epidemiologic cycles: (1) classic urban (epidemic), and (2) sylvan (jungle) yellow fever. *Urban YF* involves transmission of virus from man to man through the bite of infected *Aedes aegypti* mosquitoes. The mosquitoes become infected when they bite an infected individual during the viremic stage of disease. Infected mosquitoes remain infectious for life. When an uninfected mosquito bites an individual with viremia, the mosquito becomes infectious after an incubation period of 12–14 days, which is referred to as *the extrinsic incubation period*. During this period, the virus replicates extensively, completing its biologic cycle. Yellow fever virus cannot survive long in an urban environment unless there is a constant influx of susceptible human beings and unless the area is infested with *Aedes aegypti* mosquitoes.

Jungle YF is primarily a disease of nonhuman primates that is transmitted from monkey to monkey by jungle (arboreal) mosquitoes. Nonhuman primates are natural hosts of the virus and serve as permanent reservoirs of the virus. The disease in monkeys may be either severe or inapparent. Man, an accidental host, contracts the disease when he enters the jungle and is bitten by jungle mosquitoes carrying YF virus. The disease may also be transmitted from monkey to man by the *Aedes aegypti;* this usually takes place in nonjungle environments. All age groups are susceptible, but the disease is usually milder in young infants than older persons. Inapparent infections frequently occur. The mortality rate is about 10 percent.

7. Treatment and Prevention

If a case of YF is diagnosed in an endemic area, the patient should be kept in a mosquito-proof room during the first 5 days of illness. Vaccination is essential for all persons entering an endemic area. Vaccination with live virus of the 17D attenuated strain induces effective immunity within 7–10 days. Booster doses of the vaccine should be given every 10 years. Urban YF has been virtually eliminated in most areas of the world by reducing or eradicating the vector mosquito *A. aegypti*.

C. OTHER VIRAL HEPATIC DISEASES

Rubella virus, cytomegalovirus, coxsackievirus, herpes simplex virus, or EB virus infection may cause hepatitis as part of a more generalized infection. The neonate is highly susceptible to hepatitis with associated jaundice and hepatosplenomegaly caused by these viral agents, with the exception of EB virus. Infection of the neonate by rubella or cytomegalovirus occurs transplacentally from mother to offspring, whereas herpes simplex or coxsackieviruses infect the infant during or after parturition. Although the liver returns to normal metabolic function following recovery from hepatitis due to these "hepatitis viruses," sequelae may include cirrhosis. Other viral diseases (e.g., measles) may be accompanied by minor, nonspecific changes in liver function.

Infectious mononucleosis in adults is caused by EB virus (Chapter 29); it is usually accompanied by hepatic enlargement and altered liver functions (e.g., serum alkaline phosphatase) although jaundice is usually absent. In contrast to SH or IH, there is minimal cell necrosis, and chronic liver disease is not a sequela of infectious mononucleosis.

REFERENCES

Brechot, C., Hadchouel, M., Scotto, J., Fonck, M., Potet, F., Vyas, G.N., and Tiollais, P.: State of hepatitis B virus DNA in hepatocytes of patients with hepatitis B surface antigen-positive and -negative liver diseases. Proc. Natl. Acad. Sci. (U.S.A.), *78*:3906, 1981.

Dane, D.S., Cameron, C.H., and Briggs, M.: Virus-like particles in serum of patients with Australia-antigen-associated hepatitis. Lancet, *1*:695, 1970.

Gerlich, W.H., Goldman, U., Muller, R., Stibbe, W., and Wolff, W.: Specificity and localization of the hepatitis B virus-associated protein kinase. J. Virol., *42*:761, 1982.

Gerlich, W.H. and Luer, W.: Selective detection of IgM-antibodies against core antigen of the hepatitis B virus by a modified enzyme immune assay. J. Med. Virol., *4*:227, 1979.

Locarnini, S.A., Gust, I.D., Ferris, A., Stott, A.C., and Wong, M.L.: A prospective study of acute viral hepatitis with particular reference to hepatitis A. Bull. W.H.O., *54*:199, 1976.

Marion, P.L., and Robinson, W.S.: Hepadna viruses: hepatitis B and related viruses. Curr. Top. Microbiol. Immunol., *105*:1, 1983.

Szmuness, W.: Hepatocellular carcinoma and the hepatitis B virus: Evidence for a causal association. Prog. Med. Virol., *24*:40, 1978.

Szmuness, W., Stevens, C.E., Harley, E.J., Zang, E.A., Oleszko, W.R., William, D.C., Sadowsky, R., Morrison, J.M. and Kellner, A.: Hepatitis B vaccine. Demonstration of efficacy in a controlled clinical trial in a high-risk population in the United States. N. Engl. J. Med., *303*:833, 1980.

Tiollais, P., Charney, P., and Vyas, G.N.: Biology of hepatitis B virus. Science, *213*:406, 1981.

Zuckerman, A.J. and Harvard, C.R.: Hepatitis Viruses of Man. New York, Academic Press, 1980.

Chapter 29

Glandular Viral Diseases

A. MUMPS

Mumps was one of the earliest diseases recognized as a clinical entity. Hippocrates described its characteristic features (nonsuppurative swelling of the parotid glands) as early as the 5th century B.C. About 1790, Hamilton stressed the importance of orchitis as a complication of mumps and noted that CNS disorders sometimes accompany parotitis (Chapter 29). In 1940, physicians began to recognize pancreatitis and involvement of other organs and tissues as additional expressions of infection with mumps virus.

In 1934, Goodpasture and Johnson proved the viral etiology of mumps by reproducing the disease in nonhuman primates inoculated with filtrates of human saliva. The subsequent development of a skin test Ag for determining hypersensitivity to mumps virus revealed that mumps virus frequently causes inapparent infections that are followed by immunity.

1. Etiologic Agents

Mumps is caused by an RNA virus belonging to the Paramyxoviridae family (Chapter 13). The virus agglutinates chicken or human erythrocytes and induces parotitis and fatal meningoencephalitis in suckling mice or hamsters.

2. Clinical Symptoms

In most cases, the incubation period is 16–20 days. Swelling of one or both parotid glands is the first sign of disease. However, in adult males, pain in the testes may be the first symptom. Rarely, aseptic meningitis appears first, followed by parotitis (Chapter 27); the incidence of aseptic meningitis is variable but may be as high as 25 percent.

In severe cases of mumps, the prodromal period with fever, malaise, headache, chills, sore throat, earache, and tenderness along the region of the parotid ducts lasts as long as 2 or 3 days. Parotid swelling is usually observed below the ear. There is sharp pain if the Stensen's duct becomes partially occluded as the gland swells (Fig. 29–1). The papillae at the opening of the Stensen's duct may be reddened, but this feature is inconsistent. Symptoms tend to be mild in children and more severe in adults. The duration of swelling and fever depends on the extent and severity of the infection.

Orchitis is rare before puberty but may occur in 25 percent of adult patients. The onset is variable but usually occurs between the 5th and the 10th day of illness as the parotid swelling subsides. Inflammation of the testes is usually unilateral. Even if both testes are involved, the necrosis is commonly spotty and seldom results in complete sterility. Pancreatitis occurs in about 10 percent of all cases. More rarely, other

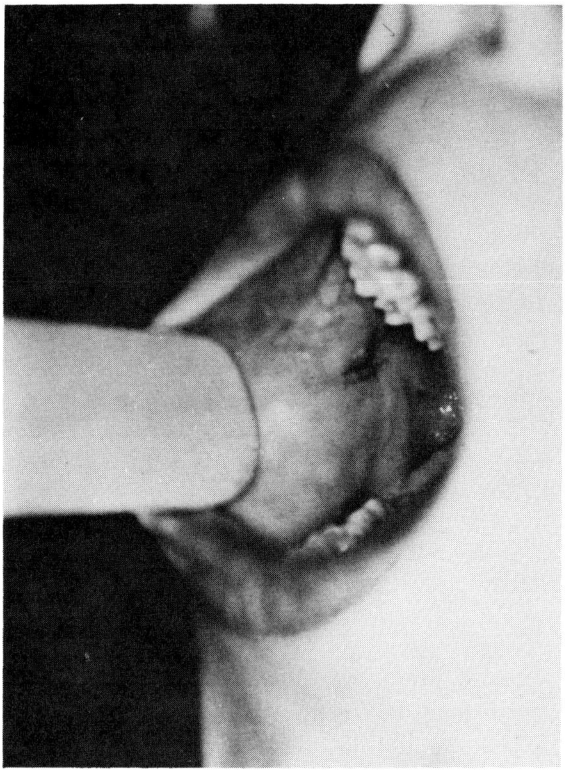

Fig. 29–1. Swollen Stensen's duct opening in mumps. Note *papillae* at the opening of the duct. Involvement of Stensen's duct may be diagnostically helpful. (With permission from Physician's Bulletin, 1959. Eli Lilly Co., p. 92.)

organs (e.g., prostate, mastoid, ovaries, thyroid, thymus, spleen, liver) are involved, particularly in adults.

3. Pathogenesis

Mumps virus enters and multiplies first in the respiratory tract. Viremia with invasion of other organs, including CNS, occurs later. The parotid glands are highly susceptible to virus infection and become markedly inflamed. Degenerative changes can occur in the testes and ovaries; lesions found in the CNS are those of postinfectious encephalitis (Chapter 27).

4. Immunity

Only one antigenic type of the virus exists, and immunity is permanent after the first infection. There is no basis for the old wives' tale that individuals who develop and recover from unilateral mumps parotitis can later develop the same disease involving the contralateral side. Mumps is rare in infants because of passively acquired maternal Ab. Skin test Ag is available commercially for determining the immune status of an individual.

5. Diagnosis

It is impossible to diagnose mumps clinically if the salivary glands are not involved. Laboratory specimens should include saliva, spinal fluid, or urine; the urine specimen should be collected 4 or 5 days after the onset of illness. The specimen is inoculated into the allantoic cavity of chick embryos. Demonstration of mumps virus is then made by hemagglutination and hemagglutination-inhibition (HI) tests using known antisera. Inoculation of monkey kidney cells and identification of mumps virus by hemadsorption and hemadsorption-inhibition tests can also be done. The skin test is of no value for diagnosis.

6. Epidemiology

Mumps infections occur worldwide, and man is the only natural host and reservoir. Children are particularly susceptible to the virus. Spread of mumps virus occurs by direct contact and droplet infection. Saliva of patients during the 6 days preceding parotitis and of persons harboring inapparent infections is highly infectious. Virus is also excreted in the urine until the 15th day of the disease. The length of the virus excretion period contributes to the highly contagious nature of the disease. The fact that man is the only natural reservoir of the virus and that virus transmission is by direct contact and by droplet nuclei explains the regional variation in the epidemiology of the disease.

7. Treatment and Prevention

The administration of hyperimmune gamma globulin prior to the development of clinical symptoms will prevent or attenuate the disease in persons exposed to mumps virus. A vaccine comprised of live virus attenuated by culture in chick embryos is available commercially. A single dose of vaccine produces an inapparent infection that is not transmissible to other individuals. The vaccine has been shown to be 96 percent effective (Chapter 23).

B. INFECTIOUS MONONUCLEOSIS (GLANDULAR FEVER)

Infectious mononucleosis (mono) is characterized by irregular fever, pharyngitis, lymph node enlargement, splenomegaly, lymphocytosis (due to morphologically atypical lymphocytes), high titers of heterophil

Abs that react with sheep erythrocytes, and the development of Abs to Epstein-Barr (EB) virus capsid Ags.

1. Etiologic Agents

In 1968, Gertrude and Werner Henle concluded that EB virus is the cause of infectious mono; their conclusion was based on seroepidemiologic data. Originally, this conclusion was not widely accepted, since EB virus is also associated with other diseases, namely, Burkitt's lymphoma, nasopharyngeal carcinoma, and possibly Hodgkin's disease (Chapter 34). Evidence supporting an EB virus etiology of infectious mono is based on the following data: (1) it occurs only in individuals lacking Abs that react with EB virus; (2) all individuals with a documented history of infectious mono have anti-EB virus capsid Abs; (3) infectious mono is a lymphoproliferative disease, and EB virus stimulates growth of cultured lymphocytes in vitro; and (4) EB virus can occasionally cause an infectious-mono-like syndrome and Ab conversion in patients following heart surgery supported by blood transfusions. Epstein-Barr virus is often present in circulating leukocytes of healthy blood donors. The infrequency of EB-virus-induced infectious mono following blood transfusions may be due to the fact that the transfused blood also contains EB virus Abs that might abort or prevent the disease. Recent evidence has shown that EB virus can be isolated from the nasopharynx during the acute stages of infectious mono, which strongly supports the EB virus etiology of this disease.

2. Clinical Symptoms

The incubation period of infectious mono varies from a few days to several weeks. Initial symptoms include malaise, fever, sore throat, headache, and severe pharyngitis. Between the 5th and 12th day, a palatal exanthem consisting of sharply circumscribed red spots can be observed. Cervical lymph node and splenic enlargement is almost always noted. Jaundice with or without hepatomegaly may occur between the 4th and 14th day of illness. A skin rash consisting of maculopapular, faintly erythematous eruptions has been described. Involvement of the CNS and peripheral nerves, heart, or lung rarely occurs. The disease course lasts about 1–2 weeks, but convalescence may be prolonged. Relapses or death are rare. Death is generally attributable to traumatic rupture of the spleen.

3. Pathogenesis

The highly varied and numerous clinical features of infectious mono reflect the wide distribution of lesions in the body. Little information is available concerning the pathogenetic mechanisms responsible for diseases due to the EB virus. In vitro experiments suggest that EB virus can produce a latent infection and perhaps persist for the life of the individual. Epstein-Barr virus can be detected in peripheral or bone marrow leukocytes derived from a wide diversity of healthy donors. The virus has the in vitro capacity to transform normal leukocytes into established cell lines.

4. Immunity

Antibodies engendered against EB virus persist for many years, in contrast to *"heterophil" Abs* against sheep erythrocytes. Up to 85 percent of the adult population have EB virus Abs despite the lack of any documented history of apparent disease. These data indicate that EB virus is ubiquitous in nature and commonly causes inapparent infections.

5. Diagnosis

Epstein-Barr virus may be isolated from nasopharyngeal swabs of patients with infectious mono, although routine isolation of the virus for diagnosis is not done. Diagnosis is usually made by examining blood samples taken during the acute stages of disease for (1) lymphocytosis characterized by the presence of atypical lymphocytes (Fig. 29–2) involving 20 percent or more of the cells and (2) heterophil Abs measured by the *Paul-Bunnell test.* Heterophil Ab titers of over 1:100 are significant. From 60–80 percent of clinically and hematologically diagnosed cases of infectious mono show a positive Paul-Bunnell test. This test will no doubt soon be replaced by CF, gel diffusion, immunofluorescence, or other specialized serologic tests for Abs specific for EB virus Ags (e.g., early Ag, virus capsid Ag, membrane Ag). The SGOT (serum glutamic oxalacetic transaminidase) and cephalin-cholesterol flocculation tests for liver damage are usually positive.

6. Epidemiology

Infectious mononucleosis is principally a disease of young persons between the ages of 15–40 years. The disease is not highly contagious and occurs sporadically. Transmission of the disease occurs principally by direct contact, especially through saliva and especially as the result of kissing. Some reports indicate that it is possible to contract the disease indirectly from fomites contaminated by persons with apparent or inapparent infections or following massive transfusions of blood.

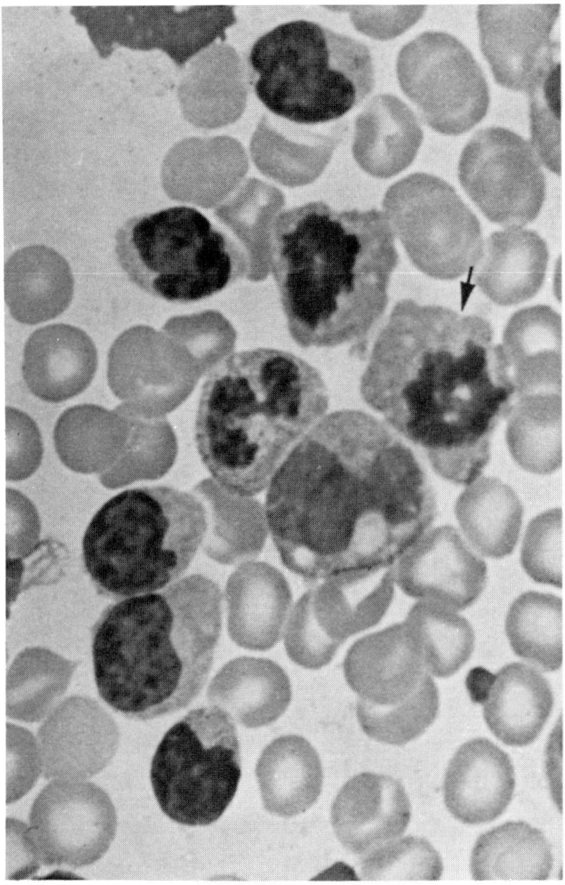

Fig. 29–2. Smear of peripheral blood showing "atypical lymphocytes" associated with infectious mononucleosis. Apart from the central monocyte (with vacuoles) and a segmented neutrophil, the nucleated cells are all lymphocytes. The mitotic figure (in telophase) shows the increased cytoplasmic basophilia often seen in the abnormal lymphocytes (↑) of this disease. Mitoses in lymphocytes are very uncommon in normal peripheral blood, but are frequently seen in infectious mononucleosis. (With permission from Hayhoe, F.G.J., and Flemans, R.J.: An Atlas of Haematological Cytology. New York, Wiley-Interscience, 1970.)

7. Treatment and Prevention

No effective infectious mono vaccine has been developed. The course of infectious mono is generally benign, and no effective therapy is available. Palliative treatment with aspirin may be used to alleviate the symptoms of the fever and pharyngitis. Rest and inactivity are recommended to prevent possible splenic rupture during the acute and convalescent phases of the disease.

C. CYTOMEGALIC INCLUSION DISEASE (SALIVARY GLAND VIRUS DISEASE)

Cytomegaloviris infection is manifested in a variety of ways, depending on age and physical condition of the patient and whether the disease is associated with primary or latent infection. Originally, the disease was recognized only retrospectively by the postmortem finding of enlarged epithelial cells bearing intranuclear and cytoplasmic inclusions in salivary glands, liver, spleen, lungs, kidneys, and occasionally the brain, and by seroepidemiologic surveys. Recently, however, the disease is being recognized more often during life and has emerged to prominence because of its association with extensive blood transfusions (e.g., for open-heart surgery), immunosuppressive therapy, congenital infections, and renal transplantation. Whereas inapparent infection is common during childhood and adolescence, most fatalities occur among children under 2 years of age. About 10 percent of infected neonates suffer some brain damage.

1. Etiology

The causative agent of cytomegalic inclusion disease is a typical herpesvirus (*cytomegalovirus*). It replicates best in human cells and is relatively unstable at 37°C (the half-life is 1 hour at 37°C). *Cytomegalovirus* grows slowly and produces a low yield of virus in tissue culture.

2. Clinical Symptoms

Cytomegalovirus infections are most often inapparent. Infections become apparent more commonly in congenitally infected infants than in infants infected postnatally. Most often the disease is associated with hepatosplenomegaly with hepatitis (Chapter 28). A maculopapular erythematous rash is sometimes seen in childhood infections. Primary infection in young adults resembles the febrile type of infection associated with infectious mono but without the pharyngitis or lymphadenopathy. Damage to the liver often accompanies cytomegalovirus mono. The atypical lymphocytes seen in patients with infectious mono are also seen in cytomegalovirus mono; however, the patient does not exhibit an elevated titer of heterophil Ab.

Cytomegalovirus mononucleosis (postperfusion mononucleosis) sometimes occurs in recipients of recent blood transfusions. The disease may be the result of either a primary infection or of reactivation of a latent infection. The febrile symptoms of the disease are particularly severe in debilitated or immunosuppressed patients (leukemic children) who require massive blood or plasma transfusions. It is now the consensus that cytomegalovirus, along with EB virus, accounts for most of the cases of posttransfusion febrile syndromes.

Observations on kidney transplant patients also suggest that immunosuppression activates latent cytomegalovirus or renders the patients susceptible to reinfection.

3. Pathogenesis

The prolonged incubation period (months) of disease caused by cytomegalovirus is paralleled by the slow replication cycle of cytomegalovirus in tissue culture. Intrauterine infection of the neonate may lead to hepatosplenomegaly with jaundice, microcephaly, mental retardation, and death (Chapter 33). Infection of the newborn presents generalized visceral involvement with viremia and viruria.

The characteristic cytopathology of infected cells includes cell enlargement and eosinophilic inclusion bodies in the nucleus and/or basophilic inclusion bodies in the cytoplasm (Fig. 29–3). A mononuclear cell infiltrate may be associated with the infected cells in many organs, including salivary glands, lymph nodes, liver, spleen, lungs, and kidneys.

4. Immunity

Complement-fixing (CF) and neutralizing Abs are present in a high proportion of the sera of healthy human beings. Young healthy children

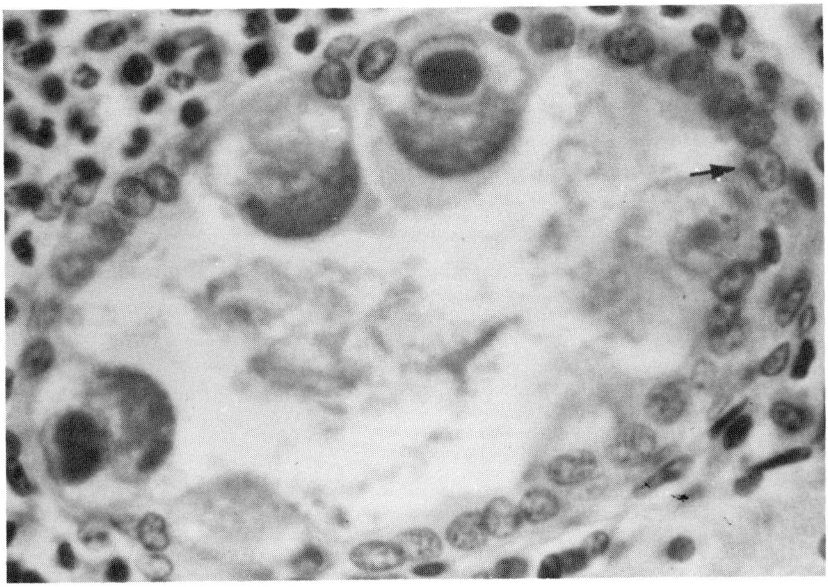

Fig. 29–3. Microscopic section of cytomegalic inclusion disease virus infected kidney. Note the swollen *"cytomegalic" cells* with intranuclear inclusions (↑) and the characteristic enlargement of the affected cells in the renal tubule. (With permission from Physician's Bulletin No. 1, 1959, Eli Lilly Co., p. 83.)

may excrete virus in the urine or saliva despite circulating Ab in the blood. This suggests that cytomegalovirus commonly produces chronic or inapparent infections. The frequency with which cytomegalovirus can be isolated from individuals decreases after the first few years of life. Evidence for reinfection is provided by a rise in serum Ab above the previous level which can be detected at the time of virus isolation from apparent disease. Furthermore, infants infected during intrauterine life are born with IgM Abs synthesized during the fetal period; titers of IgM Abs continue to rise after birth, and there is continued excretion of virus in the urine. This condition is similar to that observed in congenital rubella infection (Chapters 22 and 26).

5. Diagnosis

The presence of "cytomegalic cells" containing characteristic inclusion bodies in urinary sediments or in autopsy material from the kidney (Fig. 29–3) and other visceral organs establishes the diagnosis of cytomegalovirus infection. Virus isolation in human fibroblast cultures can be accomplished using specimens obtained from the nasopharynx or the urine. The presence of serum IgM neutralizing Abs to cytomegalovirus in newborns is suggestive of congenital infection. An increase in the titer of IgM Abs against cytomegalovirus during the first year of life also can be of diagnostic value.

6. Epidemiology

Cytomegalovirus infections are widespread and, except for infants under one year of age, are usually subclinical. Approximately 60–70 percent of adults have circulating Abs to the virus. Infants with cytomegalic inclusion disease discharge virus through the nasopharynx and urine for many months after birth. In this respect, the disease simulates congenital rubella. Cytomegalovirus may also be spread in a similar manner by asymptomatic carriers.

7. Treatment and Prevention

There is no effective treatment for cytomegalovirus infections, and no vaccine is currently available.

D. MESENTERIC ADENITIS

This disease, which involves mesenteric lymph nodes and the appendix, results from infection by *echoviruses* or *adenoviruses*. Clinically, the disease resembles an appendicitis attack. However, at the time of surgery, the signs noted are enlarged lymph nodes in the ileocecal region and a normal appendix. The etiologic agent can be isolated from stools or, in case surgery is carried out, from mesenteric lymph nodes. For further details on the biologic properties of adenoviruses and echoviruses, see Chapters 7 and 10.

REFERENCES

Charlesworth, J.A., Endre, Z.H., Pussell, B.A., Yasmeen, D., and Peake, P.W.: Complement behavior in infectious mononucleosis: possible mechanisms for the prevention of immune complex injury. J. Infect. Dis., *145*:505, 1982.

Hanshaw, J.B.: Congenital cytomegalovirus infection: A fifteen year perspective. J. Infect. Dis., *123*:555, 1971.

Henle, G., Henle, W., and Diehl, V.: Relation of Burkitt's tumor-associated herpes-type virus to infectious mononucleosis. Proc. Natl. Acad. Sci. (U.S.A.), *59*:94, 1968.

Miller, G.: Epstein-Barr herpes virus and infectious mononucleosis. Prog. Med. Virol., *20*:84, 1975.

Modlin, J.F., Orenstein, W.A., and Brandling-Bennett, A.D.: Current status of mumps in the United States. J. Infect. Dis., *132*:106, 1975.

Pereira, M.S., Field, A.M., and Blake, J.M.: Evidence for oral excretion of EB virus in infectious mononucleosis. Lancet, *1*:710, 1972.

Plotkin, S.A., Michelson, Pagano, J.S. and Rapp, F. (eds.): CMV: Pathogenesis and Prevention of Human Infection. New York, Alan R. Liss, Inc., 1984.

Plummer, G.: Cytomegaloviruses of man and animals. Prog. Med. Virol., *15*:92, 1973.

Rinaldo, C.R., Jr., Carney, W., Richter, B.S., Black, P.H., and Hirsch, M.S.: Mechanisms of immunosuppression in cytomegaloviral mononucleosis. J. Infect. Dis., *141*:488, 1980.

Rubin, R.H., Russell, P.S., Levin, M. and Cohen, C.: Summary of a workshop on cytomegalovirus infections during organ transplantation. J. Infect. Dis., *139*:728, 1979.

Sumoya, C.V.: Primary Epstein-Barr virus infections in children. Pediatrics, *56*:16, 1977.

Waner, J.L. and Weller, T.H.: Serological and cultural studies bearing on the persistent nature of cytomegaloviral infections in man. Perspectives in Virology, *8*:211, 1973.

Weller, T.H.: The cytomegaloviruses: Ubiquitous agents with protein clinical manifestations. Part II. N. Engl J. Med., *285*:267, 1971.

Chapter 30

Viral Diseases of the Eye

The eye and associated tissues may be affected during the course of many cutaneous and systemic viral diseases. Infections of the eyelid and conjunctiva that accompany pharyngoconjunctival fever, varicella, and other benign illnesses usually lead only to minor and transient clinical symptoms. However, ocular infections with viruses such as herpes simplex, vaccinia, varicella-zoster, or variola can involve the cornea and cause serious and permanent damage to the eye. In addition, congenital rubella may cause *cataracts* and *glaucoma,* and cytomegalic inclusion disease may involve the retina. Epidemic keratoconjunctivitis, Newcastle's disease conjunctivitis, and occasionally herpetic keratoconjunctivitis may occur in the absence of generalized infection and characteristically produce localized eye lesions.

A. EPIDEMIC KERATOCONJUNCTIVITIS (SHIPYARD EYE)

Epidemic keratoconjunctivitis is an acute infectious inflammation of the conjunctiva at the border of the cornea.

1. Etiologic Agent

This disease is caused by *adenovirus type 8,* which produces more serious ocular disease than those adenoviruses that cause pharyngoconjunctival fever and conjunctivitis (Chapter 25).

2. Clinical Symptoms

The incubation period of *epidemic keratoconjunctivitis* ranges from 5–10 days. The onset is usually abrupt and is characterized by a unilateral follicular conjunctivitis or an intense *catarrhal conjunctivitis* without perceptible corneal involvement. The other eye frequently becomes infected within a few days, but the involvement usually is less severe. A pseudomembrane sometimes forms, and subconjunctival hemorrhages and an iritis may develop. The preauricular lymph node on the affected side is almost always enlarged. A moderate fever, headache, and malaise are usually the only systemic manifestations. Within a few days, the cornea becomes inflamed; the corneal lesions consist of small foci of subepithelial infiltration in the form of thin corneal opacities. There is intense lacrimation, photophobia, blurred vision, and pain. The disease may last from 2 weeks to several months; repeated relapses may take place. The corneal lesions may persist for weeks, months, or years after clinical recovery. However, healing usually occurs with a few sequelae and with complete return of visual acuity.

3. Pathogenesis

Epidemic keratoconjunctivitis is characterized by marked edema and hyperplasia of the ocular mucosa and by numerous *follicles.* The pseudomembrane that may cover the conjunctivae (Fig. 30–1) contains many mononuclear cells and degenerating epithelial cells; in severe cases, there

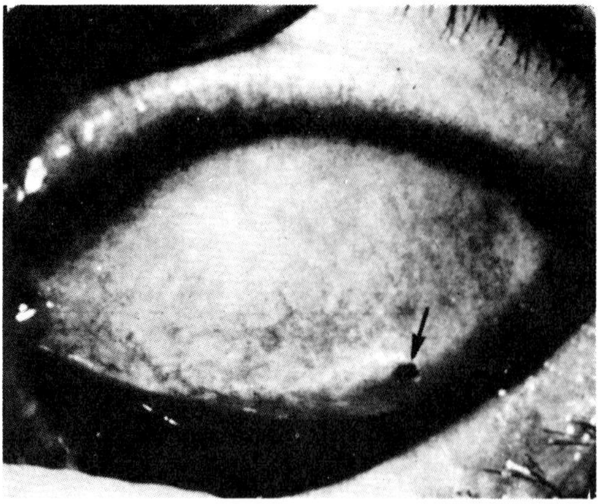

Fig. 30–1. Upper tarsal conjunctivae of patient 8 days after onset of epidemic keratoconjunctivitis. Entire tarsus is covered with an inflammatory membrane. Hole (arrow) indicates membrane's thickness. (With permission from Dawson, C.R., Hanna, L., and Togni, B.: Adenovirus type 8 infections in the United States. IV. Observations on the pathogenesis of lesions in severe eye diseases. Arch. Ophthalmol., *87*:258, 1972.)

may be conjunctival scarring. The corneal subepithelial foci are composed mainly of mononuclear cells and do not ulcerate. One of the characteristics of epidemic keratoconjunctivitis is the absence of corneal neovascularization either during the development of lesions or later. It has been suggested that the stromal opacities are caused by a cellular immune response to viral Ags that are elaborated in the corneal epithelium and diffuse to the stroma. However, the possibility that continuing viral infection is responsible for the opacities cannot be excluded. The intranuclear inclusion bodies characteristic of adenovirus infection are seldom seen in corneal cells from human patients. However, electron microscopic examination of corneal epithelium from severe infections has occasionally revealed intranuclear virus particles, indicating that the virus replicates in the cornea.

4. Immunity

Serum neutralizing Abs to adenovirus 8 develop during the course of the disease; however, there is a relatively rapid fall in titer following recovery, and immunity may be short-lived.

5. Diagnosis

During an epidemic, adenovirus keratoconjunctivitis may be diagnosed on the basis of the clinical features of the disease. However, the first cases that appear during an outbreak may be difficult to distinguish from keratoconjunctivitis or keratitis caused by other viruses or by *chlamydiae.* Since the latter infections can be effectively treated with antibiotics or chemotherapeutic agents, it is of utmost importance to establish an etiologic diagnosis.

The virus may be isolated by inoculating susceptible cells with conjunctival scrapings. The presence of virus is recognized by the characteristic cytopathic changes that occur; the virus is identified by serologic procedures (Chapter 25). The diagnosis also can be established or confirmed by demonstrating a significant increase in serum neutralizing Ab to adenovirus type 8.

6. Epidemiology

Epidemic keratoconjunctivitis is endemic in the Near and Far East where there are yearly outbreaks. The disease which was introduced into the U.S.A. during World War II, caused large epidemics among shipyard employees. For this reason, the disease was called "shipyard eye." In the U.S.A. the disease now occurs mainly as localized outbreaks that can usually be traced to eye clinics or eye specialists. Adenovirus is not inactivated by some of the routine procedures used to clean ophthalmic instruments. Accordingly, a patient with the disease can serve as a source of contamination of such instruments, which then spreads the disease to other individuals.

7. Treatment and Prevention

There is no specific treatment for epidemic keratoconjunctivitis. Topically applied steroids are effective in treating persistent corneal opacities that may cause reduction of vision. However, steroid therapy should be reserved for severe corneal involvement, since it does not appear to reduce the intensity of conjunctival inflammation.

No vaccines for epidemic keratoconjunctivitis are available.

B. NEWCASTLE DISEASE CONJUNCTIVITIS

Newcastle disease is a natural disease of fowl; the disease manifestations include a pneumoencephalitis in young chickens and an influenza-like illness in adult birds. Infections in man, who is an unnatural host, cause an inflammation of the conjunctivae.

1. Etiologic Agent

The etiologic agent of this disease is *Newcastle disease virus*, a member of the Paramyxoviridae family (Chapter 13).

2. Clinical Symptoms

The incubation period in man is 1–2 days. The onset of disease is abrupt, and the principal symptoms are an acute follicular-type inflammation of the conjunctivae, edema of the lids, profuse lacrimation, and preauricular adenopathy. The infection is usually unilateral. The cornea is not involved, and photophobia is unusual. Pulmonary involvement may occur, but it is usually inconspicuous and may not be recognized. In most cases, the temperature remains normal. The conjunctivitis lasts only 3 to 4 days, and recovery is complete within 1–2 weeks.

3. Pathogenesis

The pathogenesis of NDV infections in chickens has been studied extensively because of the economic importance of the disease. In man, the virus causes only a self-limiting conjunctivitis with no residual effects. The ocular mucosa is hyperemic and may be dark red; edema is marked, and there may be an inflammatory pseudoptosis.

4. Immunity

The limited localized involvement of the eye in man probably accounts for the low levels of neutralizing Ab in serum.

5. Diagnosis

Diagnosis can usually be made on the basis of clinical symptoms coupled with a history of laboratory exposure to the virus or contact with infected fowl.

An etiologic diagnosis can be made by culturing the virus in embryonated chicken eggs or in tissue cultures, followed by identification tests using neutralization or hemagglutination-inhibition tests. Serologic confirmation using patients' sera may be difficult because the infection does not elicit a good Ab response.

6. Epidemiology

Birds are the natural hosts of NDV. Infection in man is an occupational disease and is limited almost exclusively to poultry workers or meat processors and laboratory personnel associated with NDV research. The virus becomes airborne and is spread by droplet nuclei. Although the virus is somewhat more resistant to environmental conditions than most of the paramyxoviruses, close association with contaminated materials is required for infection.

7. Treatment and Prevention

No effective treatment of Newcastle disease conjunctivitis is available. Prevention of human infections involves control of the disease in poultry and extreme caution when working with the virus in the laboratory. Live, attenuated virus vaccines as well as inactivated virus vaccines are useful for preventing generalized systemic disease in fowl but are less effective for preventing infection of the avian respiratory tract. Slaughter of infected flocks and quarantine of infected premises can eliminate or control the disease.

C. HERPETIC KERATOCONJUNCTIVITIS

Herpes simplex virus type 1 or *type 2* are the etiologic agents of herpetic keratoconjunctivitis. This potentially serious ocular disease is discussed in Chapter 26.

REFERENCES

Dawson, C.R., Hanna, L., and Togni, B.: Adenovirus type 8 infections in the United States, IV. Observations on the pathogenesis of lesions in severe eye diseases. Arch. Ophthalmol., *87*:258, 1972.

Dawson, C.R., Hanna, L., Wood, T.R., and Despain, R.: Adenovirus type 8 keratoconjunctivitis in the United States. III. Epidemiological, clinical and microbiological features. Arch. Ophthalmol., *69*:473, 1970.

Fenner, F.: The Biology of Animal Viruses. Vol. 1 and 2. New York, Academic Press, 1968.

Fenner, F. and White, D.O.: Medical Virology. New York, Academic Press, 1973.

Grayston, J.T.: Eye Infections with Trachoma and Inclusion Conjunctivitis. *In* Principles of Internal Medicine. 8th Ed. G.W. Thorn, R.D. Adams, E. Braunwald, K.J. Isselbacher and R.G. Petersdorf (eds.): New York, McGraw-Hill Book Co., 1977.

Grist, N.R., Bell, E.J. and Gardner, C.A.: Epidemic keratoconjunctivitis. A continuing study. Health Bull. Edinburgh, *28*:47, 1970.

Nataf, R. and Coscas, G.: Involvement of the eyelid, conjunctiva, cornea, sclera, and anterior segment. *In* Clinical Virology. R. Debre and J. Celers (eds.). Philadelphia, W.B. Saunders Co., 1970.

Pavan-Langston, D.R.: Ocular viral diseases. *In* Antiviral Agents and Viral Diseases of Man. G.J. Galasso, T.C. Merigan, and R.A. Buckanan (eds.). New York, Raven Press, 1979, p. 158.

Chapter 31

Viral Diseases of the Intestinal Tract

Diarrhea and vomiting are symptoms frequently observed in many systemic diseases. Epidemic nonbacterial gastroenteritis is the second most common disease in families with young children. The most common diseases are associated with upper respiratory infections (Chapter 25). The major cause of infantile diarrhea in children is regarded as a *rotavirus. Rotaviruses* also have been implicated as the etiologic agent of diarrheal diseases in the young of a number of animal species. In developing countries rotaviruses are responsible for several million infant deaths per year. These viral agents continue to infect humans throughout life without producing clinical disease.

Other viruses causing epidemic nonbacterial gastroenteritis in humans include the *"Norwalk-agent,"* which has been identified only by immunoelectron microscopy. So far it has not been successfully cultivated in animals or cell cultures.

Finally, certain types of *echoviruses* are known to be associated with gastrointestinal disorders. *Adenoviruses, coxsackieviruses,* and *reoviruses* also have been isolated from patients with gastroenteritis but the etiologic significance of these agents has not been elucidated.

A. INFANTILE DIARRHEA

Infantile diarrhea appears to be distinct from adult epidemic nonbacterial viral gastroenteritis; it is a more severe disease, is of longer duration, and may be caused in part by a simultaneous bacterial infection. The disease usually occurs in the fall and winter in the form of explosive epidemics in nurseries, orphanages, and in large families living under crowded conditions with poor sanitation. Most infections occur in the 6-months to 2-year age groups.

The incubation period is 3–4 days. The disease is characterized by an abrupt onset with fever; a profuse watery diarrhea may cause rapid dehydration and death, particularly in newborn and premature infants.

The etiologic agent of infantile diarrhea is a *rotavirus,* which appears as spheres and has *icosahedral symmetry. Rotaviruses* are taxonomically grouped with the *Reoviridea* family (Chapter 11). The rotavirus genome consists of 10 segments of double-stranded RNA. There appear to be 8–10 structural polypeptides including a major outer shell glycoprotein that make up the virion.

Both human and animal (e.g., bovine, porcine) rotaviruses have worldwide distribution. Human rotaviruses have been transmitted to a number of different animals (calves, horses, piglets, monkeys, rabbits). The current proposal is to name strains according to the species of origin (i.e., human, bovine, porcine rotaviruses).

Rotavirus infections are restricted to the gastrointestinal tract and they do not become systemic. The virus infects epithelial cells in the small intestine on the apical half of the villi, while the undifferentiated cells in the crypt appear to be resistant. A few human rotavirus strains have been adapted to grow in human primary and secondary kidney cells.

Large amounts of rotavirus particles can be observed by electron microscopy in the feces of patients during the first 3–5 days after onset of symptoms. Radioimmunoassay (RIA) and enzyme-linked immunospecific assays (ELISAs) are becoming the methods of choice in laboratory diagnosis of rotavirus infections (Chapter 24). Treatment involves prompt intravenous administration of fluids to combat dehydration and salt imbalances, which can be fatal.

B. EPIDEMIC NONBACTERIAL GASTROENTERITIS

Epidemic nonbacterial gastroenteritis is an acute, self-limiting infection characterized by symptoms of profuse watery diarrhea, cramps, nausea, and vomiting that may occur singly or in combination. The incubation period varies from 1–5 days; the onset of symptoms is often abrupt. Systemic signs are relatively mild and include headaches, dizziness, and malaise. Fever, when present, is low and is probably related to dehydration. The acute illness lasts only a day or two; however, loose stools may continue for a week.

The disease is worldwide in distribution and affects all age groups. It is highly communicable and occurs in large epidemics as well as sporadically. The disease is transmitted by the fecal-oral route and has been passed serially in volunteers fed bacteria-free filtrates containing the viral agent.

The etiologic agent of epidemic nonbacterial gastroenteritis is called the *"Norwalk-agent."* The virus particles, which can be visualized in fecal extracts from cases of experimentally infected volunteers, appear as isometric forms, 26–30 nm in diameter having cubic symmetry. The genome of the virus has not been characterized.

A RIA has been developed which can be used to detect the Norwalk-agent or its Abs and can be used in laboratory diagnosis of the disease.

Immunity lasts for a year or longer, and there is no cross-protection among other causative agents of diarrhea. It appears likely that there is more than one antigenic type of each virus, since cases may recur in families and communities at yearly intervals.

Treatment is nonspecific and consists only of fluid replacement.

C. EPIDEMIC VOMITING (WINTER VOMITING DISEASE)

Epidemic vomiting is a benign illness of short duration characterized by abrupt onset of nausea and vomiting. Headache, diarrhea, and abdominal pain may be present; some patients develop a rubella-like rash. There is little or no fever. The disease frequently is observed in association with upper respiratory tract infections and may last only a few hours or up to 1–2 days. Recovery is rapid and uncomplicated.

The disease occurs most frequently in the winter months and is probably spread by droplet transmission. All age groups are affected. Because of its benign nature, there have been few laboratory investigations of the disease.

The suspected etiologic agent of epidemic vomiting (winter vomiting disease) is referred to as a *calicivirus*. The virus particles are isometric, about 30–35 nm in diameter with cubic symmetry.

In summary, viral diseases of the intestinal tract are sporadic and worldwide. They can occur in epidemic proportions primarily among young children. The etiologic agents are spread by the fecal-oral route. Control of these viral diseases depends on further knowledge of the epidemiology, immunology and growth characteristics of the etiologic agents in cell cultures.

REFERENCES

Appleton, H., Buckley, M., Thom, B.T., Cotton, J.L., and Henderson, S.: Virus-like particles in winter vomiting disease. Lancet, *1*:409, 1977.

Dolin, R., Blacklow, N.R., DuPont, H., Beescho, R.F., Wyatt, R.G., Kasal, J.A., Hornick, R., and Chanock, R.M.: Biological properties of Norwalk agent of acute infectious nonbacterial gastroenteritis. Proc. Soc. Exp. Biol. Med., *140*:578, 1972.

Estes, M.K. and Graham, D.Y.: Epidemic viral gastroenteritis. Am. J. Med., *66*:1001, 1979.

Faulkner-Valle, G.P., Clayton, A.V., and McCrae, M.A.: Molecular biology of rotaviruses. III. Isolation and characterization of temperature sensitive mutants of bovine rotavirus. J. Virol., *42*:669, 1982.

Flewett, T.H. and Woode, G.N.: The rotaviruses. Arch. Virol., *57*:1, 1978.

Graham, M.K. and Dimitrou, D.H.: The molecular epidemiology of rotavirus gastroenteritis. Prog. Med. Virol., *29*:1, 1984.

Greenberg, H.B., Kalica, A.R., Wyatt, R.G., Jones, R.W., Kapikian, A.Z. and Chanock, R.M.: Rescue of noncultivatable human rotavirus by gene reassortment during mixed infection with Ts mutants of a cultivatable bovine rotavirus. Proc. Natl. Acad. Sci. (U.S.A.), *78*:420, 1981.

Holmes, I.H.: Viral gastroenteritis. Prog. Med. Virol., *25*:1, 1979.

Konno, T., Ishida, N., Chiba, R., Mochizuki, K., and Tsunoda, A.: Astrovirus-associated epidemic gastroenteritis in Japan. J. Med. Virol., *9*:11, 1982.

Suzuki, H., Konno, T., Kutsuzaeva, T., Imai, A., Tazaeva, F., Isheila, N., and Sakamoto, M.: The occurrence of calicivirus in infants with acute gastroenteritis. J. Med. Virol., *4*: 321, 1979.

Thornhill, T.S., Kalica, A.R., Wyatt, R.G., Kapikian, A.Z., and Chanock, R.M.: Pattern of shedding of the Norwalk agent in stools during experimentally induced gastroenteritis in volunteers as determined by immune electron microscopy. J. Infect. Dis., *132*:28, 1978.

Tyrrell, D.A.J. and Kapikian, A.Z. (eds.): Virus Infections of the Gastrointestinal Tract. New York, Marcel Dekker, Inc., 1982.

Chapter 32

Slow Virus Infections

Persistent viral infections refers to the long period of time (years) that the infectious virus is present in the host. "Slow virus infections" refers to the long incubation period ranging from months to years and the long progressive course of the disease. There may be remissions and relapses, which ultimately lead to irreversible deterioration and demise of the host. The term "slow virus" is actually a misnomer because under optimal conditions several of these viruses replicate rapidly in vitro. Therefore, slow virus diseases are a group of persistent, degenerative, usually fatal diseases that are associated with intermittent or continuous infection throughout life. At any given time, the presence of virus may or may not result in apparent infection and disease. If infection with slow viruses results in disease, symptoms appear after an incubation period of months to years, and the pathogenesis of the disease may be mediated by the immunologic responses of the host to the infection (see Fundamentals of Immunology). Several subacute and chronic neurologic diseases of man having a virus etiology are listed in Table 32–1. Two slow infections in humans caused by atypical viruses, i.e., agents which have not been visualized, isolated or well characterized biologically or physiochemically and referred to as *viroids,* include Kuru and Creutzfeldt-Jakob disease (Table 32–1). In addition, there are several diseases in animals that also appear to be caused by "slow viruses" (Table 32–2). Recently, evidence was published indicating that the etiologic agent for "scrapie" in sheep should be referred to as a "*prion*" (for proteinaceous infectious particles) to distinguish it from "viroids" which are infectious particles composed of "naked" nucleic acid. This chapter includes in-

Table 32–1. Slow Infections In Humans Caused by Typical or Atypical Agents

Etiologic Agent	Disease	Host Range	Incubation Period	Major Findings
Typical Agents				
Papovavirus (SV40-like)	Progressive multifocal leukoencephalopathy (PML)	Humans	Months to years	Patchy foci of demyelination; gliosis; loss of memory; dysarthria; incoordination
Measles virus	Subacute sclerosing panencephalitis	Humans	2–20 years	Ataxia; mental and motor deterioration; coma; measles virus in CNS
Rubella virus	Progressive rubella panencephalitis	Humans	2–15 years	Dementia; ataxia; spasticity; rubella virus in brain
Atypical Agents				
Viroid	Kuru	Humans, chimpanzee	Months to years	Cerebellar ataxia, incoordination; status spongiosus
Viroid	Creutzfeldt-Jakob	Humans, chimpanzee	Months to years	Progressive dementia; ataxia; spasticity; cell loss; astrocytic proliferation

Table 32–2. Slow Infections In Animals Caused by Typical or Atypical Agents

Etiologic Agent	Disease	Host Range	Incubation Period	Major Findings
Typical Agents				
Visna and Maedi viruses	Visna Maedi	Sheep Sheep	Months to years	Demyelination of CNS Progressive pneumonia
Lymphocytic choriomeningitis virus (LCM)	Lymphocytic choriomeningitis	Mouse, humans	Days to months	Rapid death or persistent tolerant infection in mice; may be immune-complex disease; meningoencephalitis, meningitis or influenza-like illness in humans
Aleutian mink disease virus	Aleutian mink disease	Mink	Months	Disease of the liver and kidneys; persistent viremia; hypergammaglobulinemia; glial scarring; polyarteritis
Equine infectious anemia virus	Equine infectious anemia	Horse	Weeks	Hemolytic anemia; hypergammaglobulinemia; hepatitis, glomerulonephritis
50-nm particle	Mink encephalopathy	Mink, other mammals	Weeks to months	Extraneuronal vacuolation of gray matter; neuronal degeneration; astrocytic proliferation
Atypical Agents				
Prion	Scrapie	Sheep, goat, hamster	Months to years	Noninflammatory degeneration of CNS; incoordination; no detectable antibodies in host

formation on slow virus-host interactions and the pathogenesis of selected diseases in man and animals in which neurologic manifestations predominate.

A. SLOW INFECTIONS IN HUMANS CAUSED BY TYPICAL VIRUSES

1. Progressive Multifocal Leukoencephalopathy (PML)

This is a rare demyelinating syndrome of the CNS that generally occurs in elderly persons with a debilitating disease (e.g., leukemia, Hodgkin's disease, tuberculosis) and in patients who have recently received immunosuppressive therapy.

Etiologic Agents. Large numbers of papovavirus-like particles have been observed repeatedly in infected brain tissue. A SV40-like or wart-like virus has been isolated in monolayers of human fetal brain cells inoculated with homogenates of infected human brain tissue. Two viruses isolated from PML patients are referred to as JC virus and SV40-PML virus.

Clinical Symptoms. Gross abnormalities of motor function, vision, and speech occur. The CSF is unchanged, and the electroencephalogram shows only diffuse slowing. The time from first observation of clinical symptoms to death is about 3–4 months.

Pathogenesis. Pathologic changes include areas of demyelination that are numerous in the cerebral hemisphere, cerebellum, and brain stem. Eosinophilic intranuclear inclusions are observed in oligodendrocytes; however, there is no detectable inflammatory response to the infected cells (Table 32–1).

Epidemiology. It is not known whether the infection represents endogenous reactivation of a latent or noninvasive virus, such as the human wart virus, or recent infection by an exogenous virus.

2. Subacute Sclerosing Panencephalitis (SSPE)

This disease is relatively uncommon, the estimated frequency being 1 per million in the U.S.A. Almost all cases involve children.

Etiologic Agents. *Measles virus* (Chapter 13) has been observed and isolated in tissue culture from brain tissue and lymph nodes of patients with SSPE.

Clinical Symptoms. Early symptoms of disease are gradual deterioration followed by motor neuron dysfunction (e.g., jerks, convulsions, and incoordination). Visual difficulties and ultimate blindness follow (Table 32–1). Finally, coma and death occur 1–3 years after the onset of clinical symptoms. This disease should not be confused with postinfectious encephalitis (Chapter 27), another neurologic complication of measles that begins within a few days after the primary infection.

Pathogenesis. Brain sections show round cell infiltration, perivascular round cell cuffing, and occasionally Cowdry type A intranuclear inclusions in neuronal and glial cells.

Immunity. High titers of humoral Ab to measles virus are detectable in the patient's serum and spinal fluid. Despite the presence of high titers of specific Abs, the progression of the disease is not arrested.

Diagnosis. There is a marked increase in IgG in the CSF. Measles virus has been isolated from cell cultures of SSPE human brain tissue and lymph nodes at post mortem. A disease resembling SSPE can be produced in the laboratory by inoculating SSPE-infected brain material into ferrets.

Epidemiology. Although rare, SSPE may develop years after recovery from apparent measles. Viruses isolated from infected brain tissue have biologic and morphologic characteristics similar to those of measles virus (Chapter 13). These data suggest that in SSPE, the measles virus becomes cell-associated following initial in vivo infection of human brain cells and persists in the brain in latent form for the life of the individual. Studies of the viruses recovered from SSPE patients suggest that the agent is defective in its ability to produce matrix (M) protein and possibly the hemagglutinin protein (HA). The lack of M and HA proteins could be responsible for lack of effectiveness of the immune response and persistence of virus in the host.

TREATMENT AND PREVENTION. There is no specific treatment for SSPE.

3. Progressive Rubella Panencephalitis (PRP)

Initial virus infection may occur in utero or acquired during childhood.

ETIOLOGIC AGENT. *Rubella virus* has been isolated from infected children with symptoms of PRP. Free virus exists in the blood. Membrane-associated viral Ag could complex with Ab and circulate as Ag:Ab complexes.

CLINICAL SYMPTOMS. The disease begins with progressive neurologic impairment including dementia, ataxia, spasticity and sometimes myoclonus. Death occurs after a protracted course of several years.

PATHOGENESIS. Cell-mediated and humoral immune responses have been examined in PRP; no consistent defects which could explain the pathogenesis of disease have been observed. Some degree of host tolerance to viral Ags may be responsible for peristence of virus in the host.

EPIDEMIOLOGY. Rubella virus infection can occur in utero or during early childhood resulting in persistence of the virus in the host and progression of symptoms associated with PRP.

B. SLOW INFECTIONS IN HUMANS CAUSED BY ATYPICAL VIRUSES

Kuru and Creutzfeldt-Jakob diseases are caused by atypical agents *(viroids)* that demonstrate unusual resistance to a variety of reagents that inactivate conventional viruses (e.g., formaldehyde, trypsin, nucleases, ultra-violet light). Viroids contain nucleic acid species of low molecular weight (ca 1×10^5 daltons). Unlike viral nucleocapsids, viroids are not encapsidated.

1. Kuru

Kuru is an endemic disease of a Melanesian tribal people inhabiting a remote area in the highlands of New Guinea. Until recently, kuru had been the most common cause of death in this limited area, predominantly affecting children and adult women.

ETIOLOGIC AGENT. The putative *viroid* etiology of kuru is based on a kuru-like syndrome produced in chimpanzees inoculated with brain material from a patient who died from kuru.

CLINICAL SYMPTOMS AND PATHOGENESIS. The incubation period in man is months to years. The symptoms and laboratory findings of kuru (Table 32–1) are similar to those of scrapie (Table 32–2). Symptoms of the disease include slowly progressive ataxia with tremors of the head, trunk, and extremities. There are no febrile phases, no changes in CSF biochemical values or peripheral blood cells, and no detectable inflammation. Late in the course of disease abnormalities of extraocular movements and mental changes develop. The disease leads to death in 3–6 months. There is no detectable immune response to the disease.

EPIDEMIOLOGY. The first hypothesis regarding the etiology of kuru was that it was a sex-linked genetic disease in which a mutant gene had become widely disseminated throughout the isolated Melanesian tribe by centuries of inbreeding. Recent investigators have discredited the genetic hypothesis, since the disease has become less common. Close interrogation of tribal members revealed that since about 1910 women of the tribe had practiced cannibalism. Deceased relatives were eaten following a "nonsterilizing" cooking ritual; children were given an occasional morsel. The men played a minor role in these rituals and rarely ate any of the brain, which was considered a gourmet's delight by the women. This interesting and puzzling but unresolved disease is considered to have a viroid etiology; however, unanswered questions are: (1) what is the natural source of the etiologic agent, and, (2) what are its biologic properties?

2. Creutzfeldt-Jakob (C-J) Disease.

This uncommon disease of the CNS is a complex subacute presenile encephalopathy in which the patient becomes progressively incoordinated and demented as a result of a "spongy deterioration" of the brain. These findings resemble kuru and scrapie; however, the clinical symptoms are clearly distinguishable, in that with C-J disease there is severe dementia, myoclonic fasciculation, and somnolence (Table 32–1). Death usually follows in less than 1 year after symptoms appear. A similar disease has been produced in chimpanzees inoculated with brain homogenates obtained from patients who died of C-J disease; the experimental disease had an incubation period of about 1 year. The causative agent of this disease does not evoke a demonstrable Ab or CMI response.

C. SLOW INFECTIONS IN ANIMALS CAUSED BY TYPICAL VIRUSES

The term slow virus infection was originally used in veterinary literature to describe several transmissible diseases of sheep (Table 32–2). Two of these slow virus infections (scrapie and visna) are of particular interest because they cause slowly progressive neurologic diseases similar to those found in man. *LCM virus* and *Aleutian mink disease virus* are associated with diseases having a chronic, slowly progressive condition marked by immune-complex formation. Similar diseases of known etiology occur in humans, i.e., glomerulonephritis caused by *streptococci.*

1. Visna and Maedi

It is now well established that visna, a progressive neurologic disease of sheep, is caused by an RNA-containing virus with properties in common with retroviruses (Chapter 19). The virus can be grown in sheep choroid-plexus tissue cultures; infected tissue culture fluids can transmit the disease. These criteria fulfill Koch's postulates for assigning a virus etiology to visna. A related virus, *maedi,* causes progressive hemorrhagic pneumonia in sheep. The agent of maedi has properties very similar to those of visna and can be grown in the same tissue culture system.

In infected sheep, the incubation period of visna varies from 8 months to 4 years. During the incubation period, virus can be recovered from the CSF, blood, saliva, RES, brain, and lungs by culture in cells derived from the choroid plexus of sheep. The disease in sheep has an insidious onset. The animal first develops paralysis of the hind limbs, which progresses to total paralysis and death. The primary lesions are in the CNS. The histopathology is characterized by meningeal and subependymal infiltration and proliferation of RES cells. There is demyelination of white matter, but gray matter is unaffected. Virus-neutralizing Abs are produced in high titer by infected sheep (Table 32–2).

2. Lymphocytic Choriomeningitis (LCM)

Adult mice inoculated with LCM virus usually develop a fatal generalized infection. In contrast, infant mice inoculated with the same virus a few hours after birth or infected in utero develop a chronic infection that resembles slow virus infections. After 10 months of age, many of these mice develop a progressive disease involving the CNS. During development of disease, there is a gradual Ab response resulting in accumulation of Ag-Ab complexes in the kidneys, which causes chronic immune-complex glomerulonephritis, which is lethal (Table 32–2).

3. Aleutian Mink Disease (AMD)

Aleutian disease is contagious in mink, but the mode of spread (i.e., vertically from mother to offspring or horizontally from animal to animal) is not completely understood. The pathologic condition in diseased animals is characterized by proliferation of plasma cells and high levels of IgG. Although the Abs incited can interact with the AMD virus, they do not neutralize virus infectivity. In addition, IgG Abs against IgM develop in AMD-infected mink, suggesting that an autoimmune response involving autologous immune complexes may be involved in this disease.

D. SLOW INFECTIONS IN ANIMALS CAUSED BY ATYPICAL VIRUSES

1. Scrapie

The best-known slow infection in animals is scrapie, a natural disease of sheep (Table 32–2). The disease was first recognized in Scotland by sheep farmers and was thought to be a hereditary condition; however, in 1936, it was transmitted to healthy sheep with brain suspensions from diseased animals. The animals rub their bodies (hence the name scrapie) and nibble their skin on the lower extremities. They exhibit fatigue, weight loss, disturbed gait, tremors, and abnormal behavior. Later they develop ataxia and blindness. The disease is invariably fatal 6 weeks to 6 months after the onset of symptoms.

Scrapie is characterized by a noninflammatory, focal degeneration of gray matter that is distributed symmetrically in various parts of the brain. Perineuronal gray matter becomes spongy, astrocytes hypertrophy, and sporadic degeneration of myelin occurs. Neurons become necrotic and vacuolated. Status spongiosus and edema, associated with hypertrophy of astrocytes, are the most common lesions of scrapie. The disease can be transmitted to mice and hamsters, which show similar lesions.

A transmissible infectious agent can be isolated from lymph nodes and spleen about 1 week after infection. Several weeks later, it is present in the salivary glands, thymus, and lungs, and by 16 weeks it is recoverable from the brain. Although eosinophilic bodies are often seen in vacuoles, no virus-like particles have been observed in tissue sections of infected brain by employing electron microscopy. Recent evidence suggests that the scrapie agent *(prion)* is a small proteinaceous infectious particle whose infectivity can be reduced 99.9% by proteases (e.g., proteinase K).

2. Mink Encephalopathy (ME)

Transmissible mink encephalopathy was first recognized in Wisconsin some 20 years ago and later at Idaho mink farms. The pathologic changes in ME (Table 32–2) are quite similar to the changes in scrapie and slow virus diseases of man (e.g., kuru and C-J disease). The major pathologic lesions in ME are found in the CNS and consist of vacuolization of the gray matter and a reactive astrocytosis. There is a general lack of humoral and cellular immune reponses in ME. Clinical symptoms are characterized by locomotor incoordination, convulsions, and ultimately a semicomatose state. Death follows 3–8 weeks after the development of the initial symptoms.

E. SLOWLY PROGRESSIVE NEUROLOGIC DISEASES IN MAN OF POSSIBLE VIRAL ETIOLOGY

1. Guillain-Barré Syndrome

In several cases of Guillain-Barré syndrome, echoviruses have been isolated from the CSF. Concomitant increases in specific Ab titers suggest that the syndrome and *echovirus* infection were concurrent. Mass immunization with killed influenza virus vaccine was associated with a significant rise in the incidence of Guillain-Barré syndrome (see Chapter 23).

2. Multiple Sclerosis (MS)

Multiple sclerosis is the most common demyelinating disease of man; about 100,000 persons in the U.S.A. are afflicted. There is suggestive evidence, based on recent epidemiologic, immunologic, and pathologic data, that *measles virus* may be the causative agent of MS. In this regard, isolation of a transmissible agent in CSF of MS patients has recently been reported. These studies indicate that from 3–23 years may elapse between virus exposure and onset of clinical disease. Patients with MS have high IgG levels in the CSF. Also, inclusion bodies and small multinucleated giant cells have been found in demyelinating scarred areas of the brains of MS patients. Other reports indicate that intranuclear virus-like particles are present in mononuclear cells infiltrating perivascular areas of active myelin deterioration. These observations suggest that MS may involve activation of latent viruses in lymphocytes that interact with cells of the CNS. Ultimately, the virus induces cell injury and death. Similarly, in Marek's disease of chickens (Chapter 34), circulating lymphoid cells bearing Marek's disease herpesvirus in a latent state can incite destruction of myelin.

3. Parkinsonian Dementia

The possibility that parkinsonian dementia is a slow virus infection has been entertained. The major clinical findings include ganglia cell destruction, gliosis, neurofibrillary degeneration, bradykinesia, rigidity, and dementia.

REFERENCES

Buyukmihci, N., Rorvik, M., and Marsh, R.F.: Replication of the scrapie agent in ocular neural tissue. Proc. Natl. Acad. Sci. (U.S.A.), *77*:1169, 1980.

Diener, T.O.: Are viroids escaped introns? Proc. Natl. Acad. Sci. (U.S.A.), *78*:5014, 1981.

Diener, T.O., McKinley, M.P., and Prusiner, S.B.: Viroids and prions. Proc. Natl. Acad. Sci. (U.S.A.), *79*:5220, 1982.

Grinnell, B.W., Martin, J.D., Padgett, B.L., and Walker, D.L.: Is progressive multifocal leukoencephalopathy a chronic disease because of defective interfering particles or temperature-sensitive mutants of JC virus? J. Virol., *43*:1143, 1982.

Ikemoto, S., Minaguchi, K., Suzuki, K., and Tomita, K.: Antigenic shift of visna virus in persistently infected sheep. Science, *197*:376, 1977.

Melnick, J.L., Seidel, E., Inoue, Y.K., and Nishibe, Y.: Isolation of virus from the spinal

fluid of three patients with multiple sclerosis and one with amyotrophic lateral sclerosis. Lancet, April, 10:830, 1982.
Mims, C.A., Cuzner, M.J. and Kelly, R.E. (eds.): Viruses and Demyelinating Diseases. New York, Academic Press, 1984.
Ter Meulen, V. and Hall, W.W.: Slow virus infections of the nervous system: Virological, immunological and pathogenetic considerations. J. Gen. Virol., *41*:1, 1978.
Prusiner, S.B.: Novel proteinaceous infectious particles cause scrapie. Science, *216*:136, 1982.
Stroop, W.G. and Baringer, J.R.: Persistent, slow and latent viral infections. Prog. Med. Virol., *28*:1, 1982.
Wechsler, S.L. and Meissner, H.C.: Measles and SSPE viruses: similarities and differences. Prog. Med. Virol., *28*:65, 1982.

Chapter 33

Viral Diseases of the Fetus and Newborn

The effects of viral infections on the fetus and newborn generate continuing interest as the fields of perinatology and teratology are developing. Major congenital malformations occur in 1–2 percent of all live births. The association of maternal infection with fetal infection and defective organogenesis was first noted in 1941 when rubella infection in the first trimester was found to cause congenital ocular, cardiac, and other defects. In the past 40 years, several other maternal viral infections have been shown or suggested to be associated with ill-effects on the fetus or newborn. The effect is usually a direct result of infection of the progeny, but may also be due to maternal responses to viral infections without a virus actually infecting the fetus. The exact significance of maternal infection on the rate of abortions, stillbirths, and birth defects, however, still remains to be defined.

The epidemiology of the viral infection and what type of ill-effect the fetus or newborn will experience depend on several factors. Thus, the prevalence of particular viral agents in the community and nursery will vary according to geography and season of the year and nosocomial outbreaks. Immunity of the mother as a result of earlier infection or successful immunization will prevent such infections as rubella, chickenpox, or poliomyelitis. The time and mode of transmission of the virus (transplacental, intrapartum, or postnatal) in relation to organogenesis and immunologic competence of the developing fetus are also important considerations. Figure 33–1 presents a summary of the possible sources and modes of transmission of the most important viral infections affecting the fetus or neonate. The diagnostic, clinical and therapeutic aspects related to these agents are discussed below.

A. DNA VIRUSES

1. Cytomegalovirus (CMV)

CMV is the most common virus isolated from newborns, 0.5–2.5 percent of all neonates have virus in the urine. Only 1–10 percent of newborns with virus in the urine are symptomatic in the neonatal period, although subclinically infected infants may develop hearing and/or neurologic defects in later life.

The spectrum of clinically manifest disease is wide in that affected organs include liver, spleen, kidneys, lungs, heart, eyes, and brain. Whether or not central nervous system involvement is manifest as microcephaly, intracranial calcifications, or seizures, the large majority of neonates detected with some clinical disease will manifest later significant neurologic impairment and developmental delay. Clinical manifestations of visceral organ involvement include intrauterine growth retardation, hepatosplenomegaly, jaundice, pneumonitis, and purpuric or

This chapter was supported in part by NIH grant AI19554 from the National Institute of Allergy and Infectious Diseases.

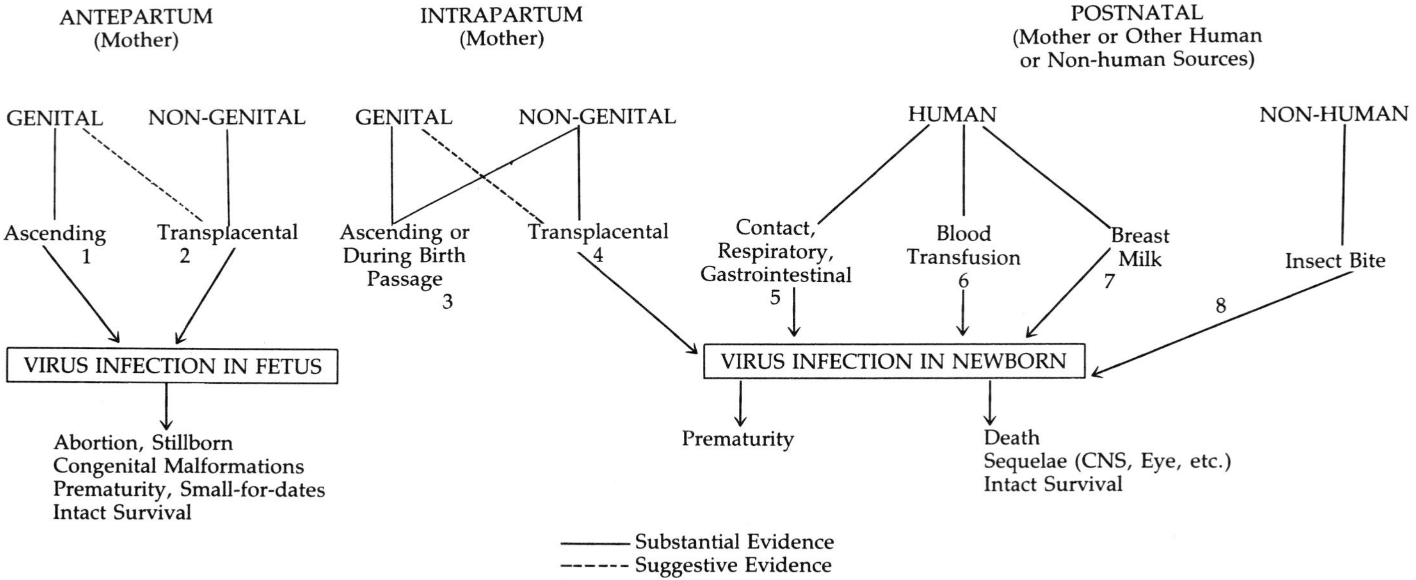

Documented transmission of the DNA and RNA viruses are listed below with suggested but not fully documented modes of transmission noted in brackets.

DNA VIRUSES

Cytomegalovirus—1,2,3,5,6,7
Herpes Simplex Virus—(1),(2),3,5
Hepatitis B—2,3,(4),5,6,(7)
Epstein-Barr Virus—(2),(6)
Poxviruses—2,5
Adenoviruses—(3),5
Papovaviruses—3
Varicella-zoster—2,4,5

RNA VIRUSES

Rubella—2, (5)
Coxsackie A and B—2,3,(4),5
ECHO—2,3,(4),5
Polio—2,3,(4),5
Hepatitis A—(3),(5),(6)
Arboviruses—(2),(4),8
Rubeola—(2),5

Mumps—(2),5
Rhinoviruses—5
Respiratory syncytial virus—5
Influenza—5
Parainfluenza—5
Rotaviruses—(3),5

Fig. 33–1. Possible sources and modes of transmission of virus infections in fetus and newborn.

petechial skin lesions. Roentgenograms of the long bones may show irregular sclerotic and lucent areas of the metaphyses referred to as "celery stalking."

Maternal infection prior to conception does not prevent congenital CMV infection, e.g., several cases have been reported of CMV infections in subsequent pregnancies. However, it appears that all infants born to women seropositive prior to pregnancy do not develop clinically apparent disease in the neonatal period. Another source of infection to the fetus is from the mother's infected cervix, in that CMV can be isolated from this site in 4–28 percent of women at the time of delivery. Approximately one-half of infants born to these mothers will become infected; virus can be isolated from infants 3 weeks to 2 months after birth. It is not clear yet whether intrapartum infections cause clinical disease.

Breast milk is probably the most common source of postnatal CMV infection although, again, clinical consequences to the infant have not been noted to date. Blood transfusions represent another method of transmission which appears to result in severe, if not fatal illness, when CMV positive blood is administered to infants born to seronegative mothers. No matter how CMV is acquired, the virus may be persistently or intermittently shed in the urine or throat for several months or years.

Diagnosis of CMV is best accomplished by viral identification methods (Table 33–1). Cytologic examinations of the urine are less sensitive, and serologic tests are primarily useful for screening purposes. The presence of serum rheumatoid factor is particularly helpful in suspecting CMV or other intrauterine infections.

Prevention of congenital CMV infection is difficult. Isolation of CMV infected individuals from seronegative women of child-bearing age is not practical in the majority of circumstances since most infants with congenital CMV are asymptomatic and older children or adults are even more commonly asymptomatic excretors. Attempts should be made to recover virus from the amniotic fluid to determine if the fetus has become infected in women with a serologically confirmed primary infection during the first trimester or women who develop a heterophile negative mononucleosis-like syndrome. If virus is recovered from the amniotic fluid, an elective abortion might be considered. Blood banks are currently developing methods to prevent transfusion-acquired CMV infection in the high risk infant who has not received maternal transplacental Abs. A live attenuated vaccine is being developed with the hope of immunizing seronegative women prior to pregnancy, but problems of oncogenicity and latency must first be evaluated, as well as the immunologic aspects of this infection. No effective antiviral chemotherapy is available at present.

2. Herpes Simplex Virus Types 1 and 2 (HSV-1 and HSV-2)

Neonatal HSV infections are not as common as those due to CMV. Since subclinical HSV cases are rare, neonatal clinical herpetic disease

Table 33–1. Diagnostic Methods of Viral Infection in the Newborn

Viruses	Method of Diagnosis	Alternative Methods of Diagnosis
DNA Viruses		
Cytomegalovirus	V (U,M,IO)	C (U,IO), SIgM
Herpes Simplex Viruses	V (S,M,E,CSF,U,B,IO)	C (S,M,E,IO), SIgM, SIgG
Varicella Zoster Virus	V (S,IO)	C (S,IO), SIgG
Hepatitis B	V (B)	—
Epstein-Barr Virus	V (B)	SIgM, SIgG
Poxvirus	V (S,IO)	SIgG
Adenovirus	V (M,R)	SIgG
Papovavirus	V,C (L), SIgM	—
RNA Viruses		
Rubella	V (M,U,B,IO)	SIgM, SIgG
Enteroviruses	V (R,M,CSF,IO)	SIgG
Hepatitis A	SIgM	SIgG
Arboviruses	V (B,CSF)	SIgG
Respiratory Viruses	V (N)	C (N,IO), SIgG
Rotaviruses	V (R)	—

Methods

V = Virus identification—preferably virus isolation, but may include electron microscopy, immune electron microscopy, immunofluorescence, enzyme linked immunosorbent assay, counterimmunoelectrophoresis
C = Cytohistopathologic methods
SIgM = Serologic test for IgM Abs
SIgG = Serologic test for IgG Abs requiring persistent or rising titers several months postnatally (to obviate problem of transplacental IgG antibodies)

Sites

U = Urine
M = Mouth
IO = Internal organs including brain
R = Rectal or stool
E = Eyes
CSF = Cerebrospinal Fluid
B = Blood
L = Larynx
N = Nasopharynx
S = Skin

may actually be more frequently recognized than clinical CMV disease. Transplacental infection with HSV appears to be rare; it is suggested by a few neonates in whom major neurologic and/or ocular involvement was detected shortly after birth. Most HSV infections are acquired intrapartum from a mother with clinically apparent or subclinical genital infection at the time of delivery. Postnatal infection from maternal nongenital sites and nosocomial transmission from infant to infant have also been documented. Neonatal acquisition of virus from nursery personnel or fathers with fever blisters is very infrequent. It has been estimated that 100–400 cases of neonatal HSV infection occur yearly in the U.S.A. Transplacentally acquired type-specific Abs have been noted in several infants to have no effect on the progression of disease in the newborn indicating that recurrent maternal infection also poses a risk. However, it is not clear if such Abs may assist in preventing some of the infants at risk from acquiring the infection. Approximately one-half of infants with HSV will be born prematurely. Although HSV-2 causes about 75 percent of clinically apparent neonatal infections, there is no difference in the spectrum of disease caused by the two types of HSV other than the more severe ocular manifestations with HSV-2.

HSV infections are particularly severe in the first month of life and can be grouped into two forms, disseminated and localized. Disseminated infection may involve a wide variety of organs including liver, adrenals, lungs, kidneys, and heart and may also affect the brain. The mortality rate is about 85 percent for untreated disseminated disease. Localized infections may affect the brain without evidence of dissemination or may be limited to the skin, eyes, or mouth. About one third of neonates with untreated localized CNS disease will die, but the majority of survivors are left with neurologic sequelae.

Although infants at risk can be suspected on epidemiologic grounds, clinical suspicion is most often aroused by the presence of skin vesicles, oral ulcers or ocular findings (conjunctivitis and keratitis). These findings unfortunately are absent in about half the cases, so that disseminated disease with hepatic involvement, disseminated intravascular coagulation, or meningoencephalitis, are often diagnosed at autopsy. Progression from the skin or mouth to more severe involvement of the brain or visceral sites occurs in about 75 percent of infected infants.

HSV is the only neonatal viral infection that has so far been found (in control trials) to respond to antiviral therapy. In a randomized controlled study, intravenously administered adenine arabinoside was demonstrated recently to reduce the mortality rate from 74 to 38 percent, being particularly effective in reducing the mortality and morbidity of neonates with localized CNS disease. Early diagnosis and prompt therapy is thus important to further decrease mortality and morbidity. Higher doses of adenine arabinoside and another drug, acyclovir (Chapter 23), are under current controlled evaluation for use in neonatal HSV infections.

The diagnosis of HSV infection in mothers or newborns is best accomplished by virologic identification methods, however, cytologic techniques in capable hands are also useful for rapid diagnosis (Table 33–1).

Mothers with a history of genital herpes or sexual contacts with penile herpes should be monitored virologically or cytologically during the last trimester of pregnancy. Delivery by cesarean section before or soon after rupture of membranes can prevent most cases of neonatal infection. HSV vaccines and hyperimmune globulin are under current investigation.

3. Varicella-Zoster Virus (VZV)

Transplacental infection with VZV in early pregnancy causes a rare congenital varicella syndrome consisting of low birthweight, cicatricial skin scars with muscle atrophy, microphthalmia, optic atrophy, chorioretinitis, hypotrophic lower limbs and associated findings of encephalomyelitis.

Infants born to mothers who develop skin lesions within 5 days of delivery or 4 days after delivery may develop severe disease with a high mortality rate. The disease may be prevented or modified if such high risk infants receive hyperimmune zoster globulin. A VZV vaccine is under current study.

4. Hepatitis B Virus (HBV)

Approximately 1 of 1,000 women in the U.S.A. are hepatitis B surface Ag (HBsAg) positive at the time of delivery, mostly as a result of being chronic carriers. Although HBV may be transmitted transplacentally during pregnancy, particularly if the mother has an acute infection, the greatest risk to the newborn appears to occur at the time of delivery during exposure to maternal blood. Close contact with the mother in the postnatal period provides another opportunity for virus transmission. The infant may be HBsAg positive only several months after birth. Infection in the infant may lead to chronic antigenemia, cause transient liver enzyme elevations or be manifest as mild to severe liver disease, which occasionally is fatal. Infection in the infant may be prevented by administering hepatitis B immune globulin (HBIG) close to the time of birth in combination with hepatitis B vaccine.

5. Other DNA Viruses

Epstein-Barr virus (EBV) infection during pregnancy, manifested as infectious mononucleosis, has been associated with a few cases of birth defects, such as congenital heart disease and cataracts. Although a few cases of transplacental transmission of EBV have been reported, the role of the virus in causing perinatal disease or birth defects remains to be ascertained.

Adenoviruses can occasionally be isolated in the early neonatal period. The infant may be asymptomatic or have respiratory signs. The infection is acquired postnatally from family contacts and occasionally in the nursery.

Some children who develop laryngeal papillomas have been found to be born to mothers with genital warts at the time of delivery. Conflicting results have been reported on the association of another papovavirus (BK) with birth defects. BK virus was originally isolated from the urine of a patient with the initials BK who was receiving immunosuppressive therapy. Interestingly, BK virus shares properties in common with SV40 (Chapter 6).

B. RNA VIRUSES

1. Rubella

Transplacental *rubella virus* infection in the early months of gestation causes varying degrees of fetal damage. More than 20,000 cases of congenital rubella occurred in the United States during the 1964 epidemic. Since attenuated rubella vaccine was introduced in the early 1970s, less than 100 cases of congenital rubella are reported yearly. However, despite the vaccine, 10–20 percent of women of pregnancy age in the U.S.A. are seronegative.

Infants with the "expanded rubella syndrome" may exhibit a variety of clinical manifestations. These include low birth weight, meningoencephalitis, microcephaly, cataracts, pigmented retinopathy, glaucoma, thrombocytopenia, anemia, jaundice, pneumonia, and hepatosplenomegaly. Cardiac defects are common and include pulmonary artery stenosis, patent ductus arteriosus and septal defects. As with CMV and HSV, radiolucency of long bones may occur.

Some infants are asymptomatic at birth and develop sequelae recognized early or many years later. These include psychomotor retardation, deafness, diabetes mellitus, thyroid dysfunction, and progressive panencephalitis.

Diagnosis is infrequently made today by virus isolation methods, serologic tests being more commonly employed (Table 33–1). Infants may continue to excrete virus for several years and are a potential source of infection for nursery personnel and the community. Prevention could be accomplished with universal vaccination and, lacking this effort, should be focused on vaccination of seronegative health-care personnel, female or male. It is currently recommended that the rubella vaccine should not be administered to pregnant women because of the potential danger of a rubella vaccine virus infection of the fetus. A therapeutic abortion should be considered in women with confirmed rubella infection in the first four months of pregnancy.

2. Picornaviruses

Coxsackie A and *B* and *ECHO viruses* have been found to infect newborns most often as a result of intrapartum or postpartum transmission, including during nursery outbreaks, and possibly also as an intrauterine infection. Most infants are asymptomatic or have mild diarrhea or meningitis. Coxsackie B virus can cause a fulminating disease characterized by encephalitis, pneumonitis, myocarditis, and hepatitis. ECHO viruses have also been found to cause meningitis, encephalitis, and diarrhea, and occasionally a severe disseminated disease involving the brain, lungs, and liver.

Neonatal poliomyelitis was rare before the introduction of vaccines and, to our knowledge, no cases have been reported since that time.

3. Other RNA Viruses

Although *hepatitis A virus* may be acquired during passage through the birth canal if the mother is excreting virus, no neonatal infections have been demonstrated as yet.

Togaviruses *(eastern equine encephalitis, western equine encephalitis viruses)* can occasionally infect newborns particularly as a result of insect bites. It appears that western equine encephalitis virus can also be transmitted transplacentally. CNS involvement may be severe causing neurologic sequelae and death.

Rubeola (measles) *virus* may cause abortions, stillbirths, or premature labor related to severe maternal infection. An association with birth defects has not been confirmed.

Some studies have also implicated maternal *mumps virus* infection with various birth defects, but the evidence is inconclusive. *Rhinoviruses, respiratory syncytial, influenza* and *parainfluenza viruses* have all been associated with nursery epidemics. The association of these viruses with abortions, stillbirths or birth defects is not well documented.

Although *rotaviruses* may be acquired by the newborn from an infected mother at the time of delivery, neonates acquire the infection postnatally most often from their nursery contacts. However, despite the fact that rotaviruses are responsible for acute enteritis in infants and children, most newborns found to excrete these viruses in their stools do not usually have diarrhea.

In summary, infections of the fetus and neonate by viruses are responsible for a significant degree of morbidity and mortality. Recognition of neurologic, ocular or hearing sequelae may not become apparent until the child is several months or years old. Some of these infections are particularly important because the fetus is at risk during the time organogenesis and immunologic development are occurring. Improved techniques of diagnosis are necessary, as are methods to prevent and possibly treat maternal and neonatal infections. The role of viral infections, other than the few already appreciated, in the etiology of birth defects requires intensive study.

REFERENCES

Estes, M.K. and Graham, D.Y.: The molecular epidemiology of rotavirus gastroenteritis. Prog. Med. Virol., *29*:1, 1984.

Hanshaw, J.B. and Dudgeon, J.A.: Viral Diseases of the Fetus and Newborn. Philadelphia, W.B. Saunders Co., 1978.

McIntosh, K. and Fishaut, J.M.: Immunopathologic mechanisms in lower respiratory tract disease of infants due to respiratory syncytial virus. Prog. Med. Virol., *26*:94, 1980.

Nahmias, A.J., Dowdle, W.R., and Schinazi, R.F. (eds.): The Human Herpesviruses: an interdisciplinary approach. New York, Elsevier-North Holland, 1981.

Nahmias, A.J., Visintine, A.M., and Starr, S.E.: Viral Infections of the Fetus and Newborn. *In* Viral Infections: A Clinical Approach. Drew, W.L. (ed.). Philadelphia, F.A. Davis Co., 1976, pp. 47–68.

Rapp, F. Persistence and transmission of cytomegalovirus. Comprehensive Virology, *16*:193, 1980.

Remington, J.S. and Klein, J.O.: Infectious Diseases of the Fetus and Newborn Infant. Philadelphia, W.B. Saunders Co., 1983.

Sever, J.L., Larsen, J.W., and Grossman, J.H., III: Handbook of Perinatal Infections. Boston, Little, Brown & Co., 1979.

Stuttgart, E.G.: Varicella-zoster virus infection in pregnancy. Prog. Med. Virol., *29*:166, 1984.

Wechsler, S.L. and Meissner, H.C.: Measles and SSPE viruses: Similarities and differences. Prog. Med. Virol., *28*:65, 1982.

Chapter 34

Viruses and Carcinogenesis

The aim of this chapter is to present an overview of some of the processes involved in the virus causation of cancer and particularly to indicate how the tools and models of basic research have aided in exploring the virus etiology of human malignancies. One proposed idea is that carcinogenesis is a multistep process involving initiation (a mutagenic event) and promotion (a nonmutagenic event). There are published data suggesting that viruses can act as both initiators and promoters of carcinogenesis.

Oncogenic viruses have the capacity to induce either *benign* or *malignant* neoplasms in susceptible animals. The oncogenic potential of most oncogenic viruses can be demonstrated also in tissue cultures, where they induce characteristic cell alterations designated as *cell transformation*. When a virus induces cell transformation in a tissue culture, small foci of cells resembling microtumors (Fig. 34–1) often appear; in most cases the number of foci is proportional to the number of infectious virus particles present in the virus inoculum.

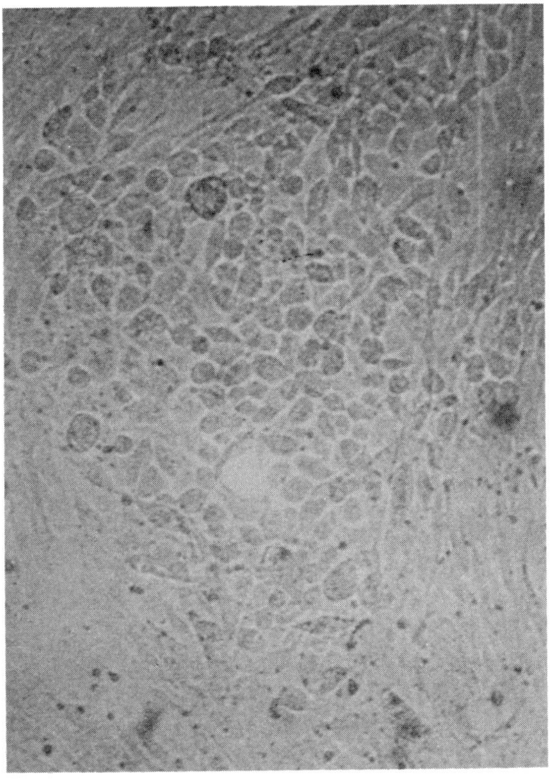

Fig. 34–1. Focus of chick embryo cells transformed by Rous sarcoma virus. × 80. (With permission from Temin, H., and Rubin, H.: Characteristics of an assay for Rous sarcoma virus and Rous sarcoma cells in tissue culture. Virology, 6:669, 1958.)

Viral oncology has advanced rapidly as the result of studies with 3 major groups of viruses: (1) The type C oncoviruses including the leukemia and sarcoma viruses of various animal species, the type B oncovirus including the mammary tumor virus (Bittner agent) and the type D oncoviruses including the primate sarcomagenic viruses. It is of singular interest that particles resembling retroviruses were detected and isolated from a human cutaneous T-cell lymphoma; however, the human retrovirus isolate apparently is not significantly related to other animal retroviruses and it is not endogenous to humans. (2) The group of small oncogenic DNA viruses, which include the polyomavirus of mice, SV40 of monkeys, and a number of adenovirus serotypes, have been isolated from man and other species. (3) Certain members of the Herpesviridae family are associated with neoplastic or proliferative lesions in a wide variety of animal species, including man (e.g., Burkitt's lymphoma, nasopharyngeal carcinoma, and cervical carcinoma). Both in vitro and in vivo methods for quantitative assay of the oncogenic activities of retroviruses, the small oncogenic DNA viruses, and herpesviruses have been devised. These achievements have resulted from the use of modern tissue culture techniques, inbred newborn animals and the application of basic biochemical, biophysical, genetic and immunologic research tools.

Because of the complexities involved, much of the current information on virus-induced cell transformation is still fragmentary and speculative. However, exciting new data are emerging regarding the possible origin of retrovirus oncogenes and the products encoded by the oncogenes of both RNA and DNA tumor viruses.

A. PROPERTIES OF VIRUS-TRANSFORMED CELLS

The transformation of cells cultured in vitro and the induction of neoplasms in animals by oncogenic viruses are related phenomena, as exemplified by the observations that transformed cells behave like tumor cells when transplanted into suitable animals, and that tumor cells from animals behave like transformed cells when cultivated in tissue culture.

Viral-induced cell transformation in vitro provides a model for determining the possible mechanism(s) by which oncogenic viruses incite tumors in vivo. During virus-cell interactions, most oncogenic viruses do not produce cytopathologic changes that permit their recognition. However, they can be detected in infected cells by other techniques, including (1) the formation of new Ags, (2) the presence of virus-specific mRNA, (3) the presence of viral DNA or RNA genomes, and (4) virus-specific enzyme activities (e.g., reverse transcriptase).

The criteria for cell transformation induced by oncogenic viruses include the following: (1) altered growth characteristics, (2) morphologic changes, (3) altered antigenic composition, (4) altered cell metabolism, and (5) the capacity to form neoplasms when inoculated into susceptible animals (Table 34–1). The most reliable criterion of true oncogenic transformation of cells is the ability of the transformed cells to produce a malignant tumor when injected into a susceptible host.

B. VIRAL-INDUCED TUMOR ANTIGENS

One of the principal properties of transformed cells is that they possess new Ags not found in uninfected cells. These new Ags can be detected by complement fixation tests (CF), fluorescent Ab, and graft rejection tests and are designated by different, albeit inappropriate, terms. For example, new intranuclear Ags detected by fluorescent Ab and CF tests in polyomavirus, papillomavirus, and adenovirus-transformed cells are referred to simply as *T antigens* (tumor Ags), whereas surface Ags responsible for tumor graft rejection are called *tumor-specific transplantation* (TST) Ags. The TST Ags are present in the cell plasma membrane but not in the virion. Although these TST Ags are nonvirion Ags, they are virus-specified in the sense that all tumors induced by the same virus carry the same antigenic specificity, irrespective of the nature of the tissue or the species of animal in which they originated. In contrast, herpesvirus- or retrovirus-induced tumors present a different picture. The TST-Ags present in the plasma membrane of these tumors may also be present in the isolated virion. The reason for this is that herpesviruses and retroviruses possess an envelope that is derived from cell membranes, whereas polyoma, papilloma, and adenoviruses are nonenveloped virions.

Transplantation experiments based on the principle of graft rejection are used to demonstrate presence of TST Ags in transformed cells. Thus, inbred mice inoculated with syngeneic polyomavirus-transformed cells develop immunity to the TST Ags and reject the transformed cells. Since the TST Ags specified by this virus in different strains of animals are the same, inbred animals immunized with polyomavirus-induced tumor

Table 34–1. Properties of Cells Transformed by Oncogenic Viruses

1. Altered growth characteristics of transformed cells
   a. loss of contact inhibition
   b. increased growth rate and saturation density in culture
   c. increased capacity to persist in culture
   d. altered metabolism
2. Changes in transformed fibroblasts
   a. cells are shorter
   b. parallel orientation of cells is lost
   c. chromosomal abnormalities are present
   d. resist superinfection by the transforming virus
3. Altered immunologic properties of transformed cells
   a. new cellular antigenic components are present
   b. virus-specified antigens (T antigens, TST antigens, gs antigens, virion antigens) are produced
4. Capacity to form neoplasms

cells of allogeneic origin will likewise be immune to syngeneic tumors induced with the same virus. The TST Ags can also be demonstrated by the fluorescent Ab, cytotoxic, and colony-inhibition test.

A substantial body of evidence suggests that viral-induced cell transformation also results in the appearance of *"embryonic and fetal Ags"* on the plasma membrane (Fig. 34–2). The specificity of these Ags is determined by the species origin of the transformed cells; their possible role in carcinogenesis is not known. It has been suggested that cellular oncogenes become turned on by viruses and may be responsible for inducing the expression of these so-called embryonic and fetal Ags.

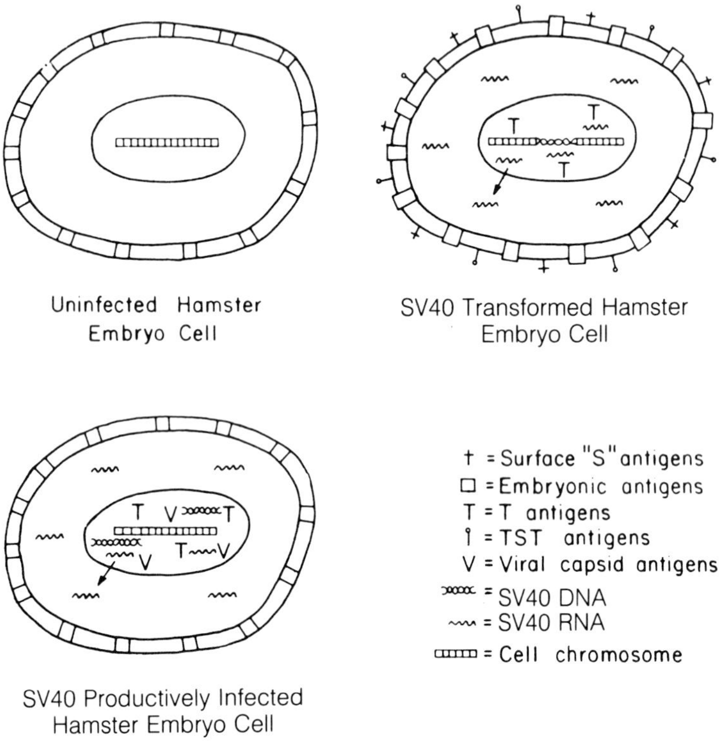

Fig. 34–2. Comparison of uninfected, SV40-productively-infected, and SV40-transformed hamster embryo cell. In transformed cells, the SV40 genome is integrated into the cellular chromosome in the nucleus. Viral mRNA is found in the nucleus and cytoplasm. T-Ags are produced and localized in the nucleus. Several changes occur at the cell surface, including the appearance of embryonic Ags, virus-specified TST Ags, and S Ags. In productively infected cells, the SV40 genome is not necessarily integrated into the cellular chromosome, and TST Ags, S Ags, and embryonic Ags are absent or masked. However, SV40-specified mRNA, T Ags, and viral capsid Ags are synthesized during productive infection. By comparison, all these changes are either absent or masked in uninfected cells.

C. RNA TUMOR VIRUSES

RNA tumor viruses (oncoviruses) are classified under the family name Retroviridae (L. retro, backward). These viruses are unique in containing an RNA-dependent DNA polymerase (reverse transcriptase) that copies single-stranded virion RNA into DNA. Various retroviruses cause leukemias, lymphomas, sarcomas and mammary carcinomas. Some retroviruses are nononcogenic.

Retroviruses are enveloped particles containing an icosahedral nucleocapsid. The envelope is derived from the host cell membrane and contains virus type-specific glycoprotein spikes. The nucleocapsid contains group specific (gag) Ags, the reverse transcriptase and a diploid 70S RNA genome composed of two 35S subunits joined together by hydrogen bonding. A low molecular weight 4S cellular RNA species is associated with the 70S RNA complex. The 4S RNA serves as a primer for synthesis of double-stranded viral DNA using the virus genomic RNA as template and reverse transcriptase. The double-stranded DNA form of retroviruses is referred to as the *provirus*. The mechanism of retrovirus replication following integration of provirus DNA into host cell chromosomes has been described in Chapter 19.

Retroviruses are classified (Table 34–2) according to their morphology in the electron microscope. A-type particles, seen only within cells, pos-

Table 34–2. Classification of Retroviruses

Genus	Subgenus	Species
Cisternavirus A Mice, guinea pigs, hamster	—	—
Oncovirus B Mammary carcinoma virus in mice		Mouse mammary tumor virus (Bittner agent)
Oncovirus C Leukemia-sarcoma viruses of animals	Avian	Rous sarcoma viruses Rous-associated viruses Fujinami viruses Reticuloendotheliosis viruses Myeloblastosis viruses Erythroblastosis viruses Myelocytomatosis viruses
	Mammalian	Mouse sarcoma viruses Mouse leukosis (Friend, Maloney, Rauscher viruses) Murine leukosis G (Gross or AKR viruses) Rat leukosis (Harvey, Kirsten, Rusheed viruses) Feline, Porcine, Bovine leukosis viruses Primate sarcoma viruses (Wooly monkey, Gibbon ape)
Oncovirus D		Mason-Pfizer monkey virus

sess a double shell with an electron lucent center (Fig. 34–3). Mature B-type particles have an eccentric core (Fig. 34–3) and C-type particles a central core (Fig. 34–4). D-type particles have a morphology intermediate between B- and C-type particles.

1. Endogenous Proviruses

Endogenous proviruses make up part of the normal genetic composition of vertebrate cells and are passed vertically from parent to offspring through germ line cells. Many endogenous viruses are defective in replication and require a nondefective helper virus to replicate. The helper virus codes for a protein(s) required by the defective virus for replication. Replication of the defective virus is made possible by complementation (see Chapter 4). Three viral gene products are required for retrovirus

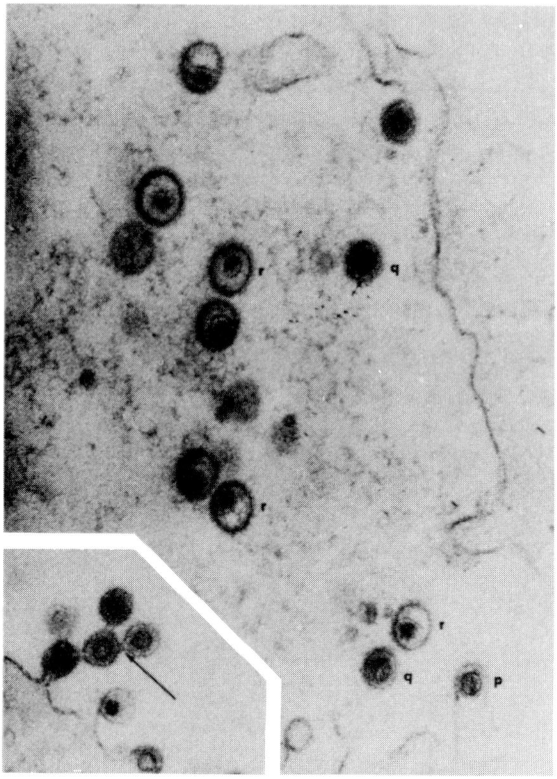

Fig. 34–3. Electron micrograph of a thin section of a mouse mammary tumor virus infected cell. q = immature type-A particles; r = mature type-B particles; p = virus particle budding through the plasma membrane. Arrow in insert indicates viruses sharing a common envelope × 100,000. (With permission from Lyons, M.J., and Moore, D.H.: Isolation of the mouse mammary tumor virus: Chemical and morphological studies. J. Natl. Cancer Inst., *35*:549, 1965.)

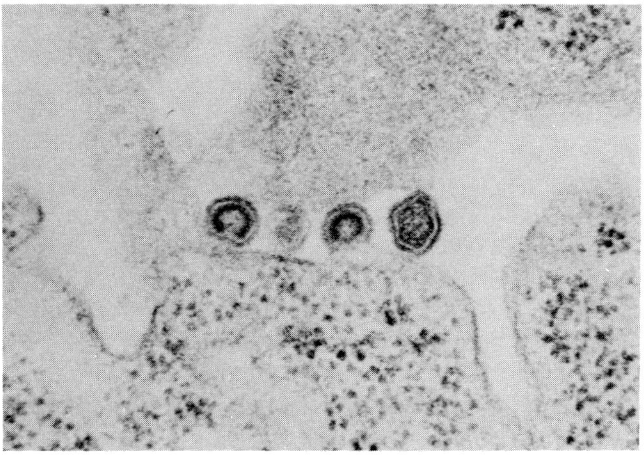

Fig. 34–4. Electron micrograph illustrating leukemia virus C-type particles. The particles are in a leukemic spleen of an AKR/Dm strain mouse that developed spontaneous leukemia. The micrograph illustrates two stages of budding and a fully developed mature leukemia virus particle. × 50,000. (With permission from Dr. Leon Dmochowski, Department of Virology, The University of Texas System Cancer Center, M.D. Anderson Hospital and Tumor Institute, Houston, Texas 77030.)

replication: gag (codes for core structural proteins or group-specific Ags), pol (reverse transcriptase) and env (envelope glycoproteins). All of these gene products are present in virus particles.

2. Oncogenes of Retroviruses

Recently it has been established that nondefective helper viruses can acquire oncogenic potential by recombination of helper virus genetic information with cellular genes called *proto-oncogenes*. The rescued proto-oncogenes code for transformation-related proteins. The transformation-related genes are found in low amounts in uninfected, nontransformed cells. For example, the avian Rous sarcoma virus (Table 34–2) oncogene called the src gene codes for a phosphoprotein of 60,000 MW($pp60^{src}$) which has protein kinase activity. The protein kinase has the distinct property of phosphorylating tyrosine residues. This counterpart for the viral src gene is present in normal cells (proto-oncogene). The proto-oncogene also codes for a phosphoprotein of 60,000 MW which is antigenically related to the avian Rous sarcoma virus src gene. The concept of recombination between retrovirus genetic information and cellular genes is intriguing. Along with this concept are other possible mechanisms to explain retrovirus induced cell transformation such as insertion of viral promoter sequences at crucial sites in the cell genome.

3. Transforming Retroviruses

Transforming retroviruses have been classified as either slowly transforming or rapidly transforming viruses (i.e., nondefective acute transforming or defective acute transforming).

a. Slowly Transforming Retroviruses. These include the avian leukosis (leukemia or lymphoma producing) virus, murine leukosis virus, murine mammary tumor virus and feline leukemia virus. All of these viruses contain the three structural genes gag, pol, env and the region called "c" located near the 3′ terminus and adjacent to the LTR (long terminal repeat) region of provirus DNA (see Chapter 19). The "c" region appears to contain the promoter site, the signal for poly A addition, the initiation site for provirus DNA syntheis and the control for oncogenicity. The slowly transforming leukemia or lymphoma viruses are probably progenitors of the rapidly transforming viruses.

b. Rapidly Transforming Viruses. These are either *nondefective* or *defective* in replication. The prototype for a nondefective rapidly transforming virus is Rous sarcoma virus. Oncogenic potential is made possible by the src gene which codes for the transforming protein ($pp60^{src}$).

The defective rapidly transforming viruses include the avian leukemia viruses, Fujinami sarcoma virus, reticuloendotheliosis, myeloblastosis, erythroblastosis and myelocytomatosis viruses, Abelson murine leukemia virus, and the murine, rat and feline sarcoma viruses of the mammalian systems (Table 34–2). All of these viruses share the common feature of deleting some internal viral gene sequences and acquiring by genetic recombination cellular gene sequences (proto-oncogene). The size of the deleted portion and the content of inserted cellular sequences varies among the defective rapidly transforming viruses.

Cellular proto-oncogenes may be inserted into the replicating unit of retrovirus genomes in at least four different ways: (1) The proto-oncogene may be inserted as an independently expressed gene that does not impose on either the structure or function of the replicating genes and is expressed from a subgenomic mRNA. The src gene of RSV is a good example of this class. (2) It may be inserted as an independently expressed gene that replaces part or all of the replicative gene and is expressed from a subgenomic mRNA. The myb oncogene of avian myeloblastosis virus is an example of this class. (3) It may be inserted as a fusion between viral src and a portion of gag that is accompanied by deletions in one or more of the replicative genes, usually pol and portions of gag and env, and is expressed as a polyprotein produced from genomic length mRNA. The myc oncogene of avian myelocytomatosis virus is an example of this class. (4) It may be inserted as two separately expressed viral oncogene domains, one used with a part of gag, the other expressed independently and the two together replacing portions of replicative genes. In this class, one gag-oncogene protein is produced from a genomic mRNA. The second oncogene protein is produced from subgenomic mRNA, for example erb A and erb B of avian erythroblastosis virus. With the exception of the src gene, the insertion of proto-oncogenes into retrovirus genomes causes genetic defects that preclude the replicating of new virus unless the defective function is provided by a nondefective helper virus.

4. Assay

Retroviruses can be assayed and identified by several methods. Sarcoma viruses can transform fibroblastic cultures. The transformed cells form foci of morphologically altered cells (Fig. 34–1). As noted above, gag Ags and env Ags can be assayed by CF, IF or ELISA tests (Chapter 24). Virions in sufficient quantity are assayed by reverse transcriptase activity. Some biologic assays use induction of leukemias or tumors, the production of spleen foci (Friend virus) or increase in spleen weight (Rauscher virus). Molecular hybridization using labeled provirus DNA (cDNA) is used to detect and measure the amount of viral RNA present in transformed cells.

5. Pathogenesis

Pathogenesis induced by retroviruses depends on the capacity of the virus to transform cells. As noted above, retroviruses can be classified as (a) rapidly transforming or (b) slowly transforming viruses. A rapidly transforming virus contains a gene (or genes) that encodes for a transforming protein. The prototype transforming gene is $pp60^{src}$ of avian sarcoma virus. Although the mechanism of transformation is unknown, the requirement for $pp60^{src}$ in the process is clearly established. In contrast, no transforming protein has been established for slowly transforming viruses. The slow development of pathogenesis (latent period 4 to 12 months) and the failure of these viruses to induce transformation in tissue culture suggest that slowly transforming viruses do not encode a protein like $pp60^{src}$. Recent results suggest that slowly transforming viruses induce the expression of a cellular gene (proto-oncogene or cellular src gene) by integrating a provirus adjacent to specific cellular sequences. The essential feature is that cellular sequences are induced by linkage to a strong viral promoter (promoter insertion).

Other factors involved in pathogenesis are genetics of the host, physiologic cofactors (e.g., hormones) and presence of physical, chemical or biologic inducers (e.g., trauma, chemical carcinogens or immunosuppression). Table 34–3 summarizes the pathogenesis of representative retroviruses.

Table 34–3. Diseases Caused By Retroviruses: Animal Hosts And Pathogenesis

Disease	Animal Hosts	Viruses	Pathogeneis
Leukemia	Avian Murine Feline	Avian leukosis; Rauscher, Friend, Gross; Feline leukosis; type C oncoviruses	Proliferation of lymphoid tissues (lymph node, spleen, bone marrow)
Sarcoma	Avian Murine Feline	Rous Sarcoma; Murine sarcoma; Feline sarcoma; type C oncoviruses	Palpable tumor with malignant cells embedded in fibrillar matrix
Mammary carcinoma	Murine	Bittner agent; mammary tumor agent; type B oncoviruses	Hyperplastic alveolar nodules; Virus transmitted in milk; Estrogens are cofactors

D. RETROVIRUSES AND THE ETIOLOGY OF HUMAN NEOPLASMS

Although retroviruses have not been shown to be clearly associated with human cancer, their role in carcinogenesis in man has been proposed because of their wide distribution among animal species and because of their demonstrated role in the genesis of certain tumors of chickens, mice, and cats. Additionally, there have been reports that typical retrovirus particles are associated with an aggressive cutaneous T-cell lymphoma of a human cancer patient. In addition, "reverse transcriptase" and 70S RNA were detected. The major gag protein was shown to be serologically distinct from all previously described retroviruses.

A singularly interesting observation is that human cells in tissue culture can be transformed to neoplastic cells by infection with avian sarcoma viruses (e.g., RSV). The mechanism of transformation by RSV requires (1) the presence of the virion, and (2) a permissive cell for transformation. The production of virions or the stable integration of "provirus" does not seem to be a requirement for the continued occurrence of cell transformation. The current opinion is that development of malignant disease is a multistep process involving initiation (a mutagenic event) and promotion (a nonmutagenic event). Retroviruses may be involved in either one or perhaps both processes depending on the oncogenic virus and the host cell.

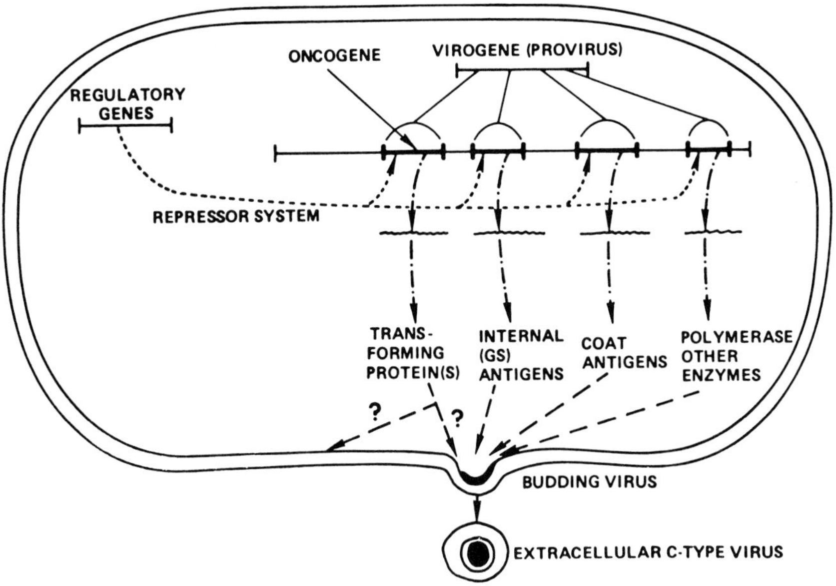

Fig. 34–5. Model for control of the virogene (provirus). The virogene or provirus is depicted as having four genes, one of which is the oncogene. These genes can be repressed or derepressed individually or in any combination. When the regulatory genes are transcribed and translated into repressor molecules, the genes of the provirus are not expressed. The repressor system can act coordinately to control expression of all genes in the provirus. Partial expression could allow virus production without transformation or transformation with virus production. (With permission from Todaro, G.J., and Huebner, R.J.: New evidence as the basis for increased efforts in cancer research. Proc. Natl. Acad. Sci. (U.S.A.), *69*:1009, 1972.)

E. THEORIES ON THE ORIGIN OF TRANSFORMING RETROVIRUSES

In the absence of experimental infection, retroviruses are thought to be transmitted vertically, that is, from parent to offspring. They are widespread among different species of vertebrates.

Transmission of transforming retroviruses depends on virogenes carried in the provirus that are associated with genetic material of the host cell. The oncogenes are considered normal cellular sequences which are carried in the provirus and they contain information needed for retrovirus transformation. Normally, oncogenes and virogenes are not expressed in host cells because of cellular repressor(s) (Fig. 34–5). The repressor(s) can act coordinately to control the expression of the whole virogene plus oncogene or independently to allow partial expression (e.g., gag Ag production or reverse transcriptase activity). Partial expression may allow for virus replication in the absence of cell transformation. Activation of endogenous virogenes has been demonstrated in primary mouse embryo cells using 5-bromodeoxyuridine or ultraviolet irradiation. During tumorigenesis, the provirus can be entirely activated by factors such as radiation, chemical carcinogens, virus infection, aging, and other alterations of genetic expression to make virus-specified products and virions.

Although the hypothesis concerned with the origin of transforming retroviruses is interesting and may have relevance to the etiology of some animal cancers, certain DNA viruses also are oncogenic and have the capacity to induce cell transformation without the intervention of retroviruses.

F. ONCOGENIC DNA VIRUSES

Of the known oncogenic DNA viruses (Table 34–4) (1) polyoma, simian virus 40, human adenoviruses, and herpesviruses have been studied in detail. Papillomaviruses have prompted renewed interest because of their association with human cancer (e.g., condylomata acuminata, laryngeal papillomas, cervical carcinoma). The possibility that herpesviruses may have tumorigenic properties is based on seroepidemiologic studies, the identification of herpesviruses in cultured tumor cells from man and animals, and the demonstration that herpes simplex virus type 2 (HSV-2) can induce transformation of mammalian cells in tissue culture using intact virus particles, purified viral DNA or specific viral DNA sequences.

Infection with oncogenic DNA viruses results in either productive

Table 34–4. Classification of Major Oncogenic DNA Viruses

- I. Oncogenic DNA viruses (about 50 different types)
  - A. *Papovaviridae*
    - 1. Genus: *Papillomavirus*
      - a. Papillomaviruses of man, rabbits, dogs, cows and other animals
    - 2. Genus: *Polyomavirus*
      - a. Polyomavirus (murine)
      - b. SV40 virus (simian)
  - B. *Adenoviridae*
    - 1. Adenoviruses of man, nonhuman primates, birds and cows
  - C. *Poxviridae*
    - 1. Molluscum contagiosum
    - 2. Yaba virus (nonhuman primates)
  - D. *Herpesviridae*
    - 1. Herpesviruses of:
      - a. Burkitt's lymphoma, nasopharyngeal carcinoma[1] (man)
      - b. Cervical carcinoma[2] (man)
      - c. Lucké renal adenocarcinoma[3] (Rana pipiens)
      - d. Marek's disease[3] (birds)
      - e. Monkey lymphoma[3] (owl monkey)
      - f. Guinea pig leukemia[3]

[1]Evidence associates this disease with Epstein-Barr virus.
[2]Epidemiologic, molecular and biologic data associate this disease with herpes simplex virus type 2.
[3]These diseases are associated with new members of the herpesvirus group.

infection with associated virus production and cell death or abortive infection with a block in virus production. Under appropriate conditions, up to 40% of abortively infected cells may undergo malignant transformation. Unlike retrovirus transformed cells, cells transformed by oncogenic DNA viruses generally do not synthesize infectious virus but continuously express certain virus gene functions. Productive infection and cell transformation by oncogenic DNA viruses are considered mutually exclusive. Usually, cells of the natural host are productively infected *(permissive cells),* whereas cells of the unnatural host are transformed *(nonpermissive cells).* It appears that nonpermissive cells either lack some essential component required for infectious virus replication or produce a repressor-like substance(s) that blocks the late virus gene functions required for infectious virus production.

1. Papillomavirus

Natural tumors induced by members of the genus *Papillomavirus* are either benign (e.g., common warts "verruca vulgaris" in man and many animal species) or are initially benign but may become malignant (e.g., rabbit papilloma and human genital warts, condylomata acuminata). The lesions of *condylomata acuminata* are warts growing on moist mucous membranes of the vagina or external genitalia of the male. The cells contain intranuclear inclusions that resemble DNA-containing inclusions described in cells of common human warts caused by human papillomavirus (Chapter 6). Condylomata acuminata presents a high proliferative tumor. Human laryngeal warts (papillomas) are mucosal proliferations which appear to be induced by a papillomavirus. A few cases of laryngeal papillomas have spontaneously converted to malignant growth after a latent period of 30 or more years. At least 18 types of human *papillomaviruses* (HPV) have now been identified in the laboratory. HPV-1, 2 and 4 induce benign papillomatous proliferations without risk for malignant growth. HPV-5 is regularly associated with specific skin warts which have an increased tendency for malignant conversion. HPV-6 and type 11 are associated with condylomata acuminata which also has been recorded to convert from a benign papilloma into a squamous cell carcinoma. Type 11 HPV is the prevalent virus found in laryngeal papillomas as well as condylomata acuminata. There is evidence that HPV-16 and type 18 virion DNA also is present in cervical carcinoma tissue. The possibility that HSV-2 is an initiator and HPV is the promoter of cervical carcinoma is being explored. Papillomavirus specific DNA has been identified in tumor cells by immunologic and molecular hybridization tests. The virus is typed by restriction enzyme analysis of purified papillomavirus DNA.

2. Polyomavirus

Considerably more information is available about the genus *Polyomavirus,* which includes *polyoma* and *simian virus 40* (SV40), the smallest known oncogenic DNA viruses. Polyomavirus can spread naturally among laboratory mice or wild house mice without apparent disease or tumors. However, when high titers of infectious polyomavirus are inoculated into newborn mice or hamsters, a wide variety of histologically diverse tumors are produced, hence the name "polyoma"; most tumors are spindle cell sarcomas. In polyoma-transformed cells, neither infectious virus nor infectious viral DNA can be detected; however, transformed cells contain polyomavirus-specific Ags, virus-specific RNA, and multiple copies of virus DNA which are not necessarily complete but are intimately associated with host cell chromosomes. The significance of multiple polyomavirus DNA copies and their location in cell-specific chromosomes remains to be determined.

SV40 was initially discovered in apparently normal cultures of monkey kidney cells being used for the production of poliovirus vaccine. Inoculation of SV40 virus into newborn hamsters revealed that this virus was oncogenic. Subsequent studies showed that SV40 resembles polyomavirus in its physical and chemical properties (Chapter 6). Polyoma or SV40 viruses induce striking biochemical alterations in transformed cells, including induction of cellular DNA synthesis, stimulation of specific enzyme activities, synthesis of virus DNA and virus-specific RNA, and the formation of early nonstructural viral Ags, T Ags, and TST Ags. Polyomavirus codes for three early proteins in productively infected and transformed cells. The proteins are referred to as large (100K), middle (55K) and small (22K) T Ags. Middle T Ag has the same apparent molecular weight as the TST Ag associated with the plasma membrane of productively infected and transformed cells. Also, middle T Ag has an associated tyrosine-specific protein kinase activity and is considered the oncogene product of polyomavirus. Its function in cell transformation is still unclear. There is recent evidence to suggest that middle T Ag forms a stable complex with $pp60^{c\text{-}src}$, the product of a cellular oncogene which also is associated with Rous sarcoma virus cell transformation. Two distinct SV40 early A gene products have been identified in transformed cells, large T (94K) Ags and small T (17K) Ags. These Ags are present in both SV40 productively infected and transformed cells. Large T Ag is a nuclear phosphoprotein with specific DNA binding properties; it is required for both SV40 productive infection and SV40 transformation. The function of large T Ag involves initiation of viral DNA synthesis in productively infected cells and stimulation of cellular DNA synthesis during transformation. It also regulates the levels of p53 cellular protein synthesis in transformed cells. The possible function of small T Ag in cell transformation is unclear.

Tumor Ags found in the cell nucleus can be demonstrated by CF or IF tests using sera from animals bearing polyoma or SV40 induced tumors. The TST Ags appear in the plasma membrane of transformed cells (Fig. 34–2), and can be demonstrated by transplantation rejection tests in syngeneic animals previously vaccinated with infectious SV40 virus. This type of immunity occurs because the vaccine virus probably induces some TST Ag synthesis which elicits a cellular immune response leading to subsequent rejection of the tumor.

The mechanism of cellular DNA induction and its role in SV40 induced cell transformation is unknown. The use of SV40 cloned DNA fragments to transform mammalian cells proved that only the early region of SV40 DNA is required for cell transformation. Viral-induced cell transformation in the SV40 system is a rare event, for even with inputs of 1000 PFU per cell, only about 5% of the cells eventually become transformed.

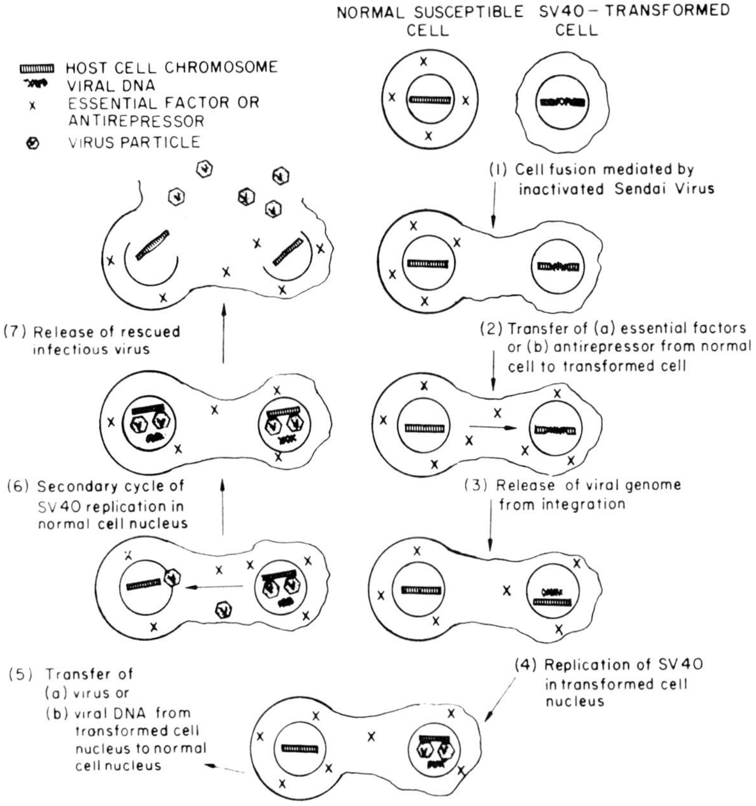

Fig. 34–6. Possible mechanism for induction of SV40 multiplication following SV40-transformed cell fusion mediated by inactivated Sendai virus with susceptible host cells. (With permission from Butel, J.S., Tevethia, S.S., and Melnick, J.L.: Oncogenicity and cell transformation by papovavirus SV40: The role of the viral genome. Adv. Cancer Res., *15*:1, 1972.)

In contrast to polyomavirus or adenovirus transformed cells (discussed below), the complete SV40 genome is often present in SV40 transformed cells, as evidenced by the production of infectious SV40 either spontaneously or as a result of artificial cell hybridization (Fig. 34–6). A notable observation is that DNA isolated from SV40 virions can transform human fibroblast cells in tissue culture; the transformed cells contain SV40 T-Ags, and infectious virus is recoverable from the transformed cells.

3. Adenovirus

The human *adenoviruses* are of special interest since they were the first human viruses shown to possess oncogenic properties in newborn rodents. Adenoviruses possess about 23 genes, but not all of them are expressed in adenovirus-transformed cells. Similar to polyomaviruses, adenoviruses can induce tumors in rodents, but no infectious virus has been detected in the tumor cells even after subjecting them to various physical or biologic manipulations, including cell fusion; these techniques are used to induce latent SV40 virus in hamster tumor cells. It is noteworthy that adenovirus-transformed cells carry the equivalent of several molecules of adenovirus DNA integrated into a number of their chromosomes; virus-specific RNA is transcribed in vivo and the transformed cells always contain adenovirus-specified T- and TST-Ags. Transformation by adenoviruses evidently involves an abortive infection of nonpermissive host cells and probably includes the following essential events: (1) Host cell DNA synthesis is induced, (2) T-Ag synthesis occurs, (3) virus genes are integrated into cellular DNA so that functions 1 and 2 are maintained, and (4) late virus gene functions (production of virus structural proteins and infectious virus) are blocked.

Although human adenoviruses induce cancer experimentally in animals, they do not appear to induce cancer in man. A survey of more than a hundred different cancers in man for the presence of adenovirus-specific RNA by RNA-DNA hybridization experiments has provided no evidence that they cause cancer.

4. Poxvirus

Several members of the Poxviridae family induce slight epithelial proliferation in their natural hosts. Two poxviruses, namely *Yaba virus* and *molluscum contagiosum virus,* which produce benign skin tumors in non-human primates and in man, respectively, are of interest and merit further discussion. Yaba tumors were first described in 1957 among rhesus monkeys kept in open-air cages in Yaba, Nigeria. The Yaba virion is morphologically similar to vaccinia virus (Chapter 9), although it may be slightly larger. It causes large subcutaneous, nonencapsulated histiocytomas in nonhuman primates. After intravenous inoculation into rhesus monkeys, hundreds of small benign tumors may develop at multiple sites, including subcutaneous tissues, heart, muscles, and lungs. Within 48 hours after inoculation, macrophages migrate into the infected area and the infected cells undergo striking morphologic changes leading to tumor formation. The nuclei and nucleoli of the tumor cells enlarge and multiply rapidly. The tumor cells contain cytoplasmic inclusion bodies, infectious virus, and many soluble viral Ags. Apparently, tumors produced in mature animals never become malignant. Yaba virus has been transmitted to man, and in vitro infection of both monkey and human cells is characterized by a prolonged growth cycle and eventual cytopathic changes. Characteristically, Yaba tumors in monkeys, either naturally acquired or induced, regress spontaneously within 1–2 months, probably as the result of in vivo cytopathic effects of the virus. Specific Abs develop but appear to have little, if any, effect on established tumors. Evidently, no antigenic relationship exists between Yaba virus and other members of the poxvirus group.

Molluscum contagiosum virus produces painless, pearly white, discrete lesions in children and young adults. Infection leads to formation of benign tumors that can occur anywhere on the body except the soles and palms. The "molluscum body" is a large cytoplasmic inclusion body composed of virus particles resembling other poxviruses. The virus has been transmitted experimentally to man. Growth of the virus in tissue culture cells has been difficult; however, the tendency to induce proliferative changes of infected cells is more striking with molluscum contagiosum virus than with other poxviruses.

5. Hepatitis B Virus

There is convincing evidence that another DNA virus, *hepatitis B virus* (Chapter 20) is the etiologic agent of primary liver cancer in humans. In areas such as Asia and Africa, it is the most common cancer in young men. Also, these individuals possess high titers of HBsAg (Chapter 20) and anti-HBc Abs. The latter is indicative of ongoing hepatitis B virus (HBV) infection. Recombinant-DNA technology combined with gel electrophoresis strongly suggests that integration of HBV-DNA into hepatocytes occurs during the course of HBV infection and precedes the onset of primary liver cancer. Potential uses of vaccines (Chapters 20 and 23) against HBV may provide further support of the role of HBV in the development of primary liver cancer.

The potential relationship between HBV and primary hepatic cancer is supported by recent observations that a virus morphologically indistinguishable from HBV has been isolated from American woodchucks which shares some common DNA sequences and Ags with HBV. Infection of woodchucks with this virus results in the development of hepatocellular carcinoma. This virus-animal system provides an excellent model for determining the possible role of virus in human primary liver cancer.

6. Herpesvirus

In contrast to many viruses discussed above, which seldom induce malignant neoplasms in their natural host, there is convincing evidence that viruses of the Herpesviridae family (Chapter 8) are linked etiologically with certain natural malignant neoplasms of lower animals (e.g., Lucké renal adenocarcinoma, Marek's disease, rabbit lymphoma, monkey lymphoma) and of man (e.g., cervical carcinoma, Burkitt's lymphoma, and nasopharyngeal carcinoma).

G. HERPESVIRUSES AND THE ETIOLOGY OF ANIMAL NEOPLASMS

1. Lucké Renal Adenocarcinoma

Of the oncogenic herpesvirus models in animals, the Lucké tumor is the only carcinoma. The other herpesvirus-induced cancer models are lymphoproliferative diseases. The Lucké renal adenocarcinoma occurs with a low frequency in the leopard frog, *Rana pipiens*. An observation of interest is that tumors from hibernating frogs (4° to 9°C) contain intranuclear inclusion bodies, herpesvirus particles and cell necrosis, whereas tumors of frogs held at room temperature (20° to 25°C) do not. Herpesvirus-containing extracts from inclusion-containing Lucké tumors (4° to 9°C) induce typical Lucké renal adenocarcinomas when injected into frog embryos. When the inoculated frog embryos develop into adult frogs, up to 80 percent of the animals develop renal adenocarcinomas.

2. Marek's Disease

This is a highly contagious disease of chickens caused by herpesvirus; it is characterized by lymphoid infiltration and uncontrolled cell proliferation in the nerves and visceral organs, resulting in tumor formation and paralysis. *Marek's disease virus* produces a productive infection in the follicular epithelium of the feather from whence they can be shed into the environment. The virus is not transmitted vertically (from mother to offspring by the ovum). Eradication of the disease in flocks of chickens has been accomplished with a live nonpathogenic turkey herpesvirus vaccine. This is the first instance in which a naturally developing neoplastic disease caused by a virus has been controlled by vaccination.

3. Rabbit Lymphoma

This lymphoproliferative disease is caused by *herpes sylvilagus,* an indigenous virus of wild cottontail rabbits (genus *Sylvilagus*). The virus is unable to cross genus lines to infect the ordinary laboratory strains of rabbits of the genus *Oryctolagus.* The disease varies from benign hyperplastic tumors to malignant lymphomas. Virus can be recovered from peripheral leukocytes and from infected immature lymphoid cells that infiltrate the lymph node and spleen. No virus-induced cytopathic effect has been observed in parenchymal cells of any organs showing lymphocyte infiltration. It is not known whether the herpesvirus genome is present in all immature lymphoid cells.

4. Monkey Lymphoma

Two unrelated herpesviruses, *herpesvirus saimiri* and *herpesvirus ateles,* were the first oncogenic herpesviruses isolated from nonhuman primates. They differ in their host cell range in tissue culture, Ab induction in animals, antigenic structure, and oncogenic potential. Herpesvirus saimiri, which was originally isolated from kidney cultures and peripheral lymphocyte cultures of squirrel monkeys, can induce leukemia and malignant lymphomas in several animal species, e.g., marmoset monkeys and rabbits. The virus remains latent in squirrel monkeys.

Herpesvirus ateles, originally isolated from kidney cultures of a spider monkey, causes leukemia and malignant lymphoma in marmoset monkeys.

H. HERPESVIRUSES AND THE ETIOLOGY OF HUMAN NEOPLASMS

1. Cervical Carcinoma

Evidence for the herpesvirus etiology of cancer in man is mainly circumstantial. Two types of *herpes simplex viruses (types 1* and *2),* differing in their biologic, immunologic, and biochemical properties, infect man (Chapter 8). Herpes simplex virus type 1 (HSV-1) is usually associated with lip lesions (cold sores), whereas herpes simplex virus type 2 (HSV-2) is commonly harbored in the genitourinary tract and is transmitted by sexual intercourse. Both viruses can be cultivated readily in tissue culture, and both types cause latent as well as active infections in man.

Exposure to HSV-1 generally occurs early in life, whereas exposure to HSV-2 usually does not occur until puberty. Either type of infection can develop into an apparent or inapparent disease (Chapter 22). Apparent disease is characterized by cell destruction, which is most often self-limiting (Chapter 26). Many herpesviruses establish latent infections following recovery from apparent disease. Latent infections often persist for the life of the individual even though circulating neutralizing Ab is constantly present.

Recent epidemiologic, serologic, and virologic evidence suggests that HSV-2 may be the causative agent of squamous cell carcinoma of the human cervix, the fourth most common cancer in women; HSV-1 is suspected of being the causative agent of lip cancer. The onset of cervical carcinoma seems to be related to sexual exposure. For example, prostitutes, women with multiple sex partners, and those who start sexual relations early in life tend to exhibit a higher frequency of the disease and higher frequency of Abs to HSV-2 than matched control groups. Other supporting evidence comes from finding viral Ags, viral mRNA and DNA in tumor cells. Also, transformation of mammalian cells is accomplished using ultraviolet or neutral red-light inactivated virus, purified viral DNA or cloned viral DNA fragments. It is clear now that exposure to restriction enzyme-generated sequences of HSV DNA (Blg II C or Blg II N DNA sequences) can induce malignant transformation (Chapter 8). Also, certain sequences of the viral DNA are frequently expressed and translated into recognizable viral Ags in both experimentally transformed cells and in human cervical carcinoma cells. One of these Ags referred to as Ag-4 has been recently identified and characterized as a polypeptide with DNA binding and ribonucleotide reductase activities. Interestingly, women with a primary diagnosis of cervical carcinoma who possess detectable Abs to Ag-4 have a good prognosis in disease following radiation or surgical therapy compared to women with disease who receive therapy and are negative for Abs to Ag-4. These latter women have a poor prognosis.

In summary, HSV infections can be linked to cervical carcinoma although not to all forms of cervical cancer or dysplasia. It is possible that HSV acts as the "initiator" of cell transformation in susceptible target cells and a second event (promotion ?) induces the cell to become neoplastic. There are some data to suggest that the promotion step may follow exposure to some environmental cocarcinogen (chemical, viral, radiation). The modern tools of molecular and cell biology should provide the needed data to definitively link HSV with some environmental cocarcinogen to the etiology of cervical carcinoma.

2. Burkitt's Lymphoma, Nasopharyngeal Carcinoma

In addition to herpes simplex viruses, a second widely distributed herpesvirus, Epstein-Barr virus (EBV), has been shown to be associated with certain human tumors, i.e., Burkitt's lymphoma in African children and nasopharyngeal carcinoma of the Chinese. The virus is also the cause of infectious mononucleosis, a common nonmalignant disease of

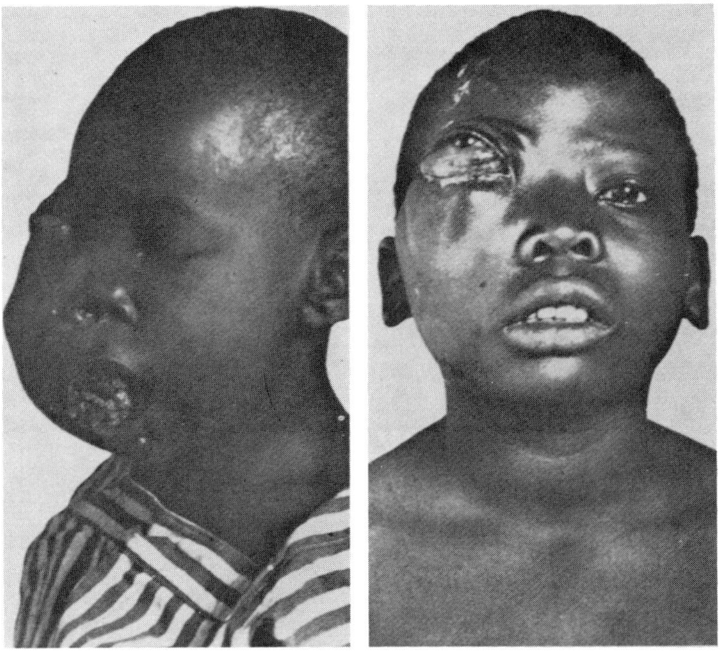

Fig. 34–7. Left panel shows a Burkitt's tumor involving the right maxilla and mandible. Right panel shows a Burkitt's tumor involving the right maxilla and extending into the orbit. (From Burkitt, D.: A sarcoma involving the jaws in African children. Br. J. Surg., *46*:218, 1958–1959.)

adolescents (Chapter 29). The incidence of Burkitt's lymphoma among African children is 1 in 50,000. In most cases, the tumor starts in the region of the alveolar process of the maxilla or the mandible (Fig. 34–7); it rapidly grows and causes distortion of the face. There may be invasion of the eyelids and orbit and metastasis to the liver, kidney, ovaries, and other visceral organs. The tumor can be successfully treated with alkylating agents such as cyclophosphamide or vincristine. If remission lasts more than one year, the prognosis is good. Patients with Burkitt's lymphoma, nasopharyngeal carcinoma, or infectious mononucleosis develop high Ab titers to EB virus.

Burkitt's lymphoma and nasopharyngeal carcinoma occur under unusual circumstances that presently are poorly understood. Recent data suggest that malaria can cause immunodepression and predispose to the development of Burkitt's lymphoma. Malignant cell transformation by EBV is correlated with a reciprocal translocation of distal fragments of chromosomes 8 and 14.

Most cell cultures established from Burkitt's tumors synthesize EB virus particles and have been shown to synthesize IgM that is bound to the cell membrane. The maintenance of this cell marker is proof that the tumor cells are derived from lymphoid tissue. The isolation and purification of the virus has been difficult; most virus particles are structurally defective, and relatively few tumor cells in culture make EB virions. The agent appears to fulfill all the criteria of an oncogenic virus: It has been identified in peripheral blood leukocytes of patients with Burkitt's lymphoma and is capable of transforming human fetal fibroblasts and both human and nonhuman primate leukocytes in vitro. Viral DNA persists in transformed cells in multiple copies as free episomal or integrated into host cell DNA. The persistence of viral DNA is paralleled by the expression of EBV-specific nuclear Ag (EBNA). More recently, DNA-DNA hybridization studies have revealed that small quantities of EB virus-DNA are present in virus-free biopsy specimens of Burkitt's lymphoma and nasopharyngeal carcinoma. Clones of somatic cell hybrids produced by fusing Burkitt lymphoblast cells with other human cells also carry the EB virus genome, which can be induced to replicate after exposure to the tumor promoter 12-0-tetradecanoyl-phorbol-13-acetate. These data indicate that the entire virus genome is present in all Burkitt lymphoblastoid cells.

In addition to EBNA other virus-induced Ags detected in EB virus infected lymphoblastoid cells include early Ag(s) (EA) in the cell nucleus, virus capsid Ag(s) (VCA) in the cell cytoplasm, membrane Ag(s) (MA) in the plasma membrane of the infected cells, lymphocyte-detected membrane Ag and an 81,000 MW DNA-binding protein. It has been demonstrated that patients with Burkitt's lymphoma or infectious mononucleosis have serum Abs against EA that appear during the acute phase of disease and will be present before the appearance of Abs to VCA and MA. Detection of Abs in the patient's serum to EA by IF or CF tests can be of diagnostic importance. The Abs to EA are not detectable within a few months following recovery from disease, whereas Abs to VCA and MA persist for years, if not for life.

The association of EBV with nasopharyngeal carcinoma (NPC) is well established. All NPC biopsies tested contain multiple copies of episomal EBV-DNA. The cancer cells express EBNA and can be induced to replicate infectious EBV particles. The virus shares biologic and biochemical properties with EBV isolates from Burkitt's lymphoma and infectious mononucleosis patients. NPC patients regularly exhibit Abs to EBV-specific EA and VCA. The obvious clustering of NPC among Southern Chinese, East Africans and Alaskan natives suggests genetic as well as geographic factors in the etiology of NPC.

Despite the regular association of EBV with NPC, nothing is known concerning the interaction between EBV and nasopharyngeal epithelial cells prior to transformation. NPC cells have not been successfully propagated in tissue culture and no animal model exists for studying NPC-like tumors. Further studies require careful study of naturally developing NPC in humans.

I. A HYPOTHESIS FOR VIRAL CARCINOGENESIS

There is substantial evidence in the scientific literature to support the hypothesis that carcinogenesis is a multistep process involving independent steps. Support for this idea comes from epidemiologic studies indicating that cancer increases in proportion to a multiple power of elapsed lifetime. Pathologic studies indicate that transformed cells progressively acquire new phenotypes by passing through a series of stages such as anaplasia, dysplasia, metaplasia and neoplasia. In the mouse skin induction of malignant papillomas seems to require three distinct types of stimulation such as initiation (a mutagenic event), promotion (a nonmutagenic event) and a second mutagenic event which causes malignant disease.

Other evidence suggests that oncogenic viruses that cause malignant transformation by themselves possess two genes encoding distinct functions, i.e., immortalization of cells and oncogenic transformation functions. Both of these functions must be expressed sequentially to convert cells into tumor cells.

Other precedents support a model of multiple cooperating independently-activating genes, i.e., a virus gene or chemical carcinogen required for initiation of cell transformation and the concomitant expression of a cellular oncogene for final conversion to malignant transformation. Again the central theme for the hypothesis is that carcinogenesis is not a single event but may involve both viral and cellular oncogene expression to realize the full malignant potential of a cancer cell.

REFERENCES

Bishop, J.M.: Retroviruses and cancer genes. Adv. Cancer Res., *37*:1, 1982.

Clanton, D.J., Jariwalla, R.J., Kress, C., and Rosenthal, L.J.: Neoplastic transformation by a cloned human cytomegalovirus DNA fragment uniquely homologous to one of the transforming regions of herpes simplex virus type 2. Proc. Natl. Acad. Sci. (U.S.A.), *80*:3826, 1983.

Courtneidge, S.A. and Smith, A.E.: Polyoma virus transforming protein associates with the product of the c-src cellular gene. Nature, *303*:435, 1983.

Erikson, J., Ar-Rushdi, A., Drwinga, H.L., Nowell, P.C., and Croce, C.M.: Transcriptional activation of the translocated c-myc oncogene in Burkitt lymphoma. Proc. Natl. Acad. Sci. (U.S.A.), *80*:820, 1983.

Friedman-Kien, A.E. and Laubenstein, L.J. (eds.): AIDS The Epidemic of Kaposi's Sarcoma and Opportunistic Infections. New York, Masson Publishing USA, Inc., 1984.

Gilden, R.V. and Rabin, H.: Mechanism of viral tumorigenesis. Adv. Virus Res., *27*:281, 1982.

Gissmann, L., Wolnik, L., Ikenberg, H., Koldovsky, V., Schnurch, H.G., and Zur Hausen, H.: Human papillomavirus types 6 and 11 DNA sequences in genital and laryngeal papillomas and in some cervical cancers. Proc. Natl. Acad. Sci. (U.S.A.), *80*:560, 1983.

Griffin, J.D., Spangler, G., and Livingston, D.M.: Protein kinase activity associated with simian virus 40 T antigen. Proc. Natl. Acad. Sci. (U.S.A.), *76*:2610, 1979.

Groff, D.E. and Lancaster, W.D.: Evidence that integration of virus DNA may not be necessary for maintenance of cell transformation. Prog. Med. Virol., *29*:218, 1984.

Groffen, J., Heisterkamp, N., Reynolds, F.H., Jr., and Stephenson, J.R.: Homology between phosphotyrosine acceptor site of human c-abl and viral oncogene products. Nature, *304*:167, 1983.

Hampar, B.: Transformation induced by herpes simplex virus: a potentially novel type of virus-cell interaction. Adv. Cancer Res., *35*:27, 1981.

Huszar, D. and Bacchetti, S.: Is ribonucleotide reductase the transforming function of herpes simplex virus 2? Nature, *302*:76, 1983.

Hynes, N.E. and Groner, B.: Mammary tumor formation and humoral control of mouse mammary tumor virus expression. Curr. Top. Microbiol. Immunol., *101*:51, 1982.

Johnson, P.J. and William, R.: Of woodchucks and men: The continuing story of hepatitis B and hepatocellular carcinoma. Br. Med. J., *284*:1586, 1982.

Kucera, L.S., Furman, P.A., and Elion, G.B.: Inhibition by acyclovir of herpes simplex virus type 2 morphologically transformed cell growth in tissue culture and tumor-bearing animals. J. Med. Virol., *12*:119, 1983.

Land, H., Parada, L.F., and Weinberg, R.A.: Tumorigenic conversion of primary embryo fibroblasts requires at least two cooperating oncogenes. Nature, *304*:596, 1983.

Lebowitz, P. and Weissman, S.M.: Organization and transcription of the simian virus 40 genome. Curr. Top. Microbiol. Immunol., *87*:43, 1979.

Levine, A.: Transformation-associated tumor antigens. Adv. Cancer Res., *37*:75, 1982.

Maupas, P. and Melnick, J.L.: Hepatitis B infection and primary liver cancer. Prog. Med. Virol., *27*:1, 1981.

Newbold, R.F. and Overell, R.W.: Fibroblast immortality is a prerequisite for transformation by EJc-Ha-ras oncogene. Nature, *304*:648, 1983.

Rapp, F. (ed.): Oncogenic Viruses. Vols. I and II. CRC Press, Inc., Boca Raton, Florida, 1980.

Reiss-Gutfreund, R.J., Dostal, V., and Novotny, N.R.: Enhancement of Lewis lung carcinoma by the concomitant infection of the host with herpes simplex virus type 1 and type 2. Oncology, *40*:46, 1983.

Spector, D.H. and Spector, S.A.: The oncogenic potential of human cytomegalovirus. Prog. Med. Virol., *29*:45, 1984.

Zur Hausen, H.: The role of viruses in human tumors. Adv. Cancer Res., *33*:77, 1980.

Glossary

This Glossary includes a selected list of terms and their definitions that are commonly used for communication in basic and applied virology. The list is not intended to be inclusive of all terms.

Albumin: A protein widely distributed throughout the tissues and body fluids of animals and plants.

Alkylating agents: Chemicals or drugs that induce substitution of an alkyl radical for a hydrogen atom in a cyclic compound such as a purine or pyrimidine.

Anamnestic response (recall phenomenon, memory phenomenon): Accelerated immune response to an antigen that occurs in an animal that has previously responded to the same antigen; it may involve humoral or cellular immunity.

Anaplasia: Loss of cell structural differentiation as seen in malignant cancer.

Antibody (Ab): A substance (commonly, if not always, a gamma globulin) that can be incited in an animal by an antigen or by hapten combined with a carrier and that reacts specifically with the antigen or hapten. Some antibodies occur naturally without known antigen stimulation.

Antigen (Ag): A substance that can react specifically with antibodies and, under appropriate conditions, can incite an animal to form specific antibodies.

Aseptic meningitis: An infection of the meninges caused primarily by viruses; occasionally other microorganisms may be etiologic agents of this disease but they are not detectable in spinal fluid.

Asthenia: A syndrome characterized by increased susceptibility to fatigue.

Astrocyte: A large nonnervous cellular element of the central nervous system.

Ataxia: A loss of muscular coordination.

Attenuated: A reduction or weakening of virulence (in a microorganism) obtained through selection of mutants that occur naturally or are obtained experimentally.

A-type virus: A morphologic classification of the retroviruses. The immature form of B-type retroviruses.

Australia antigen: A virus-like antigen associated with serum hepatitis (hepatitis B); the antigen was originally discovered in an Australian aborigine, hence the name.

Autoantibodies: Antibodies produced by an animal that react with the animal's own antigens. The stimulus is usually not known but could be the animal's own antigens or cross-reacting foreign antigens.

Autoimmune disease (autoallergic disease, autoantibody disease): A disease which by appropriate criteria is judged to result from the reaction of antibodies or from cellular immunity against antigens of the affected individual.

Avian: Pertaining to birds, e.g., chickens.

B-cells: Refers to bone-marrow-derived lymphocytes.

Bradykinesia: A condition characterized by extreme slow movement.

B-type virus: A morphologic classification of the retroviruses. The mature form of mouse mammary tumor virus.

Burkitt's lymphoma: A cancer of the lymphoid system associated with a herpesvirus (Epstein-Barr virus) first described in children living in certain areas of Africa.

Capsid: The protein coat that surrounds the nucleic acid of the virus.

Capsomeres: Substructures of virus particles composed of aggregates of polypeptide chains that interact to form the basic structural units of the capsid.

Carcinogenesis: The development of cancer.

Carcinoma: A cancer derived from epithelial tissue.

Cell-mediated immunity (cellular immunity): A state of immunity mediated by specifically immune lymphocytes (T-lymphocytes) coupled with lymphokine-directed participation of macrophages. It can be transferred with living immune lymphocytes but not by immune serum.

Cell strain: A cell culture derived from a primary culture by selection or cloning of cells having specific markers or properties.

Chlamydia: Bacteria belonging to the family Chlamydiaceae that cause the diseases psittacosis, lymphogranuloma venereum, trachoma, and inclusion conjunctivitis.

Choroid plexus cells: Cells that are derived from a network or interjoining of nerves and blood vessels.

Communicable: A property that relates to the ease with which infection is transmitted from one individual to another.

Complement: A multifactorial system consisting of about nine interacting components of normal serum; these components are characterized by their capacity to participate in certain antigen-antibody reactions.

Complement fixation (CF): The orderly fixation or participation of complement components with an antigen–antibody complex; some complex molecules (e.g., endotoxin) can fix complement by the alternate pathway from C3 to C9 in the absence of antibody.

Continuous cell line: A cell culture that possesses the potential of being subcultured indefinitely *in vitro.*

Coproantibodies: Antibodies occurring in the intestinal tract. They consist primarily of the IgA class.

Coryza: Inflammation of the nasal mucosa characterized by nasal discharge and watery eyes.

Councilman bodies: Inclusion bodies, found in hepatic cells, that are of diagnostic importance in yellow fever.

C-type virus: A morphologic classification of the retroviruses. The mature form of leukosis-sarcoma retroviruses.

Cytomegalic cell: A large swollen cell that is pathognomonic for cytomegalic inclusion disease (salivary inclusion disease).

Cytopathogenic effect (CPE): This effect consists of morphologic alterations of host cells that usually result in cell death.

Delayed hypersensitivity: A specific sensitive state characterized by a delay of many hours in onset time of reaction following antigen administration; it peaks 24 to 36 hr after elicitation with antigen. It is transferable with specifically sensitive T-cells but not with serum.

Dementia: A general mental deterioration due to organic or psychological factors.

Demyelination: Destruction or loss of myelin from the medullary sheath of Schwann cells.

Deoxyribovirus: A term employed to designate those viruses that possess a DNA genome.

Dermatotropism: See *Tropism.*

Desquamation: The shedding of cells from the outer layer of the epidermis.

Diploid cell: A cell having the 2x number of gametic chromosomes.

Down's syndrome: The concurrence of symptoms related to mongolism.

Dysarthria: Disturbance of speech due to paralysis, uncoordination, or spasticity of the muscles used for speaking.

Dysplasia: Abnormal cell development that may progress to cancer.

Dyspnea: Labored breathing, "shortness of breath," usually associated with serious diseases of the heart or lungs.

Endemic: The continuous presence of a disease in a community, usually with low incidence.

Endosymbiotic: A mutually advantageous association between two or more organisms and a host.

Envelope: A host-cell-derived membrane, containing virus specific antigens, that is acquired during virus maturation.

Epidemic: An outbreak of a disease that simultaneously affects a significant number of individuals in a community.

Erythema: An eruption of reddish patches in the skin.

Feline: Pertaining to cats.

"Fixed" virus: An attenuated variant of the virulent "street" rabies virus.

Gamma globulin: A fraction of serum based on electrophoretic mobility, composed of a number of molecular classes and subclasses of immunoglobulins and other nonantibody globulins.

Genome: A set of genes.

Giant cells: See *Syncytium.*

Gliosis: A condition marked by overgrowth of the neuroglia (nonnervous cellular elements of nervous tissue).

Guarnieri bodies: Acidophilic intracytoplasmic inclusion bodies in epidermal cells infected with smallpox virus.

Hemadsorption (HAd): The attachment of red blood cells to the surface of host cells.

Hemagglutination (HA): Aggregation of red blood cells.

Hemagglutinin (viral hemagglutinin): A nonantibody protein on the outer surface of some viruses (e.g., orthomyxoviruses) which reacts with a surface determinant(s) on red cells to cause agglutination of the red cells (hemagglutination).

Herpangina: Vesiculopapular lesions about 1 to 2 mm in diameter that are present above the pharynx. These lesions are associated with coxsackievirus infections.

Heterophil antibody: An antibody having heterogenetic distribution.

Heteroploid: Denotes chromosome numbers deviating from the normal chromosome number.

Hyperplasia: An increase in the size of a tissue or organ due to increase in cell numbers.

Hypoxia: A condition characterized by reduced levels of oxygen in the blood.

Icosahedron: A geometric figure composed of 12 vertices, 20 triangular faces and 30 edges.

Icterus (jaundice): A condition characterized by excess of bile pigments in the blood and tissues that leads to a yellow color of the surface integuments.

Immune complex disease: A vasculitis mediated by antigen-antibody complexes, complement, and neutrophils.

Immunity: The state of being able to resist and/or overcome harmful agents or influences.

Active: Immunity acquired as the result of a natural experience with an infectious agent or due to vaccination.

Passive: Immunity due to acquisition of maternal antibody or injection of preformed antibody.

Immunogen: An antigen that stimulates the production of protective antibodies.

Immunoglobulins (Ig): Classes of globulins to which antibodies belong, e.g., IgG, IgM, IgA, IgD, and IgE.

Inclusion body: Acidophilic or basophilic heterogeneous masses of new material in the nucleus or cytoplasm of cells that are associated with some virus infections.

Infection: The presence of microorganisms in parenteral tissues.

Viral infection: The presence of a virus particle or its genome inside a host cell.

Infectious dose (ID_{50}; $TCID_{50}$): That amount of virus required to cause a demonstrable infection in 50% of the inoculated animals or tissue culture cells.

Inflammation: The vascular and tissue responses to injury; these are important in host defense and healing.

Insidious onset of infection: Onset of infection without readily apparent symptoms.

Kupffer cells: Macrophages lining the hepatic sinusoids.

Latent infection: A persistent infection with intermittent acute episodes of disease. The virus is not detectable between episodes.

Leukemia: Proliferation of abnormal leukocytes in the blood in increased numbers.

Leukemic cell: The malignant cell of leukemia; may arise from monocytic, lymphocytic, or granulocytic stem cells.

Leukocytosis: An abnormally large number of leukocytes in the circulating blood.

Leukopenia: A condition associated with a smaller than normal number of circulating leukocytes.

Lymphadenopathy: Any disease process affecting a lymph node or lymph nodes.

Lymphatic leukemia: A malignant disease characterized by uncontrolled proliferation of malignant lymphocytes and conspicuous enlargement of lymphoid tissue (e.g., lymph nodes, spleen).

Lymphokines: Biologically active substances elaborated by stimulated lymphocytes.

Lymphoma: A neoplastic disease of lymphoid tissue, e.g., lymphosarcoma, lymphocytic leukemia, Hodgkin's disease.

Maturation: The final step in the production of new virus particles.

Metaplasia: The transformation of differentiated tissue to another type of tissue.

Molluscum bodies: Large eosinophilic inclusion bodies found in epidermal cells infected with molluscum contagiosum virus.

Morbidity: The ratio of sick to well individuals in a community.

Mortality: The ratio of the number of deaths to a given population in a defined situation.

Mutation: Heritable changes in the genome that do not result from the incorporation of genetic material from another organism.

Myalgia: Muscular pain.

Necrosis: The death of cells or tissues resulting from irreversible damage.

Negri bodies: Intracytoplasmic inclusion bodies in nerve cells associated with rabies virus infections.

Neoplasm: An abnormal collection of cells characterized by more rapid cellular proliferation than the surrounding normal tissue.

Benign: A neoplasm that remains localized and does not spread to other parts of the body.

Malignant: An invasive type of neoplasm that spreads to other parts of the body.

Neuraminidase: An enzyme on the outer surface of some viruses (e.g., orthomyxoviruses) that splits off the terminal N-acetylneuraminic acid on the mucoprotein receptors on cells (e.g., red blood cells).

Neurotropism: See *Tropism.*

Nosocomial infection: An infection acquired in a hospital.

Nucleocapsid: The virus structure composed of the nucleic acid surrounded by the capsid.

Nystagmus: The rhythmic oscillation of the eyeballs in a horizontal, rotary, or vertical motion.

Oncogenes: That portion of the virogene (RNA genome of retroviruses) responsible for cell transformation.

Oncogenesis: The mode of development of a neoplasm.

Opsonin: A serum substance, usually an antibody, that adsorbs to the surface of microorganisms and promotes their phagocytosis.

Orchitis: Inflammation of the testis.

Pandemic: A disease attacking the population of a large geographic area(s); may be worldwide.

Papillomas: An epithelial tumor growing on the surface of the skin and consisting of neoplastic cells.

Parenchymal cells: The specific cells of a gland or organ that are supported by and contained in the connective tissue framework or stroma.

Pathogenesis: The mode of development of a disease or morbid process.

Paul-Bunnell test: A serologic test used in the laboratory diagnosis of infectious mononucleosis. The test measures the titer of heterophil antibodies that will react with antigens on the surface of sheep red blood cells.

Penetration: The entrance of the virion or virus nucleic acid into host cells.

Peplomeres: See *Spikes.* Projections extending from the outer surface of a virus envelope.

Peplos: See *Envelope.*

Phagosome: A cell vacuole resulting from phagocytosis of particulate materials.

Plaque: A defined area of cell destruction resulting from *in vitro* virus infection.

Plaque-forming unit (PFU): A fundamental measure of infectious virus particles. One infectious virus particle is equivalent to one plaque-forming unit.

Pleomorphism: The potential for morphologic variation.

Pleurodynia: A painful affection of the tendinous attachments of the thoracic muscles.

Pock: A discrete pustular lesion found on the chorioallantoic membrane or skin following infection with certain viruses.

Polyvalent vaccine: A vaccine that contains more than one immunogen.

Postinfectious encephalitis: Disease invoked by an allergic reaction following recovery from an acute virus infection.

Preicteric: See *Icterus.* Before the onset of jaundice.

Primary cell culture: A culture started from cells, tissues, or organs taken directly from living tissue.

Prodromal: Relates to an early sign or symptom that precedes overt signs of disease.

Prognosis: A forecast of the course and final outcome of a disease.

Provirus: The integrated or latent form of some viruses; e.g., the integrated DNA form of retroviruses.

Receptor sites: Specific areas on the surface of host cells that serve as points of attachment for viruses.

Repressors: A special group of molecules coded by DNA that block the synthesis of one or more proteins.

Rhabdomyosarcoma: A malignant neoplasm derived from skeletal muscle.

Rhinitis: Inflammation of the nasal mucous membrane.

Riboviruses: A term used to group together those viruses that possess an RNA genome.

RNA-dependent RNA polymerase: An enzyme that catalyzes the formation of RNA from an RNA template.

RNA-directed DNA polymerase (reverse transcriptase): An enzyme(s) associated with retroviruses that catalyzes the synthesis of DNA from an RNA template.

Sarcoma: A solid tumor growing from derivatives of embryonal mesoderm such as connective tissue, bone, muscle, and fat.

Sequela: A morbid (abnormal) condition that develops as a consequence of a disease.

Secretory IgA: A special molecular form of the IgA Ab composed of 2 Ab molecules plus a secretory piece; it is found in secretions and functions against microorganisms that colonize on mucosal surfaces.

Serum: The fluid portion of the blood obtained after removal of the fibrin clot and blood cells.

Acute phase serum: Serum recovered during the early stage of a disease.

Convalescent phase serum: Serum recovered during the recovery stage of a disease.

"Slow" virus: A virus that causes subacute or chronic diseases having long incubation periods lasting several weeks to years before the onset of clinical symptoms.

Spikes: Surface projections of varying lengths spaced at regular intervals on the virus envelope.

Status spongiosus: A condition in which multiple cavities are formed in the white matter of the brain due to degeneration of axons.

"Street" virus: See *"Fixed" virus.* Used to describe the virulent type of rabies virus isolated in nature from domestic or wild animals.

Structural units of a virus: The repeating polypeptide chain(s) that comprise the morphologic unit or capsomere of viruses.

Syncytium: A multinucleated protoplasmic mass formed by the fusion of originally separate cells.

Syngeneic: Relates to tissue transplant between identical twins or individuals who are identical with respect to histocompatibility genes.

T-cells: Lymphocytes processed through the thymus.

Tachycardia: A condition characterized by rapid beating of the heart.

Teratogen: Any agent that may induce abnormal development of the fetus.

Thrombosis: The formation of a blood clot in the blood circulatory system.

Titer: The assay value of an unknown measured by volumetric means.

TMV: Tobacco mosaic virus. A single-stranded RNA virus that infects and multiplies in tobacco leaf cells.

Transformation, viral: A permanent heritable change induced in a cell by an oncogenic virus or viral nucleic acid.

Tropism, infection: Infection preferentially supported by a particular cell, tissue, or organ.

Neurotropism: Infection preferentially supported by the cells of the CNS.

Dermatotropism: Infection preferentially supported by cells of the skin.

Tumorigenicity: Capacity to induce tumors in a susceptible animal.

Uncoating, virus: An intracellular event in which all or part of the virus capsid is removed from the virus nucleic acid; this process allows the viral genome to be transcribed and replicated.

Vaccination: The administration of a protective (immunogenic) antigen(s).

Vaccine: A suspension of dead or living microorganisms or their products administered for the purpose of producing active immunity.

Viral interference: A phenomenon whereby one intracellular virus inhibits the replication of a second virus.

Viremia: Presence of virus particles in the blood.

Virion: The mature infectious virus particle.

Virogene: See *Oncogene.* The complete RNA genome of retroviruses.

Viroid: Infectious subviral particles consisting of nucleic acid without a protein capsid.

Viruria: Presence of virus particles in the urine.

Virus: A small obligate intracellular parasite that depends on a living host cell for energy, precursors, enzymes, and ribosomes to multiply. It consists of a single type of nucleic acid, either DNA or RNA, and a protein coat surrounding the nucleic acid.

Zoonoses: An infection shared by humans and lower vertebrate animals.

Index

Page numbers in *italics* indicate figures; page numbers followed by "t" indicate tables; abbreviations are in **boldface.**

psychic experiences:
e.s.p.
investigated

psychic experiences: e.s.p. investigated

By Sheila Ostrander
and Lynn Schroeder

STERLING PUBLISHING CO., INC. NEW YORK
Oak Tree Press Co., Ltd. London & Sydney

OTHER BOOKS BY THE SAME AUTHORS

Executive ESP
Handbook of Psychic Discoveries
Psychic Discoveries Behind the Iron Curtain
The ESP Papers

Two Park Avenue, New York, N.Y. 10016
Distributed in Australia and New Zealand by Oak Tree Press Co., Ltd.,
P.O. Box J34, Brickfield Hill, Sydney 2000, N.S.W.
Distributed in the United Kingdom and elsewhere in the British Commonwealth
by Ward Lock Ltd., 116 Baker Street, London W 1
Manufactured in the United States of America

Library of Congress Catalog Card No.: 77-79512
Sterling ISBN 0-8069-3092-6 Trade Oak Tree 7061-2572-X
3093-4 Library

psychic experiences:

e.s.p.

investigated

Table of Contents

This book is dedicated to

Grace Y. Schroeder

1899–1977

a teacher whose help, encouragement and wit are remembered by many students.

Stranger than We Can Imagine

We are, it seems, growing into a world in which things once considered as strange and rare, and even as impossible as a unicorn, will come to brighten everyday life. This book is written for young people to give some ideas and to ask some questions about what is happening on a new and real frontier. You are the ones who will see the flowering of many things mentioned in these stories. Yet, no matter what your chronological age, you can join in this broadening of human abilities and possibilities—if you are aware of it.

Do you ever have the feeling you know what a friend is thinking? Has something you dreamed ever come true? Or when you're daydreaming, have you ever thought you could tell what was happening in a faraway city you've never visited—or even in the next room? If you have, you're in good company. If not, maybe it will happen to you one day.

The people in the following happenings went through such experiences and much stranger ones —like the ex-CIA man who talked to his plants and found they answered back. Or the writer who kept in contact with an explorer at the North Pole. What's so strange about that? He didn't use a radio—just his mind. There's a college where scientists worked out a system of *mental* Morse code to communicate with astronauts on the far side of the moon.

Today, scientists and even governments in many countries are looking into the expanding world of psychic research. The scientific study of psychic happenings is called parapsychology or paraphysics. The famous American Association for the Advancement of Science has accepted parapsychology as a recognized field of research. Hundreds of college courses are offered in parapsychology and countless schools offer others in related subjects.

Psychic happenings are often called ESP, which stands for Extra Sensory Perception. However, "perception" is only a small part of the psychic world, so scientists use a new neutral word, psi (pronounced "sigh"). Psi is short for psychic, an umbrella term for all things paranormal. The 23rd letter of the Greek alphabet, it is the up-to-date word used internationally.

Conclus

ESP has been around for centuries, as some of the stories in this book show. In the past, when people had startling ESP experiences, the best they could do was to try to get them documented thoroughly. Today, there have been enormous changes in psi research. Today psi can be made objective, trainable, usable. Today we have a machine that shows on a graph exactly how telepathy affects your blood circulation; a camera that photographs your aura (energy coming from you); detectors that show how your body reacts to hidden water or minerals; machines that reveal when you're dreaming.

Once researchers discovered a way to measure psi, they discovered something else: almost everybody has these psychic abilities to some degree. With proper training many can develop their psychic powers. Nobody had told people before that their bodies were aware of all sorts of things happening around them. Through awareness training, people are discovering what these incoming signals feel like, and are learning to use the body's built-in detection system to enhance their lives. Different kinds of training systems are helping us to sense psi signals, from meditation/relaxation to biofeedback training. A whole new world is beginning to open up in which we can explore and develop human potential.

Another big change in modern psychic research is that all over the world, whether psi is fully understood or not, it is being put to work. As you will see in many of the following reports, psi is helping to solve crimes; find water; diagnose illnesses; overcome handicaps; locate missing persons; find lost objects. Psi is helping professionals from archeologists to engineers, farmers to businessmen. Many scientists feel psi can contribute to all fields of knowledge and should be investigated from the viewpoint of each of the many different arts and sciences.

This book offers a collection of genuine psychic experiences, explorations, experiments. So many exciting new things are happening in the psi field, that often we can give only samples. Some follow-up reading is listed at the back of the book. The stories and accounts are designed to entertain and inform, and all of them have been authenticated by scientists or other responsible observers.

A British scientist who studied peculiar happenings once said the world is not only stranger than we imagine, it is stranger than we *can* imagine. But we can always try—try to imagine more about what might be possible in this world. Exciting possibilities began to open up for a lot of the people in these stories. Maybe it can happen to anyone of us. Maybe it will happen to you.

1 Telepathy

Telepathy

Almost everyone has heard about telepathy, the ability to communicate mind to mind, to send someone a thought, to pick up a thought. What many people don't realize is that a new wave of research into telepathy has rolled around the world in the last few years. Scientists have turned to the latest technology to explore the ancient mysteries of telepathy. They are taking a new look at long ignored accounts of ESP experiences. They are helping people to develop telepathy and put it to work.

Telepathy now means a lot more than just an unexpected hook-up between your mind and my mind. Telepathy can happen when you're dreaming. It can be used to track your whereabouts. It might even be used to knock you out. And you don't even have to be human to get in on the telepathic connection. As we learn more, it seems that telepathy may be telling us something about an invisible but very real way we are connected to each other and the rest of the living universe. Like any communication channel, it's got great possibilities.

But that's getting ahead of the story. Before people could explore telepathy, they had to believe it existed. In the 1930's at Duke University, Dr. J. B. Rhine began his famous ESP card tests. He asked people to tell him which of 5 symbols was being sent telepathically. If, in a series of tests, the subjects answered more questions right than they should have by the laws of chance, then he assumed that telepathy was at work. Rhine's thousands and thousands of card tests began to crack scientists' resistance to telepathy. The tests proved telepathy existed, but they didn't tell much more.

In the 1960's, American research began to move away from the "does-it-or-doesn't-it exist" approach. People were startled to hear that the Soviet government was pumping money and effort into psychic research. We spent months in the USSR and Eastern Europe in 1968 to check out Soviet findings first-hand. We saw what they were doing in telepathy, aura photography, research in mind over matter and many other things. Finally, we reported this fascinating yet solid Communist research in our book, *Psychic Discoveries Behind the Iron Curtain*, and you'll find some information about it in this book. The Soviets wanted to learn how telepathy works, how to control it and use it. People in many other countries, we learned, had the same idea and were seriously researching psi—the word scientists use to cover everything in the psychic field (pronounced "sigh"). (See page 10 for further explanation.) The time was ripe to bring telepathy more into scientific labs and more into many people's experience.

ESP to the North Pole

While Dr. Rhine was running ESP card tests, another American, the writer Harold Sherman, was trying to show telepathy exists in a more spectacular way. It all began when Sherman struck up a conversation with a stranger in a New York club. They had something unusual in common. Both were intrigued with the idea that people could communicate mentally over vast distances. The stranger turned out to be the famous Arctic explorer, Sir Hubert Wilkins. As the evening wore on and their conversation grew more excited, they decided to test their ideas about the possibility of mental communication.

Shortly after, a rare opportunity came up. A plane had crashed near the North Pole. Sir Hubert was being sent to find the missing craft and any possible survivors. This was in 1937 when communications weren't as advanced as they are now and Sir Hubert's trip would be a hard one. He would have to fly in short hops through northern Canada and then go on by dog sled to the Pole. They decided that one night a week Sherman would tune in mentally to Sir Hubert and write down what the explorer was doing.

Faithfully, every week, Sherman sat in his New York apartment and tried to connect with Sir Hubert thousands of miles away. Sometimes he must have wondered if he was getting the right mental picture, especially on one November evening. Sherman saw Sir Hubert's plane tossing about in a heavy blizzard, the snow so blinding that the explorer turned back to a Canadian city. Then Sherman received puzzling impressions—music, a large, well lit hall and people whirling around. It seemed to be a ball. Sir Hubert was there, too, dancing in evening dress. Was Sherman imagining things? He couldn't believe that Sir Hubert packed a dress suit to take to the North Pole. But trusting in his own psychic abilities, Sherman sent his notes to Dr. Gardner Murphy, head of the Psychology Department at Columbia University. He was keeping the records of what Sherman picked up mentally to compare with Sir Hubert's diary when he returned. This way no one could say that the two friends changed or made up their stories once they were together again. They were serious about trying to prove beyond question that mental communication is possible.

After an adventure lasting almost six months, Sir Hubert returned to New York. Dr. Murphy discovered that Sherman had mentally contacted Sir Hubert 70 per cent of the time. The friends certainly had proved their point. "ESP can be learned," Sherman says. He has continued to develop his ability and help others do the same. Why? You may not need to get in touch with someone at the North Pole, but there are other times, plenty of times, when it would be handy to get through to someone telepathically.

What was Sir Hubert doing when Sherman got the unlikely picture of an Arctic explorer in a fancy dress suit? Dr. Murphy looked at the explorer's diary. He had written that a heavy blizzard kept his plane from its destination and he'd been forced to land in the Canadian city of Regina. "Was met at the airport by the Governor of the Province, who invited me to attend an Armistice Ball being held there that night. My attendance at this ball was made possible by the loan of an evening dress suit."

Terrorism and ESP

As lab scientists work to understand telepathy, others are learning by doing. What good is ESP anyway, some people ask. There are probably hundreds of ways it can help. For starters, we might look to areas where regular ways of doing things aren't working well. For example, top psychics in many countries have turned their talents to crime busting.

On October 5th, 1970, terrorists struck in Montreal, setting off a two-month crisis that attacked the roots of the Canadian nation itself. Disguised as delivery men, terrorists kidnapped James Cross, the British Trade Commissioner, from his Montreal home. Five days later they struck again, turning an international crisis into an even more serious national one. Terrorists snatched Pierre LaPorte, Labour Minister of the Province of Quebec, second in rank only to the Premier of the province.

So serious was this case that on October 16th, the Canadian Federal Government placed the whole of Canada under martial law. People's civil rights were suspended. The government believed the kidnappings were part of a conspiracy to separate the province of Quebec from the rest of Canada. Civil war? Revolution? No one knew what was going to happen. Armies of police and undercover agents searched everywhere for the two kidnap victims. In some neighborhoods, they searched house to house. They found nothing. Were the hostages still alive?

In Prince George, British Columbia, a radio announcer at station CJCI, Robert Cummings, got an idea. Maybe a psychic could provide some clues. He called a famous Chicago psychic, Irene Hughes, who often worked with police and federal agents to help solve difficult cases. On October 14th, Cummings tape-recorded and broadcast the psychic's impressions about the kidnap victims.

"I feel the case will drag on for two or three months," Irene said, "But eventually James Cross will be released unharmed." She felt that he was sick but alive. Cross was being held she thought, "about five miles northwest of Montreal. It seems that the place he is in is about three stories high. I feel that it is red brick, a kind of old place, and it actually could be an apartment building."

Irene said she feared for LaPorte's life. On a part of the tape the station did not air publicly, she reported a psychic impression that LaPorte would be assassinated, stabbed to death by the terrorists.

When would the authorities get a break in the case? Irene thought there would be an arrest, something striking that would lead later to the solution of the kidnapping. "I would pinpoint November 6th."

As the search went on, the radio station continued to tape information from the American psychic. Canadian authorities asked that none of the tapes be aired until after they had heard them.

Three days after Irene's prediction, on October 17th, police opened the trunk of an old green car and found the body of Labour Minister Pierre LaPorte. First reports said the assassins had strangled him. Later reports revealed he also had been stabbed to death.

On November 6th, the date Irene had predicted, police captured one of the Quebec terrorists. He eventually fingered the others.

Finally, on December 3rd, the remaining terrorists agreed to trade Cross for an airplane that would fly them to safety in Cuba. After 60 terrifying days, James Cross was a free man, very weak, but basically unharmed. He'd been captive in a three-story brick duplex, northwest of Montreal, as Irene had said.

Radio station officials were astounded by the accuracy and detail of Irene Hughes' psychic sleuthing. They never found out whether the government acted on any of her psychic clues. If they didn't, they should have, for it was time to put such human ability to work. "Undoubtedly," the radio people concluded, "this endeavor represents an impressive documentation of ESP at work in a 'now' manner in modern history."

The Government Agent and ESP

Patrick Price, former police commissioner and vice-mayor of Burbank, California, leaned back in a comfortable leather chair at Stanford Research Institute. A solidly built Irishman, Pat Price talked into a tape recorder. Now and then, he sketched pictures.

"I see a little boat dock along the bay. I see little boats, some motor launches, some little sailing ships." Pat described the dock area and the positions of the masts on some of the sailboats. Then he said, "Funny thing—this flashed in—kinda looks like a Chinese or Japanese pagoda effect. It's a definite feeling of Oriental architecture that seems to be fairly close to where they are."

What was the former police commissioner doing? He was hot on the trail of a couple of scientists. He was tailing them with ESP. He was also in a tough test called remote viewing. When Pat was shut in the room, even his quarry didn't know where they were headed. It could be any place within several hundred square miles. The men were given a packet of instructions chosen by chance. They read where they were to go. When they arrived, Pat was to click on his mental TV and describe where they were, what they were seeing.

Impossible? A lot of people think so, until they try it. Pat Price knew it was possible. He used ESP in his police work. Could he do it in a laboratory? The scientists were walking around the Redwood City Marina, a dock along the bay. And yes, a Chinese restaurant of oriental design stood on the dock.

Trail a suspect—drop in on a hideout—you don't have to be chief of police to realize just how handy this remote viewing could be. Electronic bugging presents many problems. What about psychic bugging? One strong psychic at Stanford Research Institute was given a location simply by longitude and latitude. He supposedly dropped in mentally (and invisibly) on a completely restricted defense base. He drew an accurate ground plan of the base. This is another "benefit" of remote viewing.

Stanford Research Institute is a world famous think-tank research center often used by the U.S. government. Dr. Hal Puthoff and Russel Targ, physicists who have worked there on many projects, designed remote viewing. Unlike the older scientific approach, this new system creates a set-up in which the subject can learn and improve telepathy. After all, you don't grab people off the streets to test their piano-playing ability. You show them how to play, you have them practice, then you test. Telepathy seems to be a talent or skill like playing the piano or winning at tennis.

Hal and Russ often greet government agents who drop in to see what's up—or far out—at the ESP lab at Stanford. That's when they have some fun. At first, they thought only people like Pat Price who knew they were psychic could follow people with telepathy. To their delight, they found that many people can do it, at least a little bit, in the same way that most people can hum, though they would have to practice to sing well.

"I don't believe what you're doing. It's impossible."

Hal and Russ answer that kind of remark with, "Fine, you're it!" Their guest has to do the remote viewing personally.

Not too long ago, a government man arrived. He couldn't stomach the idea of telepathic television. Hal and Russ challenged him. On his first try, he mentally saw Hal "standing on a wooden walkway with a railing before him. The ground falls away underneath." Later, Hal and Russ took him to the target spot. It was a wooden bridge over a stream in a nearby park. He admitted there were many similarities between the spot and what he'd seen in his mind. But he was sure there had to be a catch somewhere.

He asked to be left alone in the sealed room. This time he mentally saw triangles, squares and more triangles and some sort of electrical shielding. He also spoke of a building with a small movie theatre. Triangles and squares didn't sound right, but it turned out that's what Hal was seeing when the man tuned in. Hal was lying on his back staring up at the open frame of a tall electrical transmission tower. A small movie theatre stood close by.

The agent didn't figure out that Hal was gazing at a transmission tower. He simply got the gist of the image. That's why subjects sketch impressions. People often get correct visual information, but it gets confused when the rational mind tries to tag a name to the image.

One man correctly described a solid rectangular structure. "Maybe it's a ticket counter at the airlines." It was a church altar. He was right—but wrong.

Why did the government man want to be alone in his telepathy test? He decided the scientist in the room might be giving him clues by body movements. Or, he thought, there might be a loudspeaker piping him information just below the range of conscious hearing. This time, he huddled in the corner, all alone, with his hands over his ears. But he still could gather what was happening at a distance. He did it himself, but he wasn't convinced.

"Ah," he said, "I know. You hear my tape, then switch targets and drive me to a place that is similar to the one I described." This is clearly impossible from the way the test is set up. But Russ and Hal decided to humor their increasingly upset friend.

"Okay. This time we'll both go to a place and record where we are. When we return, we'll swap tapes before anyone opens his mouth."

The spot, chosen by chance, was a merry-go-round in a children's playground 4 miles away. Returning, Hal and Russ handed over their tape and looked at the man's sketch. They smiled but didn't say a word. He'd drawn a pretty good sketch of the target. They drove to the playground to let him decide for himself. As they crossed the parking lot, he spotted the merry-go-round. "That's it, isn't it!" the man exclaimed. He finally admitted, "It really works!" Now he agrees with Russ and Hal that remote viewing "must show the existence of an astonishing, hidden human potential."

The Case of the Stolen Wallet

"My friend's wallet was stolen!" a woman's voice said over the phone. "The wallet had a number of personal things in it that she'd like to get back. Can you help her find it?"

The man who answered the call was a psychic who goes simply by the name of DeMille. That day, April 3, 1964, he'd just checked in at his hotel in Yakima, Washington. He was to give a lecture that evening to an executives' club. "I'll do my best to find the wallet," he told the caller. "But you'll have to wait until after my speech. Call me about 10:30."

DeMille, who does not specialize as a psychic crime buster, decided to get a few minutes' sleep before his lecture. As he relaxed, instead of dozing off, he began to pick up psychic impressions. He saw two teenage boys following a woman shopper. He saw one of them bump into her, while the other reached into her handbag and removed her wallet. In his mind's eye, he saw them run out, climb into a beat-up Ford and drive to a used car lot. They looked through the wallet, pulled out a roll of bills and divided about $46 between them. Next, the boys seemed to be looking at a checkbook. Hazily, he saw a number that looked like "2798301" and "First National Bank of Washington." He also got something about a "meat-packing" firm.

Unable to sleep, DeMille called the secretary of the club where he was to lecture. After hearing his story, the secretary suggested, "Call the police."

Soon DeMille was recounting his impressions to Sergeant Walt Dutcher of the Yakima Police Department and to Frank Gayman, a reporter for the *Yakima Herald*. Gayman was skeptical. The sergeant was both disbelieving and hostile.

DeMille felt that if he could check out some of his psychic impressions, he might be able to locate the wallet. "Why not call the First National Bank?" said DeMille. "Maybe a meat-packing company has Account #2798301. Then it would be easy to contact the company and see whether or not a woman employee has been robbed."

The bank officials checked. "Account #2798301 is not a meat-packing company," they reported. In fact, they added, the bank had no meat packers as customers.

That did it. Sergeant Dutcher threatened DeMille with arrest for turning in a false crime report.

The reporter waited. It looked as if the psychic was way off base. The phone rang. It was the bank again.

There *was* an account #2798001 belonging to Cub Scout Pack #3. Could that be the one? Instantly, DeMille felt that it was. Often psychic impressions appear hazy and confused, with details mixed-up and pictures that are difficult to interpret. His unconscious mind had mixed up the words "cub *pack*" with "meat-*packing*."

DeMille felt he was on the trail again. If a few of the impressions he'd picked up about the theft of the wallet were right, then maybe the picture he'd seen of the two teenage thieves parked at a used car lot and throwing the wallet away, was also right. If they could make the rounds of the handful of used car lots in Yakima, DeMille felt he had a good chance of finding the wallet.

The president of the local executives' club offered to drive him about. Following DeMille's psychic impressions, they drove through the city to an outlying area and pulled up in front of a used car lot which seemed to fit the picture he'd seen mentally. After a brief search, there, in the bushes in front of a nearby house, they found the stolen wallet.

That's the big point—he found it. Some of his psychic information was wrong, but there was enough to solve the case. It is a rare person who gets the whole picture right psychically. That's why some researchers ask psychics to work in teams on certain cases. Between them they may pick up enough bits of information to piece together the whole puzzle.

Psychic House Call

Picking up someone else's thought or sending a message mentally: that's telepathy. But sometimes things happen to make researchers think that telepathy is not quite as straightforward as a radio broadcast you purposefully click on. Sometimes a telepathic message seems to come into a person's unconscious and push him into action, even though the message was not aimed specifically at him. Such a strange experience happened to Dr. Edwin Boyle, of the Miami Heart Institute.

In January, 1968, Dr. Boyle flew to New York for a medical convention at a large hotel. Sitting at a meeting, he began to think about the Reverend Arthur Ford. He had never met Rev. Ford, but he'd heard of the famous psychic who years before began exploring ESP with Sir Arthur Conan Doyle, the creator of Sherlock Holmes. Since then, Ford had worked with many famous people, including kings and queens. He'd written books and been on television. Though he was old now, he was still considered one of the finest psychics in the country. That's about all Dr. Boyle knew.

Well, it would be interesting to look up Ford sometime, he thought. Right now he'd better pay attention to the medical discussion.

The thought of Rev. Ford would not go away, no matter how Dr. Boyle tried to push it out of his mind. He felt a growing urge to meet Ford, right now. Where did Ford live anyway? Dr. Boyle dimly remembered something about Philadelphia. Finally, like a man propelled by an unseen force, Dr. Boyle left the meeting and went to the phone booths in the hotel corridor. He opened the Philadelphia phone book and found Rev. Ford's number. Wondering why he was doing it, the doctor dropped money in the slot and dialed the psychic. Maybe he could see him on the way home to Florida. No answer. Well, that was that.

But it wasn't. Dr. Boyle, the reasonable and well-known heart specialist, began to feel haunted. An inner voice kept urging him, "Go to Arthur Ford. He needs you . . ." Feeling desperate by this time, instead of flying home to Florida, Dr. Boyle went to Penn Station and boarded a train to Philadelphia, a hundred miles away. He was a man with a mission. He almost forgot to wonder why he was doing such a crazy thing.

Hurrying off the train in Philadelphia, Boyle went straight to the Westbury Hotel where Ford lived. He rushed to the psychic's room and knocked on the door. Dead silence. He knocked and knocked again. There was no answer. Boyle was not to be put off at this point. He had to get into that room. He went downstairs and somehow talked the hotel clerk into coming up with the passkey.

They swung open the door. There on the floor lay Arthur Ford in a coma. He'd had a very bad heart attack. Dr. Boyle, the heart specialist, gave him emergency treatment while the hotel clerk called an ambulance. If the psychic had stayed alone much longer, he would have been dead.

What or who sent Dr. Boyle off on his strange life-saving mission? Had Ford sent him a psychic message while he was in a coma? If so, he wasn't aware of it. When Ford finally opened his eyes and looked at his rescuer, Dr. Boyle, he said, "Who the heck are you?"

The Dream Shop

Arthur Ford managed to send out a telepathic SOS when he was in a coma. Could you do the reverse and send a telepathic message to someone who was unconscious, to someone asleep? If there's anything that seems to be ours alone, it's our dreams. It doesn't occur to most of us that someone might be able to send us a dream, a sort of "night letter" that acts itself out while we sleep.

Sending people dreams is exactly what a group of people have been doing for years at the Maimonides Hospital in Brooklyn, New York. Dr. Montague Ullman and Dr. Stanley Krippner started this dream shop that is officially known as a sleep laboratory. Can you telepathically plant an idea for a dream in another person's mind? That's what the two doctors decided to find out.

Spending the night in a sleep lab is quite an experience. After you settle down on a cot, technicians paste small wires to your head. These run to machines in another room that let the scientists keep track of your brain waves and your eye movements. The lights go out and you fall into a deeper and deeper sleep. The doctors see your brain waves slow to a nice steady rhythm. They observe even though your eyelids are closed, that your eyes are moving. That is what's called REMs (rhymes with "stems")—rapid eye movements—and it means you're having a dream. As soon as your eyes stop moving, a buzzer sounds and a loudspeaker voice says, "Please talk into the tape recorder. Tell us your dream." That's how scientists capture dreams before they fade away. But how do they send you a ready-made one?

Often they use striking pictures by famous artists. One is chosen by chance from a group of pictures and given to the telepathic sender, the dream maker. One night the choice was *Dempsey and Firpo* by the artist George Wesley Bellows. The dream maker studied this bold, realistic picture of the two famous prize fighters. It's a dramatic moment: sweat glistening on his biceps, Firpo has just knocked Dempsey through the ropes. The front row crowd leans back as the huge Dempsey tumbles into their laps. The referee, an older man, stands counting at the side of the rectangular ring. The ropes and posts of the ring make a strong pattern.

The telepathic sender concentrates on the picture. Now and then, he sketches something that reminds him of boxing. He even gets up and shadowboxes a moment. Down the hall, a young man begins to dream. Awakened, he reports, "Posts standing up from the ground. There's some kind of a feeling of moving. Ah, something about Madison Square Garden and a boxing fight. It's as if all these things that I see were in a rectangular framework. It's connected with a Madison Square boxing fight . . . I had to go to Madison Square Garden to pick up tickets to a boxing fight, and there were a lot of tough punks, people connected with the fight."

In another dream that night, he spoke of two or three figures in the dream, in the presence of other people. They'd come together for a purpose, "Two younger men and one older man."

The dream idea came through, mind to mind. In hundreds of dream transmissions, the doctors found that two subjects most easily catch our dream attention—what else but sex and violence. These telepathic dream tests, done hundreds of times, are classic, breakthrough experiments in psi research that show two major things. First, telepathy does exist. Second, you can project a thought telepathically into another person's mind even though he is not consciously cooperating.

The young man sleeping in the Brooklyn hospital knew that someone was trying to get into his dreams. But what if you didn't know? Could you be influenced unawares? Other dream shops in sleep labs have sprung up in universities around the country. Six students went to sleep in one of these. Three were in a telepathy test. The others supposedly were being checked for other things. *All six* picked up the telepathic dream broadcast and wove it into that night's dreams.

In the future, will some devilish dictator set up dream broadcasts to try to sway the minds of sleeping people? Will some advertiser decide that dream broadcasting can get people to buy more soap? Pipe dreams? Probably. But psi researchers are not only trying to find out how to encourage ESP, they're also trying to discover what shields us from ESP, what keeps it from occurring.

If one person can send you a dream, how about 2,000 people? The audience at a rock concert tried it. They were about 45 miles from the lab where a young English photographer, Malcolm Bessent, slept peacefully. The music stopped. A picture flashed on the theatre screen. "Now everybody, let's really concentrate and send that guy in Brooklyn a dream!"

The crowd looked at a picture called "The Seven Spinal Chakras." A man sits crosslegged, meditating. Seven very brightly colored circles are shown on his spine. These are supposed energy centers of the body. People meditate to get this natural body energy flowing. It ignites and shoots up from the base of the spine through the solar plexus to the top of the head.

Back in the lab, doctors noticed Malcolm's eyes were moving rapidly. In a while, they woke him. "In the dream I was very interested in using natural energy. I was talking to this guy who said he'd invented a way of using solar energy and he showed me this box . . . to catch the light from the sun which was all we needed to generate and store the energy. He was suspended in mid-air or something . . . I was thinking about rocket ships . . . I'm remembering a dream I had about an energy box and . . . a spinal column."

If you kept sending someone the same dream, would it affect how he acted when he was awake? Can you find out about people by the way they make dreams out of ideas you send them? Do two sleeping people ever send dreams back and forth to each other?

The House that Pikki Built

Soviet scientists knew something about telepathy before most of the rest of the world did. They knew that the minds involved didn't have to be human. In the 1920's, a psychic named Pikki helped start Soviet psi research.

Pikki was a star, but he never let it go to his head —maybe because he was a wire-haired terrier. Pikki often toured Russia in a circus act, dancing, counting, playing the clown with his trainer, Durov. Durov taught every kind of animal you can think of, from aardvarks to lions to mice, and he came to believe something that's occurred to a lot of people who have pets. You can communicate mentally with an animal. He found he could think a message to his dog, Pikki. "Nonsense," is what most scientists of the time said. You can't do that with a person, let alone with a dumb animal. This made Durov all the more determined to prove, in a laboratory test, that he and Pikki had a telepathic connection. But he needed powerful friends to back his work.

One night he saw his chance. A scientist, famous for his discoveries about the workings of the brain, invited Durov and Pikki to his home. He knew there were mysteries about the brain waiting to be discovered. But could a dog show him one?

Pikki resisted the urge to sniff the heavy furniture, the plush sofas and the piano in the scientist's living room. Soon he had his chance to leap all over the room, when the scientist thought of things for him to do and Durov sent him the commands mentally. Once, Pikki bounded to the piano stool, hopped up and pounded away on the high notes with his paw. A dog would hardly play the piano by accident. The scientist looked almost as pleased as Durov. He seemed to be seeing the impossible: the dog was picking up their exact wishes mentally. Or was he? The scientist frowned. Maybe Durov was sending him signals somehow. The only way to be sure would be if he himself sent a mental message, one unknown to the trainer. Now it was Durov's turn to frown. He and Pikki had worked together for years. Could the dog pick up the thoughts of a stranger? If he couldn't, instead of a powerful friend, Durov might have an enemy who'd say mental communication was a fake. It was a big responsibility to be put on a wire-haired terrier.

Durov crossed his fingers as the scientist stared into the dog's eyes for a long minute. Pikki ran to a round table and began circling and circling it as though he were confused. The scientist shook his head. Pikki had failed. But then the scientist called the dog back. As he wrote later, "Maybe the mistake was mine. I wanted the dog to jump on a chair, hop on the table and sit. Yet, all I'd thought about in my mental command was the table. This time, I pictured, as clearly as I could, the dog going from me to the chair to the table and sitting. I released him and almost before I knew it, there he was sitting on top of my table!"

A dog taught a great scientist something—the right way to send a telepathic message—at least to a psychic dog. Thanks to Pikki, the influential scientist helped Durov move into the scientific labs. Working with Pikki and his other dogs, Durov proved telepathy with animals. He also began one of the most important centers for animal studies in the world, which still flourishes today in Moscow. It's called The Durov Institute. You might call it the house that Pikki built.

The Cursed Plant

Thought can connect minds, human minds and animal minds. The more people look into it, the more it seems that all of nature responds to thought. That's what a group of people in California proved to their own satisfaction not too long ago.

"*You never do anything right. You look terrible. Oh, why try anyway, the world's too hard!* These are the kinds of negative remarks we too often make," the Reverend Franklin Loehr said one Sunday to his congregation. Over six feet tall, standing there in the pulpit, dressed in his long dark robe, Rev. Loehr was an imposing figure, as he talked about the power of positive and negative thinking. But inwardly, he sighed to himself. His audience probably didn't understand what he was saying. They probably supposed he was just talking about being nice. He was talking about real power. Maybe it wasn't as obvious as a punch in the nose, and maybe it didn't happen as fast as flipping on an electric switch, but thought, he believed, has real force.

Rev. Loehr is a trained chemist as well as a Congregational minister. The scientist in him wanted to demonstrate the power of thought. He wanted his congregation to see it with their own eyes, touch and measure it with their own hands. But how? He came up with a plan. If it worked, people would be able to see how their thoughts influence living things. They would have the evidence on their own window sills.

"We're going to take up gardening," he told 150 volunteers. "We're going to see if we can grow bigger and better plants by adding one magic ingredient—thought power. We're also going to see if we can curse a plant. Yes, curse it and stunt its growth with thought alone."

The volunteers found the instructions for this new adventure simple. Fill three flower pots with earth. Plant the same number of bean seeds in each one. Make sure the three pots get the same amount of sun and water.

For 15 minutes in the morning and at night, send the plants in the first pot warm, positive, loving thoughts. Don't think about the second pot, let the seeds grow normally. The seeds in the third pot are to be cursed, or at least sent discouraging thoughts twice a day.

Pretty soon, even though some of their relatives looked at them a little oddly, 150 men and women were sitting down each morning, beaming thoughts to newly planted bean seeds.

Some prayed and blessed them. A few tried to send mental pictures of strong, leafy plants. Others thought and said such things as, "I love you. You're special. You're the most wonderful bean plants in the world. You can't help growing strong and beautiful."

When it came to cursing, most people tried to think of a pet hate. Then they aimed their bad feelings at the seeds. Others just fed the plants a steady diet of negative remarks. "Boy, are you ugly! There's no use growing. Even if you break ground you'll find it's cold and nasty. No one wants you. Don't grow."

Does mental gardening pay off? About two weeks later, Rev. Loehr's group measured the height, fullness and roots of all the plants. They weighed them. They took pictures. In the months that followed, they sent thoughts to 27,000 seeds. As the people measured the plants, they began to see with their own eyes what their minister was preaching: thought has power.

The plants treated with loving thoughts did grow better than the ones left alone. Like all power—the power of electricity or atomic energy—it seems the power of thought can be used for good or ill. The cursed plants did not grow as well as the normal ones. Some seeds never even broke ground.

But should you curse a plant? "Well," the Rev. Loehr says, "the Bible tells us that Jesus cursed a fig tree and it withered. There are things we don't want to grow, like certain diseases. Perhaps, as we study the power of thought to stop growth in plants, we can learn something about using thought to help the sick."

Even so, some people don't have it in them to hate a plant. Morning by morning, one woman to her dismay watched her cursed plants sprouting up full and lush. They almost outgrew the ones she was loving. She didn't know what to think. She confided in her elderly father who lived with her. Seeing how upset she was, he solved the mystery. He couldn't bear to think of her sending such hateful thoughts to the poor defenseless little seeds. So, every night he came quietly downstairs and gave those seeds a lavish dose of love, affection and grandfatherly encouragement!

Freud's Trick on Einstein

There are Olympic gold medal winners and prima ballerinas, and there are psychic superstars, too. Wolf Messing is one such psychic star. His career began in earnest when he was "discovered" in one of the most unusual telepathy try-outs on record.

Wolf thought he'd never seen so many books; they lined the walls all the way to the high old ceilings of the apartment. Then the teenage boy remembered he had something else to think about. Here he was, a runaway, barely making his living in a sideshow. Yet two of the world's most famous men, the great psychiatrist, Sigmund Freud, and the even greater scientist, Albert Einstein, were waiting to meet *him*. Both men were familiar with mysteries; Freud uncovered those of the human mind and Einstein was revealing the mysteries of the universe itself. They wanted to see for themselves if this so-called "Wonder Boy" had any mysterious powers of mind.

Every Friday in the sideshow, Wolf put himself into a trance. Then he was sealed inside a "crystal" coffin. He would lie there all weekend, still as a corpse, while curious onlookers crowded close, trying to catch sight of the slightest movement to see if he were dead or alive. When he wasn't in his coffin, people said Wolf could read their minds, that he could sometimes send thoughts to them and make them think what he wished. Working in the sideshow was one thing, but coming face to face—or mind to mind—with Einstein and Freud was another.

Now in the apartment, Freud would send Wolf a mental message. Wolf was supposed to do whatever he thought Freud was thinking about. Einstein stood back, watching Wolf intently. Could the boy read the great doctor's mind? Silently, Freud began to concentrate. After a few moments, Wolf turned and walked into the bathroom. Seeming to forget he was a guest, he opened the cupboard, looked around and picked up a pair of tweezers. He came out and strode straight to Einstein. He hesitated, and then said boldly, "I beg your pardon, Dr. Einstein, may I pluck three hairs from your moustache?"

How could he ask the great Einstein such a question? What if he'd mixed up the mental command? Einstein just smiled and offered up his moustache for the sake of science.

Holding three prize hairs, Wolf realized, as Freud nodded yes, that he'd sailed successfully through one of his most important psychic demonstrations. Perhaps, among many other things, Einstein remembered this incident when he wrote, "The most beautiful experience we can have is the mysterious . . . Whoever does not know it and can no longer wonder, no longer marvel, is as good as dead, his eyes are dimmed."

Years later, Wolf had to go through an even more intimidating telepathy test dreamed up by another famous man—Joseph Stalin.

Stalin's ESP Test

"You are Wolf Messing!" the Nazi officer shouted. "The man who predicted the death of our leader!" He smashed his prisoner in the mouth, knocking out six teeth with one blow. ("A master torturer," Wolf called him later.) The Nazis had just invaded Poland and Hitler himself had offered a large reward for Wolf's capture. Hitler, says Wolf, always took psychic things very seriously. And Wolf, who had become a famous psychic—someone people listened to—predicted at a huge public meeting that Hitler would die and never rule the world.

Lying beaten in the hall of a police station, Wolf realized that if ever he had mental power, this was the time to use it. Concentrating as hard as he could, he sent a thought to the Nazi soldiers around him. "Go to the room at the end of the hall, go, go, it's urgent!" For one reason or another, they did. Wolf, suddenly alone, escaped. Travelling undercover he made his way from the country and, hidden in a hay wagon, crossed into Russia. Safe at last from the Nazis, now he had another problem.

At this time it was forbidden even to talk about

psychic things in Russia. People who thought they had psychic abilities and the scientists who studied them were often hounded into jail and sometimes shot. Yet this was the only kind of work Wolf knew. He began to give his mental demonstrations in small towns. One night, two uniformed men walked onto the stage where he was performing. "The show's over," they told the audience. Wolf thought he was finished too, for these men were members of the feared secret police. They drove through the night and took him into a room where a man was waiting. Wolf stood face to face with Joseph Stalin, the ruler of all Russia. Stalin asked about Wolf's supposed ability to send thoughts into other people's minds, to cloud their minds and make them do what he wished. Obviously, a dictator would like to corner the market on such a talent. Stalin would see about this supposed psychic. He made up a seemingly impossible test for Wolf. There was nothing to do but try. Nobody said no to Stalin and lived.

Stalin asked Wolf to come see him at his country home outside Moscow. But there was a trick to it, a big one. He must get in without a pass, without permission. Guards stood thicker than trees around Stalin's home. Even the most important people were checked and re-checked when they arrived—except for one man, Lavrenti Pavlovich Beria, the head of the secret police, a man whose mere name inspired terror. The guards were Beria's men and they wouldn't dream of stopping him.

As evening came a car drove up to Stalin's house. A slim, curly-haired man got out and nodded at the guard who opened the gate. He strolled past the sentries lining the drive. A member of the secret police opened the door and welcomed him. Then the man walked down the hall past yet more guards, until he came to a room where Stalin sat reading. Stalin looked up. Wolf Messing stood smiling before him.

"How did you do it?" asked the amazed Stalin.

"Easy," said Wolf, "I simply thought *I am Beria, I am Beria*, and sent this thought into the minds of the guards who for a moment did think I was the head of your secret police."

Wolf and Beria looked about as much alike as an apple and a banana. Stalin was impressed. We don't know if the dictator ever put Wolf's talents to use secretly in affairs of state. If the scientists we met in Russia knew, they weren't talking, but they did think the story was true. Years after Stalin's death when Wolf wrote up the incident, it was passed by the government censor and printed in an official journal. We do know that after his meeting with Stalin, most of Wolf's problems were solved. He was permitted to give psychic demonstrations and become famous throughout the Soviet Union at a time when other people hardly dared whisper "ESP."

Many people might object to Wolf's account, because even though they think telepathy is possible, they don't think it's possible to project a thought so strongly it can cloud someone's mind. But the Soviets have the best tested reasons in the world for believing you can, at times, put a psychic whammy on someone.

The Psychic Whammy

A Russian chemistry teacher held up a foaming test tube. He was discussing a chemical reaction with a young woman college student. He saw her eyelids drop shut, then flutter open. She started to say something when, in mid-sentence, her eyelids closed again. She was out cold.

It worked! The teacher could hardly believe his own eyes. The young woman was hypnotized. But not by anything he did. She'd been hypnotized by a teacher who wasn't there, who sat concentrating in a different part of the building. That man had managed to put a psychic whammy on her. He put her in a trance telepathically. She didn't know anyone was trying to do anything to her.

Telepathic hypnosis, the ability to put somebody into a trance at a distance, is the classic Russian ESP experiment. For 30 years in a secret laboratory, a team of scientists, with Stalin's permission, worked on this odd, somewhat frightening power. When the dictator died, the internationally known Soviet scientist, Dr. Leonid L. Vasiliev, revealed that his team had perfected telepathic hypnosis. In hundreds of tests, they successfully sent telepathic commands from room to room, building to building and even over a thousand mile distance. They sent telepathy from regular labs and also from special structures that blocked out all known electromagnetic waves. The Russians found that telepathic hypnosis takes a special subject and a highly trained hypnotist. Fortunately, it isn't easy to do. (It also probably isn't very nice to do.) But for all their work, the Russians do not claim to have discovered the psychic whammy. That honor goes to a group of distinguished European professors. One of their more famous adventures toward the end of the last century shows that scientists, like everyone else, can be funny without realizing it.

One evening after dinner in the French city of Le Havre, Dr. M. Gibert looked thoughtfully at the five guests around his table. He himself was a well-known physician, yet he considered it an honor to have such guests. From France, England, Poland, these men were all leaders in the new field of studying the human mind.

"Will you do it?" one asked Dr. Gibert. They wanted him, using mind power alone, to try to put one of his patients in a trance and summon her across town to his house. The patient's name was Leonie. Well, why not, Gibert thought. He'd been able to control Leonie mentally before. It would be interesting entertainment for the evening. Besides, with such a scholarly group, something new might be learned.

The six men synchronized their pocket watches. Then Dr. Gibert went into his study to concentrate. His guests headed across town to Leonie's cottage ¾ of a mile away. When they arrived, the five scholars surrounded Leonie's house, hiding behind bushes and trees. It was time for Gibert to start sending his mental commands. Right on schedule, the door opened; Leonie stepped out and walked to the garden gate. One scientist, peeking around the corner of the house, saw with delight that Leonie's eyes were shut tight. Then he was disappointed. She turned and went back inside. Why? According to the records, at this point Dr. Gibert "because of the strain of thinking, had fainted—or dozed off." But he came round and started concentrating again.

Walking rapidly, Leonie came out the door again and bumped smack into one of the scientists who'd crept up the walk to get a closer look. But she kept going at a good clip. The five scientists trotted along behind her, taking notes. For ten minutes, she hurried through town towards Gibert's house "successfully avoiding street lamps and traffic." One scientist noted that none of the other people on the streets seemed to notice anything strange about Leonie even though she barrelled along with her eyes shut. They may have been too busy looking at the pack of high-collared, distinguished-looking gentlemen following this rather dishevelled middle-aged woman. After all, Leonie had just been finishing dinner and didn't have the slightest idea she ought to fix up to go out on the town.

Suddenly, Leonie paused at a corner. She seemed confused. She was. Dr. Gibert had decided the whole thing was useless and started playing pool. He changed his mind and began concentrating once more.

Leonie was off again, the scientists jotting and jogging at her heels. At last, she reached Gibert's house. Just then, the good doctor, wondering where everybody was, dashed out of the door. He and Leonie collided head on and fell down. Leonie struggled up, climbed over Gibert and rushed on into the house. She ran from room to room, saying sorrowfully, "Where is he? Where is he?" She searched upstairs and down, surrounded by six nimble-footed scientists trying not to be tagged by the hypnotized woman. Finally, Gibert sank into a chair and *mentally* called her to him.

"She takes him by the arm," one of the group recorded. "She is seized with great joy!" So were all the scientists, for they were convinced you surely can influence people at a distance.

Telepathic Emotions

Mikhail Kuni, a Russian art student, awoke with a start. A sickly feeling of dread surrounded him. He tried to shake it, but couldn't. Finally, he woke up the two other students who shared his Moscow apartment.

"I've just had a terrible dream," he told his sleepy friends. "I dreamed my mother was bitten by a rat and is near death."

"It was just a dream," they said. To get his mind off it, his friends made Mikhail breakfast. They had just about finished eating when a knock sounded on the door. "Telegram for Mikhail Kuni," the messenger said.

It was from his family in Vitebsk, a city a few hundred miles west of Moscow. Mikhail read, "MAMA DANGEROUSLY ILL. COME HOME."

When he arrived, Mikhail learned that a rat had bitten his mother some days before. Unexpectedly, gangrene developed. Her leg would have to come off, but because of her other problems it wasn't likely she would survive.

"As the doctors talked, one phrase stood out vividly in my mother's mind," Mikhail recalls, "that her condition was hopeless. It was a painful jolt!"

Mikhail learned that at the time the doctors were consulting in his mother's room in Vitebsk, he was asleep in Moscow. "Images and information about what was happening came to me at the moment of the most intense emotional experience for the telepathic sender—my mother," says Mikhail.

He realized there is more to the mind than people think. Mikhail left art school and became internationally known for his mental feats: he trained himself to calculate huge sums almost instantly, to use ESP and supermemory.

"My mother's emotion propelled the telepathy," Mikhail Kuni says.

Emotion also propelled some unusual Russian telepathy experiments. A big, barrel-chested, red-haired man, Karl Nikolaev, got the telepathic message from his more scholarly-looking "telepathic twin," Yuri Kamensky, a biophysicist. The Russians trained them together hoping they would develop some of the invisible rapport that real twins have.

Karl and Yuri starred in many important telepathy tests. They sent images of screwdrivers, coils and other objects over hundreds of miles. Once again, they proved that telepathy is a fact. And they proved something more. The scientists connected them to machines that showed what was happening in their bodies and brains when telepathy was active. They had sent images; but now the scientists wanted to know if they could send their feelings to each other telepathically—strong emotions like joy, anger, fear.

How about using ESP to send someone a punch in the nose? That was Yuri's job one night. He was supposed to try to beat up Karl, something he wasn't likely to be able to do, except, maybe, mentally. After the clock in the Kremlin tower struck the hour, Yuri in a Moscow lab began his telepathic fist fight.

"I saw myself punching Karl in the nose, hitting him in the stomach, wrestling him to the floor." Yuri thought about it, he felt it, he summoned as much violent emotion as he could, trying to project his attack to Karl.

Meanwhile, the good-natured Karl was all wired up in a military lab in Leningrad, 300 miles away. Scientists checked readings from his brain, heart, muscles and other functions. Karl expected a message from his telepathic twin, but he didn't know when or what would come through. As Yuri attacked in Moscow, the scientists in Leningrad saw Karl's brain wave pattern changing to a less harmonious one. Other physical changes started, and then Karl began to get the idea of attack consciously.

"I really don't like this kind of emotional telepathy," Karl told us. "It sometimes makes me feel sick for days." But he also said nothing could keep him from the lab, even though he works full-time as an actor.

"For years," he went on, "nobody thought I could

do it; they thought ESP was a trick." Karl explained that once, as a Russian soldier on leave in Hungary, he saw a stage mentalist do ESP. It's almost impossible to tell if someone on stage is using ESP or not, but Karl was electrified. He decided, "If that guy up there can do it, so can I."

One reason scientists think Karl can do it is that the impartial machines show that something is indeed happening to him at the moment a telepathic message arrives. Use of equipment makes telepathy more visible. It gives clues to the ways it works and doesn't work. The Soviets didn't invent this approach to ESP. Even before we went to Russia in the late 1960's, we noticed that in *Pravda* and in the Communist youth papers, Russians were writing about Douglas Dean in America. He was hooking people to machines for telepathy tests.

Not many people in the United States knew of Dean's work. At that time, scientists weren't exactly crowding to follow in his footsteps. It took the Soviets to kick off an American breakthrough—or perhaps we should say an Anglo-American breakthrough.

Psychic Bodies

"The redcoats are coming! that's the kind of message we could send from the planet Jupiter with this new ESP communication system." Douglas Dean's eyes twinkled as he spoke, for he himself is a "redcoat," or at least an Englishman who is working in the U.S. Doug, a chemist, investigates psychic abilities, hoping to make them useful to society.

He heard stories like this one about a New Jersey salesman. During rush hour, the salesman was driving his car along a crowded highway, fighting his way to work, when he began to feel sick. The further he drove, the worse his stomach hurt. Something's terribly wrong, he decided. He couldn't go to work. At the next crossover, he swung around and headed home. Driving along, he felt as though he had swallowed a magic pill. All of his pains vanished. He was fine. "All right, let's go to work," he said to himself and once again turned toward the office. He had gone only a little way when the pains hit, if anything worse than before. Again he turned around; again the pains vanished. This time he kept going until he pulled into his own driveway.

Opening the door, the salesman saw a clutter of suitcases in the hall. His wife had planned secretly to take the children and leave him while he was at work.

Coming home unexpectedly, he stopped her and was able to settle their disagreement.

A lucky coincidence, many people would say. But the pains never came again and the doctor could find nothing wrong. Other people like Doug Dean wonder if it wasn't something more than coincidence at work, something like a very personal kind of communication system. People say they have a "gut feeling" about something. Was this man's gut trying to tell him something? We think of telepathy as communication from mind to mind. But Doug Dean wondered if perhaps your whole body picks up ESP messages. Doug began a new approach to ESP—physical testing—which was to become the new trend in research. In the early sixties he got his big break. The New Jersey Institute of Technology, the fourth largest engineering school in the country, agreed to let him start a telepathy lab.

Much to Doug's surprise, Dr. John Mihalasky, a professor at the college, was one of the first volunteers for telepathy. John had agreed to be the faculty supervisor for the project. "I was open-minded about the possibility of ESP," he says. "Still I thought I better get right into that lab. I wanted to see for myself that this guy Doug Dean wasn't some kind of nut or fraud."

"Give me the names of five people that really mean a lot to you," Doug said to John. He wrote these on cards and mixed them with other cards that had names of people Doug knew, names from the phone book and some cards with nothing on them. Next, Doug connected John's hand to a machine that measures changes in blood volume in a person's fingers. Then he left John alone, lying on a cot in a dim room.

In another room, Doug and his helpers mixed the cards. They were going to try to send the names to John telepathically. At specified times the telepathic sender turned up a card and concentrated on the name. He thought of a blank screen if the card was blank.

John knew one thing: he wasn't getting any telepathic messages. He didn't feel anything except rather pleased to stretch out for a few minutes on the cot. At the end, John carefully checked the recordings from the machine. He was shocked. When the telepathic sender concentrated on names that meant nothing to John, the blood volume in his hand remained the same. But, when the sender thought about people John knew and cared about, the blood volume changed very definitely. Though John was completely unaware of it, his body had been neatly picking up telepathic messages.

"I was more psychic than I thought," John says and adds that probably many of us are. Impressed, he teamed with Doug to do hundreds of ESP tests. One man's name always got a big reaction from John's body. This was a man he worked with, someone he didn't like, a man in a position to do him harm. People seem to have a sort of radar, John and Doug say, that constantly scans the mental airwaves. There's a blip in your body when it picks up a strong thought about someone you're emotionally attached to, particularly if that person is a threat. If we can learn to make those ESP blips conscious, we'd certainly know a lot more about what's going on.

"The body's ability to pick up telepathy can be turned into a communication system," Doug told a

NASA space conference. We can use this ESP system when astronauts are on the far side of the moon and radio is blanked out. It could be used when we get to Jupiter where there will be an 8-minute lag between sending a radio message and its reception on earth. It could be used by submarines far under water and at other times when a radio is not practical or not working. You don't need regular psychics for this set-up. Anybody—or at least many bodies—can do it.

The ESP communication system is as simple as Morse code—dots and dashes. If, in a certain time period, your body reacts, that's a dash. If it doesn't react, that's a dot. This is how you send messages. From room to room is one thing. But could your body really react to a thought that was far, far away?

Doug set up tests between New York and Florida, 1200 miles apart. The signals came through. He ran communications tests from Newark, New Jersey, to France. The signals came through, spanning thousands of miles. Finally, he cooked up a test with a woman who is an expert skin diver. Underwater, off the coast of Florida, she concentrated on names. A person on shore in a lab picked up her ESP message.

The redcoats may not be coming, but it seems that ESP has landed. Today, it can be used as a reliable if very slow communication system. Of course, John and Doug say, this is just the beginning. In the future, will we have sophisticated ESP systems for communication? How about entertainment? If we do, unlike radio or television, the most important part of the hook-up will not be a glass tube or a cool transistor. It will be a nice warm human body.

Telepathic Suicide?

In the lab, Doug and John tested for telepathy happening along an unconscious emotional connection. Some people seem to have a much more direct telepathic link.

Joan lived in Cape Cod, Massachusetts. Her identical twin, Joy, lived thousands of miles away, far across the United States in the state of Washington. One fall day, Joan began complaining about very bad pains in her lower back. It felt as though something was terribly wrong with her kidneys. Her husband helped her into the car and drove fast to the local hospital.

If you'd been watching the scene from a high cloud, you might have thought you were seeing double. As Joan was wheeled away for examination in Cape Cod, her twin Joy was being examined at the hospital in Washington for sudden severe back pains. Who had the kidney problem? It was Joy, as it turned out. The doctors could find nothing wrong with Joan. It seems, unknown to herself, Joan was suffering telepathic sympathy pains.

The records are full of examples of conscious and unconscious telepathy between members of a family and particularly between twins. Recently, doctors found out in tests that if one twin's brain waves change, so do the other's, though she may be rooms away.

Perhaps the strangest proof that people can and do connect with each other over a distance in a very real, physical way is the rather bloodcurdling case of Bobbie Jean and Betty Jo Eller. Identical twins, the girls grew up in Purlear, North Carolina. They did everything together. They picked the same hobbies, liked the same clothes. Bobbie Jean seemed to be the leader. Whatever she did, Betty Jo would mirror.

After the girls were graduated from high school, the family began to notice that they were acting oddly. They wouldn't go out. They didn't want to have anything to do with anyone else. Bobbie Jean took to sitting for hours and gazing into space. She wouldn't answer when people spoke to her. Betty Jo did the same thing. They stayed in their room. Finally, they had wandered so far off into their own world that the only thing you could notice about these two silent, identical figures was their tears. Off and on, both would start to cry. They didn't seem to be able to stop.

Bobby Jean and Betty Jo went to the state mental hospital at the beginning of 1961. The specialists tried drugs; they tried talking; they tried all kinds of thing to reach into the twins' private world. Nothing worked. After a year, in desperation, the doctors decided that the only thing left to do was to totally separate them. If they didn't have each other, they might find their world lonely and come back to this one.

They were placed in separate wings of the hospital as far from each other as possible. They were allowed no contact whatsoever—or so the doctors thought. But it seems they had developed their own invisible communication system.

One night in spring, Bobbie Jean had a seizure. She stiffened up. After midnight, checking the beds, the head nurse found Bobbie Jean dead on the floor. She phoned the other wing of the hospital. Would the nurses there check on Betty Jo? They found Betty Jo lying on the floor. She was dead. Both girls died in the same position, curled up like unborn babies, lying on their right sides.

Dr. John Reece performed autopsies on their bodies. He found no evidence of suicide. Or anything else. For once, Dr. Reece couldn't fill in "Cause of death" on the death certificate. He left it blank. "I found no evidence of injury or disease that could cause death," he reported.

Do You Have an Astro-Twin?

Other kinds of twins are beginning to interest researchers. You probably have one, too, though you haven't met. Your astro-twin was born on the same date, at the same time and around the same geographical location that you were born.

Astro-twins are of interest to a new branch of science, *cosmo-biology*, that has grown up recently, particularly in Russia and Europe. This is the study of how the ever-changing patterns of the planets affect life on earth. That sounds like astrology, you're probably saying. And there are similarities, though not every scientist will admit it.

Psi researchers are interested in cosmo-biology, because the influence of the planets on the force fields of the earth (and therefore on people) seems to have something to do with psychic abilities like psychokinesis, healing and telepathy. Over the years at the dream lab in Brooklyn, for instance, they've found you're most likely to transmit a telepathic dream successfully at full moon or new moon.

We're unconsciously connected to each other telepathically more often than we know. In ways we rarely realize, we're also connected to what's happening in outer space. The study of astro-twins might tell us more about how these connections work.

Researchers have found that if you have an astro-twin, the same events tend to happen in your lives at the same time. Astro-twins, though not related in any way, tend to have accidents on the same date; they may share the same interests and hobbies, receive promotions at the same time, get married, have an identical number of children at the same time and even die on the same day of similar causes.

Edna Hanna met Edna Osborne in a hospital in Hackensack, New Jersey. Both women had given birth to babies at the same time. To their surprise, they discovered they had the same birthdate and the same number of brothers and sisters. Both Ednas had married men named Harold exactly $3\frac{1}{2}$ years before on the same day. Both Harolds were in the same business and owned identical cars. The two Harolds had also been born on the same day, month and year. They were of the same religion. Edna Hanna and Edna Osborne were both blue-eyed brunettes of identical height and weight. The Hanna and Osborne families each had a dog. It turned out they'd bought their dogs on the very same day. The dogs were the same age, same sex, same size, same breed. Both families christened their dogs "Spot." Incidentally, the babies that Edna Hanna and Edna Osborne gave birth to that day in Hackensack were both girls of the very same weight. Ahead of time, each mother had chosen a name for her new baby—Patricia Edna.

John Brown from Ontario, Canada, had the eerie experience at age 41 of encountering his astro-twin in the office where he worked. The two men shared the birthdate of July 22, 1914. Both worked in the accounts department of the company; both had taken secretarial courses; both had worked in army orderly rooms in World War II; and both had been employed at various times on the very same street in downtown Toronto. They discovered they had each married a woman from another province. Both women had been born in February. The two men weighed the same and owned similar cars. Their last names both started with "B".

In July 1972, John Brown died. A few weeks later, John Brown's time-twin died "suddenly." Brown's widow noted that funeral services for both men were held at the same funeral home.

The odd parallels in the lives of astro-twins show up even when they come from totally different social levels in life. In 1762, King George IV of England was born the same day and hour as a commoner who was to become a chimney sweep. George IV became known for his vices and follies, for being a gambler and a spendthrift. The chimney sweep was known for the same vices, but on an entirely different scale. George IV kept horses and ran the best horse races in England. The chimney sweep kept a stable of donkeys and ran the best donkey races of his time. George IV was kicked by a horse and injured. At the very same time, the commoner was kicked by a donkey and injured. Both took the same amount of time to recover. On the day that George IV went bankrupt and lost everything, so did the commoner. When all the King's horses were being sold, across town the commoner was selling all his donkeys to pay his debts.

King George III also had a commoner astro-twin born June 4, 1738, at almost the same minute and in the same locale. The commoner's name was Samuel Hemmings. When Hemmings became ill, George III did, too. When Hemmings had an accident, George III did also. Major events in the life of one were mirrored in the life of the other. The day Hemmings founded his own business as an ironmonger, George III became King of England. The day Hemmings married, September 8th, 1761, King George got married. On January 29th, 1820, Samuel Hemmings died. At that very same hour and on the same date, King George III died too, and of the same cause.

2 Psychic Energies

Thought Power That Moves Objects

The plump, dark-haired young woman frowned intently as she sat gazing at a group of objects on the table in front of her. There were matches, a silver salt shaker, and cigarettes lined up inside a plastic cube. She bent her head forward as if her eyes were beckoning some unseen presence. Her pulse raced, her heartbeat speeded up. She felt weak and dizzy. Suddenly, the objects in front of her began to move, slowly at first, and then more rapidly. The matches spun out in many directions at once, some away from her, some toward her. The silver salt shaker marched sturdily forward. The cigarettes moved along vertically without falling over.

The woman was Nelya Mikhailova of the Soviet Union, world famous for her strange powers of mind over matter—psychokinesis—or PK for short.

Nelya was exhausted from this brief demonstration. Sometimes she loses as much as three pounds in a half-hour of tests, as if she were somehow converting the substance of her own body into energy. But even though she often felt sick for days after taking part in the tests, she felt a patriotic urge to help her countrymen unravel the riddle of her thought power.

After all, as a young girl in World War II, she helped defend her native city of Leningrad during the terrible 3-year siege. Now she felt that by being tested, she was working on a new front, one of scientific discovery. And test her, they did.

For years, one major drive has powered Soviet psi research. Russian scientists want to learn more about the energies involved in psychic events and try to use them. Dr. Leonid Vasiliev, the founder of psi research in Russia, said, "The discovery of the energy behind psychic events will be as important, if not more important than the discovery of atomic energy." His government took him seriously.

Since the early 1960's, Nelya has been the focus of high-level political, military and scientific attention. The list of investigators, including Nobel Prize winners, reads like a "Who's Who" of Soviet science. They have trussed up Nelya like an astronaut. They have monitored every type of physical function as her thought power makes an object move. And they report that Nelya has used her PK powers in startling ways: she has stopped a frog's heart; she has caused burns on human skin; she can affect solids, liquids and gases; she has separated the yolk from the white of an egg floating in a sealed aquarium six feet away; she has moved heavy crystal bowls by thought power; she has produced psychic pictures on film; she has interfered by PK with medical and scientific equipment.

Russian scientists invented special detectors to try to track Nelya's psi force. Dr. Harold Burr, professor of Neuroanatomy at Yale University, established in 1935 that all living matter, from a seed to a human being, is surrounded by an electromagnetic energy field. This is a sort of a "cocoon" or "aura of energy" around the body. These energy force fields change with different thoughts and emotions. Soviets developed Sergeyev detectors to monitor the energy field changes.

Researchers set up the detectors all around Nelya. They found something new! When objects moved, the energy fields around Nelya's body began to pulse in rhythm with her heartbeat and brain waves. She could mentally direct her energy fields toward the objects.

"The pulsations in the fields around her body may act like magnetic waves," Dr. Sergeyev of Leningrad suggests. "The magnetic waves may cause non-magnetic objects to act as if magnetized—to be repelled from her or attracted to her."

Kirlian photography—photographs of energy coming from the body—showed changes in the energies coming from Nelya's hands during PK. Some Soviets called the energy "bioplasma." They believe there's a whole energy circulation system in the body—and that it is the same one that the Chinese talk about as the basis of acupuncture (healing with needle therapy). This energy system, they feel, could be the key to many psi events. Through it, we may sense oil or water deep in the ground, as in dowsing. The energy also seems to be involved with psychic healing.

The Soviets believe that everyone possesses these psi energies and it's a matter of learning conscious control of them in order to use them. From what they discovered about Nelya, they developed programs to train people to do PK. Alla Vinogradova learned to move light objects without the stress Nelya experiences. Boris Yermolaev can make objects hang in the air.

Westerners found out about Nelya through the report in our book. When we brought the first film footage of Nelya to the West, some television stations were afraid to broadcast it, because they'd never heard of mind over matter. The film of Nelya was shown in lectures or on television all over the world from Australia to Japan, Europe and the U.S. People began phoning the television stations to say they could do the same thing. They'd just never thought of trying.

Dozens of Western researchers have now journeyed to Russia to see and test Nelya. Elsie Sechrist, who travelled with a group from the Edgar Cayce Foundation, recalled that having dinner with Nelya caused some surprises in a Soviet restaurant. She didn't have to ask for the food to be passed. She could make it jump toward her!

Among the Westerners who tested Nelya, were Benson Herbert and Manfred Cassirer from the Paraphysics Lab in England. They first met Nelya in the spring of 1972 and discovered for themselves that this strange PK force not only moves objects but can also affect people.

Nelya is known to be able to heal people. She can also use the same energy in a negative way. "If she could stop a frog's heart," Benson mused, "could she affect me in some way?" He rolled up his sleeve and said to Nelya, "Try to use your PK to harm me. Put your hand on my arm for five minutes!" Benson began to feel such acute physical pain, he had to clench his teeth. Nelya removed her hand. Benson had a bright red, swollen burn on his arm that took eight painful days to heal. The test was repeated on Manfred, who instantly complained of the searing, burning heat.

The team returned to the USSR in 1973 bringing their own special lab equipment from England. They found that Nelya could physically affect many scientific instruments, including a hydrometer and a radiometer. They measured her PK force at 8340 dynes (unit of force in physics). After exhaustive tests they reported that they had amply confirmed Soviet findings. "She can move objects at will without the employment of any known force."

An unknown energy force, an energy everyone has, an energy known or unknown that your mind can direct, an energy that can affect the body—no wonder interest in PK is running high in many scientific circles. It's more than moving matches or bending a key.

"PK could be the ultimate weapon," say some researchers. Imagine what would happen if an enemy could aim this force at a nuclear plant, a missile base or an electrical station. Psychics Ted Serios and Uri Geller have been known to break electric circuits, to scramble data stored on magnetic tape and in computers. Increasingly, computers are the work horses of our society. An invisible, unstoppable force that could interfere with computers and other sophisticated machines is an alarming idea, to say the least.

It's not surprising that the Soviets are not the only ones trying to get a handle on exactly how this human energy force works.

Poltergeist

A call came to a German telephone company. An answering machine picked up. "The correct time is now 2:45 p.m." The phone rang again. "The correct time is now 2:46 p.m." This was the 60th call in a single hour that had come from the same phone to the number that gave the time. Who was making all these calls? The telephone company traced them to a firm of lawyers, Siegfried Adam, in Rosenheim. The firm didn't need the phone company to tell them something funny was going on. They had soaring phone bills to prove it.

"We know what time it is. We didn't make those calls," the office employees insisted. "None of us were even near the phone when some of those calls went through."

Still, day in day out, the company billing equipment kept registering calls, all to the *time* number. The phone bills grew larger. "If someone was making calls at such an astonishing rate," the law firm personnel said plaintively, "surely one of us would have seen the prankster dialing." Or would they?

This wasn't the only odd thing going on at the unhappy law firm. One day, a secretary, Annemarie

Schaberl, walked down the hallway and all the overhead light bulbs blew out. That's not too strange. But it is when it keeps happening. Next, neon tubes seemed to unscrew themselves from their sockets. Pictures turned 90° on the wall as if tilted by invisible hands. Exploding light bulbs sent glass splinters flying into the walls.

Soon the employees began whispering about "poltergeists." Poltergeist literally means "noisy spirit." The activities of these supposed spirits, who enjoy playing practical jokes on the living, have been recorded through history. Today, most scientists don't think it is a spirit who moves pictures or unscrews light bulbs. They think it's done by some sort of energy *unconsciously* shooting out of a person who's mixed up and frustrated.

Meanwhile, the phone bills kept mounting. Physicists and technicians were called in from Germany's most famous science centers. They filmed the strange goings-on. Hanging light fixtures swung wildly by themselves with no visible means of propulsion. Loud explosions were recorded on tape. Physicists conducted tests and asserted, "Some unknown form of energy is at work." The head of the electric company's maintenance division admitted, "Our experts cannot explain these extraordinary occurrences. It is all very strange." So strange, it was fast becoming the best scientific evidence of poltergeist activity on record.

The law firm wasn't interested in the niceties of physics or proving psychic energies. They were desperate to stop the phone and repair bills. At this point, in December 1967, Dr. Hans Bender, a parapsychologist from the University of Freiburg, arrived on the scene. "This seems to be an extraordinary example of mind over matter," Dr. Bender said. He was sure the person who was "doing" it was just as scared as the rest of the employees. The unconscious psychic was just blowing off steam, Bender explained. But as always in genuine poltergeist cases, the person had no idea he or she had chosen such a strange way to get rid of inner pressures. "The disturbance never happens on weekends when there are no people in the office," Dr. Bender pointed out.

By a process of elimination, Dr. Bender traced the poltergeist activity to the young secretary, Annemarie Schaberl. She disliked her job; she didn't like her boss. She was having emotional conflicts at home. "I didn't do anything!" Annemarie said. And it was hard to believe she had the power to unscrew bulbs and affect the phone when she wasn't near them.

But the scientists found she did have some unknown ESP power. Later it came out that she was having trouble with her boyfriend, too. He wasn't sure he wanted to keep seeing her. He thought she was some kind of a witch. Whenever they went bowling and her score was low, the automatic pin-setting machine stopped.

What should the law firm do? Dr. Bender suggested, "Transfer Annemarie out of the office and things will settle down." She was and they did.

Uri and the Unbendable

A tall, dark-haired, handsome young man gently strokes a large silver spoon held by TV interviewer, Barbara Walters. Slowly as if the metal were turning to spaghetti, the spoon begins to bend. Many people all over the world have seen the famous Israeli psychic, Uri Geller, perform on television. Seemingly, with psychic power, he has bent keys, silverware, even screwdrivers; he's made broken watches run and picked up telepathic messages.

Supposedly, Uri has even interfered mentally with electrical equipment; stopped an escalator in a big Munich department store; halted a German cable car; knocked out electric lights in Oslo, Norway. Often, when Uri appears on television, people claim strange events happen in their homes, spoons and forks bend and broken clocks start ticking. Could Uri be for real? People wondered.

There is really no way to be 100 per cent sure that a stage demonstration of psychic ability is genuine. But there is a way to separate real from fake in a laboratory. Millions have exclaimed or argued over Uri on TV, but few people know about his even more extraordinary feats in the laboratory under the strictest scientific controls.

At Stanford Research Institute, for instance, Uri was shown a sealed magnetometer, an instrument that measures how strong a magnetic field is. Uri summoned his powers, and without touching it, he made the gauge change on the complex instrument. "I was as surprised as anyone when the needle moved sharply," said Uri. "They told me this was scientifically impossible."

Uri received ESP messages sent to him in tests even when scientists sealed him inside a refrigerator-like box made of thick, massive steel.

At the University of London, scientists decided to see if Uri's supposed invisible force could do anything to a Geiger counter that measures radiation. Uri interfered with the working of the machine. Then he caused all sorts of metals to bend and crystals to break up.

Whatever the energy force is that Uri lets loose, if it can bend metal and disrupt electricity, it is of interest to the nation's defense department. Secretly, Uri has been investigated and tested by defense scientists. Here is one of the recently released discoveries.

At the Naval Surface Weapons Center in Silver Spring, Maryland, scientists have developed a new miracle metal called Nitinol. This metal has a very strange property: it cannot be bent! When Nitinol is heated, it instantly jumps back to its original straight shape. What could Uri Geller's metal-bending powers do to the *unbendable* metal? Surely now, the defense experts reasoned, they had a metal which could screen out Uri's psychic powers.

Uri began the tests in 1973. He gently touched the Nitinol metal wire and mentally willed it—"Bend, bend, bend." The wire bent to a right angle. The scientists heated the metal, expecting it to return to its original straight shape. They got a surprise. The metal continued to curl up. They heated the wire to a glowing 900° Fahrenheit (482° Celsius), but still the "Geller bend" would not leave. Tests were run again and again through 1974. No known chemical or energy ever had this effect on Nitinol wire.

Say the baffled weapons experts, "We can't offer any scientific explanation of how this could happen. Uri obviously has some very special power."

Photographing the Human Aura

"It's indescribable! Some lights glitter constantly, others come and go like wandering stars. It's fantastic, alluring, a mysterious game—a fire world!"

"An unseen world opened before my eyes—flashing, twinkling, flaring."

"A spectacular show of colors, whole galaxies of lights, blue, gold, green, violet, all shining and twinkling!"

Scientists, including Soviet ones, don't usually talk in such rapturous terms. When we landed in Russia, and saw what they were exclaiming over, we had to admit they had reason. They were looking at pictures of the human aura, shifting fields of light and color surrounding the body. Painters have long shown religious figures with glowing halos or auras. For centuries, mystics and psychics have said they could see such beautiful color-filled force fields around people. They claimed they could tell a person's mood and the state of his health from the aura. Now, thanks to the Russian invention *Kirlian photography*, anyone can see at least part of what psychics see. Anyone can see the aura.

Working in a small, overcrowded house in southern Russia years ago, Semyon and Valentina Kirlian, a husband-wife research team, began photographing in an unusual way. They used electricity instead of light to make pictures. Their pictures were lovely. A hand or leaf was described in light. The inside of the hand was filled with sparkling light and canals of color. A flaring, glowing aura surrounding the edges extended into space.

To make a photo, Valentina might put her hand inside a lightproof bag. Her fingers rested on unexposed film on top of a plate hooked to an electric device. Semyon would send a burst of high frequency electricity through the plate. Valentina felt her fingers buzz and that was it. They developed the film.

The pictures held more than beauty. The Kirlians found they could learn things from the ever-changing auras around bodies and plants. They tried to interest scientists in their discoveries, but most of them couldn't believe anything important could come from a makeshift home lab. Finally, farming experts agreed to check the Kirlians' invention. They brought leaves that looked like twins: the same kind, the same age, the same freshness.

The Kirlians photographed the leaves and their spirits began to sink. The sparkling pictures of the twins were very different, as though they were only capturing some chance effect. As the Kirlians' spirits sank, the scientists' interest rose.

"You've got it!" the scientists finally exclaimed. One leaf was from a healthy plant; the other had been experimentally infected with a plant disease. "Your invention caught the difference," the experts told the Kirlians, "even though there were no symptoms and no known test could have revealed the disease." Kirlian photography, it seems, could diagnose *ahead of time*, before a problem could be detected in a leaf or body. Psychics had always said that they could see diseases starting in the aura before they showed in the body. After that, more people became interested in the Kirlians' home-grown invention.

Psychics also claim the aura has something to do with PK, with psychic healing, even with out-of-the-body experiences. These ideas and more buzzed in our heads as we had coffee with an outstanding young physicist in a Moscow café. As a boy in southern Russia, he had often dropped in at the Kirlians' for cookies and a look at their fascinating lab. Today, he devotes his time to developing his old friends' invention. After a long, excited talk, he said, "Now, I'll show you the best picture of all, the picture of a ghost."

We didn't exactly expect to see a glowing bed-sheet dragging chains, but we were surprised to see just a leaf. It was the usual glowing leaf; the far right side wasn't as bright as the rest, but we could still see the pattern and edge.

"That's a picture of the leaf that wasn't there—the picture of a ghost." He pointed to the dimmer part of the leaf. "We cut that piece of the leaf off beforehand. But see, *the energy body of the whole leaf shows up!*" our friend said triumphantly. It may lack the chills of a horror show, but it was the most exciting ghost we ever saw. It could change many ideas about how the body is made. In psi research, it set off a lot of bells.

Psychics also have claimed that all of us have a second body, an energy body of light or high frequency fields. Your energy body, they say, is involved with psychic happenings. This is the "body" one uses in an out-of-body experience. Some psychics also explain that it is really the edges of this energy body which they see when they speak of the aura. The physical body, they believe, eclipses the energy body. It's like the moon getting in front of the sun. The bright body of the sun is blocked, but you can still see the corona or aura of the sun around the edges of the moon. Such ideas have been handed down for centuries. We wondered if the Russians were really on the way to finding scientific proof of things psychics had long maintained.

When we returned home, we wrote about Kirlian photography and tried to tell every scientist we could find about it. Most of them wouldn't listen, but some did. Dr. Thelma Moss of U.C.L.A. saw such promise in it that, armed with her heaviest fur coat, and at her own expense, she travelled to the far eastern part of the Soviet Union to a university with specialists in Kirlian photography.

Thelma Moss is a psychologist, not an electronics expert. Back home, she asked experts to help build the equipment and met the same kind of response we had found. "The Russians are fooling." "We don't have the parts." "The plans don't make sense." "If there was anything to it, we'd have known about it." Thelma was at a low point one night when a student in her parapsychology class came up with a funny grin on his face. Ken Johnson handed his teacher a picture. Thelma looked at the swirling lights of a Kirlian photo.

"I made it in my garage," Ken said. He'd managed to put together a crude Kirlian machine. Ken went to work with Thelma, developing the Kirlian process.

Some of his aura pictures of people and leaves are so beautiful that they were shown at a special exhibit at a major art gallery.

Thelma set out to see just what these Kirlian auras might show about a person. They seemed to show changes in mood and strong emotions. The aura changed when the person took a drug. One student (with doctors standing by) volunteered to drink 18 shots of vodka. Beforehand, the aura around his finger was a well-defined shape of bluish light. After he'd downed the vodka, the aura changed to a great rosy glow. It makes you wonder if it was a psychic who first said of a drinker, "He's got a glow on."

The Russians tested healers, people who do laying-on of hands to help the sick. Dr. Moss tested them, too. Some of the first patients were plants. She poked a couple of holes in two identical leaves and took pictures. They glowed, but dark swirls were starting, lights had begun to go out like in a city late in the evening. A healer held her hands over one leaf. Thelma took another picture. After "treatment," the glow was brighter; lights seemed to be coming on again. This could be called the "green thumb effect," and it may be a clue to why some people are good at growing plants.

Thelma asked the healer to direct negative energies to the second leaf, to harm it. When that leaf was photographed, the glow was gone; only a few lights were left, like a great city blackened out. Maybe this is the kind of energy "brown thumb" gardeners send out.

Other scientists put Kirlian photography and healers together. Douglas Dean of the New Jersey Institute of Technology worked with the well-known healer, Ethel De Loach. One of the patients was a scientist with a growth on his arm. Before the healing began, there was a fairly small light blue aura around Ethel's fingers. She began to practice her healing and it was like the sun coming up bright and red out of a blue ocean. The photo showed a great fireball with streamers of light radiating from it. This does not prove she was healing the patient. It does show that something more than suggestion was happening, that there was a great change of energy. (The scientist's growth disappeared by the next day and has never returned.)

Recently, the official Soviet medical journal wrote that Kirlian photography should be ranked equal to X-ray in its usefulness to medicine, and the largest scientific supply house in the U.S. makes sophisticated Kirlian equipment. Its most important function is medical diagnosis. But, just as they use tape recorders and brain-wave machines, psi researchers use Kirlian photography to explore the psychic world.

Most of the results the Russians claimed for Kirlian photography were soon verified by scientists in the U.S., except for the "ghost." Then one day in New York we saw something familiar on a magazine cover. It was *The Sciences*, published by the New York Academy of Sciences. It had a picture of a leaf, a sparkling leaf, but with one strip a little dimmer. It was our old friend the "ghost," yet it wasn't quite the same. This "ghost" was captured in the lab of Dr. Thelma Moss of the University of Southern California. Not every ghost makes the cover of a prestigious scientific journal. Kirlian photography had arrived in the U.S.A.

Mind Over Nails—and Pain

Healing involves energies. What if we could learn to control these energies to live without pain and heal super-fast? If you're an "ordinary person," maybe you can.

Out of the limelight, Vernon Craig of Wooster, Ohio, is a slightly chunky, pleasant-looking, ordinary person in his 40's. The only thing different you might notice about him is a slight air of excitement, as though he had something important he wanted to tell you. He does and he's figured out some mind-boggling ways to do the telling.

In the spotlight on stage, Vernon Craig becomes Komar, holder of three Guinness world records. Before an audience, he stands barefoot on a bed of nails. "Most of you could do this," Komar pauses. "But I don't think you could do this!" he says and he leaps high in the air, landing again on the upturned nails. Instead of twisting his face in pain, he smiles. No blood, no wounds. "But you could learn to do it," he continues, after the applause. "I'm just an ordinary man."

Komar picks up a long curved sword and with one

blow easily slices through a head of cabbage. He does the same with other swords to prove they're sharp. "I started on the stage when I was asked to do performances to help retarded people. Soon, it occurred to me that in one sense we're all retarded. We don't develop the powers that we have. You could all learn as I have, not to feel pain and how to protect and heal your body."

Komar puts the gleaming swords in a rack so they make a mean-looking ladder. Balancing carefully on the blades, he shows a special kind of mind over matter. He climbs up the ladder of swords. "If I slip, I'll have four feet instead of two."

Like many yogis in the far East and some psychics in the West, Komar has learned to control body functions that usually work automatically. For a long time, most people didn't believe the yogis or the psychics. It had to be a trick, they said. But now, doctors at a number of medical clinics have tested Komar and others like him. They find that people can control their heartbeat, their temperature; they can stop bleeding, can turn off pain. They can somehow make their skin resistant to burning even when intense heat is applied to it. The *Guinness Book of World Records* has documents and film to show that Komar in his bare feet walked 25 feet through coals burning at a temperature of 1,494° Fahrenheit (812° Celsius). Flesh should not only burn; it should be consumed at that heat. But doctors examining Komar's feet found not even a blister.

"Are there men in the audience who weigh more

than 200 pounds?" Komar asks. Six large volunteers go onstage. On his back, Komar stretches out on a large bed of nails. (The *Guinness Book of World Records* reports that Komar can lie on a bed of six-inch nails longer than anybody else. He did it for 25 hours and 20 minutes.) On stage, the two largest men step up and stand on Komar's stomach and chest. At a signal, they both jump up in the air and come down heavily—shoes and all—on Komar. They look upset; Komar looks happy. He gets up to show no harm was done.

He stretches out on the nails again with a large slab of concrete on his chest. A reluctant volunteer is convinced finally to bring a sledge hammer smashing down on the stone. The slab cracks in two. The smiling Komar rises and says to the volunteer, towering over him, "What a pussy cat! I've been hit much harder."

Komar lies down again and a second bed of nails is put on his chest and stomach. He looks like the meat in a nail sandwich. Volunteers call out their weights. All together they are 1,610 pounds (731 kg.). A 32½-pound (14.7 kg.) board is placed on top of the nail sandwich. All 1,610 pounds sit on the board pressing the beds of nails into Komar. After what seems like a long time, they rise and lift the top bed of nails. Up pops the cheerful Komar. Deep marks from the nails show in his back, but they haven't punctured his skin. He doesn't seem to notice and shortly the marks disappear. "I'm just an ordinary man," he says again. "What I can do you can learn." He's just had 1,642½ pounds (745.7 kg.) sit on his nail sandwich. It's unbelievable, and another Guinness world record.

Dr. Norman Shealy, a leading expert on pain, is one of the doctors who has examined Komar. At the Pain Rehabilitation Center in LaCrosse, Wisconsin, Dr. Shealy tested to see if Komar is one of the rare people who because of some disorder feel no pain. In a normal state, Komar can and does say "ouch." Dr. Shealy found he has average sensitivity to pain.

Asking Komar to control his pain, Dr. Shealy tied a tight tourniquet around his arm. Usually this becomes extremely painful in a few minutes. Nothing happened. Dr. Shealy held Komar's hand in freezing water. He drove a large needle right through Komar's arm. He prodded him with electrical shocks from a device used to measure sensitivity to pain. He turned it up as high as it would go. Komar sat there happy and relaxed. Dr. Shealy concluded that Komar can indeed turn off pain by altering or changing his state of consciousness. As you can imagine, this ability could be of tremendous benefit to millions suffering chronic pain, if they would learn.

Learning is the key word. You would not barrel down a speedway in a car if you'd never sat behind a wheel before. You wouldn't sail off a high diving tower if no one had taught you to dive or swim. It would be just as foolish to decide suddenly to jump on nails or walk on swords. You wouldn't survive to learn how to control pain. It's worth learning, as Komar proved one night when by accident he showed the audience something that's been demonstrated by others in the lab.

Komar cut his foot. A doctor friend and another doctor, a stranger, came up from the audience to have a look. It wasn't serious, but there was a definite cut. Almost instantly, Komar stopped the bleeding and went on with the show. About fifteen minutes later, the doctors asked to check his wound. First one looked, then the other. They seemed to be pulling at the sole of Komar's foot as if they were trying to open a seam. But they couldn't do it. The flesh had joined; the cut had disappeared completely.

"Scientists say we only use about 10 per cent of our minds," Komar remarked. "Through training, you can learn to use more of the wonderful powers of your mind. You can learn to do what I do."

Is Komar a living psychic prediction—the "ordinary" person of the future?

Plants and Energies

Plants "talking" and "reacting!" It seemed like an absurd idea until Cleve Backster published his most important basic plant experiments in 1968. Backster, a lie-detector expert, hooked up plants in his lab to lie-detectors and recorded their electrical activities on graphs. Then he set up a totally automated experiment in a locked—and peopleless—lab. Machines in a distant lab room would at random dump tiny brine shrimp into boiling water. The instant these tiny shrimp died, the charts showed that the plants responded with sharp changes in their patterns. The deaths of living things seem to send out an energy signal to which plants reacted. A life energy signal may connect all living things, Backster says. He calls this basic communication network "primary perception."

Today, researchers all over the world are investigating how plants perceive and react. Plant communications has become a new frontier. Using all kinds of equipment, from lie detectors to Kirlian photography to brain-wave detectors, they found: plants react to thoughts about their well-being; they react to the deaths of living things; they establish a communication link with their owners that extends over hundreds of miles; they "faint" or close off signals during disasters; they can be used as detectors or "watch plants"; they can be hooked up to equipment to turn it on or off. One researcher even has a plant open his automatic garage door on a mental command.

If you would like to see plants "talk" for yourself, the how-to for hooking up simple equipment for plant talk and experiments is covered in our *Handbook of Psychic Discoveries.*

The bright colorful energy flares that can be seen coming from plants in Kirlian photos, may hold the key to the important plant communication energy signal. These same plant signals also have been monitored with ultra-weak luminescence detection equipment. Soviet physicist Dr. Victor Adamenko remarks that plants communicate over hundreds of miles and no known methods of screening seem to be able to block this signal.

The next accounts highlight some of what is happening with plants in labs in the U.S. and the USSR.

The Witness Had Roots

A murder trial is in progress. The crowd starts to whisper as the doors at the back of the court swing wide and the star witness is carried in. The witness isn't sick. The witness can't walk because instead of legs it has roots.

People hunch forward to get a look at the large green plant placed at the front of the court. Experts hook the plant to electronic equipment. The accused is brought before the plant. Is this the man who murdered the plant's owner?

Far-fetched? Yes. But one day such a scene may not be quite so impossible, thanks to a discovery made by Cleve Backster.

Cleve is a hard man to keep a secret from. If you met him, honesty would be the only policy. He just might hook you to his lie detector equipment. The graphs on the machine tell him what is happening inside your body. When you lie, there are small changes in your body that can reveal you're not telling the truth. Cleve is so expert that he set up the lie detector system for the U.S. Central Intelligence Agency—the famous CIA—and also for major police departments in several countries. But the biggest secret he ever uncovered was not pried from a shifty-eyed burglar or a cool counterspy. The biggest secret Cleve ever found, a secret that has led to a widening mystery, came from a tall, broad-leafed plant in his office.

One evening Cleve remembered to water his office plants. He decided to hook up his lie detector to see what happened to the electrical activity in a plant as it soaked up water. He noticed changes. Then he wondered idly, "What would happen if I burned the plant?" Before he could reach for a match, the machine registered a huge reaction, as though it were saying, "Alarm! Alarm!"

The graph was similar to that given by an alarmed person. "How could the plant possibly know what I had in mind?" Cleve thought. "In mind!" It hit Cleve that he hadn't done anything. He simply *thought* about hurting the plant. Plants can't read minds. Or can they?

Cleve now has hooked his machines to hundreds of plants. All of them seem to be telling him the same extraordinary secret. Plants are a lot more wide awake than they look. They react and respond to people's acts and thoughts. They react to dogs, cats, and even to shrimp. Plants, it seems, are part of the great big communications system that connects all living things. Like most of us, they give a big reaction when they "sense" they're going to be attacked.

Cleve, the hard-boiled crime fighter, now insists that plants be treated lovingly. He apologizes if he damages a leaf. But because his work is crime detection, he did once break his rule of kindness and plot a plant murder. He wanted to see if a plant at the scene of a violent crime could be used to give detectives a clue.

Six people came to help. Cleve folded six pieces of paper and shook them up. One was marked X; whoever picked it would do the terrible deed and murder a plant. Silently each person drew a slip. Not one changed his expression. No one except the murderer knew who held the fatal X. One by one, each went into a distant room where two (unsuspecting?) plants sat rooted in their pots. After everyone had a turn, Cleve opened the door to the room. At his feet lay the victim, a mashed—mugged—vegetable. Someone had ripped up the plant by the roots, twisted and torn it and smashed its pot. A real whodunit situation.

Cleve connected his equipment to the second plant, the witness. Each suspect was to return to the scene of the crime separately. The first man entered. Cleve studied the graph showing the plant's electrical activity. Nothing much happened. It recorded a steady line. Another suspect entered, then another. Cleve wondered if his witness would be any help. Suddenly the recording line shot right off the graph. Cleve gazed at the man in front of him. He didn't look any more or less guilty than the others.

After all the suspects filed through, Cleve decided on a showdown. The plant had only reacted to one of them. Cleve went up to the man. "You," he said, "are the murderer!" After a long moment, the man produced a slip of paper. It was marked with a large black X.

The Telltale Geranium

There is an old saying that daisies don't tell, but maybe geraniums and other kinds of plants, maybe even daisies, *will* tell on you. They might give away one of your secrets. But if a plant ever does tattle, don't be too angry. It will be only because the plant is sympathetic to your feelings.

If you think that's an odd idea, you agree with the people who worked near Dr. V. I. Pushkin. He's a scientist in a teachers' college in Moscow.

Everyone in the building heard that something weird was going on in his laboratory. If you peeked through a crack in the door, at first you'd see what looked like a homey scene. Tanya, a college student, sat in a comfortable armchair. Nearby, on a table, stood a big, healthy, red geranium plant. But then you'd notice wires hooked to the geranium leaf and running to complex equipment in the corner. You'd see Tanya begin to look upset. Her face was clouding over. She shivered and started to breathe faster like someone frightened.

Tanya *was* frightened. A hypnotist was talking to her. He held her in trance. For the moment, his words were her only reality.

"You're outside, a cold wind is springing up, you have no coat. Clouds, low, black clouds are gathering, shutting off the sun," he told her. "There is nothing around, no place to run for shelter. Now a man, a dangerous, desperate-looking man appears. He's coming toward you. Closer . . . closer . . ."

Tanya shrank back into the chair, growing more upset. But the scientists in the room weren't paying the least attention to her. Their eyes were glued on the geranium—or actually to the recordings coming from the machine hooked to the plant.

As Tanya became more distraught, so did the electrical activity inside the geranium. It seems hard to believe a plant would "sympathize" with a person. Still, the scientists had to admit that something very strange was happening.

Between sessions, the geranium always showed a steady recording of normal plant activity. As soon as Tanya, with the help of the hypnotist, began to feel strong emotions—joy, sorrow, anger—the geranium reacted. It mirrored her feelings as if an unknown bond existed between them. That's just what the scientists sought, a bond between plant and person. They decided they'd found one. They couldn't say what this invisible connection is, but if it exists, there are dozens of ways to use it. Could you, for instance, find out a person's secrets without her knowing you were checking on her?

The Russian scientists attached a lie detector, not to Tanya, but to the geranium. They didn't bother to hide the set-up, but they could hide it in a real interrogation.

"Tanya," they said. "Pick a number from one to ten. No matter how much we ask, keep it a secret. Don't let us know what it is until the end of the experiment." This time Tanya was her own normal self and not hypnotized. She chose a number.

"Is it four? Is it eight? Five?" They questioned her from one to ten, over and over.

Each time, Tanya said firmly, "No." Finally they stopped. Did they get a hint from Tanya? The scientists agreed they hadn't the slightest idea which number she was thinking of. They checked the recordings from the geranium. Had it picked up some burst of emotion from Tanya when she lied? Something they hadn't noticed? The only number the plant reacted to was five.

She confessed. Tanya's secret choice was—"Five."

3 Knowing the Future

Knowing the Future

In ancient times, every king, queen, general and anybody else worth his salt had a seer, a person who tried to peer into the future and tell us what was going to happen. Some of these so-called prophets were, no doubt, just very clever talkers. One, on the eve of a battle between his nation and another, said, "When the battle is over, O King, a great empire will have fallen." The king took it to mean he would win. When he lost, he couldn't say his seer was wrong. A great empire had fallen, his own.

Other seers probably were worth their pay and did get a glimpse of the future. When psi investigators collect ESP experiences, they find, strangely enough, that the most common sort is a flash from the future, particularly in dreams. Scientists call this "precognition," which literally means to know ahead of time, to know directly without the use of reason. The experts can't explain precognition, but they're finding out some things about how it works. Most correct predictions don't involve the crash of a jumbo jet, the cure for a terrible disease or any such headline event. Most involve small, personal glimpses of the future, often unimportant ones.

A classic case concerned an English Lady who dreamt that a pig waddled into the parlor during prayers. The next morning, as the Lady and her staff said their prayers, for the first and only time, an escaped pig wiggled into the house and burst into the parlor. It was a funny little look at the future that didn't do anyone much good. But sometimes these quick glimpses might be of great help. Not long ago tragedy hit a mining town in Wales when a huge mountain of slag from the mines fell on the schoolhouse and killed many children. Later, it was found that before the disaster, a number of people dreamed or had a quick vision of some detail of the tragedy. A single glimpse wasn't enough to let anyone realize what was coming and shout the alarm. But what if they had all been put together? Since then, premonition bureaus have started in London, New York, Toronto, Los Angeles and other cities. If someone thinks he's had a glimpse of the future, he mails it in. This puts him on record as having made the prediction. It also allows people to look over many future glimpses and see if any pattern appears, if some big future event is casting its shadow on present minds.

If ESP suddenly pops into your life, it's likely it will involve precognition. On the other hand, this area of psychic research has, until recently, faced the most hard-nosed resistance from scientists. Why? The problem lies in our idea of time. St. Augustine said, when asked to define time, "If no one asks me, I know; if I want to explain it to a questioner, I do not know." He had a point. When you really think about time, you see that it is not a simple thing. There's clock time, there's inner time—the kind that goes fast when you're enjoying yourself and slowly when you're doing something unpleasant—and there are a great many other kinds of time. In fact, time from a philosophical and scientific point of view is probably the most complicated thing you can talk about.

In the first half of the century many scientists would not admit the possibility of precognition, because it did not fit in with the way they thought the world was put together. But the way scientists see the world changes. With new theories of physics and consciousness, the ability to know the future is not ruled out. That's one reason that the current research in precognition has gone ahead. The other is that, theories or not, it seems that some people—psychics, business people, inventors, in fact, all kinds of men and women—do once in a while get a look at the future. And that's a very handy thing.

Shadows of the Future

"My son, my son!" the young mother shouted herself awake in the middle of the night. Still half in the grip of her nightmare, she tried to explain to her groggy husband, "I dreamed I was on a long, desolate beach. No one was there, just sand dunes and the waves rolling in one after the other. But our son was *buried* under the sand. I knew he was. I kept digging and digging, but I couldn't find him in all that sand . . ." and she broke off sobbing.

"It was only a dream, a bad dream," her husband reassured her. "It's okay." Finally to calm her, he led her into the room where their three-month-old baby boy lay sleeping gently, safe and snug under his covers.

For days the dream haunted the young woman's thoughts. It had been so real, so vivid. She'd heard of dreams coming true. Was something going to happen to her son? But the baby thrived. Eventually the dream faded. Once in a great while the mother remembered it and wondered at her fears. After all, her boy was growing up now and going to school.

Years later, the mother had something much worse than a dream to worry about as World War II mushroomed across Europe. Like millions of other young men, her son left home for the battlefront. He was a soldier in the German army. At first, his parents heard from him; then his letters stopped coming and they learned he was a prisoner somewhere in France. When the war ended, they hoped at last they would get some news. Was their boy dead or alive? Still, no word came.

Finally, the parents located two of their son's friends who'd been with him in the war. He was dead. They had buried him themselves in a deserted place in Northern France. They drew a picture of the spot—a strip of beach, an X where they'd left him. He lay buried under the sand dunes overlooking the sea. He lay in all that sand, just as his mother had seen him 26 years before!

Dr. Hans Bender of the University of Freiburg collected this account and many like it. He wanted to see if people had foreseen the terrible war that almost destroyed Europe. Apparently, many of them did but in small, personal ways. Nothing in the mother's dream would make her think of war. Bender wonders what could have been learned, and perhaps avoided, if her dream was put with others that showed a husband in uniform, a brother peering from behind bars, a family stuffed into a railroad car.

Dreaming of a White Future

Dreams of the future have been reported throughout history and a few are so remarkable that it's hard to think they are mere coincidence. Still, scientists like to prove things in the lab. But how do you corral a dream of the future in the lab? One good thing about psi research is it makes you call on all your creativity to come up with a way to catch ESP at work. In the dream lab at Maimonides Hospital in Brooklyn, Drs. Ullman and Krippner and their team found a clever way to see if people really dream ahead.

They wondered if Malcolm Bessent, a young English photographer, could dream, say on Tuesday night, something that would happen to him the next day, on Wednesday. Could he pick up a piece of his future?

Malcolm went to sleep in the laboratory. No one tried to send him telepathic dreams. They just let him sleep and have his own dreams. All through Tuesday night, whenever the machines showed Malcolm was dreaming, the scientists woke him and asked him to tape record a report of his dream.

In the morning, Malcolm got up in the lab and had breakfast. The scientists sealed the records of Malcolm's dreams and mailed them to another place. Then they went home and a fresh team that did not know Malcolm's dreams took over.

Now what? The researchers could follow Malcolm around all day recording his experiences, and then later, check them against the dream records of the night before. But that wouldn't have been a very good test. After all, Malcolm remembered his dreams, and he might have made one come true, consciously or unconsciously.

What if the scientists thought up some odd experience and sprang it on Malcolm? Later they'd be able to determine if he'd gotten wind of their plot ahead of time in his dreams. This idea was on the right track, but they would have to come up with a plot by chance—something they didn't consciously choose—something they didn't know either until that moment. If they'd had a plot in mind beforehand, some people would say that Malcolm might have read their minds in his dreams. That would certainly be an extraordinary demonstration of telepathy if it ever happened, but it wouldn't help prove that you can dream the future.

The researchers turned to an interesting little book, which lists things found in dreams. Using a complicated method, they chose, by chance, one of the items. It turned out to be *parka hood*, the sort of hood found on snow jackets. This is a very uncommon thing to see in dreams, but parka hood would be the theme for the experience the scientists were plotting for Malcolm.

Next, one researcher sorted through several hundred pictures kept in the lab to find one that he thought went with parka hood. He chose "Walrus Hunter." It showed a sculpture of an old eskimo man. The researcher had just an hour to rig up "an experience" for Malcolm that would relate to parka hood and walrus hunter.

Malcolm walked into a room completely draped in white sheets. Two blue fans representing icebergs blew cold air at him. A color organ provided "northern lights," and he heard the national anthem of Finland, a very cold country. The researcher draped white towels over Malcolm. He fashioned him a white hood like that worn by the old walrus hunter.

Suddenly, Malcolm felt ice cubes rolling down his neck. The researcher grabbed his hand, plunging it into ice water. "Smell the ice, taste it," he ordered.

Malcolm had an experience, all right. Did his dreams Tuesday night give him a clue to what was coming on Wednesday, something no one in the world knew?

Awakened after his first dream the night before, Malcolm had said, "I was just standing in a room, surrounded by white. Every imaginable thing in that room was white." In another dream he had reported, "I was talking to a black man . . . White-grey hair . . . He had an old face. Blues. Ice blue and white . . ."

The series of tests with Malcolm seems to indicate that you *can* dream ahead at times and glimpse at least a bit of your future experiences.

Do you ever dream ahead? If you want to know, keep a pen and paper by your bed. Before you go to sleep, tell yourself you're going to remember your dreams. When you wake, immediately jot down any dreams while they are fresh in your mind. After a few weeks, go back over your records. Did any bit of the future show up in them? If so, try to figure out if there was anything different about those dreams compared to the others you had. Remember, not all dreams come true. Most of them have nothing to do with precognition.

If predicting the future generally is your interest, beware of what's called a self-fulfilling prophecy. Watch out if you dream, or if someone predicts, that you'll have trouble learning French next year. You may have trouble with French—not because it's your fate—but because you took the negative suggestion to heart and worked at making it come true. Make sure you're not unconsciously making a dream come true—unless it's a good one.

The King's Murder

Florida psychic, Anne Gehman looked at the photograph she'd been handed by a group of business executives. It was a portrait of a man and there seemed to be nothing unusual about it. Suddenly Anne's hands went to her throat. Her face twisted in pain. She touched her lower jaw, her neck and her chest. "I'm sorry," she gasped. "This man is threatened by forces over which he has no control. This man is going to die—and very soon."

The group of executives looked stunned. The photograph was of their King, Saudi Arabia's King Faisal. These high-ranking businessmen from the oil-rich Middle East regularly consulted Anne Gehman for her amazing psychic powers in helping them find new oil wells. She'd had an extraordinary record of finding oil for numerous companies which had greatly impressed them. They certainly hadn't expected any sort of chilling prediction like this about their King.

"It was March 18th," a cousin of Faisal's recalled.

"How could she feel this?" the others wondered.

The King was not ill in any way, and he had no symptoms in any of the parts of the body she had pointed out. Anne Gehman felt that the pains were coming from the King's picture to her body. There were several sharp, agonizing pains and then burning sensations, she said.

One week later, on March 25th, 1975, King Faisal was shot and killed by his own nephew who had gone mad.

Later, it was learned that the areas of the body that Anne Gehman had mentioned, the throat, face, neck and chest, were the exact location of King Faisal's fatal bullet wounds.

Was the king destined to die? If we can predict the future, does that mean it's fixed and can't be changed? Most psychics think it is partly but not completely fixed. Psychics seem to see what's going to happen if things continue as they are. Good psychics say that if you change your actions or your thinking, you can change the future.

The future seems more fixed when it involves events on a grand scale, like world war or famine. Even then, most great psychics insist that the future is not laid out, and then laid upon us by some cackling Fates. It comes from us, from the conscious and unconscious behavior of masses of people. Change is possible, they say, if improbable.

Weird Warning

Sometimes it's hard to take a hunch about the future seriously. It can be worth fighting for, however, as a cabful of people found out. They sped through the night darkness of Palm Beach, Florida, late one evening in January, 1968. Manufacturing executive, John Myers, a passenger in the cab, suddenly experienced a strange, eerie sensation. There was no sign of anything wrong, but abruptly, from nowhere, he heard a voice calling his name, "John, stop . . . danger."

The other two passengers in the car, his co-workers, were talking animatedly and seemed to sense nothing unusual. Everything was quiet.

Suddenly, an invisible force seemed to push Myers right out of his seat, and he began shouting, "Stop, stop, there's danger ahead! A train is coming!"

The others gaped at him in astonishment. "What train?" They seemed to be on a clear open road. "We don't hear anything." The driver didn't hear anything, either.

But the conviction grew in Myers' mind that their lives were in immediate danger. He shouted again at the driver, "Stop!" The cabbie looked at his strange passenger with confusion but kept going. Myers leaned over from the back seat and grabbed the steering wheel from the driver. Startled, the cabbie applied the brakes.

They stopped with a jerk that lifted them right out of their seats. Just in time! A freight train whooshed out of the darkness and clanged across the road directly in front of them. There was no warning signal from the train. The train gates on the road that should have gone down were jammed. The red crossroads warning lights were broken. Had it not been for their split-second stop guided by the strange voice John heard, they all would have been killed.

Across the Time Barrier by Chair

How would you feel if you decided on the spur of the moment to go out somewhere—to a sports event or a concert or a lecture—and after settling into a seat, you discovered that weeks in advance a psychic had foretold that you would sit in that exact seat? What if the psychic had described all sorts of personal details about you in advance? Well, that's been the uncanny experience of hundreds of people all over Europe and the U.S. who've been involved with breaking the time barrier by "chair."

One of the most fascinating scientific tests ever developed for psychically probing the future comes from the world's first government-supported ESP lab in Holland. It's called the "chair test," and it was created by Professor W. Tenhaeff, head of the parapsychology department at the University of Utrecht.

Scientists pick, by chance, the numbers of several unreserved seats for a future event—a lecture, concert or play. Then they give the seat numbers to a psychic.

"Tell us on tape, something about the people who will sit in these seats at the lecture next week," they say to the psychic.

Does it work? The chair test has given astounding results in hundreds of repeat tests. Here's what happened at a recent transatlantic chair test. The famous Dutch clairvoyant, Gerard Croiset, was the psychic. Often called the man with the "radar brain," Croiset is sought throughout Europe for his extraordinary ability to solve crimes and to track missing persons. Police in Europe and the U.S. have used his help frequently.

The American chair test was organized by scientists in Denver headed by Dr. Jule Eisenbud of the University of Colorado.

On January 6th, 1969, scientists in Utrecht, Holland, made a film of Croiset. Who would sit in certain seats at a lecture in Denver three weeks later? Croiset's psychic impressions began to flow. When the psychic finished talking, the Dutch sealed the film and airmailed it to Denver. It would not be opened until the night of the meeting.

At 8 p.m. on January 23rd, about a hundred people milled about in the lobby of the International House in Denver. The doors to the auditorium opened and the crowd went down the aisles sitting where they pleased. They were going to see a film, the film made by Croiset three weeks before, thousands of miles away in Holland. Dr. Eisenbud looked at the crowd sitting there expectantly. Would any of them get a surprise?

The film rolled. People sitting in specified seats suddenly began hearing themselves talked about by this strange man in Holland.

"The first person is a lady about 5 feet 6 inches tall," said Croiset on film. The first person *was* a woman, 5 foot 6.

"She is dark-haired or wears a dark beret." The woman was dark-haired and she wore her hair in a style that did resemble a beret.

The woman had recently viewed her newborn grandson at the hospital nursery. Croiset accurately described how she'd pressed her nose to the glass.

Croiset went on. "The lady recently experienced some emotion connected with page 64 of a book."

The woman recalled this incident vividly. She had bought a book called *The Cat You Care For*. When she came to a section on putting old cats to sleep, it reminded her of a very upsetting day when she'd been forced to have her daughter's favorite cat destroyed due to age and sickness. It remained a sore point between her and her daughter. The offending page was page 64.

Croiset described a man in the audience, not only revealing personal details, but also his notes involving important research. "The man is a physicist with a firm doing atomic energy work for the government," said Croiset on film. He described his height, coloring and a single gold tooth in his lower jaw. "This man is partly in science, partly in business, and often works in a laboratory." This was confirmed by his personnel office and co-workers.

The Dutch psychic continued. "He had something to do with green material, perhaps paint or a chemical, which has made a stain on his coat." The physicist worked with a greenish yellow radioactive chemical that often stained his lab clothes and had a greenish cast under the lab's fluorescent lights. Again co-workers, two physicists and a chemical engineer, confirmed.

Croiset went beyond exteriors. "I see something like a roulette wheel. I see a ball rolling—it is on number 24." For his hobby, the study of intra-atomic forces, the physicist kept a special notebook in which he'd drawn a sketch on January 17th. If he'd finished the sketch it would have formed a complete "roulette wheel," he explained. The rolling ball and number applied to a sketch drawn on the previous December 19th. His family later produced the notebooks as proof. There were the very sketches described by Croiset.

How does Croiset do it? Working weeks ahead of time in Holland, Croiset revealed information about a physicist and a grandmother in Colorado. Neither knew Croiset, and neither knew they would be targets in an experiment. What led those people to sit in those particular seats? Do elements of one's self extend beyond the space-time coordinates of earth? Again, what do we mean by time?

Churchill's Hunch

Night after night, the enemy airplanes swarmed over the city. Burning buildings glowed everywhere against the black horizon, as far above, beams from huge searchlights crossed the sky. Now and then, the lights would catch the planes swooping like evil bats to drop their terrifying bombs on the seemingly helpless people below. But this was London during World War II and the people weren't helpless. Rallied by Winston Churchill, their Prime Minister, the people swore to withstand this nightly pounding.

One evening, Churchill was conferring over dinner with three other government leaders at his official house, No. 10 Downing Street. No one was surprised when the air raid sirens sounded. The government leaders kept right on talking, and in the large, unusually built kitchen with a 25-foot glass wall, the staff kept on preparing dinner. Government emergencies—bombs, or not—the Prime Minister was known to like a good meal.

Suddenly, Churchill excused himself. He called his butler. "I want the staff to bring dinner into the living room immediately. They can use hot plates to finish it and serve."

Shaking their heads a bit at this odd request and grumbling about the extra mess, the staff nevertheless quickly did as Churchill instructed, they brought everything into the living room. Three minutes later, after it was all arranged there, a great whistling roar cut off everyone in mid-sentence; a deafening explosion rocked the house as a bomb smashed directly into the kitchen, blasting everything to smithereens.

Churchill had saved his staff from certain death and perhaps his friends and himself from injury.

How did Churchill sense the danger ahead of time? His staff didn't know; they were just thankful their Prime Minister took time out from pressing government concerns to listen to a hunch urging him to *get those people out of the kitchen, right now!*

Many national leaders were to some extent psychic and interested in the paranormal, such people as Franklin D. Roosevelt, Mackenzie King, Prime Minister of Canada during World War II, and Abraham Lincoln, to name a few. Generally, highly successful people seem to have a special interest in ESP. Chester Carlson, the lawyer who invented Xerox, the copying process, said that ESP helped in his work. He wanted to know more about it and put up the money for the psi lab at the New Jersey Institute of Technology. It's not surprising that the New Jersey team began to explore the supposed ability of super-successful people to get a glimpse of the future.

Psychic Tycoons and their Hunches

Jesse Livermore—before the turn of the century, most people in America knew that name. Jesse Livermore made millions in the stock market on Wall Street. Many feared him. It wasn't his wealth that frightened them, but his unnerving habit of knowing what was going to happen ahead of time. He seemed to know when to expect the unexpected.

One day, Livermore abruptly interrupted a vacation and rushed to the nearest telegraph office. He sent his brokers an urgent message. "Sell short all of my stock in Union Pacific Railroad. Sell immediately!" Railroads were thriving in those days. The Union Pacific was bringing a fine profit. Maybe Livermore has finally lost his touch, the brokers thought. It was such an unreasonable request. Nothing could hurt the railroad in the near future; the stock was sure to go up. But they followed Livermore's orders and sold everything.

A few days later, the great 1906 San Francisco

earthquake struck. Miles of track belonging to Union Pacific were twisted like spaghetti; the railroad was almost wiped out. Everyone connected with it lost great sums of money, everybody but Jesse Livermore who took his profit of over a quarter of a million dollars and calmly returned to his vacation.

Livermore is far from the only American tycoon who used his hunches or intuitions to help him get to the top. Many highly successful executives use their ESP when in need, though they often don't call it that. Conrad Hilton, who built the famous international chain of hotels, often spoke of his hunches. He used them to submit winning figures in sealed bids, to buy hotels and stocks even when the facts pointed in the opposite direction.

Not everyone who puts intuition to work is as famous as Hilton. In the spring of 1971, the owner of a Brooklyn toy factory ordered his employees to increase greatly the production of stuffed panda bears. The toy man couldn't tell his employees why he wanted more pandas; he didn't know why, he just had a hunch. Over half a year later, people everywhere were surprised to hear that President Nixon would go to China. The United States had had no dealings with China for years; no American president had ever gone there. During his visit, the Chinese government gave Mr. Nixon two giant panda bears. With much fanfare, the animals were put in the Washington zoo and, of course, the demand for toy pandas soared. The Brooklyn toy maker was ready. He made a lot of extra money and finally understood what his hunch was all about.

Maybe the extra added ingredient that leads to great success is intuition or ESP. This idea occurred to Doug Dean and Dr. John Mihalasky at the New Jersey Institute of Technology. John is an expert in business management. He knew all too well that major decisions cannot always be made on the basis of facts alone. They must be made by what businessmen call "a good guesser." How do you find a good guesser who can pick up what's going to happen ahead of time? Doug once worked for a firm that gave employment tests. These worked fine to a point, but they didn't help choose people who might become top executives, the good guessers. Successful presidents of some large companies tried out the tests themselves. According to the exams, some were only qualified to be janitors; most of them wouldn't have been hired at all. Obviously, something was missing. There must be some X factor that goes along with training and experience to add up to smashing success. Let's find out if it's ESP, John and Doug decided.

"We'd like each of you to choose 100 numbers in a row," Doug and John said to a gathering of top business executives. "*Next week*, a computer is going to choose numbers for each of you. We want you to guess ahead of time what the computer will choose." It sounded like trying to win an impossible lottery, but the businessmen gave it a try.

John and Doug played their guessing game with more than 5,000 top level executives. They got a couple of surprises. The biggest surprise came when many of the business people said, "It's about time someone looked into hunches." These executives, the cream of the American business world had much more confidence in the existence of ESP than the average public. Why? Because, they answered, we use it!

Many executives did show ESP ability and guessed more numbers than they should have by chance alone. John and Doug tested the presidents of 25 companies. Twelve of them had more than doubled their companies' profits in the last five years. Eleven of these super successful men scored high on the ESP test. Those who did not increase their firms' profits at all did not show any ESP ability.

"If you're so psychic, why aren't you rich!" goes the old line. It looks like a lot of people are, only they don't call themselves psychic. They call themselves things like "president," or "chairman of the board." ESP, it seems, can help you succeed, not just in making dollars, but in anything you want to do. It wouldn't be a bad idea for all of us to pay more attention to how our hunches work—and how they don't work. Some business people are trying out the ESP guessing game as an employment test. They hope it may help choose future executives. Who knows? One day when you go in for a job, they just might check out your psychic ability.

The Time-Traveller

If psychic ability can give us a glimpse of the future, can it help us get hold of a piece of the past? Anyone can learn about the past in many ways, by reading, talking to people, visiting places. That's why it's difficult to prove that someone has used ESP to pick up information about it. Yet, there are enough cases on record to make you think that psychic abilities can be sighted toward the future, the present—or the past.

George McTavish is a time traveller. And he's already proved a boon to modern archeologists who search out the past. George visits, in a way that science cannot explain, ancient villages and sites. The places can be 500 or 10,000 years old. As though he were actually there, George "sees" how people lived in these long vanished times. Is he just imagining?

"George has been at least 80 per cent right on hundreds of tests that pitted him against archeological knowledge," says Dr. Norman Emerson, a leading archeologist of the University of Toronto.

One of the biggest problems in digging up the past is to know where to start shovelling. Recently, George stepped slowly through the stubble and bush of a piece of land staked out in Southern Canada. Archeologists believed that the remains of a Huron Indian camp lay underground and they wanted to find the palisades—a fence-like fortification. "You'll find it over there," George said, pointing to some low-lying land.

"That doesn't make any sense," one scientist objected. "Palisades are always built on the brow of a hill."

George, who knows nothing about archeology, insisted.

The men began digging. "We hit it within five inches!" they reported excitedly. "It's following the direction George said it would." Normally, they would have had to dig a 200-foot trench 5 feet wide to find one of these things, the archeologists admitted.

"This village is about 800 years old," George told them. After they did scientific dating of the things they'd found, the experts again had to admit George was right.

George has another psychic talent that is a big help to archeologists and a lot of other people. This is called "psychometry." It's the ability to hold an object and gather impressions about its past. It's as though everything from a watch to a battle ax carries an invisible record of its life history. People who do psychometry seem to be able to get this record to play back to them.

One evening, Dr. Emerson handed George a fragment of an Indian clay pipe. What could George tell him about it?

"This piece was thrown away. It was taboo—no good." The clay it was made of came from the American side of Lake Ontario. "They couldn't use it, but they valued it, and kept it in a well-thought-of place." George went on to describe the Indians who had made the pipe. Then he added that the pipe had a top on it resembling a human head. "There was a cap, a plug that went on top for a head." He drew a picture of the missing piece. "The lips stuck out toward the smoker."

Dr. Emerson confirmed it was indeed a fragment of a Huron Indian pipe from around Orillia, Ontario, dating from the 1600's. The pipe bowl was actually a fragment of a "pinch-faced human effigy pipe," widely used by the Huron Indians. George's sketch was absolutely accurate, though he personally knew nothing about these Indian cultures.

Says Dr. Emerson, "This marriage of intuition and science may bring us into a whole new era."

The Murderer Struck Again

Like our regular senses—seeing, hearing, touching—psychic abilities tend to operate together. Scientists go to great lengths to filter out one ability so they can study it. But in the hustle of everyday experience, it's often impossible to say which kind of psi is at work. What would you say about the following case—is the psychic seeing the past, reading an object, or reading someone's mind?

"Thirteen people murdered. Each alone at home. And we haven't one clue to help us," the police officer said. "The murderer always burns the house to the ground with the victim inside."

In a small town in Sweden, people were afraid to stay home alone, afraid to answer a knock on the door. A madman was on the loose. When would he strike again? Today? Or maybe tomorrow? No one knew and the police were desperate for leads.

Olaf Johnson, an engineer, had come to help, not as an engineer, but as a psychic. With Olaf were his sister and the two newspapermen who had pleaded with him to try to use his psychic ability to track the murderer. As he walked towards the police station,

Olaf wondered if he would get a friendly welcome. Psychics in many countries have helped solve crimes, yet some people still think of them as magicians who only trick people. Olaf didn't know then that years later, when he'd moved to America, he'd be chosen to receive ESP from Astronaut Edgar Mitchell in outer space. Now he had something closer to home to worry about, a killer who might strike at any moment.

"I'm so glad you've come," Officer Hedin greeted Olaf. The psychic felt relieved. Here was a young, friendly police officer assigned to guide him around. "I've read a lot about your psychic demonstrations and your work with scientists," the officer said as he started up the car. "It must be wonderful to have such abilities." He headed toward the first murder site.

Olaf stepped carefully through the fire-blackened ruins of the victim's house. Slowly, he began to pick up psychic impressions. In his own body, Olaf felt the same zing of terror that had seized the victim. He felt the bullet wounds and the hot rush of flames.

"Can you see anything?" Officer Hedin asked.

"I'm just getting feelings," Olaf told him.

"What about the murderer, can you get any feeling about him?"

In all honesty, Olaf had to say, "No, not yet."

For two days, Officer Hedin led Olaf to murder sites. The young officer did his best to cooperate with Olaf. Yet, each time he asked, "Do you get any psychic impression of the murderer? Any clue?" Olaf would shake his head, "No."

Finally Officer Hedin drove Olaf and his friends to

the last murder spot out in the country. He handed Olaf a charred rifle. "We found this here. We don't know if it belongs to the victim or the murderer." Standing in the crumbled remains of the house, Olaf took the burnt rifle. Suddenly he turned very pale. He started to sway.

"You see something!" Officer Hedin exclaimed. "What is it? Can you see the murderer?"

"I'm sorry," Olaf said, straightening up. "I didn't see anything clearly. It's just that I'm feeling sick. I want to go back to my hotel."

Olaf had reason to look pale. He'd seen something he couldn't tell the others. Alone, in his hotel room, he tuned in mentally to all the murders. It was as though he were watching a television series. He saw victim after victim open the door and invite in someone they trusted. Then suddenly the supposed friend raised his rifle and began shooting. In all thirteen cases, the same man raised the gun, the same finger pulled the trigger.

"He did it because he desperately needed money. He burned the houses to cover up the robberies," Olaf explained to his friends. "We've got to tell the authorities, even though they may not believe me." The authorities agreed to meet Olaf. But they also had another matter to look into. Officer Hedin was missing. Had the murderer struck again? In a way, he had.

"He would have killed us too, if I'd said anything out in the country," Olaf said. The person he mentally saw committing all those murders was Officer Hedin. The young officer's body was found in the river. He left a note admitting his guilt and saying he knew that sooner or later Olaf would find him out.

The Inventor and the Psychic

You can tune in to a picture of the past and you can discover telepathically people's thoughts about the past. Some psi researchers claim that you may also pick up thoughts about the past from dead people. It's hard to say just how the psychic in this case got her information.

Only a short time after the helicopter had been invented, a pilot from Rhode Island was hovering in a craft above a swamp. Suddenly something went wrong. The helicopter crashed and both the pilot and his assistant were killed.

Arthur Young, the man who had invented and developed Bell Aircraft's first helicopter, was very upset. Was his helicopter at fault? Was there something wrong with the design of his machine? Was the crash caused by some piece of equipment that had broken during flight?

Arthur Young went to the swamp where the crash had occurred and searched patiently for pieces of the helicopter. "What killed the pilot?" he wondered, as he hunted. He and his staff went through the crash fragments carefully, but they never found what had broken. "We couldn't find the answer," says Young. He had a feeling of tremendous responsibility about the new type of aircraft he'd invented. If there was something wrong with the design of his helicopter, it should be corrected before more people were killed.

Could a psychic help unravel the mystery? Arthur Young recalled that a famous medium from Britain, Eileen Garrett, who had been tested by many scientists, had been able to get ESP information about what had gone wrong when a newly designed British dirigible crashed in France on October 5th, 1930. The pilot, the only person who could have told the designers what had gone wrong with the aircraft, had been killed in the crash along with a crew of 45. Two days later, in a scientist's lab, Mrs. Garrett, who personally knew nothing about the design and mechanics of dirigibles, described in detail the technical problems that had caused the crash. "The whole bulk of the dirigible was . . . too much for her engine capacity. Useful lift too small; gross lift computed badly; elevator jammed; oil pipe plugged . . ." Her comments had also included all kinds of closely guarded military secrets that only the pioneer aircraft designers knew about. Experts at the Royal Airship Works called it "an astounding document." Moreover, the final reports on the crash investigation coincided with what Mrs. Garrett had said.

Could Mrs. Garrett help Arthur Young find out what had gone wrong with his helicopter? By now, Mrs. Garrett had become a prominent businesswoman in New York and the owner of a publishing company.

Arthur Young arranged a meeting with Eileen Garrett and brought with him a piece of the crashed helicopter's broken blade. She could often tell something about the history of an object just by touching it. She held the broken blade in her hand. "The machine was all right," she said. The crash, she felt, was due to pilot error. "The pilot was on the verge of a nervous breakdown and was about to enter a monastery when the crash occurred," she revealed.

A year later, Arthur Young went to Providence, Rhode Island, to see the man who operated the helicopter company that had hired the pilot. The conversation turned to the personality of the pilot. "At the time of the helicopter crash," the operator of the helicopter firm told Mr. Young, "that pilot was having a nervous breakdown and was about to enter a monastery."

4 Dowsing, Radionics and Eyeless Sight

Dowsing, Radionics and Eyeless Sight

Dowsing consists of psychically sensing at a distance concealed water, oil or minerals deep in the earth. This sensing is generally done by means of a dowsing rod—often simply two L-shaped metal rods. Modern research reveals that when people walk over underground water or minerals, their bodies react, and we can chart these physiological changes with modern medical instruments. Dowsing rods are, in a way, like biofeedback devices; they reveal outwardly the changes that are taking place inside you.

Dowsing has been known and used for centuries. Today, it's gone scientific. Utah State University has done much scientific dowsing research; Marines used dowsing in Vietnam; major corporations use dowsers regularly to locate lost items and buried cables; a course in dowsing is taught at Edison Community College, Fort Myers, Florida.

In the USSR, geologists call dowsing the "biophysical effect," and it is widely used in mining explorations. Soviet engineers think the human body is sensitive to changes in weak, electromagnetic fields and this is the means by which many of us can sense concealed water or minerals.

Dowsing can be enormously useful in many areas,

from agriculture to geology. One of America's foremost dowsers was Verne Cameron of California, and you'll find an account of one of his numerous dowsing experiences later on. In Europe, decades of research into dowsing led to the development of new forms of dowsing rods. (For how to make one new type, see our *Handbook of Psychic Discoveries*.)

When many dowsers pooled their findings, they came to the conclusion that each substance has a specific energy emanation or magnetic pattern. This led to the development of instruments that could be tuned to different magnetic frequencies to aid in detecting various hidden substances—a kind of higher-level dowsing. These devices are called Radionics or "Psionics" equipment, and they have opened up a whole new frontier for exploration. Researchers discovered that the radionics detectors could also sense things about plants, animals and humans, at a distance. It is possible that another energy spectrum exists, beyond the energies we already know.

George De la Warr was one of the pioneers in the development of Radionics, and you'll read about some of his experiments in one of the following reports.

Hidden substances and colors can be sensed not only with dowsing rods, but also with your skin, as you'll discover in the report on "eyeless sight." Each color has a different electromagnetic frequency and with practice, thousands of people have learned to sense the "feel" of colors without seeing them. Once people become aware of the subtle changes in their bodies caused by color and hidden substances, they frequently develop general ESP.

Eyeless sight was pioneered in Russia. After decades of research, the Soviets developed programs which trained blind people to become aware of these sensing abilities within them. The blind learned to "see" or perceive with their skin, and this transformed their lives. In America, a full-scale training program for the blind has been developed in Buffalo, New York, to teach eyeless sight. These new discoveries are not just for the blind; we all have vast hidden potential that can be developed with training.

The Motel Mutiny

It was the big Decoration Day weekend in 1957 and the city of Elsinore in California was jammed with guests who'd come to enjoy bathing in its famous, health-giving hot mineral waters.

But soon the sound of loud grumbling voices began to be heard in various hotel lobbies.

"This isn't mineral water!" they complained.

"This is plain old river water. I could have stayed at home for river water," grouched one discontented old man.

"They say the mineral wells went dry. They're pumping in Colorado River water," said another.

The news spread. "There's nothing health-giving about this low-grade water," they complained.

"I will neither drink it nor bathe in it."

There was a mass mutiny on the part of the guests. Within days, the tourists and their baggage filled the hotel lobbies—checking out. They packed up and left the hotels, motels and bathhouses in droves, never to return.

"What are we going to do?" Hotel owners were in a panic. "We could have had a million dollar hot springs. We could have become a resort like Palm Springs. But without the mineral baths, we'll have no tourists at all." Resort owners were facing bankruptcy. Other businesses were also affected. Without the tourists, everyone was losing money.

Three years went by. The city spent millions piping in water from other places. Finally, the Elsinore Valley property owners got together. They planned to get to the bottom of this water problem. Engineers were brought in. After months of surveying, the engineers reported that the wells had gone dry because the source that fed them had gone dry, though they didn't seem to know what that source was.

In desperation, the property owners turned for help to the famous dowser, Verne Cameron. Cameron had found scores of wells for people. Over several decades, everyone from movie stars to military experts had consulted him to find water for them.

Cameron had a unique theory on the source of water, and he had been expounding it for years, but the authorities never listened to him. This was a chance to prove his idea. "California doesn't ever have to be short of water," he insisted. "You see," he told a meeting of motel and hotel owners, "I believe that water from the ocean and lakes seeps down through faults in the earth deep into the interior and rises up again as steam. There's a vast underground supply of water beneath us."

"Go ahead and find it," they told him.

In the next twelve months, Verne Cameron did just that. Using his years of experience as a dowser and a dowsing device he'd invented called the Aurameter, he found eleven different hot mineral water wells in the area.

"This hot mineral water comes from condensed steam-water mixed with volcanic gas," he explained. "The source will never run dry."

Whatever his theory, the wells were there. Cameron checked out the city's original well which the authorities claimed had run dry.

"There's nothing wrong with the water supply," he announced. "But the well has a broken shaft." After the well was repaired, it pumped enough water to supply a major part of the city's needs.

Cameron went on to find fourteen more cold water wells for Elsinore. Four years later the wells he'd found were supplying the whole city with more hot and cold water than it had had in the 70 years of its existence. The property owners were delighted. In a short time, the tourists returned. The owners were out of the red and enjoying prosperity once more.

Photo Rapport

One day in August, 1965, in Oxford, England, at the Delawarr Laboratories, a 17-year-old boy called Rex was hooked up to a special machine—a Reflexograph—that would monitor moment-by-moment changes in his body. The machine plotted sound being transmitted through Rex's knee. Though he didn't know it at the time, Rex was going to take part in a strange transatlantic telepathy experiment with his own photo!

George De la Warr, the British researcher who organized the test, believed that a person has "rapport" with his or her own photo. Could a photo be a communication link with you even over thousands of miles?

Before the test began, Rex's photo had been sent to scientist John Hay, 3,000 miles away in Fairfield, Connecticut. They were going to see if shining a bright light on the boy's photo would cause any physical change that could be picked up by the Reflexograph.

A transatlantic call signalled the beginning of the experiment. At some randomly selected point after that, the scientist would shine a light on the boy's picture for 20 minutes. At the end of the test, the details were placed on record and filed in New York.

What happened?

At exactly the moment when the scientist began shining the light on Rex's photograph, far away on the other side of the Atlantic, Rex's body showed a clear-cut, dramatic change, as monitored by the Reflexograph. Rex himself didn't know what the experiment was about, or even when it would begin or end.

Says George De la Warr, "The shining of the light on the boy's photo apparently triggered energy releases in his body thousands of miles distant."

The Delawarr Labs carried out numerous tests with people and their photos separated by varying distances. Not only people's photos can be used, they found. The photos of plants or even of gardens also were found to be in "rapport" with the actual plant or garden. If the pictures were "treated" with magnetism or sound, the plants seemed to grow faster and bug-free. This discovery has found a practical application, as you will see in the next report.

Boxes and Bugs

Ed Hermann looked out the front window of his New Jersey house at the beautiful wild cherry tree on his lawn and began muttering under his breath. "Those darn caterpillars," he groaned. The tree was loaded with caterpillars, busily munching away. "There'll soon be nothing left of my tree."

For several years, despite all the pesticides he and his neighbors had used, every wild cherry tree in the district was being gobbled up by legions of tent caterpillars.

Ed Hermann was an engineer, who worked at the McGraw-Hill Publishing Company. He'd heard about an incredible new psi development—extermination of insects—at a distance and without pesticides. It was done in a very strange way by "treating" a photograph of the trees or crops with a magnetic broadcast from a small box-like apparatus the size of a radio. The photo was in some way a link with the real trees and crops. Some people called it a kind of ESP in instrument form. In the U.S. and Europe, scores of people were claiming that the "boxes" controlled the bugs on their crops and they no longer needed pesticides. In addition, the crops seemed to grow better.

"I'm ready to try anything," Hermann said, "if it'll save my tree."

He contacted Brigadier General Henry Gross, a well-known dowser and pioneer in long-distance insect extermination using magnetic broadcasts from patented "psionic devices."

"Put a few of your tree leaves and caterpillars in a box and mail them to me," Gross told Hermann. "Be sure to send a photograph of your tree, and you must include the negative."

It sounded weird, but Hermann posted his odd package to General Gross who lived 300 miles away in Harrisburg, Pennsylvania.

A few days passed, and then one day as he drove up to his home after work, Hermann jammed on the brakes with astonishment. There was a commotion in the grass on the edge of his property. It was marching hordes of caterpillars *leaving* his lawn. As he drove closer toward his tree, he couldn't believe his eyes. A full circle of dead caterpillars lay under the branches and leaves of the tree. Further off, swarms of caterpillars were scurrying away from the tree as fast as they could go.

Hermann rushed to report back to Gross, "Whatever you did to the cherry tree on our lawn was good. We don't have a caterpillar in sight!" This was during June and July. The previous year at the same time, they'd had to burn off the caterpillars with flaming kerosene torches. However these magnetic broadcasts operate, Herman concludes, they work.

Ed Hermann is one of many people who have tried the new techniques of Radionics/Psionics as a means of bug control without pesticides for garden or crops. Today, many thousands of acres of farmland in the U.S., from Arizona to Pennsylvania, are under pest control by Radionics. The Pennsylvania Farm Bureau reports good results on the crops they tested and notes that Psionics is not only effective, but far less expensive than the pesticides that also harm our food.

Dr. William Hale, former chief of Dow Chemical's Research Division, was highly impressed with Psionics, and wrote a report on its application in agriculture. He considered it the science of the future.

Brigadier General Henry Gross set up a Homeotronic Research Foundation in Pennsylvania and successfully exterminated insects at a distance on more than 90 farms in Cumberland Valley. Some of General Gross' research and equipment is now available from Mankind Research Unlimited in Washington, D.C.

ESP—A New Way for the Blind to See

The middle-aged woman walked purposefully through the ladies'-wear section of a large department store in Buffalo, New York. She stopped at a rack of dresses, moved her hands quickly past each of them, and picked out a few. She separated the dresses into two groups: patterned fabrics and single colors. Then she selected several dresses from each group that would match. It was a simple thing that anyone might do. Except this case was different. The woman was totally blind. Lola Reppenhagen had learned to "see" with her hands!

"I was very skeptical at first," Lola recalls, when she was first invited in 1973 to attend a Blind Awareness Project along with 19 other blind people. Psychic Carol Liaros planned to teach them her regular course in ESP. If the blind could develop their psychic abilities, she thought, they could overcome some of their visual handicaps; they could develop new ways of perceiving.

Carol Liaros stumbled onto the idea while teaching psychic development at the Human Dimensions Institute in Buffalo. One man had responded unusually well in describing things he "saw" psychically. Afterwards, Carol was astonished. "I've been blind since the age of two," the 41-year-old man told her.

In Project Blind Awareness, Lola and the others started with basic exercises in relaxation and meditation. Then came training in identifying colored paper, photos and small objects. Two scientists supervised testing. Finally, the blind would learn "mind travelling" to places they'd never visited. This would decrease their fear of new places and help them get around more easily. It was with "mind travelling" that Lola was later to amaze television and radio audiences.

The first results of Project Blind Awareness were extraordinary. Totally blind people learned to "see" colors with 70 per cent accuracy in over 2,000 tests. Professor Douglas Dean of the New Jersey Institute of Technology reported that the odds are millions to one that this could happen by chance. Psychologist Dr. S. Zieler revealed, "Some of the things that happened were mind-boggling."

After seven weeks of ESP training, some of the blind persons discovered they were "seeing" in a new way that they could hardly explain or understand.

Barbara Engel, 26, recalled, "One night at home I went into my bedroom and to my amazement the whole room was lit up in chartreuse (yellowish-green). I could see the outlines of the bed and dresser in the glow. And it wasn't coming through my eyes—it was coming through my head." She added excitedly, "The experiment might be the start of something . . . there could be a whole new world opened to blind people."

Another student reported, "I can walk down the street now without a cane because I can 'see' the plate glass windows, lamp posts and other obstacles 'out of the side of my forehead'."

Bill Focazio, who is blind and runs a men's-wear shop, was sitting in his store one day "When I realized I could 'see' things. Not clearly of course—just blurred outlines of the doorway and desk beside me." He covered his eyes with his hands and the scene was still there. "The images were coming to me through the side of my head." Perhaps we all can perceive in ways we don't fully understand yet.

The blind students could place their hands over a picture and describe it. "This is a woman with her hair pulled back . . . But what is that round ball on the top of her head?"

The picture was of Adelle Davis, the author of nutrition books, who wore her hair with a bun on top.

Other blind students suddenly could make out people across the room even in the dark. They saw them as patterns of light—as "auras."

Most fascinating of all were the "mind travelling" experiments devised by Mrs. Liaros. Several blind people were taken with other people to a building where they'd never been before. In a relaxed state, one of them "mind travelled" to a room in the building and described the room and furniture to the others. Then the whole group went up to the room to see if the description was accurate.

Classes for the blind have continued. Instructors' courses are given at Niagara University. In 1976, some of Lola Reppenhagen's "mind travel" journeys were filmed, taped and documented by Buffalo television and radio commentators, with some startling results.

Before going on the air with Jeff Kaye at WBEN radio, Lola was to travel mentally to his home and get an impression of the interior layout and decor. He was given her typed comments at the time of the interview. "The results staggered me," says Kaye. Her accuracy in placing objects in their proper place in the various rooms was nothing short of incredible.

"She even identified a picture of a small thistle on the wall of the living room and felt that this had a particular meaning for me. It does. It's a pen and ink drawing done by one of my children and presented to me on a birthday some years ago. It, by the way, is one of several pictures hung on that wall.

"I was astounded at the accuracy she demonstrated and quizzed my wife to determine if any strangers or even friends had been in the house and commented on that picture and the answer was negative.

"I'm totally convinced that her impressions were genuine and not gathered in any clandestine way," says Kaye.

5 Are the Dead Trying to Tell Us Something?

"Those who hope for no other life are dead even for this," said the great German writer, Goethe. If there's one question that everyone asks at some point, it's "Is there life after death?"

That question was in the minds of some scientists and scholars who gathered in 1882 at Cambridge University in England. Men of brilliance and wide knowledge, they started scientific investigation of psychic things when they founded the Society for Psychical Research. They would investigate telepathy, PK and many other psi experiences, but at the heart of their interest was the question of survival after death.

A few years before, two American sisters had astonished their friends, their relatives and eventually the world when they claimed that they regularly communicated with the dead. The girls provided information that questioners felt could only have come from their dearly departed. Communicating with spirits became a fad and, for some, a serious religion. Hordes of people consulted mediums. A dead person supposedly takes control of a medium's body and mind and speaks through her or him.

The spirits outdid themselves to convince people they were really there. They gave intimate, secret information to questioners; their pictures showed up on old-fashioned glass photographic plates; they sent heavy tables and large trumpets flying across scores of darkened rooms. It turned out that much of the spirit "proof" was rigged by clever but phony mediums, as still happens today.

As is also true today, there were mediums then

who were not fakes. A medium may be genuine, but that still doesn't mean she's putting us in touch with the dead, researchers decided. The medium could be getting the information telepathically. A medium might unconsciously be using her own PK to move objects. A so-called spirit who spoke regularly through a medium might be a split-off part of her own personality of which she is unaware. All of this is true and all of this continues to make it difficult to prove a medium is in contact with the dead.

Yet, some communications are hard to explain away, like the famous cross-correspondence cases that began after the turn of the century. A highly respected classical scholar, R. W. H. Myers, helped start the Society for Psychical Research. Not long after his death, outstanding mediums in America and England began to receive messages, supposedly from Myers. The mediums were disappointed; the messages didn't make sense. One might get a snatch of a poem, another a quote in Greek or Latin, a piece of a rhyme or a reference to a literary work. It finally dawned on investigators that somebody or something was playing a very complicated game of hide-and-seek with them. When the pieces were put together from the different mediums, they did make sense. Here were quotes, puns, clever allusions to classical literature. Sometimes, it took a couple of years to trace the original sources.

Myers kept communicating, and other dead scientists supposedly joined the act. Increasingly complicated psychic puzzles came from mediums widely separated from each other. Some people maintain that the mediums used unconscious telepathy to come up with these extraordinary puzzles. Others, after studying the massive records, decide it's less far-fetched to believe that the organizing intelligences behind the puzzles were the dead R. W. H. Myers and his friends.

From the 1930's, American psi researchers tended to steer away from the question of survival after death. They couldn't figure out a way to prove it, so they decided to tackle things they might prove, like telepathy. Today, there is once again a growing interest in investigating life after death and not just among psi researchers. People in many fields are beginning to explore consciousness and the many forms it can take, including its existence beyond the physical body.

As in other areas of psi research, people are turning to equipment. Some, as you'll see, claim they've already made the breakthrough and are getting messages from the so-called dead on tape recorders. Using machines to get through to the dead may sound odd, but it isn't new. Thomas Edison supposedly spent years trying to invent a sort of telephone to talk to the dead.

The study of the possibility of "life after life," as Dr. Raymond Moody has so nicely put it, is the oldest and newest area of psi research. For what it's worth, famous psychics have been predicting for over a century that we would learn in one way or another how to communicate with the dead between now (1977) and the year 2000.

People Who Came Back

Did you ever begin to drift off to sleep and suddenly feel that you were up around the ceiling looking down at your body on the bed? It's not something most people do every day. Yet, there are quite a few people who feel that on occasion they've slipped out of their bodies and wandered around as invisible beings. They don't usually talk about it. A special group of people, however, do talk. They want to share what they consider good news. These are people who have "died" and come back, people the doctors said were dead or near dead, but who revived and live to tell their story.

The wife of a Harvard professor lay in a coma, dying of pneumonia. The doctor and her husband left the room for a moment. Just then the woman felt herself slip out of her body. Rather surprised, she decided to follow her husband. He crossed the hall and paced back and forth in the library. Finally, he selected a book and read a few lines. The woman drew close and looked at the page. Her husband obviously wasn't aware of her presence.

It occurred to the woman that if she tried hard, she could get back in her body and live. She didn't really want to. She felt so free and healthy. A wonderful warm glow of happiness was beginning to enfold her. But her husband seemed so very sad. She decided to try. The woman recovered. When she was out of bed, she went to the library, took down the book and turned to the page she'd seen her husband read. "Remember reading this?" she asked him. He was astonished to find that she knew all his actions at a time when her body lay motionless in a coma across the hall.

A 19-year-old man drove through an intersection. He just had time to see the glare of headlights bearing down on him at high speed. Then he felt himself shoot through space. The next thing he knew, he was floating about five feet above the sidewalk a little way from his car. "I heard the echo of the crash. I saw people come running and could see my own body in the wreckage with all those people trying to get it out."

A middle-aged woman felt herself slowly rising upward. She had just suffered a bad heart attack in the hospital. "I drifted past the light fixture, then stopped right below the ceiling." She looked down and saw a dozen nurses come running. She heard a nurse exclaim, "She's gone!" Another started mouth-to-mouth resuscitation. "I was looking at the back of her head from above while she did this. I'll never forget how her hair looked." She watched them roll in a machine and begin giving her body electric shocks while frantically thumping on her chest. "I thought, why are they going to so much trouble? I'm just fine now."

Medical people have started collecting the experiences of those who have "come back." Interviewing hundreds of people, they've found something striking. Whether they are 5 or 95, the people report similar experiences. They suddenly find themselves out of their bodies. They have no pain. They feel free and mentally alert. Some are not quite sure where, if anywhere, they should go. They seem to have a sort of body—or at least a kind of form to get around in. If the experience lasts long enough, often a wonderful, glowing, loving light begins to enfold them. Those who reach this point say they honestly did not want to come back, but they did or we wouldn't be hearing their stories. All agree on one thing: the fear of death has left them.

Scientific exploration of death is a new field. Not much is known yet. The most impressive finding so far has been the similarity of the numerous reports of feeling whole and on one's own, out of the body. Are there different routes we might take to see if there are other worlds? Johnnie Duncan of Oklahoma had an experience—"As real as any in my life," she says—that may be a clue. Johnnie was almost thirty when she developed a bad infection after an operation. Her doctors said they'd never seen anyone pull through who'd been so close to death. In this borderland state, Johnnie had her experience.

She was boarding a plane for the White City. Her Uncle Jake was there and tried to get aboard with her. But he was turned back at the door. As they were about to take off, another man she knew, a Mr. Swyden, rushed up the ramp and got on. Flying high, Johnnie felt increasingly joyful.

On landing, she got off by a small stream. Johnnie ran over a little bridge to hug and kiss her Aunt Dove who waited on the other side. Johnnie knew she'd been dead for ten years.

After they talked a minute, she asked her aunt, "Who are the children with you?"

"Why they're my babies, Ima and Cameron," she said.

Johnnie didn't remember them. Aunt Dove pointed out a woman wearing a long white dress with ruffles around the collar and the sleeves. She wore an old-fashioned white bonnet. "That's your grandmother," she said. Johnnie turned to get her bags by the stream—and with a jolt she felt herself come back awake in the hospital.

Johnnie told her strange dream to those at her bedside. Her mother got so excited she started to cry. "Why your Aunt Dove did have two babies, Cameron and Ima! That was long ago and I didn't think you knew of them. We buried your grandmother years before you were born. But Johnnie, you saw her in the clothes she wore to her grave."

Then Johnnie got a surprise. Mr. Swyden, who hopped on the plane with her, had died a few days before. Her Uncle Jake was now recovering from a serious illness. Johnnie had been too sick to be told about either of them.

Deathbed Visions

The room was silent except for the heavy breathing of the sick man on the bed. Thomas Edison, the great inventor, was close to breathing his last. Suddenly, Edison roused himself and sat up on his own. He opened his eyes and gazed straight ahead for several moments. His wife, close to his bedside, looked at him with amazement, but said nothing. He seemed so intent. At length, Edison turned to her. "I'm surprised!" he exclaimed. "It's beautiful over there!" And then he died.

What did Edison see? Whatever it was, he isn't alone. A great number of dying people exclaim with delight over a beautiful place that they're suddenly able to focus on. But even more dying people insist they see something else. They see friends, dead friends.

That's what happened to 77-year-old John Thompson. His daughter Margret and a woman friend sat by his bed. John leaned back on a bank of pillows. He knew he was dying, but his mind was clear. To pass the time, he chatted about this and that with the women. Eventually, his eyes travelled toward the foot of his bed. John's face lit up like the morning sun. "Why, there's Emmy!" he said. "There's Bill!" A little later he died with a great big grin on his face. Years and years ago, the long dead Emmy and Bill were his closest friends. She was the bridesmaid and he was the best man at John's wedding.

Was John just imagining that his old friends had come for him? Or was something a little more interesting going on? Not long ago, Dr. Karlis Osis decided to see what he could discover about deathbed visions. He found that a great many dying people who are wide-awake and alert at the very end see dead friends and favorite relatives. As you can imagine, every one of them is overjoyed. The dead seem so real to them that a few people have even tried to introduce them to others in the room.

Imagination? Certainly, some kinds of illness and drugs can cloud your mind. But Dr. Osis found that when your mind is confused, you tend to imagine all kinds of strange things. You may think you're in the middle of a hot poker game or an emotional political argument. You don't usually imagine scenes having to do with dying—with going on to somewhere else. When you are mixed up by sickness and drugs, if you imagine people in the room, they are almost always people who are alive, not long gone dead people.

No one has yet figured out a way to prove beyond a doubt that your dead friends will drop in and pick you up when you go. But reports come in that at least make you think there are possibilities. Two girls, Jennie and Edith, both 8 years old and best friends, fell seriously ill with diphtheria at the same time. Around Wednesday noon, Jennie died. The family and doctor decided not to breathe a word to Edith; it might be too much for her. Saturday afternoon, Edith picked out a few of her favorite pictures. Would someone take them to the sick Jennie? She appeared to have no idea her best friend was dead.

At supper time, Edith woke from a feverish nap and talked of dying. She wasn't afraid. She spoke about several dead people she knew. She could see them, Edith said. Then with surprise, she turned to her family. "Why, I am going to take Jennie with me! You didn't tell me Jennie was here!" She leaned forward and reached out her arms to what looked to the others like empty space. "Oh, Jennie," Edith exclaimed, "I'm so glad you're here." Then she died.

Lady Barret, a London surgeon, watched another case that makes you wonder. A young mother was dying of heart trouble after childbirth. Her husband, her mother and medical attendants surrounded her. She looked toward an empty space in her room. "Oh, lovely, lovely," she said smiling. Then she noticed her dead father. "He's so glad I'm coming, so glad," she told the others. "Oh, and there's Vida," she said referring to her sister. Again she began telling about the place she was seeing. "You can't see as I can. It's lovely and bright."

Her father wanted her to come. Looking puzzled, she exclaimed again, "He has Vida with him." She turned to her mother and said strongly, "Vida is with him!" Finally, she said, "You do want me, Dad; I'm coming." And died.

A typical deathbed vision you might say. But there's something more: that's sister Vida. Vida, an invalid for years, had died three weeks before. They hadn't told the young woman for fear it would wreck her health further. The family even opened her get-well cards to make sure nobody slipped in a mention of the death of her sister. Yet, she saw Vida in the "right" place, so to speak.

Reality or imagination? It's probably the biggest question in the world. If in your lifetime, human understanding doesn't come up with an ironclad answer, don't worry. You'll find out for sure, one fine day.

Ghostly Literary Agent

The young woman and her husband strolled slowly along a wintry street in downtown Buffalo one January evening. They were feeling discouraged and depressed. A publisher who had previously agreed to bring out the young woman's novel, had rejected it the day before. The pair felt that their fortunes were at their lowest ebb. As they walked, they happened to pass the Statler Hotel and saw a notice—there would be a psychic demonstration that evening at the hotel by a medium from England called Charles Nicholson.

Neither of them had ever heard of him and they didn't believe in spiritualism, but as a distraction from their gloom, they decided to join the crowd inside. Charles Nicholson was on the platform, giving messages from "the beyond" to various people in the audience.

"All fake," they were thinking to themselves.

"Did anyone here have a father named Arthur?" Nicholson asked. "I seem to be getting a message from him."

The young woman slowly raised her hand. "My father's name was Arthur," she said.

The medium continued. "Don't be discouraged. That's what your father wants to tell you. He knows you've had a bitter disappointment about the manuscript. But he wants you to know that the manuscript you have written will be published and that your book will be a great success."

She listened in disbelief. The medium went on to give more specific details. "Your father wants you to know that the manuscript will be sold on April 2nd this year to another publisher. It will make you famous as a writer all over the world. In addition," he continued, "your father says that in another year, after the book is published, it will be made into a motion picture. You will go to California to work on the movie script."

How could the medium know she was a writer and how could he know the details of her recent manuscript? Still, the young woman thought, it sounded outlandish.

On April 2nd, 1938, the young woman did sign a contract for her novel with another publisher. Her book, *Dynasty of Death*, made its author, Taylor Caldwell, world famous. The novel became a best seller and was translated into many languages.

One year later, Hollywood decided to turn the book into a movie. Taylor Caldwell and her husband found themselves in California, working on the screenplay of her novel.

The Dead Man Kept Writing

As she sat at her desk in her London home, Grace Rosher's thoughts followed the great ocean liner, the *Queen Mary*, gliding across the Atlantic to North America. On its return, the ship would carry the most important person in Grace's life, Gordon Burdick. We are really going to be married, Grace told herself, really. It was hard to believe. They had waited forty years to become husband and wife. All those years ago, Grace, as a young woman, took a trip to Canada, to Vancouver on the Pacific coast. There, she met Gordon and they fell in love. Then, family and business duties forced Grace to return to London and Gordon to stay thousands of miles away in Vancouver. Now and then, they got together, but for most of the forty years they kept their love alive and shared each other's days through letters. Now, when Grace thought of a funny story she wouldn't have to run to her desk and start writing to Gordon. She would be able to turn and tell him.

They were going to be together—that made the unexpected telegram she received doubly harsh. Gordon had died in his sleep before he was able to board the ship. Now she wouldn't even have his letters to look forward to.

One afternoon after Gordon's death, a strange thought popped into Grace's mind as she finished some correspondence. "Leave the pen on the paper," a voice seemed to say. It sounded ridiculous, but Grace did. Her hand began to move. It moved automatically, without any conscious thought. At first, the pen made scrawls. A few minutes later Grace looked down. She'd written words. Almost to her horror, she read, "With love from Gordon."

Her subconscious mind was playing tricks on her! She'd heard of this automatic writing, supposedly a direct pipeline from the unconscious, but it took most people a long time to get the hang of it. Here, it suddenly came upon her, full blown. In the days that followed, even though she was sure she was fooling herself, Grace couldn't resist letting her hand move automatically across a blank page. Soon, she saw something even more appalling. Sentences were flowing indeed, not in her own bold handwriting, *but in Gordon's handwriting*. Once again, she was getting long chatty letters in Gordon's familiar, small, back-slanted script! As always, he wrote about where he was.

And where was Gordon? "Right beside you, when writing," he said. "I put my hand over yours to guide the pen." Why couldn't Grace see him or feel him?

Scientists will tell you that your whole world is made up of vibrations, Gordon explained. So is my world. Only we are on a much faster frequency than you are and you cannot usually sense us. This idea is similar to looking at an electric fan. When it's on, the blades move so fast they become invisible; we seem to see right through them. In a few more decades, Gordon believed, we will be able to get in touch with the dead more easily. "It won't be psychics that bring the breakthrough, but scientists interested in ESP."

It's comforting to get letters from Gordon, Grace thought, but I'm probably well on the way to being crazy. Such things don't happen. But what if she were actually hearing from a dead person? That would be important to say the least. Grace had always found spiritualist groups that tried to contact the dead rather distasteful. Still, she was religious, a member of the Church of England.

"Have courage!" she told herself and took her story to the churches' Fellowship of Psychical Studies. This interfaith group that included twenty Anglican bishops agreed to investigate. The more that people looked into them, the more puzzling the letters became.

The London Daily Mail got news of the ghostly letter writing. One day, Grace opened the door to a reporter and a photographer. By this time, she'd asked Gordon if he could think of something that would make it clear that she wasn't doing the writing. The reporter stared, the photographer managed to keep filming, as Grace's hand moved across the paper. She wasn't holding the pen. She made her hand into a fist so the upright pen could lean against it as it moved. Yet, the reporters swore they saw sentences flow, seemingly in Gordon's handwriting. They were overwhelmed.

The Churches' committee asked Grace to work with one of England's top handwriting experts, F. T. Hilliger. He compared samples of the real Gordon's and the ghostly Gordon's handwriting. "They are not similar," he announced. "They are identical." Even in expert forgeries, he explained, there are usually some slight differences between the real and the fake. Grace's letters were in Gordon's handwriting. "Scientifically, this is impossible!" Hilliger declared.

The Churches' committee, not an easy group to impress after their long experience in the psychic field, was impressed by Grace's unusual letter writing. Eventually, they published a long report with a glowing introduction by Sir Victor Goddard, Air Marshal of Britain.

It was a marvel. But still, couldn't it be a marvelous demonstration of the powers of Grace's subconscious mind? Mr. Hilliger, who'd become fascinated by the impossible letters, heard of an American woman from Virginia doing similar automatic writing. She, supposedly, was getting letters in the handwriting of her dead husband, a minister.

Grace and the American woman met in London. Hilliger, perhaps feeling rather foolish, asked the ghostly letter writers to please switch correspondents.

The women began to write automatically. They finished and Hilliger, the handwriting expert, picked up their letters. Grace's was in writing that looked like the dead minister's hand. The writing of the American woman's letter was identical to Gordon Burdick's handwriting when he'd been alive.

Did two dead men really switch correspondents to prove a point? If not, by what weird connection of mind did Grace and the American woman automatically produce handwriting unfamiliar to themselves?

Possession or Body Rental?

A pale February sun rode in the morning sky over Watseka, Illinois, as Mrs. Vennum stood at the bedside of her sleeping daughter. Fourteen-year-old Lurancy breathed peacefully. Thank goodness, there was no sign of the pain and restlessness that so often kept her tossing and pulling at her covers. The summer before, the usually healthy Lurancy suddenly started falling into strange trance-like states. She would be semi-paralyzed for hours.

"There are people in my room. They're talking to me," Lurancy said after her first attack. No one else could see anything. They decided poor Lurancy must be imagining things. From then on, with no notice, trances would overwhelm Lurancy. In this odd state she would talk endlessly, talking—she said—to spirits, invisible people, angels and heaven knows what. When she came to, grinding pains spread through her body. Night after night, the exhausted Mrs. Vennum would sit by the bed of her groaning daughter.

They didn't want to think about it, but eventually the family had to admit to each other that their Lurancy was insane. They could find no hint as to why she'd suddenly gone mad, but the evidence of madness was all too real. Finally, they decided that before everyone else in the house went mad too, they would have to send Lurancy to the state institution for the insane, an ill-kept, unhappy place.

Yesterday, they'd given Lurancy a last chance. A famous doctor from out of town came to look at her. "Bless him," Mrs. Vennum thought, he does seem to have helped. Watching her in trance, the doctor began giving Lurancy soothing suggestions. "Why do you trouble yourself so? You say there are spirits around you, why don't you let a happy, cheerful spirit come near you?"

Mrs. Vennum began to hope her daughter had been returned to her. Shifting in bed, Lurancy opened her eyes and gazed at her mother. Lurancy said, "Who are you?" Mrs. Vennum felt her hope depart.

From that moment on, one of the most mysterious questions on record became: Who was Lurancy Vennum? It was a question to everyone but the 14-year-old girl. "I am Mary Roff," she said firmly. She knew of the Vennum family, the girl said, but she didn't know Lurancy. Then she described to Mrs. Vennum where the Roff family lived in town.

Mrs. Vennum knew where the Roffs lived. She knew they had a daughter Mary. She also knew Mary Roff had died twelve years before at the age of eighteen. Her own daughter was not yet two at the time.

Lurancy never wavered in her insistence that she was now Mary. When the Roffs came to see her, she greeted them with delighted hugs and started talking about all the things they used to do. With the Vennums she was formal, as one would be if she suddenly found herself living with slight acquaintances. Finally, the Roffs agreed to let the girl move in with them. The only hope Mrs. Vennum had, as her daughter went down the walk, was that "Mary" said, "The spirits tell me I can stay until sometime in May."

As she went from the Vennums to the Roffs, Lurancy-Mary became known as the "Watseka Wonder." Well-known scholars came to study her. They agreed the townspeople weren't exaggerating. It was a wonder, but they couldn't explain how such a thing could happen.

Lurancy, now Mary, was delighted to be home again, as she put it. When old friends came by, she called them by nicknames. She seemed to remember everything that had gone on in Mary's life, no matter how people tried to trick her.

"Mother," she asked one day. "Do you still have that old box of my letters?" Mrs. Roff recalled where she put the box and dug it out. The girl opened it and found along with the letters a tatted collar. "Oh, I made this." She told how she'd done the sewing and picked the pattern. Mrs. Roff admitted she was right. After the first few weeks, the whole Roff family couldn't help thinking that Mary was back. She had the same personality, used the same gestures as Mary, and she remembered.

"Remember," Mrs. Roff once asked, "when the stove pipe fell and burned your brother Frank?" "Yes," said the girl, and described the exact spot on the arm where Frank was burned.

One morning, the girl said to Mary's sister, "Right there by the currant bushes is where Allie greased the chicken's eye!" An odd remark. But the sister recalled that years ago, on that spot, their cousin Allie treated a sick chicken by rubbing oil in its eye. Cousin Allie now lived in another town. Lurancy never knew her.

When the Vennums visited, the girl treated them politely; that was all. Then one day in late May, Lurancy's brother Henry dropped in. She threw her arms around him, greeting him just like a long lost brother. It seemed to her he was her brother, for she was Lurancy again and wanted to go home. For a day or so, Lurancy was like a person waking from a long dream.

"What happened?" she asked. Then she answered herself, "Oh, I know, Mary told me."

Lurancy regained her health. Eventually she married and moved to the country where she lived a routine life. No one knows where Mary went.

Voices that Shouldn't Be There

Peter Bander patted his dog Rufus' head and absentmindedly hoped the Great Dane would behave while he was out. That was the least of his worries, as he dressed to meet a very important person. The VIP was Sir Robert Meyers, a crusty old businessman who headed many English companies, including the publishing house where Peter worked. It was one thing to entertain the chairman of the board, a very different thing to do what Peter hoped to do. He couldn't quite believe it himself. Peter was hoping *to show Sir Robert that dead people can communicate with us through a tape recorder.*

Rubbish, nonsense—well, of course. That's what he'd said, too. Peter thought back to when a book called *Breakthrough* appeared in his office. It was written by a professor in Germany. This man, Dr. Constantine Raudive, claimed that he regularly got communications from the dead, in many languages, on his tape recorder.

At last, he said, with modern technology, the dead have a way to talk with us.

"And that certainly would be one heck of a breakthrough," Peter thought, "if it were true." But he didn't think it was true, in the least. He'd heard vaguely about "voice taping" as it's called. Suddenly popular in Europe in the late 1960's, it began with a Swedish film maker.

One summer day, the film maker taped some bird songs in the woods. Later, playing back his tapes, he heard the birds, but he also heard voices, voices that shouldn't have been there. They weren't just any voices, but men's and women's voices talking to him. They were voices he recognized, and strangest of all, they were voices of people who had died. Thunderstruck, the Swede started making more tapes. The voices that shouldn't have been there kept recording. He had made a documentary film for the Vatican. He'd been decorated by the Pope. Finally, the Swedish producer took his story and his tapes to Rome. The Pope and his advisors were interested. Eventually, a Swiss priest was allowed to devote his time to voice taping. Others began taping: scientists, psychologists, technical people. Perhaps the best known was Dr. Raudive, who wrote *Breakthrough.*

Well, Peter thought, a lot of people seemed to be getting voices that shouldn't be there. Maybe they were picking up radio waves. Maybe they were misguided, who knew? Tape recording of the dead—the whole thing seemed far removed from everyday life, and, as any sensible person can tell you, it was impossible. He didn't think the book should be published and forgot it.

Some days later, Peter's boss, the president of the publishing house, called him to his office. Peter probably groaned inwardly as he heard him bring up the book on voices. An enterprising man, he'd tried some taping. Peter started to pay more attention when he heard his boss say, "And I've gotten voices, too, voices that shouldn't be on the tape." Would Peter listen and give his opinion? Some of the voices spoke in German, a language the boss didn't know, but Peter did.

As the tapes wound round, Peter listened intently. All of a sudden, he was shocked into super attention. A woman's voice said in German, "Why don't you open the door?" His mother's voice. She spoke in her native German dialect. It was unmistakable. He knew her voice on tape because in years past they often corresponded that way. Peter's mother was dead.

Peter went home and plunged into voice taping. He found it demanded a great deal of time and patience. Days would go by without any voice appearing on the tape. Other times, the voices were blurred or spoke very fast. Peter noticed that his dog, Rufus, barked and his fur bristled whenever a voice sounded on the tapes. Rufus had never paid attention to anything else on tape, radio or television. Was Rufus, as dogs can, hearing a sound with too high a frequency for humans to hear? People often say that dogs react to ghosts. Was that what was disturbing Rufus?

Peter still found it hard to accept what was happening. So many people believed the voices were simply radio broadcasts picked up by mistake. No doubt, some of the 100,000 messages Dr. Raudive recorded were from the radio. Others are hard to explain as a wandering radio beam. The radio rarely addresses you by name. It rarely says, "Ah, you have on a blue sweater," when you do have on a blue sweater.

Peter felt a little more confident about meeting Sir Robert this evening because of what had happened recently at the Radio Frequency Screened Laboratory. Part of the British defense system, this lab was constructed so that no radio beams could enter. England's top man in radio shielding supervised a voice taping session. The engineers put brand new tapes on their recorders, probably thinking it was a waste of time. They knew no waves could enter the room, so nothing could be recorded. The engineers played back the tapes. Voices spoke, voices that not only shouldn't, but from their point of view, *couldn't* be there.

Only one thing seemed clear: there were voices. But what were they? Some researchers believe that they are examples of the psychic phenomenon of PK. They think a person at a taping somehow uses psychic energy unconsciously to impress a message on tape. Others think the voices might be from outer space. What do the voices say about themselves? A theme recurs in voices picked up around the world. One of the voices—supposedly that of a recently deceased woman—put it well. "Just think. I am!" Others speak more formally. "We are the dead, but we live."

Peter's company decided to tell the story of the voices. The deputy prime minister of England heard about it. He called the chairman of the board, Sir Robert Meyers: "You can't print such a book. It will ruin your reputation."

Sir Robert agreed, no book. It seemed this was one breakthrough that wouldn't become known to English-speaking people. Fortunately, however, Sir Robert thought seeing and hearing was believing. The book had a last chance.

That night, Peter went to a voice-taping session put on for Sir Robert. An independent recording company set up the equipment. Peter noticed there were about two dozen people gathered when Sir Robert and Lady Meyers appeared and were led to their seats of honor. The sound engineers solemnly explained that they were using factory fresh tape on four recorders that were shielded and grounded. They assured the audience that no sound could be picked up. They clicked on the machines.

The room was silent. Sir Robert sat stoically. His wife looked serene. Peter noticed a baffled look on the engineers' faces, at one point, when the dials revealed that impressions were being made on the tapes.

After 18 minutes, the tapes were played back. Sir Robert, Lady Meyers, Peter, the engineers and everyone else heard over 200 paranormal sounds. Of these, 27 were very clear voices. And Peter realized with relief that they weren't just any old voices.

"I am Artur Schnabel," a man's voice said in German. He gave some personal information to Sir

Robert and Lady Meyers. "The great pianist, Schnabel, was a dear friend of ours!" Sir Robert exclaimed. It sounded like his voice and they had always spoken German together—when Schnabel was alive.

Other voices called other people in the room. Everyone stayed late listening again and again to the tapes. Finally, Sir Robert said that he was 92 years old, not a time when you are apt to change opinions. But, he admitted, from that moment on, he would have to revise his ideas about life and death. "I'm relieved," he added, "that eternity does not mean being condemned to eternal inactivity."

Lady Meyers was glad that her husband had at last seen the light. The elderly lady smiled and said, "I knew it all along."

The book about the impossible voices was published. In England and now in America too, patient people seem to be getting more tape recordings of voices that shouldn't be there. The voices are happening. What they are remains to be proved. What do you think?

6 UFO's

UFO's

Do UFO's exist? All over the world, an increasingly large number of people report seeing UFO's. The reports come not only from Europe, the Americas and Australia, but from the Soviet Union and other Communist countries as well. Astronomers, pilots—even astronauts—say they've seen UFO's.

Today, some of the closed files on UFO's are being re-opened and serious scientific research is being carried out at such places as the Center for UFO Studies, P.O. Box 11, Northfield, Illinois 60093. This major center is headed by Dr. Allen Hynek, former Director of the Lindheimer Astronomical Research Center of Northwestern University. His book, *The UFO Experience*, is a good one for finding out what's been going on in the UFO world, and it reveals things about UFO's that the authorities have been trying to cover up.

The two accounts which follow, one from America, the other from Russia, have been thoroughly documented.

The UFO Souvenir

A couple, returning home from a vacation in Canada, drove carefully along a deserted New Hampshire road. It was the night of September 19th, 1961. Suddenly, Betty and Barney Hill noticed a large glowing object in the sky above their car. Barney stopped the car and observed the craft more closely with binoculars. "We could see windows in it, lit up. We were certain we could see figures in the windows peering out. It was definitely a craft and huge."

They got back in their car and sped home to Portsmouth, New Hampshire. When they got out of the car, they noticed strange markings on the car trunk. They noticed something else that was very peculiar. Two hours they could not account for had mysteriously vanished from their lives between the time of the sighting and their arrival home.

For a while they thought nothing more of it. Betty was busy as a social worker and Barney was active in civic affairs. Both were highly regarded in their community, and as they'd generally been skeptical about UFO's, they never discussed what had happened. But then they both began having terrible nightmares about their UFO contact. Barney developed an ulcer from worry. Finally they turned for help to Dr. Benjamin Simon, a Boston psychiatrist.

What had happened to the Hills during those mysterious missing two hours? Could Dr. Simon use hypnosis to remove their memory blocks about the experience?

Betty and Barney were each hypnotized separately and their reports taken down. Independently, each told almost the same incredible story. They had been abducted by space beings, into a UFO.

"They were about five feet tall," Betty reported. "They had grey, metallic-looking skin. No noses, just nostrils, and their mouths were slits. They had eyes that seemed to extend around the sides of their heads. They communicated with us by ESP and some English."

Betty and Barney were each given a physical examination while aboard the craft. They opened Barney's mouth and were fascinated to discover that his false teeth could be removed. Betty had to explain that her own teeth were real and wouldn't come out. They did some strange test which involved putting a needle into Betty's navel. Before she left the craft, Betty asked the space beings where they came from. They gave her a UFO "souvenir." They showed her a star map marked with trade and exploration routes. Later she drew a sketch of this map.

Dr. Simon was concerned mainly with clearing up the couple's nightmares. "Probably just a vivid dream that somehow communicated from one to the other," he said. But was it just a dream?

Months later, Pease Air Force Base in New Hampshire confirmed that on the night of this strange incident, military radar had actually picked up a shimmering mass, a "UFO" in the very area of New Hampshire where they said they'd seen the space vehicle.

Even more curious, astronomers recently used computers to check out Betty Hill's star map. They discovered her map shows the way things look from the star Zeta Reticuli. Moreover, it happens to be one of the stars in the vicinity of our solar system that scientists believe could support life.

Considering the incredible number of stars in the galaxy, scientists say the odds that Betty Hill's star map could match this pattern by chance are about 1,000 to one.

Big Bang in Siberia

Something had blown up; something had exploded with unbelievable power. Something had happened in the deep forests of Siberia that was truly out of this world. Early on a June morning in 1908, a tremendous, deafening blast jolted the townspeople in a Siberian village. The ground rocked. Wind rushed through the trees. The terrified people looked toward the deep forest where the blast seemed to be. They saw a gigantic, light-colored, mushroom-shaped cloud rising slowly into the blue sky.

Shock waves from the blast circled twice around the world. A glow of light from the explosion spread through the atmosphere and across all of Europe. It was so light that for three nights people thousands of miles away, in London and in Paris, were able to read without turning on their lamps.

What was it that exploded? What uprooted ten

million forest trees in one instant? At that time, nothing in this world could have caused such a blast. Therefore, scientists and everyone else decided it must have been a meteor or comet from outer space that crashed into the earth.

Research teams set out, but the going was difficult. The forests were unexplored and unmapped. The few woodspeople who might have led the scientists refused to go near the blast site. "The Gods are angry," they said. Any living thing that went near the spot would weaken with a strange sickness. All the scientists could do was shake their heads at such superstitious tales. Eventually, a few did struggle through the wilderness. They saw countless giant trees torn up by the roots, now lying on their sides, all pointing in the same direction. They saw other trees that stood like crooked telephone poles, shorn of all their branches. But they didn't see much that moved or flew in the forest. And they never found a meteor or even a huge hole in the ground to show them the center of the blast.

Revolution and wars came to Russia and people had more urgent things to think about than the unexplained explosion in the deserted forest. Then, in 1945, the United States set off the world's first atomic explosion. Suddenly, the old stories about the Siberian blast began to have a strange new ring.

Since then, the Soviets, using the latest scientific equipment, have worked hard to make sense out of the mysterious Siberian explosion. The more they learn, the more the mystery. All the evidence indicates that whatever caused such a bang that June morning was not a meteor. It was not a comet. What was it? Checking all their data, the scientists have to admit that a nuclear explosion as big as a 10-megaton hydrogen bomb destroyed the forest. Yet, not one person on earth knew there was such a thing as atomic energy in 1908.

The explosion didn't take place on the ground; it went off a few miles *above* the forest. To make matters even more peculiar, the scientists find that the thing that rocketed across Europe to Siberia slowed down and changed its course shortly before it blew up.

Was someone trying to send the earth a guided message from outer space? Did a UFO pilot make a last desperate manoeuvre to save his ship? If not, what did cause an atomic explosion on June 30, 1908?

7 Other Lives and Lost Civilizations

America's Most Famous Psychic

The slim, stoop-shouldered man lay on the couch, apparently asleep. Beside him, a woman waited, note pad in hand. The sleeping man spoke. "Yes, we have the body, and here are the conditions as we find them . . ." In his "invisible housecall," he proceeded to outline in technical terms, a medical diagnosis and treatment for a patient living thousands of miles away whom he had never seen.

The sleeping man was Edgar Cayce (*Kay*-see), destined to become one of the most famous psychics of all time. Twice a day, until his death at age 67 in 1945, "the sleeping prophet" went into trance and dictated diagnoses and cures that proved to be amazingly accurate. The Kentucky-born Cayce, a photographer by trade, had only a 7th grade education and no medical knowledge. In his lifetime, he dictated more than 14,000 trance "readings" and earned the gratitude of thousands of people he'd helped.

Today, over 500 U.S. health practitioners use Cayce remedies and cures, and in Phoenix, Arizona, a Cayce medical clinic treats patients according to Cayce methods.

Today, at the Edgar Cayce Foundation headquarters in Virginia Beach, Virginia, the Association for Research and Enlightenment (A.R.E.) has preserved and catalogued the seer's documented trance "readings" in an impressive, modern library.

The Cayce health diagnoses had proved highly accurate for decades. Could his trance predictions on other topics also be right? That was the grim question people around him asked when he awakened from trance one August day in 1941. Cayce never knew consciously what he'd dictated in trance. He shook his head as he read the notes of that session. He'd predicted Los Angeles, San Francisco and New York were all to be destroyed. "I hope it's wrong," he said.

But the Cayce material held other startling prophecies along the same lines. The period from 1958 to 1998 would be a time of great earth changes. "The earth will be broken up in the western portion of America." He predicted a shift in the earth's axis which would cause many severe earthquakes around the globe. There would be volcanic eruptions in the torrid zones.

He foresaw upheavals in the Arctic and Antarctic; the sliding of much of Japan into the sea; rapid land changes in northern Europe; new land rising off the eastern U.S.; earthquakes from top to bottom of South America.

Cayce predicted a drastic reversal of global climates and with this, lean times of food shortages. Find ways to grow your own food, he urged. The signal of coming earth changes in America, he said, would be volcanic eruptions of Vesuvius or of Pelée on Martinique in the Caribbean. However, he added, if there was a change of consciousness these disasters need not occur.

By the winter of '76, people were beginning to notice climatic shifts. Snow fell in Florida, Alaska had balmy rains. Food crops were damaged on both coasts. By the mid 1970's, severe earthquake activity seemed stepped up around the globe. Guatemala, Turkey, Italy, Romania, China, suffered quakes with massive death tolls. An increase in volcanic activity also was recorded, some of it on Martinique.

However, all of Cayce's predictions weren't about such negative events. Cayce's foresight often made fortunes for others, although he and his family lived very modestly. Investors profited hugely from his predictions of real estate booms, his finding of oil wells, his accurate stock market and business advice.

The readings contained some other puzzling material for Cayce. In his sleep state he was talking about reincarnation, the idea that some part of an individual has lived before as another person in other centuries. Cayce, a devoted Christian Sunday school teacher, couldn't swallow that. But as the years went by and more and more information in the readings proved valid, Cayce came to believe that he and all of us have lived before and that we've come back for a reason. A more modern way of expressing it, dictated through *The Seth Material* by Jane Roberts, suggests that each individual is a part of a larger Self, and that it is different aspects of this larger Self that take bodily form to develop further, and not exactly the same Jean or Joe.

Cayce gave readings on thousands of people, describing in detail their previous incarnations. Once when Cayce was sitting in a Virginia Beach barber shop, a small boy suddenly climbed onto his lap. The boy's father pulled him away. "You don't know that man."

"I do know him," the child said. "We were hungry together at the river."

Cayce was startled. Only his family knew of a secret reading about himself in a previous life in which he and others were on a raft on the Ohio River fleeing Indians. Weak and starving, they were finally caught and massacred.

Reincarnation?

Nine-year-old Robert Dunbar talked excitedly at the dinner table about his latest adventures—in chemistry. He'd just gotten his first chemistry set that afternoon and he'd already done every experiment listed. He seemed to know all the technical terms already.

His parents listened with amazement. "I can't wait to get my hands on the chemicals to make gunpowder," he said. His enthusiasm was electric. "I'm going to make my own firecrackers and fireworks."

The Dunbars looked at each other. They remembered the extraordinary predictions they'd had about their son at the time of his birth from America's most famous prophet, Edgar Cayce. In a sleep-like trance state, Cayce gave "readings" for new-born babies, listing their previous "incarnations" and the various talents and abilities they would have in this life as a result.

"This boy lived before in India," the sleeping Cayce said of new-born Robert Dunbar. "He developed a combination of chemicals that produced explosives which were used against enemy tribes." Cayce said little Robert had previously lived in Germany, Egypt and Atlantis. In all of them he was involved in chemical, electrical and mechanical developments. Cayce told the parents their new son would have brilliant abilities in all these areas. Then he gave them a warning. Guide his talents into positive areas, he advised. If you don't, Robert's talents again will be used to bring death and destruction to other people. This continued use—or mis-use—of his abilities would not be good for Robert's development in the long run, according to the sleeping prophet.

Robert's parents watched with surprise and then concern, as each of their boy's new interests seemed to follow the pattern Cayce predicted. From chemicals and explosives, Robert went on to develop an intense interest in cars and all mechanical things. In high school, electricity consumed his attention. He later got a degree in electrical engineering. Robert's parents showed him his Cayce reading. He became interested in it, and on many occasions, he dug it out when he had important choices to make. World War II erupted. The gifted Robert was asked to work in defense research. Would he help develop new and more effective ways to kill and destroy the enemy? Robert remembered his Cayce reading. He decided to serve his country doing radar work.

After the war, the immensely talented Robert was showered with job offers, both military and civilian. All involved work on destructive devices, missiles and bombs. The military wanted him to do secret work with captured German scientists on rockets and atomic bombs.

Robert remembered the sleeping prophet. Finally, he decided not to spend his life figuring out new and better ways to destroy his fellow man. He took a job in the field of electrical utilities. Robert feels that as a result his life has been unusually happy.

What destiny put so many offers in Robert's path to use his abilities for destructive purposes? Robert did turn out to be a gifted scientist with a real talent for developing explosives and destructive devices. Was this know-how a result of previous incarnations or just a coincidence?

Atlantis, Archeology and the Occult

A legendary lost land—a superior culture of super-beings—the myth of Atlantis has intrigued people for centuries. The Lost Continent of Atlantis was said to be a vast land in the Atlantic, stretching from the Bahamas to the Azores. It sank thousands of years ago, leaving no trace, only rumors and stories of incredible secrets and lost civilizations.

Edgar Cayce woke up from a trance one day and read the beginnings of an account some of his family wished he'd never mentioned. For twenty years this data poured through. His followers in the 1920's thought he'd gone into science fiction. Out of the thousands of past life readings came an account of Atlantis that some feel may prove extremely important for our era.

Atlantis had the highest level of civilization and scientific knowledge that was ever attained on earth, Cayce said. It existed between 50,000 and 10,000 B.C. and then vanished into the sea in a spectacular holocaust. The Atlanteans destroyed their own land, Cayce said, through misuse of scientific discoveries.

In the 1920's people could hardly follow what Cayce was talking about when he described such Atlantean devices as television and tape recording, submarines and commercial aircraft, atomic explosives, lasers, and death rays, long before we had any widespread use of these inventions. They had every convenience, he said, from electricity to all sorts of mechanical appliances. Cayce said some of their aircraft were like the one described in Ezekiel, a shiny, whirring four-sided flying craft that went up and down by means of wheels spinning within wheels. Others were like huge boats that could fly and go on or under water.

They knew how to overcome gravity itself and how to make stone float. They could levitate stones weighing tons to construct their buildings, Cayce said. They had equipment that rejuvenated the body, so they remained youthful through lifespans of hundreds of years. They could photograph at a distance right through walls. They had harnessed solar energy and possibly etheric or cosmic energy from the sun by means of massive crystals. According to the Cayce account, they may have known forms of energy we have not yet discovered. They were highly developed psychically, knew how to use "supernatural energy," and regularly practiced out-of-body travel.

Through genetic manipulation or in some other way, they developed animal/human "Things." These were their slaves and did all the work. Were the "Things" humans or machines? Atlanteans asked each other this question and became embroiled in bitter disputes over slavery, Cayce related. They split into two groups: one advocated the use of advanced knowledge for selfish purposes, the other, for more spiritual goals. They fought over ethics and the concentration of knowledge in the hands of a few. There was a moral collapse, says Cayce. Then something went wrong with the cosmic crystal power stations. Scientists adjusted them incorrectly and the energy set off earthquakes and volcanic eruptions that caused massive destruction. With the first disaster, the continent split into five islands, said Cayce. There were two more upheavals, the last, about 10,000 B.C. when Atlantis finally sank totally beneath the sea.

The Atlanteans were warned in advance. Many had time to escape to other countries, taking with them records of much secret and technical knowledge. They fled *by airship,* says Cayce, to Egypt, Central and South America, the western U.S., and other areas. In these new lands, concealing their records, they built pyramids and set up sophisticated civilizations among the natives. They brought the arts of medicine, embalming, and architecture.

Folklore and religious texts from very ancient cultures in India and South America do describe a flood, disasters, flying machines, and the arrival of highly civilized refugees. Nevertheless, Cayce's account sounded like one more addition to the fanciful fables that have been told about Atlantis since Plato wrote of it 2,500 years ago.

Then, Cayce made a very strange prediction. "Atlantis will rise again!" By 1968, he said, it might be sighted near Florida. The sealed records of Atlantean knowledge about cosmic crystals would be found under the slime of sea water, in temples off the coast of Bimini in the Bahamas. The race to find Atlantis was on.

Why are so many people in the psi field so interested in Atlantis and archeology? In part, because many psychics claim that Atlanteans had developed psi forces to a very high degree to the benefit of their civilization. For centuries, occult (hidden) groups around the world claimed they had preserved the true lost knowledge, the secret techniques for psychic powers which the Atlanteans knew: how to levitate heavy stone; the keys to astral travel; secrets of rejuvenation; clairvoyant photography; shapes that generated energy. The secrets were so well concealed, it was hard to separate possibility from legend.

The records from Atlantis containing all this knowledge are hidden in three places in the world, said Cayce. One of them is Egypt. He predicted that in the next twenty years, these records will be found. The Temple of Iltar will rise again, he said, and it will be found near Yucatan.

Why had so much data on Atlantis been dictated through Cayce? He said it was because many Atlanteans were reincarnating at this time in history to meet the same problems. We could blow ourselves up again, perhaps even unintentionally, if earthquakes occurred under nuclear installations or if excessive nuclear testing brought about earthquakes. In genetics, we've already created a man/plant hybrid, man/mouse hybrids, and "cloned" (duplicated) animals, and we are close to constructing test-tube humans. Will someone come up with an Atlantean "slave" being? Will the newly re-discovered psychic powers of humans be used for evil purposes, as one Soviet researcher fears?

Would Cayce's strange Atlantis prophecy be fulfilled? With a sense of growing excitement, people waited for archeological bulletins. Over the last ten years, they have been coming in. Incredible underwater "buildings" have been discovered near the Bahamas: walls, causeways, streets, plazas, harbors, temples. There are so many, they suggest a small city, says expert Charles Berlitz. Near Bimini, a huge underwater stone wall has been located. Radio-carbon dating of mangrove roots—still clinging to other underwater walls—showed they'd grown around the year 10,000 B.C. Others claim sighting two pyramids, one near the Grand Bahama Banks, the other a "step pyramid" 12 fathoms down. There is even a report of a marble citadel—five acres in size, with roads leading from it—and a sword-shaped underwater building. Another researcher claims to have recovered what might be Atlantean sculptures.

Much exploration is being done secretly to protect

the finds, but an Atlantis Museum has already been set up in Nassau.

While underwater archeology opened new vistas on the past, aerial archeology also made startling findings. In the Nasca Plain of Peru, ground pattern markings, *visible only from the air,* have been discovered that some people feel are markers or landing strips for aircraft. Were they landing fields for the Atlantean airships Cayce described?

In the following accounts, you'll find more information about evidences of highly-developed past civilizations that hint we may be close to unravelling the mystery of Atlantis, and with it, many mysteries of the mind.

The Mystery of the Ancient Maps

Way back in 1513, the Grand Admiral of the Turkish Navy concocted a grand mystery for the 20th century. With a few twists of his pen, he left us clues that are revolutionizing our ideas of the past and ourselves. The Grand Admiral, who once roamed the seas as a pirate, was named Piri Reis. One day, he laid out a very fine piece of animal hide and set to work drawing a map of the world's oceans and continents.

For help, he turned to the Turks' collection of very, very old maps. One of these, he noted, was used not too long before, by that Italian, Christopher Columbus, when he sailed across the ocean and discovered the new world of America. Looking at the maps, the Admiral knew that one should say Columbus "rediscovered" the Americas. Obviously, some people knew North and South America well enough to draw them in detail long before Columbus pulled up his sails. The Admiral said he took his information from maps which were used at the time of Alexander the Great in 340 B.C. These in turn were evidently copied from even older maps.

With a sure hand, Admiral Piri Reis put together his grand map of the world. We don't know if Turkish sailors venturing into unknown seas ever used his charts. We do know the map was kept and prized—and then forgotten.

In 1929, after the Sultan of the Turks was overthrown, the old palaces were being swept clear. Looking through a batch of crumbling, faded records, modern Turks came across the Admiral's map. Obviously, they had much better maps now. But the Turks saw they were holding an historic memento, since the Admiral had written on the document that he'd drawn information from an old map also used by Columbus.

As a friendly gesture, the Turkish government gave a copy of the map to the government of the United States. Eventually, it came into the hands of another sailor, Captain Arlington Mallery, of the U.S. Navy Hydrographic Office in Washington. Captain Mallery was an expert in map making, old and new. He turned to the old Turkish map with interest.

In a few seconds, he realized he was not looking at an historic curiosity. He was staring at a time bomb with a slow fuse that was finally about to go off and shake up current ideas of history and the progress of civilization. There, clearly detailed on the Admiral's map were a number of things supposedly unknown until recently.

Amazingly, the Admiral's map correctly showed the coastline of Argentina. In 1513, New World explorers had not yet sighted Argentina. More amazingly, the Admiral's map showed correct longitude markings. These lines are drawn lengthwise on the globe to help pinpoint location. Yet it was not until the late 18th century that people knew enough to calculate longitude correctly on a global scale. Or so it had been thought until that moment.

Captain Mallery looked down the Admiral's map to the southern part of South America. It was distorted—an understandable mistake on the part of the ancient map makers. But it was not a mistake. This is how a land mass on a spherical shape, like the globe, would look when projected on a flat surface, or *when seen from the air*. In our day, satellites orbiting 60 miles above the earth have taken photos showing the same distortion.

Beyond all these improbable features, there was something on the Admiral's map that gave Captain Mallery definite chills. He saw detailed charting of the coastline of Antarctica. The chart even showed inlets, bays, and rivers, some of which had not yet been mapped when the document was discovered. It wasn't until 1949 that an international team came up with a complete, accurate map of Antarctica. And it agreed with the chart put together so long ago by the Turkish admiral from much older maps.

Even more astonishing was what the map did *not* show. It showed no ice on Antarctica. The original seemed to have been carefully drawn before Antarctica was sealed over with ice, and it has been covered with ice for thousands of years! When the Admiral's map was discovered, scientists had just begun to chart the coastline of Antarctica through a one-mile-thick ice cover!

If someone had unearthed the Admiral's map a hundred years before, he might have thrown it away as a fairytale drawing. Experts of the time did not have the knowledge to understand that it was right. How many other old drawings are waiting on dusty shelves for us to learn enough to understand their messages?

Other ancient maps began to come to light after 1929. They agreed with the Admiral's. But modern experts noted they did have one outstanding "error." The old maps showed a waterway through the middle of Antarctica. They drew it as though these were two large islands instead of one big continent. In the International Geophysical Year of 1958, using newly developed equipment, scientists found that if it were not fused with a layer of ice, Antarctica would indeed be two large islands.

It seems that before what we call the dawn of history, people somewhere on this planet knew a great deal more than we think they did. If that's so, how did we lose so much knowledge?

Impossible Objects

Imagine that you were lucky enough to find the unopened tomb of an Egyptian Pharaoh. Finally, you manage to work your way to the burial chamber crammed full of glistening treasures. Gold, silver, jewelry, furniture—suddenly the bright beam of your flashlight picks up another large gleaming object. There sits a low-slung, high-horse-powered sports car. "That's impossible!" you say. "People didn't know anything about automobiles until now." You're right. It's impossible. No one has found a sports car in a Pharaoh's tomb. But people have been turning up equally impossible objects.

In the late 1930's, Wilhelm Konig, a German engineer, worked on a construction project for the ancient city of Baghdad. Wilhelm was interested in how people lived in ancient times. He went to the Baghdad museum. Walking the long halls, he found himself wondering about some peculiar-looking objects on display. Vases, that's what people said they were. The museum simply classified them as "Ritual objects." But what kind of object, for what ritual, no one cared to guess.

One day in the course of his work, Wilhelm picked his way through an old ruin of a building. It had been constructed about 1,700 years before in 200 A.D. Wilhelm stumbled across an object. He picked up another of the so-called vases he'd seen in the museum. But this vase, still had something in it, a cylinder of copper coated on one side with asphalt. A plug of asphalt was fixed into the upper end of the cylinder. In the center of the plug was a piece of iron. Because he was an engineer, Wilhelm realized he was looking at an electric battery, an electric battery from 200 A.D. An impossible object.

Was electricity "simply" used to goldplate jewelry in long ago Baghdad? Was it used more widely? Did the mighty have some sort of lights in their palaces? No one knows. All we know is that some people understood quite a bit about how electricity worked a very long time before Ben Franklin flew his kite.

Another museum halfway round the world has an equally interesting group of things from vanished civilizations. This is the national collection of the country of Colombia in South America. Many lovely models, trinkets, and ornaments of gold are on display. One lucky day, Ivan T. Sanderson, a world-renowned biologist, naturalist and archeologist was touring the museum. His eye caught on a triangular little gold object. It was very interesting and very old, made around 1000 A.D. The museum said the object represented a moth, or perhaps some other flying creature, or maybe even a sea animal, like the triangular-shaped stingray. Ivan knew that other representations of creatures in the museum were quite exact. Ivan also knew his insects, birds and sea animals; this was his lifelong speciality. The little gold object did not have the features of any known animal. It was indeed an impossible creature. Living things were not put together the way this model was. Closely examining the little gold object, a different impossibility occurred to Ivan. Using technical devices, he made an enlarged scale model of the object. The more he studied it, the more Ivan felt sure he was right. Here was a model of a piece of mechanical equipment. It was obvious to him just what piece of equipment had been copied over 1,000 years ago.

Without saying a word about where it came from or when it was made, Ivan gave a model to several engineering experts. Could they please tell him what this object was? The experts didn't seem to think Ivan's question was very difficult. They all came to the same conclusion. One of these engineers was J. A. Ulrich, who was also a pioneer of German rocket planes.

"It's a plane," he said of the little gold object. He thought it might be an F-102 fighter. The triangular or delta-shaped wings have a special curve. "This spells out one thing—jet," Ulrich explained. He noted that a regular rudder system was depicted. He thought the model must be a very new one because it showed special speed brakes at the rear instead of the usual systems. These speed brakes were brand new at the time; they'd just come out on the Swedish Saab jet.

"Don't be ridiculous. That's no moth. That's a

good little model of a jet plane." The experts agreed with the thought that had hit Ivan. Ivan never did figure out who picked up a lump of gold a thousand years ago and crafted a little model of a jet. What did the artist copy the trinket from?

Ivan was interested in other impossible things from long ago. He looked into an odd piece of glass found during digging in 1957 in Israel near the city of Haifa. No one even realized that this old slab—11 feet long, 6 feet wide and almost 2 feet thick—was a piece of glass until 1963. Scientists determined the glass was made sometime before 700 A.D., and there's reason to believe that it might have been made several centuries before that date.

According to Ivan, this is the third largest single piece of glass in the world. The two larger pieces are mirrors for modern telescopes made in our times. Experts at the Corning Museum of Glass said that to make such a slab one would have to heat 11 tons of material and keep it at 2226° Fahrenheit (1050° Celsius) for five to ten days. Just who managed to do that so long ago when they weren't supposed to have any modern machinery or techniques?

On the subject of glass, Ivan came up with another choice tidbit, a tantalizing clue that makes one wonder just how much we do know about the history of this planet. American scientists tested the first atom bomb in New Mexico. They found the tremendous heat of the explosion fused large areas of desert sand into green glass.

Archeologists were digging to uncover an ancient civilization in the historic Euphrates Valley, sometimes called "the cradle of civilization." They uncovered a farming culture 8,000 years old. Then they dug further and found a much older culture of herdsmen. Going even deeper into the earth, they found what they considered the leavings of people at the very beginning of history, the so-called cavemen. For one reason or another, they kept digging. Finally they uncovered another layer. They discovered an area of fused green glass.

A coincidence? Perhaps. Lightning can, on rare occasions, cause a jagged line of fused green glass to appear in sand. But areas of fused green glass are not found in nature. If we looked further into the green glass, would we find the ghosts of people like ourselves looking back? Have we come this road before? Have civilizations risen to sparkling heights only to explode and start the struggle up the spiral again?

Every year impossible objects are being retrieved from the depths of the earth and the deeps of the oceans. If nothing else they are telling us that we still have a great deal to learn about our past.

Pyramid Power

The pyramids in many countries, from Mexico to Honduras, from Egypt to China, have aroused curiosity and awe. Are they huge time capsules of information waiting for future discoverers who will have enough knowledge to interpret them? The Great Pyramid at Giza in Egypt has aroused speculation for thousands of years. Legend has it that secrets were built into its structure. It covers more than 13 acres and stands about 45 stories high. It contains $2\frac{1}{2}$ million blocks of granite weighing $2\frac{1}{2}$ to 12 tons each. Each of these blocks was fitted into the pyramid and adjusted like a jewel in a precision watch. The Great Pyramid was thought to be the burial tomb for King Cheops, but no remains were ever found in it.

When we visited Czechoslovakia in 1968, we discovered another pyramid—a small cardboard pyramid modeled on the Great Pyramid—set up inside a well-known scientist's home. Inside it, there was a razor blade balanced a third of the way up on a matchbox.

"What's that for?" we asked.

"Would you like to know one of the secrets of the pyramids?" our hosts asked.

"Of course."

"Well," they began, "one of the secrets of the pyramid is its shape! It generates energy."

These pyramid models, which could be found all over Czechoslovakia, were the result of the research of Karel Drbal, a Prague engineer, who had pioneered radio and television in Czechoslovakia. He'd begun his pyramid experiments years ago and in 1959, he had received Czech patent #91304 on the Cheops Pyramid.

"How could anyone patent the pyramid?"

"He patented the pyramid as a razor-blade sharpener!" The Czech scientists showed us the patent. Could the Czech Patent Office be confused?

Later, we met Mr. Drbal, a man in his 70's. Years ago, living in France, he came across the work of Antoine Bovis. Bovis claimed to have made a strange discovery while visiting the Great Pyramid in Egypt. He'd stopped to rest in the King's Chamber which is a third of the way up the pyramid. He noticed trashcans containing animals that had strayed in and died. In spite of the humidity, the animals had not decayed. They were dehydrated and mummified. His reading on pyramids gave him an idea. Could the shape of the pyramid and its precise north-south alignment have something to do with it?

Bovis returned to France and began experimenting with exact cardboard replicas of the Great

Pyramid. He put a dead cat in one, aligned it north-south, and the cat mummified. He tried other substances and various foods. They all became preserved or mummified. Antoine Bovis revealed more in his book, *On the Radiations of all Substances* published in the 1930's.

In Czechoslovakia, Drbal wondered if this strange pyramid power would work on other things, razor blades, for instance. He knew the edge of a razor blade has a crystal structure which is "alive." After use, this sharpness becomes deformed, but some materials can return to their original form after a time. Drbal shaved with a new blade, then put it in a small cardboard pyramid. To his amazement, the blade stayed sharp, shave after shave. For years, he did thousands of tests. Some blades stayed sharp for 50 to 200 shaves, thanks to pyramid energy. At this point, he patented his discovery and soon plastic pyramids were on the market in Czechoslovakia.

We tested out pyramid power for ourselves, and after getting good results, wrote about it in our book, *Psychic Discoveries,* and introduced the Czech model pyramids on television shows across the U.S. Soon a pyramid craze swept America. Model pyramids sprang up everywhere. Every day somebody, somewhere, discovered another intriguing thing about pyramids. Some of the claims were highly exaggerated, but at the same time, scientific research was being done by many specialists, among them, Dr. Ottmar Stehle of NASA and Dr. William Tiller of Stanford. Dr. Eric McLuhan felt the pyramid chambers were "resonant cavities" like hi-fi speakers, and pyramids formed an energy communication network.

On a U.S. federal government grant, Dr. Boris Vern, a medical doctor working with Mankind Research Unlimited, a Washington research and development firm, did many controlled tests on pyramids in the mid-1970's. His findings were "revolutionary," he said, "a breakthrough in the physical and biological sciences." He reported that seeds sprout faster, food decays slower and human blood pressure is lowered after extended periods inside pyramids.

"We are dealing with some staggering discoveries that may have been lost to man for thousands of years," says Dr. Carl Schleicher, president of Mankind Research Unlimited. He believes the pyramid findings could have benefits for many areas of human life. "They indicate the need for radical changes in the way we grow and package food and design our homes and cities."

In the growth experiments, black-eyed peas and lima bean seeds were placed under a pyramid, under a cube-shape and in open space. None were watered. The peas under the pyramid grew at 1.5 times the speed of the uncovered seeds and 1.129 times as fast as those under the cube. Hamburger in a pyramid decayed at half the rate of decay of meat under a cube or in the open.

Les Brown constructed a three-story pyramid greenhouse in Bancroft, Ontario. He believes the pyramid may be the answer to the world's food problems. Plants and vegetables grew so fast and so big in his pyramid that he believes a few pyramid greenhouses could supply a large population. Cucumbers were 21 inches (53 cm.) long and weighed $3\frac{1}{2}$ pounds (1.6 kg.) as against the usual 14-inch (35 cm.), 1-pound (454 gm.) standard. Cabbages exceeded the 3 pound (1.4 kg.) limit and grew to 12 pounds (5.4 kg.). His 4-inch (10 cm.) wide radishes looked like turnips. "Pyramid energy increases the size of growing cells in the early stage," he says.

Thousands of pyramids have burgeoned over plants in Toronto, thanks to the enthusiasm of the Indoor Light Garden Group. Gilbert Milne, a photographer, grew a beautiful plant in his darkroom, inside a wire frame pyramid. Plants under pyramids in his basement greenhouse never attract bugs.

Pyramids seem to enhance life energy in growing things. Some researchers believe they harness bio-cosmic energy. Others say that pyramids magnify the energy of the earth's magnetic field. Others have found an ionizing effect from pyramids.

People report that pyramids improve the flavors of juices, wine, and cigarettes; energize water for beverages and plants; keep perishables; lengthen battery life; sharpen razor blades; improve gas mileage; and improve ESP. Some people are more successful than others with pyramid power. Some days are better than others.

Many volumes have now appeared on pyramid

power. There's a magazine for buffs, "Pyramid Guide." There are international pyramid conferences. Countless pyramid companies have sprung up—some making highly dubious claims.

If you would like to experiment with pyramids, you'll find instructions in our *Handbook of Psychic Discoveries*.

8 Clairvoyance or Out-of-the-body Experience?

Clairvoyance or Out-of-the-Body Experience?

A college freshman sat listening to a Sunday sermon. Then she saw something in the front of the church, in the left corner, that she knew wasn't there. Instead of the stone walls, she was watching a large plane, an American war plane, flames shooting out of the tail, falling toward the earth. The sight was so vivid, brighter, and so much more real than a movie, that the young woman sat staring when the congregation rose to sing the next hymn. Months later, she learned that a childhood friend, a tail gunner on a B-17, had gone down with his plane the day she'd seen the flaming apparition in church. This is probably an example of clairvoyance, the psychic ability to tune into events or objects at a distance, to see what's going on without the use of the known senses.

It's clairvoyance "probably" because, as we've noticed, it's hard to separate psi abilities. Just possibly, the young woman experienced telepathy and tuned into the mind of someone watching the crash. In the lab you can freeze the fabric of life and separate out strands of psi to study. Dr. J. B. Rhine did just that with clairvoyance during his famous card tests at Duke University. He showed it's a distinct ability. Some people can mentally see things at a distance without using another mind as a relay station. Rhine used fresh decks of cards, shuffled by machine or by independent researchers. No one person knew the order of the cards in the deck. There could be no question of telepathy. Yet, people with ESP were able to tell Rhine the correct order of cards.

Working in Czechoslovakia, Dr. Milan Ryzl created a hypnotic system to help people develop ESP, particularly clairvoyance. Josephka was a student who caught on quickly. "Use your clairvoyance, go down the street and tell me what's happening," Dr. Ryzl would say to Josephka in trance. "Go up the stairs to your apartment. Look around. What's cooking on the stove?" Often, using her clairvoyance, the young woman could see across town and give Ryzl a description of what was happening.

"I've lost my keys," she complained once before her training session.

"I want you to see the keys clairvoyantly," Dr. Ryzl said when she was in trance. In a few moments she said, "They're on a closet shelf in the kitchen." Later, that's where she found them, exactly where her grandmother put them that morning. Josephka also demonstrated her new-found clairvoyant ability in lab tests. She correctly saw which cards had been placed in doubly sealed envelopes. Odds were a trillion to one against doing as well as she did by chance.

Travelling clairvoyance is what Dr. Ryzl called it when he had students mentally go to distant places to give an on-the-spot report. Some of his students, however, may have been edging into a much stranger ability than clairvoyance. They may have had out-of-the-body experiences. As you know, at times of great crisis, some people report suddenly finding themselves out of their bodies, hovering above the frantic scene below. They don't try to do this, it's an unexpected experience, just as you may unexpectedly pick up a telepathic message from a distant friend. But there are people who have trained themselves to switch into telepathy when they want to. There are also those who say they've learned to leave the body when they wish. They claim there is a great difference between going somewhere in your mind's eye clairvoyantly and slipping out of your body and going somewhere as an invisible, full-bodied person. Scientists are just beginning to investigate the real differences between clairvoyance and out-of-the-body experiences.

Out-of-the-body experiences are astonishing to us. Before the invention of movies and television, plain old-fashioned clairvoyance was just as extraordinary. A clairvoyant experience that happened in 1759 seemed worthy of investigation by some of the best minds of the day.

TV Dinner, 1759

A handful of prominent citizens in the Swedish city of Gothenburg gossiped and talked excitedly among themselves. They had been lucky enough to be invited to a dinner in honor of one of the most famous men of their times. These chosen few were going to be able to spend an entire evening with the great man at close quarters. It promised to be an extraordinary night. They didn't know that events at the dinner party would be so extraordinary that they would go down in the history books.

The great man they awaited was Emanuel Swedenborg, scientist, philosopher, theologian, and apparently, clairvoyant. It was 1759 and, after a trip to England, Swedenborg was journeying home to Stockholm. On the way, he stopped off in Gothenburg. William Castell, a town dignitary, gave a grand party in the philosopher's honor.

Suddenly, in the midst of the formal dinner, Swedenborg abruptly left the table. A few minutes later he returned, looking pale and excited. "A terrible fire is this moment raging in Stockholm," he announced.

If he hadn't been quite so eminent, his fellow guests, numbering more than a dozen, might have laughed and made some unkind remarks. How could he know what was happening in Stockholm, some 300 miles away on the opposite coast of Sweden? Speedy communications in those times meant two or three days between Gothenburg and Stockholm.

Swedenborg proceeded to give a more or less blow-by-blow description of the conflagration. "My house is in danger!" he exclaimed. Finally, he reported, "It's under control now. The flames stopped but three doors from my own."

The next day, the Governor questioned Swedenborg about the alleged fire. The day after that, a royal messenger arrived from Stockholm. "A terrible fire has devastated Stockholm," he reported. He gave detail after detail that coincided with what Swedenborg had narrated.

Immanuel Kant, the well-known German philosopher, investigated Swedenborg's strange clairvoyant experience. He even hired a scholar to check out minute details. Kant concluded, "Swedenborg's extraordinary gift is proved beyond all possibility of doubt."

Psychic X-Ray

As a young woman in the Middle East, Shafica Karagulla may have had a feeling that one day she would be internationally known as a doctor specializing in a difficult part of the human anatomy—the brain. She had no hint that her brain research in Europe, England, Canada and the U.S. would finally lead her to study that mysterious thing people called psychic ability. But it did.

Today, Dr. Karagulla heads the Higher Sense Perception Research Foundation in Los Angeles. She investigates creative, usually highly successful professional people with these now not-so-mysterious abilities. One of the more dramatic cases she traced involved a colleague, to whom she has given the pseudonym "Dr. George."

Dr. George is one of many physicians who have a high degree of ESP. Throughout a long and distinguished career as a surgeon with patients from all over the world, he frequently used his powers of intuition to make accurate diagnoses. He could often foresee events that would take place in his patients' lives. He had a special sixth sense that told him whether or not a patient could survive an operation. He also knew when someone would not survive, even if all medical evidence showed the patient to have an excellent chance.

One day Dr. George scrubbed up for a serious and tricky operation. The patient was already anesthetized. Dr. George came through the operating room doors just as one of his colleagues was preparing to make the first incision. "Wait!" he ordered his surprised associate. "I've changed my mind as to where we should make that incision."

The other doctors and the nurses were astonished. "We must make the cut . . . here!" Dr. George said as he carefully marked another line on the patient's skin. "We'll run into an artery in the area originally marked out," he announced.

"That's absurd," said the other doctors. "You know there isn't any artery there." The assistant surgeon began to wonder about Dr. George. He was bewildered. How could this world-famous surgeon be confused about basic anatomy?

Dr. George would not be swayed. He took a scalpel from the nurse and started to make the incision in the new area, ignoring standard procedure. When the patient was opened up, Dr. George stood back and let the others look. They saw a one-in-a-million peculiarity. One of the patient's major arteries was in the wrong place. If they had gone ahead and made the incision where they'd originally marked it out, they would have sliced into the artery and probably killed the patient.

Psychic Astronauts

On a clear night you sometimes can catch sight of the planet Mercury twinkling in the sky far away, 57 million miles away, to be exact. Only very recently have we had the equipment and know-how to find out something about this small planet circling close to the sun. Or so almost everybody thought. A strange experiment seems to hint that we always may have had the "equipment" to explore the solar system. It hints that perhaps a great many of us could learn to explore outer space, in a very unusual way.

On March 24th, 1974, the American spacecraft, Mariner 10, was scheduled to fly close by Mercury and radio back information. This would be the first time that human beings would have a close-up view of Mercury. Or would it? Two and a half weeks *before* the spacecraft got to Mercury, two men, Harold Sherman and Ingo Swann, were asked to see if they could mentally visit the planet. It sounds like a mad request, but the two agreed. The descriptions of what they mentally saw on Mercury would be kept by research groups to be compared with the facts that the spacecraft would later radio back.

Psychic lift-off was set for March 11th, 1974. The spacecraft still had 18 more days to go speeding through space before it got to Mercury. Swann took off mentally from New York City and Sherman from Arkansas.

In a few minutes, Swann reported, "I'm out past the moon, heading toward the sun, trying to zero in on Mercury." Then, both psychic astronauts reported they'd arrived mentally at the planet. What did they see?

"There are rainbows," said Swann. "Huge rainbows leaping up in all directions."

"It's a fiery spectacular," said Sherman. He saw "blinding, brilliant, many-colored lights waving in the atmosphere." A beautiful sight, but could it possibly be real?

Scientists who study planets did not expect to find any rainbows on Mercury. They didn't think conditions were right for such a show of lights.

Weeks later, the spacecraft began to radio back information. Decoding it, scientists came up with an unexpected finding. There is a strong glow of rainbow-colored lights, like Northern Lights, on Mercury!

As they explored Mercury, both psychic astronauts found an envelope of magnetism around the planet. Swann reported that this magnetic envelope was rather flattened in on the sun side and formed a sort of tail on the opposite side. That really seemed wrong. At the time, not one textbook on space predicted that there would be a magnetic envelope around Mercury.

After the facts from the spacecraft were sorted, scientists announced a major discovery, a discovery, they said, that would have great impact on the study of the solar system. What was it? *They found a magnetic envelope around Mercury. It formed a sort of tail on the side opposite the sun,* just as Swann reported.

How did two men sitting in their own homes, hundreds of miles apart, come up with the same correct information about Mercury 57 million miles away? How did they come up with facts that not even the most knowledgeable scientists had guessed?

Sherman and Swann "sighted" other complicated things that later appeared to be right. Not one single thing they reported on their mental flight was found wrong compared to the spacecraft information. Fantastic as it sounds, it seems that one spring night, the two psychic astronauts took off, passed the speeding spacecraft and beat it to Mercury. The craft, Mariner 10, is one of the genuine wonders of modern technology. Yet it took months to go from Earth to Mercury. Sherman and Swann made their entire "flight" in less than an hour.

Both Sherman and Swann think that almost anyone can learn to use their minds in the many astonishing ways that they have. And Sherman got an interesting idea from his Mercury adventure about how one's mind works.

Science fiction writers have long told stories about fantastic civilizations and other planets. They thought they were making them up. But who knows, says Sherman, maybe they have been tuning in mentally on actual conditions somewhere and then coloring them with their own ideas. Who does know? Next time you get an idea for a fantastic story, you might just wonder where it's coming from.

Feeling Half There

The psychic astronauts seemed to explore Mercury's rainbow world in the mind's eye, clairvoyantly. But there are times when psychics feel they've left their physical bodies behind, that they are on-the-spot reporters in a second body—a kind of "energy body"—which is usually invisible to others. Sometimes, it's hard to tell if the psychic is using clairvoyance or is, perhaps, having an out-of-the-body experience. Often, even the psychic doesn't know for sure.

Eileen Garrett, one of the very best and most intelligent psychics of this century, never gave up trying to understand her powerful talents. As a child living with her aunt in Ireland, she talked to the "little people." She saw auras; she knew when people would die and often what they were thinking. Her relatives weren't impressed. They feared her powers might come from the devil. At this point, Eileen developed a longing that never left her. She burned to understand the how and why of psychic abilities.

Eileen grew up to be an even better psychic and a successful businesswoman in Europe and America. She started a publishing house, ran a magazine, owned restaurants. But she still wanted to understand. For years, Eileen let scientists practically take her apart in the laboratory as they tried to learn how she did what she did. She was the guinea pig. She was also the person who made it possible for many scientists to explore ESP. Eileen started the Parapsychology Foundation in New York, a world-wide clearing house for information relating to psychic things. For decades, Eileen's foundation funded much of the worthwhile psi research in America and other lands. And it still does.

Scientists might call something Eileen did telepathy. She wasn't always so sure they were right. One evening, she kept an appointment with a New York psychiatrist in his apartment. He had set up a test. Could Eileen telepathically communicate with a friend of his? Eileen had never met the friend, a doctor, who lived hundreds of miles north, in Newfoundland.

Eileen went into trance. She found herself outside a house. She felt a damp ocean atmosphere curl around her. She saw little flowers growing beside the path leading to the door. Then, she was beside a staircase in the house. A man came down. As though he knew she was there, he said aloud, "I think this will be a successful experiment!" His voice sounded close, but toneless to Eileen.

She couldn't seem to shift her eyes from him. Suddenly, a table stood beside him with objects on it.

She described them aloud and they faded. She heard the man's voice again. "Make my apologies to the group. I have had an accident and cannot work as well as I hoped." She felt forced to examine his head carefully. He pointed to a bandage around his brow.

Eileen watched the man walk to a bookcase. Before he reached for a book, she knew it would be one by Einstein. As he opened it, the sentences he would read flashed through her mind.

As Eileen came out of trance, the New York psychiatrist told her that 15 minutes had gone by. While he checked the records to send to Newfoundland, a telegram arrived. The doctor, they read, had accidentally hurt his head shortly before the experiment began. Eileen did correctly describe the house, the objects, the book by Einstein. The psychiatrist called it telepathy.

Eileen pointed out that she seemed to be doing more than just linking to the doctor's mind. She seemed to be using all her body senses, as if she was partially there, at least as far as feeling went.

Part-Time Ghost

An actress in the city of Warsaw lay sleeping soundly in her bedroom. Her house was locked for the night; the drapes were drawn; everything was still. Suddenly, she woke with a jump. "Stephan!" Her friend Stephan stood at the foot of her bed looking at her. How had he gotten in? What was the matter? He didn't say a word. "Stephan," she cried again, and then more loudly, "Stephan!"

Slowly, Stephan began to fade, seemed to go out right through the wall of the room. The actress didn't sleep any more that night. She was positive she wasn't dreaming. But if not a dream, what had she seen? Stephan couldn't be a ghost. He was still alive. Then a frightening thought hit her. Perhaps he was dead, had died that very night and somehow had come to bid her goodbye. Nervously, she waited for morning so she could try to find out if anything was wrong.

Miles across town, Stephan Ossowiecki lay unconscious on a couch. He very well could have been dead from the way he looked. A worried doctor worked over him. He massaged Stephan's heart, he gave him a shot and kept checking his pulse. It was so weak that sometimes for long moments the doctor couldn't detect any heartbeat at all. What if Stephan should die? He was a good engineer and a famous psychic too, in the prime of his life. Not only would the doctor lose a friend, he would also be hard put to explain the circumstances of Stephan's death. At long last, Stephan's eyes opened and life seemed to come flooding back into his body. "I did it," Stephan said wearily. "Just as we planned, I left my body and went to see the actress. I went in through the wall by the window. She saw me, too; she sat up in bed and shouted my name three times." Stephan and the doctor had been trying what is called an out-of-the-body experiment.

As soon as possible, they went to see the actress, who was greatly relieved to find that her friend was only a part-time ghost. She confirmed his story in detail.

Many people do not think you can leave your body. They say that on such supposed trips, psychics are actually using clairvoyance. Something remarkable about Stephan's case, though, makes one wonder. The actress, who knew nothing about the experiment, saw Stephan exactly where he said he was, when his body lay miles away under the watchful eyes of a doctor. This is one of the few well-documented cases in which someone has seen a living person who claimed to be out of the body.

Dr. Charles Tart, psychologist at the University of California, points out that out-of-the-body experiences have been recorded throughout history. They were accepted as real in other societies from ancient Egypt to China and India. The thing that makes you wonder is that the reports are the same whether from a farmer in Kansas today or from a scribe in the time of the pharaohs. All say they separate from their physical bodies and float around in "second" bodies. They insist they are wide awake and not dreaming.

For the first time, Western science is starting to investigate out-of-the-body experiences at UCLA, at the Society for Psychical Research in New York and in England. Perhaps the people who report they're out of the body are really travelling in their mind's eye and experiencing ESP. On the other hand, maybe they know what they're talking about. If so, a fascinating new realm of human experience is open for exploration.

What About the Future?

Though more and more people seem to be having psychic experiences, these are not yet everyday experiences for most of us. But what about the future? The Czech scientist, Dr. Milan Ryzl, thinks one day we'll use our sixth sense as normally as our other senses and pay no more attention to it than to hearing and seeing. Is he right? Will many of us learn how to travel around in our mind's eye or even out of the body to explore the impassable jungles of South America or the rings of Saturn? Will we learn to be "ordinary" like Komar and turn off pain and heal super fast? Is there something in the area of psi energies that just might help us out of the energy crunch? Or?

A few hundred years ago, when electricity was first discovered, about the only way people could think of using it was to cure rheumatism and power lighthouses. Today, even the greatest psychic probably couldn't give us a full-color picture of how psi might be used in the future. Great psychics, however, have told us something worth remembering: we create our own future. And if you're in school now, that future, as far as the psi field goes, is just opening up.

What should you study if you want to get in on this new frontier? Just about anything that interests you. You might study biophysics or engineering; you might learn medicine, business or agriculture. You could study anthropology or psychology. People in these fields and many others are involved with psi, and their special knowledge makes them extremely valuable investigators. The American Society for Psychical Research publishes a growing list of colleges that give courses in psi, called "Courses and Other Opportunities in Parapsychology." If you are interested, send a letter-size, self-addressed envelope to: The Education Department, The American Society for Psychical Research, 5 West 73rd Street, New York, New York 10023. The price is $2.00 now, but that may change.

You may not want to take a college course or do professional research, but there are reasons why it is still a good idea to learn at least a little about psi. For one thing, it can help you understand what's going on if you do have a psychic experience. For another, it can help you spot the difference between the genuine and the fake. Certainly, not everyone who says, "Ah, I see your aura, it's purple with a tinge of yellow . . ." knows what he's talking about. Thought reading and moving objects with so-called mind power can be done by trickery on the stage and even in people's living rooms. The more you know, the less likely you are to be fooled by fakes, and the more likely you are to benefit from genuine psychic happenings.

There is another more serious reason we think people should know about psi. When we were in

Russia, one of the leading researchers said to us, "Please, please tell the people in the West that psi must be used for good, it must not be used as a biological weapon." This scientist, who knows a great deal about ESP, had reason to worry. Some people will tell you psychic abilities can be used only for good. But if you think about it, every ability we have, including intelligence and creativity, has been used for good—and for bad. Psychic talent is no exception. And, if we are on the way to discovering a new form of energy, a psi energy, no doubt it too could be used to help or harm, like any energy. You can fry an egg with electricity; you can also electrocute someone.

Around the world, a number of governments and special interest groups are funding psi research. In part, they are trying to find things to help society. They are also exploring how psi might be used in spying, warfare, mind control and brainwashing. Not just the Soviet government was listening when Dr. Leonid Vasiliev made his famous statement. "The discovery of the energy associated with psychic events," he said, "will be as important as, if not more important than, the discovery of atomic energy." Obviously, anyone who learned how to control such an energy would have a great advantage over the rest of the world.

Should psi research be stamped out? You couldn't do that if you wanted to. The best way to insure that psi is not misused is for people to realize that it does exist and to understand some of its possibilities.

The time seems right for such understanding. Dr. Willis Harmon, Director of the U.S. Education Policy

Research Center, remarked that until recently a large part of our experience, including psychic experience, was hardly investigated at all. Today, he thinks, we're finally at a point where we can develop a science of subjective or personal experience. What does that mean? It means, Harmon says, that a real revolution is in the making, "the consequences of which may be even more far-reaching than the Copernican, Darwinian and Freudian revolutions."

As the psychic revolution progresses, scientists may develop sophisticated ways to use psi. But no matter how automated it becomes, one basic fact remains. Psychic experiences are very personal. Psi seems to weave through the fabric of our bodies and the stuff of our minds. You are far more intimately connected when you pick up a telepathic message than when you answer the phone.

This revolution is not just something happening on the outside, at the barricades, or at the borders of scientific knowledge. This revolution is going on inside too, opening new possibilities for all of us.

In the world of psi as in everything else, the wave of the future is upon us. We can try to stand against it. We can ignore it. Or we can do something a great deal more exciting. We can ride the wave and choose the direction we take.

Index

About the Authors

Sheila Ostrander and Lynn Schroeder have travelled extensively to look into the wide range of psychic research currently being carried out in the U.S., Europe and the Soviet Union. After several years' investigation, they published *Psychic Discoveries Behind the Iron Curtain*. The book became an international best-seller, translated into eleven languages, and set off a wave of contemporary research into psychic phenomena. Since 1970, the authors have appeared on more than a thousand TV and radio shows, and have lectured to professional and ESP groups across the country. Their other collaborative books include the "how-to" *Handbook of Psychic Discoveries*, *Executive ESP*, *The ESP Papers*.

Lynn Schroeder, a native of New Jersey, was graduated from Skidmore College and also studied at New York's New School for Social Research. She has written on many subjects, including a book on the U.S. Supreme Court. Her poetry has appeared in numerous literary journals, including the Borestone Poetry Award Anthology.

Sheila Ostrander is a graduate of the University of Manitoba, Canada, and has also studied in Europe. She lived in Paris and London for several years and also travelled widely in the Middle East, Europe and Soviet bloc countries. She has written several successful children's books including *Festive Food Decoration* and *Etiquette for Today*. Ms. Ostrander speaks several languages (including Russian).